Bacterial Fish Diseases

Edited by

Gowhar Hamid Dar

Assistant Professor (Environmental Science), Department of Environmental Science, Sri Pratap College, Cluster University Srinagar, Higher Education Department (J&K), Jammu and Kashmir, India

Rouf Ahmad Bhat

Department of School Education, Government of Jammu and Kashmir, India

Humaira Qadri

Assistant Professor (Environmental Science), Department of Environmental Science, Sri Pratap College, Cluster University Srinagar, Higher Education Department (J&K), Jammu and Kashmir, India

Khalid M. Al-Ghamdy

Head, Department of Biological Sciences and the Vice Dean in the Faculty of Science, King Abdulaziz University, Jeddah, Saudi Arabia

Khalid Rehman Hakeem

Professor, King Abdulaziz University, Jeddah, Saudi Arabia

ELSEVIER

ACADEMIC PRESS

An imprint of Elsevier

Academic Press is an imprint of Elsevier
125 London Wall, London EC2Y 5AS, United Kingdom
525 B Street, Suite 1650, San Diego, CA 92101, United States
50 Hampshire Street, 5th Floor, Cambridge, MA 02139, United States
The Boulevard, Langford Lane, Kidlington, Oxford OX5 1GB, United Kingdom

ISBN: 978-0-323-85624-9

For Information on all Academic Press publications
visit our website at https://www.elsevier.com/books-and-journals

Publisher: Mica Haley
Acquisitions Editor: Anna Valutkevich
Editorial Project Manager: Lindsay C. Lawrence
Production Project Manager: Swapna Srinivasan
Cover Designer: Miles Hitchen

Typeset by MPS Limited, Chennai, India

Dedication

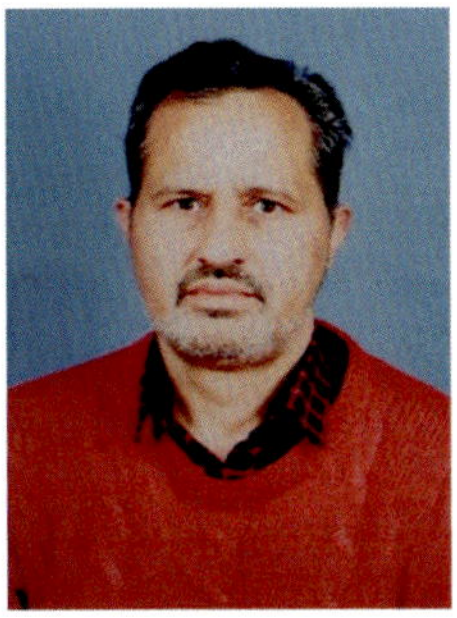

*Dedicated to the loving memory of the late Abdul Hamid Dar,
the father of the lead editor, Dr. Gowhar Hamid Dar.*

Contents

List of contributors

Ismail Abiola Adebayo
Department of Microbiology and Immunology, Faculty of Biomedical Sciences, Kampala International University, Kampala, Uganda

Tariq Oluwakunmi Agbabiaka
Department of Microbiology, Faculty of Life Sciences, University of Ilorin, Ilorin, Nigeria; Microbiology Unit, Department of Science Laboratory Technology, School of Science and Technology, Federal Polytechnic, Damaturu, Nigeria

Toyin Olayemi Agbabiaka
Department of Microbiology, Faculty of Life Sciences, University of Ilorin, Ilorin, Nigeria

Ishtiyaq Ahmad
Fish Nutrition Research Laboratory, Department of Zoology, University of Kashmir, Hazratbal, Srinagar, India

Sheikh Bilal Ahmad
Division of Veterinary Biochemistry, Faculty of Veterinary Science & Animal Hisbandry, SKUAST-Kashmir, Srinagar, India

Imtiaz Ahmed
Fish Nutrition Research Laboratory, Department of Zoology, University of Kashmir, Hazratbal, Srinagar, India

Kamoldeen Abiodun Ajijolakewu
Department of Microbiology, Faculty of Life Sciences, University of Ilorin, Ilorin, Nigeria

Md. Niamat Ali
Centre of Research for Development, University of Kashmir, Srinagar, India

Shafat Ali
Centre of Research for Development, University of Kashmir, Srinagar, India

Mohammad Yasir Arafat
Fish Biology and Limnology Research Laboratory, Department of Zoology, University of Kashmir, Hazratbal, Jammu and Kashmir, India

Yahya Bakhtiyar
Fish Biology and Limnology Research Laboratory, Department of Zoology, University of Kashmir, Hazratbal, Jammu and Kashmir, India

Abdul Baset
Department of Zoology, Bacha Khan University Charsadda, Charsadda, Pakistan

Pradyut Biswas
Department of Aquaculture, College of Fisheries, Central Agricultural University, Lembucherra, India

Monica Butnariu
Banat's University of Agricultural Sciences and Veterinary Medicine "King Michael I of Romania" from Timisoara, Timis, Romania

Candy Carranza-Álvarez
Professor of Faculty of Professional Studies Huasteca Zone, Autonomous University of San Luis Potosi, San Luis Potosi, Mexico

Luis Jesús Castillo-Pérez
Multidisciplinary Graduate Program in Environmental Sciences, Autonomous University of San Luis Potosi, San Luis Potosi, Mexico

Tanmoy Gon Choudhary
Department of Aquatic Animal Health, College of Fisheries, Central Agricultural University, Lembucherra, India

Hilda Guadalupe Cisneros-Ontiveros
Student of Graduate Studies and Research Center, Faculty of Engineering, Autonomous University of San Luis Potosi, San Luis Potosi, Mexico

Núbia Alexandre de Melo Nunes
Institute of Biological Science, Federal University of Minas Gerais, Belo Horizonte, Brazil

Diksha
Division Zoology, Department of Biosciences, Career Point University, Hamirpur, India

Mir Zahoor Gul
Department of Biochemistry, University College of Science, Osmania University, Hyderabad, India

Shefali Gulliya
Department of Zoology, DPG Degree College, Gurugram, India

Shabana Gulzar
Plant Reproductive Biology, Genetic Diversity and Phytochemistry Research Laboratory, Department of Botany, University of Kashmir, Srinagar, India

Younis Ahmad Hajam
Division Zoology, Department of Biosciences, Career Point University, Hamirpur, India

Afrozah Hassan
Plant Reproductive Biology, Genetic Diversity and Phytochemistry Research Laboratory, Department of Botany, University of Kashmir, Srinagar, India

Tasaduq Hussain Shah
Division of Fisheries Resource Management, Faculty of Fisheries, Sher-e-Kashmir University of Agricultural Sciences and Technology of Kashmir, Srinagar, India

Cesar Arturo Ilizaliturri-Hernandez
Professor of Multidisciplinary Graduate Program in Environmental Sciences, Autonomous University of San Luis Potosi, San Luis Potosi, Mexico

J. Immanuel Suresh
Department of Microbiology, The American College, Madurai, India

Swapnali Jadhav
Government Institute of Forensic Science, Aurangabad, India

Arizo Jan
Division of Fisheries Resource Management, Faculty of Fisheries, Sher-e-Kashmir University of Agricultural Sciences and Technology of Kashmir, Srinagar, India

Hanan Javid
Plant Reproductive Biology, Genetic Diversity and Phytochemistry Research Laboratory, Department of Botany, University of Kashmir, Srinagar, India

Rajeev Kumar
Department of Forensic Science, School of Basic and Applied Sciences, Galgotias University, Greater Noida, India

Rajesh Kumar
Department of Biosciences, Himachal Pradesh University, Shimla, India

Breno Luiz Sales Lemos
Institute of Biological Science, Federal University of Minas Gerais, Belo Horizonte, Brazil

Moline Severino Lemos
Institute of Biological Science, Federal University of Minas Gerais, Belo Horizonte, Brazil

Juan José Maldonado-Miranda
Professor of Faculty of Professional Studies Huasteca Zone, Autonomous University of San Luis Potosi, San Luis Potosi, Mexico

Junaid Malik
Department of Zoology, Government Degree College, Bijbehara, India

Nahum Andrés Medellin-Castillo
Professor of Graduate Studies and Research Center, Faculty of Engineering, Autonomous University of San Luis Potosi, San Luis Potosi, Mexico; Multidisciplinary Graduate Program in Environmental Sciences, Autonomous University of San Luis Potosi, San Luis Potosi, Mexico

Maibam Malemngamba Meitei
Department of Aquaculture, College of Fisheries, Central Agricultural University, Lembucherra, India

Debashri Mondal
Department of Zoology, Raiganj University, University Road, College Para, Raiganj, India

Sandip Mondal
School of Pharmaceutical Technology, School of Medical Sciences, ADAMAS University Kolkata, India

Tamal Mondal
Department of Botany, Hiralal Mazumdar Memorial College for Women, College Para, Dakshineswar, Kolkata, India

Irshad A. Nawchoo
Plant Reproductive Biology, Genetic Diversity and Phytochemistry Research Laboratory, Department of Botany, University of Kashmir, Srinagar, India

Nighat Un Nissa
Department of Zoology, University of Kashmir, Srinagar, India

Bilal Ahmad Paray
Department of Zoology, College of Science, King Saud University, Riyadh, Saudi Arabia

Kapil Parihar
State Forensic Science Laboratory, Jaipur, Rajasthan, India

Amauri Ponce-Hernández
Student of Graduate Studies and Research Center, Faculty of Chemistry, Autonomous University of San Luis Potosi, San Luis Potosi, Mexico

Raksha Rani
Division Zoology, Department of Biosciences, Career Point University, Hamirpur, India

Muneeb U. Rehman
Department of Clinical Pharmacy, College of Pharmacy, King Saud University, Riyadh, Saudi Arabia

Andrea Guadalupe Rodríguez-López
Faculty of Medicine, Autonomous University of San Luis Potosi, San Luis Potosi, Mexico

Mahipal Singh Sankhla
Department of Forensic Science, Vivekananda Global University, Jaipur, India

Preeti Sharma
Division Zoology, Department of Biosciences, Career Point University, Hamirpur, India

Podduturi Sindhura
Telangana Social Welfare Residential Degree College for Women, Bhupalapally, India

Soibam Khogen Singh
Department of Aquaculture, College of Fisheries, Central Agricultural University, Lembucherra, India

Ngasotter Soibam
Department of Fish Processing Technology and Engineering, College of Fisheries, Central Agricultural University, Lembucherra, India

Swaroop S. Sonone
Government Institute of Forensic Science, Aurangabad, India

R. Sowndharya
Department of Microbiology, The American College, Madurai, India

M.S. Sri Janani
Department of Microbiology, The American College, Madurai, India

Uzma Sultana
Telangana Social Welfare Residential Degree College for Women, Kamareddy, India

Fernanda Maria Policarpo Tonelli
Institute of Biological Science, Federal University of Minas Gerais, Belo Horizonte, Brazil; Pitágoras College, Divinópolis, Brazil

Flávia Cristina Policarpo Tonelli
Estácio de Sá University, Ribeirão Preto, Brazil

Podduturi Vanamala
Telangana Social Welfare Residential Degree College for Women, Kamareddy, India

Tanaji Vasavilatha
Telangana Social Welfare Residential Degree College for Women, Nizamabad, India

Rohit Kumar Verma
Dr APJ Abdul Kalam Institute of Forensic Science & Criminology, Bundelkhand University, Jhansi, India

Gusheinzed Waikhom
Department of Aquaculture, College of Fisheries, Central Agricultural University, Lembucherra, India

Aadil Farooq War
Department of Botany, University of Kashmir, Srinagar, India

Leticia Guadalupe Yánez-Estrada
Professor of Faculty of Medicine, Autonomous University of San Luis Potosi, San Luis Potosi, Mexico

Ali Mohd Yatoo
Centre of Research for Development/Department of Environmental Science, University of Kashmir, Srinagar, India

Tabasum Yousuf
Fish Biology and Limnology Research Laboratory, Department of Zoology, University of Kashmir, Hazratbal, Jammu and Kashmir, India

Zarka Zaheen
Centre of Research for Development/Department of Environmental Science, University of Kashmir, Srinagar, India

About the editors

Dr. Gowhar Hamid Dar (Ph.D.) is currently working as an assistant professor in environmental science, Sri Pratap College, Cluster University Srinagar, Department of Higher Education (J&K). He has a Ph.D. in Environmental Science with specialization in environmental microbiology (fish microbiology, fish pathology, industrial microbiology, taxonomy, and limnology). He has been teaching postgraduate and graduate students for many years at the Postgraduate Department of Environmental Science, Sri Pratap College, Cluster University Srinagar. He has more than 50 articles (*h*-index 10; *i*-index 12; total citation >300) in international and national journals of repute and a number of books with international publishers (Springer, Elsevier, CRC Press, Taylor and Francis, Apple Academic Press, John Wiley, IGI Global) to his credit. Moreover, he supervises a number of students for the completion of degrees. He has been working on the isolation, identification, and characterization of microbes, their pathogenic behavior, and the impact of pollution on development of diseases in fish fauna for the last several years. He has received many awards and recognitions for his services in science and development. In addition, he serves as a member of various research and academic committees.

Dr. Rouf Ahmad Bhat (Ph.D.) pursued his doctorate at Sher-e-Kashmir University of Agricultural Sciences and Technology Kashmir (Division of Environmental Science) and is presently working in the Department of School Education, Government of Jammu and Kashmir. Dr. Bhat has been teaching graduate and postgraduate students of environmental sciences for the past 3 years. He is an author of more than 50 research articles (*h*-index 14; *i*-index 19; total citation >591) and 40 book chapters, and has published more than 28 books with international publishers (Springer, Elsevier, CRC Press, Taylor and Francis, Apple Academic Press, John Wiley, IGI Global). He has his specialization in limnology, toxicology, phytochemistry, and phytoremediation. Dr. Bhat has presented and participated in numerous state, national, and international conferences, seminars, workshops, and symposia. In addition, he has worked as an associate environmental expert in the World Bank-funded Flood Recovery Project and also on the environmental support staff in Asian Development Bank (ADB)-funded development projects. He has received many awards and recognitions for his services to the science of water testing and air and noise analysis. He has served as editorial board member and reviewer of reputed international journals. Dr. Bhat is still writing and experimenting with diverse capacities of plants for use in aquatic pollution remediation.

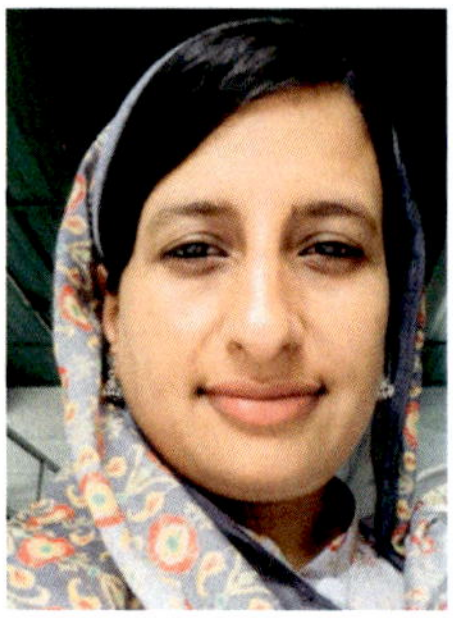

Dr. Humaira Qadri (Ph.D.) has been actively involved in teaching postgraduate students of environmental science for the past 10 years in Sri Pratap College Campus, Cluster University Srinagar, J&K, India. She also heads the Department of Environment and Water Management. A gold medalist at the master's level, she has received a number of awards and certificates of merit. Her specialization is in limnology, nutrient dynamics, and phytoremediation. She has published a number of papers in international journals and has more than 10 books with national and international publishers. She is also the reviewer of various international journals and is the principal investigator for major projects on phytoremediation. She guides research students in Ph.D. programs and has supervised more than 60 master's dissertations. She also has been on the scientific board of various international conferences and holds life memberships in several international organizations. With a number of national scientific events to her credit, she has been an active participant in national and international scientific events and has organized a number of conferences as well as seminars of national and international repute.

Prof. Khalid M. Al-Ghamdy is a professor of biology (major specialization in biological control of insects). He obtained his Ph.D. degree from McGill University, Montreal, Canada in 2006. He is currently head of the Department of Biological Sciences and the Vice Dean in the Faculty of Science, King Abdulaziz University. He has more than 100 research publications published in peer-reviewed journals.

Dr. Khalid Rehman Hakeem, Ph.D., is professor at King Abdulaziz University, Jeddah, Saudi Arabia. After completing his doctorate (botany, with specialization in plant eco-physiology and molecular biology) from Jamia Hamdard, New Delhi, India in 2011, he worked as a lecturer at the University of Kashmir, Srinagar, for a short period. Later, he joined Universiti Putra Malaysia, Selangor, Malaysia, and worked there as a postdoctorate fellow in 2012 and fellow researcher (associate professor) from 2013 to 2016. Dr. Hakeem has more than 10 years of teaching and research experience in plant eco-physiology, biotechnology, and molecular biology, medicinal plant research, and plant–microbe–soil interactions, as well as in environmental studies. He is the recipient of several fellowships at both national and international levels; also, he has served as a visiting scientist at Jinan University, Guangzhou, China. Currently, he is involved with a number of international research projects with different government organizations.

So far, Dr. Hakeem has authored and edited more than 65 books with international publishers, including Springer Nature, Academic Press (Elsevier), and CRC Press. He also has to his credit more than 135 research publications in peer-reviewed international journals and 62 book chapters in edited volumes with international publishers.

At present, Dr. Hakeem serves as an editorial board member and reviewer for several high-impact international scientific journals from Elsevier, Springer Nature, Taylor and Francis, Cambridge, and John Wiley Publishers. He is included on the advisory board of Cambridge Scholars Publishing, United Kingdom. Prof. Khalid has recently been elected as a fellow of the Royal Society of Biology, London, United Kingdom. He is also a fellow of the Plantae group of the American Society of Plant Biologists, a member of the World Academy of Sciences, a member of the International Society for Development and Sustainability, Japan, and a member of the Asian Federation of Biotechnology, Korea. Dr. Hakeem was listed in *Marquis Who's Who in the World* in 2014–19. Currently, Dr. Hakeem is engaged in studying plant processes at eco-physiological as well as molecular levels.

Foreword

Fish form one of the important food sources for human nutrition, as they are rich sources of proteins, fats, minerals, and vitamins. The need for food security, interest in a healthier diet, and meeting the demands of the rising population worldwide have led to the promotion of aquaculture as well as captured fish techniques. However, equally important is the production of healthy and pathogen-free fish, as they are vulnerable to microbial infestation by bacteria, viruses, and fungi. Among these, fish bacteria have been studied and documented extensively. A literature review provides an understanding of the various bacterial species infecting fish, based on the identification of bacterial genomes concerning virulence, virulence factors, host adherence, colonization, structural components, extracellular factors, secretion systems, iron acquisition, and quorum sensing mechanisms. Fish are found in seawater, freshwater, and also cultured in man-made aquaculture facilities. All these water sources are prone to contamination by toxins, industrial effluents, anthropogenic pollutants, pesticides, and other wastes, which affect the health and the immune system mechanisms of the fish. From this perspective, there is a need in the research and academic fields to understand the nature, mode, consequences, and control of bacteria-related disease in fishes. This book, *Bacterial Fish Diseases*, is a detailed account of all the common bacterial pathogens, their mechanisms related to disease phenomenon in fish, and the strategies in their alleviation. Furthermore, it highlights the problems associated with the culturing of fish. This book will be a useful resource to ecologists, environmentalists, nutritionists, and scientists in general for understanding the bacteria-related diseases in fish.

The book includes 19 chapters, contributed by different authors from different origins worldwide. The hard work of the authors provides a wide-ranging view of bacterial diseases of fish fauna and possible treatment to control these dreadful diseases. The book has also provided enough space for discussing the economic losses that occur due to fish diseases.

Chapter 1, Aquatic Pollution and Marine Ecosystems, has been contributed by scientists from Romania, who report that the contamination of the marine ecosystems has gained sizeable proportions and also elaborate on how human interventions have induced huge contamination of marine ecosystems. Chapter 2, Heavy Metals as Pollutants in the Aquatic Black Sea Ecosystem, was also written by the Romanian authors and explains the role of heavy metals as toxic pollutants that enter the biogeochemical circuits and accumulate in natural and artificial ecosystems, including in the

Black Sea ecosystem. In addition, on a global scale, there is evidence that anthropogenic activities have polluted the environment with heavy metals from the poles to the tropics and from the mountains to the depths of the oceans. Chapter 3, Effects of Heavy Metals and Pesticides on Fish, presents the severe damage caused by heavy metals and pesticides in fishes. The pesticides that cause several behavioral and migratory changes in fishes that lead to their decline have been addressed in Chapter 4, Pesticide Toxicity and Bacterial Diseases in Fishes, by authors from India. These authors stress the enforcement of strict environmental laws for checking and controlling the excessive use of pesticides. Chapter 5, Impacts of Aquatic Pollution on Fish Fauna, illustrates the role of molecular, cellular, and biochemical biomarkers in the monitoring of the environment and aquatic pollution. The bacterial diseases in fishes due to close contact between the marine environment and the fish pathogens that disrupt the stable supply of fishes around the world caught the attention of Indian authors, as illustrated in Chapter 6, Bacterial Diseases in Fish With Relation to Pollution and Their Consequences—A Global Scenario. Chapter 7, Common Bacterial Infections Affecting Freshwater Fish Fauna and Impact of Pollution and Water Quality Characteristics on Bacterial Pathogenicity, expounded by authors from India and Saudi Arabia, highlights the impact of bacterial infection in fish populations and corroborates the role of stress factors, that is, poor diet, inadequate husbandry procedures, and polluted environmental conditions around the host body (fish). Chapter 8, Global Status of Bacterial Diseases in Fish in Relation to Pollution and Their Consequences, has been presented by scientists from India. In this review, the authors discuss the role of different bacteria, that is, *Aeromonas*, *Mycobacterium*, *Streptococcus*, *Edwardsiella*, and *Flavobacterium*, causing diseases in fishes at the global level. Chapter 9, Understanding the Pathogenesis of Bacterial Diseases in Fish, elucidates the role of the pathogenesis of bacterial diseases caused by gram-positive and gram-negative bacteria in fishes, to expound the possible targets for treatment and disease prevention measures. Chapter 10, Evaluation of the Fish Invasiveness Scoring Kit (Fisk v2) for Pleco Fish or Devil Fish, is a case study exemplified by authors from Mexico. Furthermore, they stress FISK v2, a fish invasiveness detection tool to detect the potential risk of invasion of the pleco fish introduced into the aquatic systems of the Huasteca Zone of the State of San Luis Potosi, Mexico. Chapter 11, Profiling of Common Bacterial Pathogens in Fish, contributed by authors from Nigeria, Yobe State, and Uganda, is an extensive and critical review of aquaculture practice; its impact; its current status of production in Nigeria and major sub-Saharan Africa; and three major bacterial fish pathogens, *Aeromonas hydrophila*, *Flavobacterium psychrophilum*, and *Flavobacterium columnare*. The review considers the genome, pathogenesis, and control of the pathogens, immune response of the host, host–pathogen interaction, and recent advances in vaccine development. *Furunculosis*, a worldwide severe, contagious, principal, and economically the most significant disease, caused by *Aeromonas salmonicida* in fresh and seawater fishes, has been addressed by an author from Pakistan in Chapter 12, "Status of Furunculosis in Fish Fauna." Chapter 13, Bacterial Gill Disease and Aquatic Pollution: A Serious Concern for the Aquaculture Industry, focuses on the impact of water pollution on fishes, especially the emergence of bacterial gill diseases along with their etiological agents, pathogenesis, epizootiology, diagnosis, and control, which are significant for successful management. Chapter 14, Common Bacterial Pathogens in Fish: An Overview, has been written by Indian authors, who have compiled specific dispersed literature published about several bacterial pathogens associated with fish. Chapter 15, Bacterial Diseases in Cultured Fishes: An Update of Advances in Control Measures, addresses the recent updates of the control tools for bacterial infection. The development of fish vaccines, antimicrobial peptides, probiotics, prebiotics, synbiotic concepts, and herbal biomedicines derived from plants are covered in this chapter. Chapter 16, Ulceration in Fish: Causes, Diagnosis, and Prevention, provides an exhaustive review of the dreadful disease of ulceration syndrome in fishes. The authors discuss major causes, diagnostic methods, and some preventive measures, hence providing an insight into the early detection of this disease at the hatcheries level and in farms. The antibiotic resistance in fishes and impedance of the beneficial microbiota of the gastrointestinal ecosystem drew the attention of Chinese and Indian authors, who addressed this issue in Chapter 17, Application of Probiotic Bacteria for the Management of Fish in Aquaculture. Furthermore, this review deals with the analysis and corroborates that probiotics could act as growth promoters, increase disease resistance, enhance the immune response in fish, and improve the water quality for better survival. Chapter 18, Efficacy of Different Treatments Available for Bacterial Pathogens in Fish, is aimed at summarizing the various bacterial diseases and effective treatments available to combat the pathogenesis of these bacterial diseases. The authors have addressed many fish infections ("columnaris referred to as cottonmouth, gill infection, itch, swelling, tail and fin rot, fungal disease, pop and cloudy eye, swim bladder disease, lice and nematode worm infestations, water quality-induced disease, alimentary stoppages, anorexia, chilodonella, ergasilus, TB, glugea, henneguya, hexamita, head and lateral line erosion disease, injuries, leeches in an aquarium, lymphocystis, marine velvet, and neon tetra disease") that are most prevalent in fish farms. Different methods/treatments employed for controlling fish diseases, including antibiotics, bioremediation (improving water quality), disinfectants, prebiotics, and synbiotics, have been addressed by the authors in this chapter. Chapter 19, Summary of Economic Losses due to Bacterial Pathogens in Aquaculture Industry, addresses the economic losses incurred due to the incidence of bacterial diseases in fishes. The authors summarize the numerical data on production losses worldwide, according to pathogenic bacteria, with a particular focus on the Mexican region.

The book structure is designed to cover essential aspects of bacterial fish diseases concerning the aquatic environment. It is unique and provides an ideal source of scientific information for research scholars, faculty, and scientists involved in fish and fish health, fish diseases, and the mechanism of controlling bacterial infections in fish fauna. I greatly appreciate the efforts expended by the editors for efficaciously bringing together this informative and comprehensive volume.

Karuna Rupula

Department of Biochemistry, University College of Science,

Osmania University, Hyderabad, India

Preface

Fish supply people with long-term benefits, including food and the direct financial benefits of providing employment, profits, and saving money. More indirect but equally valuable benefits of fish and aquatic ecosystems include recreational boating, sport fishing, swimming, and relaxation. Consumption of seafoods has taken place from ancient times. Fish are the most essential source of proteins for humans and have high nutritional value. Fish are rich in omega-3 fatty acids, with much medical evidence behind their efficacy. On the other side, fish health is affected due to close contact between the marine environment and the fish pathogens, which disrupts the stable supply of fishes around the world as well as affects fish health. Fish diseases are caused by bacterial and fungal pathogens and aquatic environmental factors, such as poor water quality, are accountable for mass mortalities of fishes. Environmental and industrial pollutants are a neglected source that impacts the aquatic ecosystem. Pollutants can have direct and indirect impacts on the behavior of aquatic organisms. The fish biota may be affected by pollution by altering their biochemical, respiratory, and population structure, as well as developmental and structural functions. Many types of organic and inorganic contaminants including plastics, pharmaceuticals, pesticides, and metals are released by humans into the environment, both aquatic and terrestrial. Contaminants from agricultural activities, such as pesticides, are highly toxic to nontarget organisms, including fish.

Diseases and infections inside the host body or in its physiology have a tendency to take place only when the fish is stressed due to insufficient diet, poor husbandry procedures, and/or polluted environmental conditions. Therefore, in order to understand the various bacterial diseases of fish species, it is equally important to comprehend the correlation between the bacteria and its host as well as its environment. The upsurge of bacterial gill diseases in fishes is believed to occur when there is deterioration of ecological conditions of the aquatic habitat. The triggering of diseases in fishes by pollution reveals that human-induced attributes are widely responsible for the onset of pathogenic disease in them, because these attributes are directly proportional to the vulnerability of host species in their habitat. Furunculosis is a worldwide severe, contagious, principal, and economically very important disease in fresh and seawater fishes. The major bacterial fish pathogens, *Aeromonas hydrophila, Flavobacterium psychrophilum*, and *Flavobacterium columnare*, are severely affecting fish species worldwide. Fish infections ("columnaris referred to as cottonmouth, gill infection, ich, swelling, tail and fin rot, fungal disease, pop and cloudy eye, swim bladder disease, lice and nematode worm infestations, water quality-induced disease, alimentary stoppages, anorexia, chilodonella, ergasilus, TB, glugea, henneguya, hexamita, head and lateral line erosion disease, injuries, leeches in aquariums, lymphocystis, marine velvet, and neon tetra disease") are common in fish farms and can cause "mass mortalities"; and the treatment of these requires the intensive use of chemicals and antibiotics.

Recent progress in aquaculture production parallels the growing number of disease incidences that harm global aquaculture production, profitability, and sustainability. Bacteria-related diseases can cause severe mortality in both wild and cultured fish. Bacterial diseases and infections associated with them are very widespread and are one of the most challenging health problems to resolve effectively. An array of bacterial pathogens causes major losses to aquaculture, comprising around 34% of total diseases. Antibiotics are frequently used as a control measure against bacterial diseases in fish, but the risk of developing antibiotic-resistant bacteria strains is increasing. Research on various options other than antibiotics is being carried out and therapeutic alternatives are being developed. Fish vaccines, antimicrobial peptides, probiotic, prebiotic and synbiotic concepts, and herbal biomedicines derived from plants are often regarded as beneficial developments to control bacterial diseases in aquaculture. However, despite the strong impact this has had on production in the aquaculture industry, there are still a number of environmental and health factors that can affect and trigger fish diseases, causing significant economic losses. Among these diseases are those generated by the attack of pathogenic bacteria, more specifically by the genera of gram-negative bacteria and, to a lesser extent, gram-positive bacteria.

This book has examined in detail the ill consequences of the rising pollution levels in our aquatic systems on fish species. The starting chapters give an elaborative outline of pollution and its detrimental impacts on the fish fauna of different ecosystems. A considerable space has been given to highlighting the major bacterial pathogens that are having an impact on fish health and triggering different types of diseases in them. Furthermore, sufficient attention is given to the different types of treatments available to deal with these rapidly increasing bacterial infections in fish. The book also contains a detailed description of the economic losses suffered from bacterial infections in fish.

In general, the book is a valuable source of information on different bacterial diseases in fish and highlights advanced research on the theme. This book addresses the diverse challenges faced by fish fauna and suggests some unique future action plans to deal with them. The content of the book is therefore diverse and addresses the needs of students, researchers, ichthyologists, microbiologists, and scientists worldwide.

Gowhar Hamid Dar
Rouf Ahmad Bhat
Humaira Qadri
Khalid M. Al-Ghamdy
Khalid Rehman Hakeem

Acknowledgments

The editors would like to acknowledge with gratitude the efforts of all the authors who contributed to this volume on bacterial fish diseases. Acknowledgment is also extended to Principal Sri Pratap College, Higher Education Department (J&K) and Cluster University Srinagar for their encouragement and support in bringing out this volume. The editors would also like to take this opportunity to thank the faculty and nonteaching staff of the PG Department of Environmental Science and the Department of Environment and Water Management, Sri Pratap College, Cluster University Srinagar for their kind guidance and support. We would also like to acknowledge the help of all those friends and colleagues who encouraged us to edit this volume. Exclusive thanks to Mr. Asif Ahmad Bhat and Mr. Shahid Bashir Haji for their gracious support while designing this book.

CHAPTER

Aquatic pollution and marine ecosystems

1

Monica Butnariu

Banat's University of Agricultural Sciences and Veterinary Medicine "King Michael I of Romania" from Timisoara, Timis, Romania

1.1 Background

An aquatic environment is considered to be polluted when it has been permanently altered by the intake of excessive quantities of more or less toxic compounds from natural sources or from human activities. These contaminants can cause different types of pollution, some of which can increase mortality for certain species of animals and plants, change physiological capabilities, or degrade water physicochemical properties to the point of making the water unusable by humans. Not all contaminants pose the same hazards to ecosystems (defined as the ensemble of biotope and biocenosis, or all living things of a biocenosis), as some are biodegradable.

Ecosystems are rich in animal and plant species and microorganisms, and thus are naturally able to transform or eliminate biodegradable compounds in part or in whole, thus maintaining the natural balance as well as the natural water properties. However, if the amounts of these compounds exceed a critical threshold, self-purification capacities become insufficient. If the contaminants cannot be removed quickly enough, they accumulate and disturb the natural dynamic balance of the aquatic environment and may even become toxic, allowing a state of aquatic pollution to be identified. Other contaminants, such as plastics, metals, and certain pesticides, are not biodegradable, or are only partially biodegradable; the self-purification process described is thus inoperative in this case and consequently the compounds can accumulate in the ecosystem and possibly become toxic to living species that ingest them.

Certain compounds, such as heavy metals or pesticides, accumulate in organisms and build up in tissues and organs at concentrations that are sometimes much higher than those measured in water, due to the phenomenon of *bioaccumulation*.

Population growth, increased urbanization and industrialization, and more intensive agriculture are inevitably accompanied by an increase in water pollution. Under these conditions, it is not always easy to identify the sources of contamination or to estimate the effects, which depend on both the nature and the concentration of the contaminant and the ecosystem in which it acts; the circumstances surrounding contamination are thus very complex. Consequently, it is often difficult to determine for each toxic compound a maximum acceptable concentration for an ecosystem, as there is no consensus on the threshold not to be exceeded, and in the rare cases where rules and regulations are in place, they tend to vary by country (Alimba & Faggio, 2019).

Bacterial Fish Diseases. DOI: https://doi.org/10.1016/B978-0-323-85624-9.00009-9

As water flows, it seeps through different layers of soil and rock in the subsoil, dissolving and absorbing the compounds it encounters. Inorganic minerals, fertilizers applied in agriculture, and toxic products used and discarded by factories and household consumers are absorbed by water, polluting it and making it unsafe, sometimes even nonpotable.

In large cities with water distribution networks, polluting compounds can also be found in pipes, old tubing, tanks, or in other equipment that has contact with water along its route to the consumer. Impurities can also enter the water through small cracks in the pipes. Contaminants such as gasoline, fluoride, lead, drugs, pesticides, asbestos, nitrates, and radioactive waste all have negative effects on the water they come in contact with. Many of these cannot be detected with our natural senses; we cannot, for example, detect dissolved lead, mercury, or trihalomethanes in water. And sometimes even a glass of clear water can contain one or more harmful compounds, even though the water looks perfectly clean.

Within our planetary ecological system, the presence of water is an indispensable condition of life, and for human civilization it represents that natural resource on which any domain of economic activity depends. It plays different roles in the body; without water all biological responses become impossible. Lack of water or contaminated water both have multiple negative repercussions on humans and their health. Water shortages have increased as a result of population growth, urban sprawl, rising living standards, and industrialization, and there is a profound danger of deteriorating water wells.

Most contaminants that find their way into aquatic ecosystems come from human activities, both along coastlines and far inland (see Fig. 1.1).

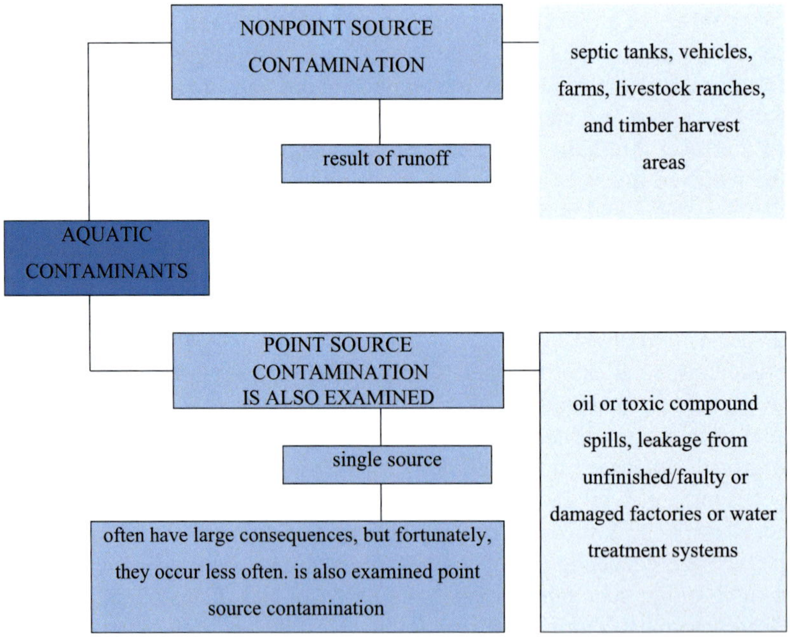

FIGURE 1.1

Types of contaminants of aquatic ecosystems.

The increase in maritime oil delivery from global buying and selling, has the excessively accompany the trends of oil exhaustion and fabrication, testing every year delivery volume increases, from approximately 1100 million tonnes (Mt) in 1970 to approximately 2000 Mt in 2006. Even more impressive progress has been shown in the seaborne trade of petroleum products, which has grown from approximately 200 Mt in 1970 to approximately 700 Mt in 2006. Nautical delivery of oil and petroleum articles has hold, for the last time intervals, specially last four decades, a constituent that can lead to a finishing degree of hazard in terms of aquatic contamination escapades, notwithstanding the fact that oil construction technology has made significant progress during that time. After 1970, maritime transport of liquefied natural gas also increased significantly. Thus, in recent years, a sizable global fleet of specialized liquefied gas tankers has been developed, employed in delivery of two distinct product types: liquefied petroleum gas (LPG) (flammable mixture of hydrocarbon gases) and liquefied natural gas (LNG) (predominantly methane, CH_4, with a mixture of ethane, C_2H_6). This fleet today totals approximately 27 Mt. The extraction and processing of offshore oil and gas have developed rapidly in recent decades.

Note that over 80% of aquatic contaminants of marine ecosystems come from land-based activities as described in Fig. 1.2.

Physicochemical characteristics of the aquatic environment are identified in Fig. 1.3.

The progression of offshore drilling and manufacturing has been rapid. At the start of 1974, approximately 192 fixed platforms and 239 mobile platforms were in operation globally. The

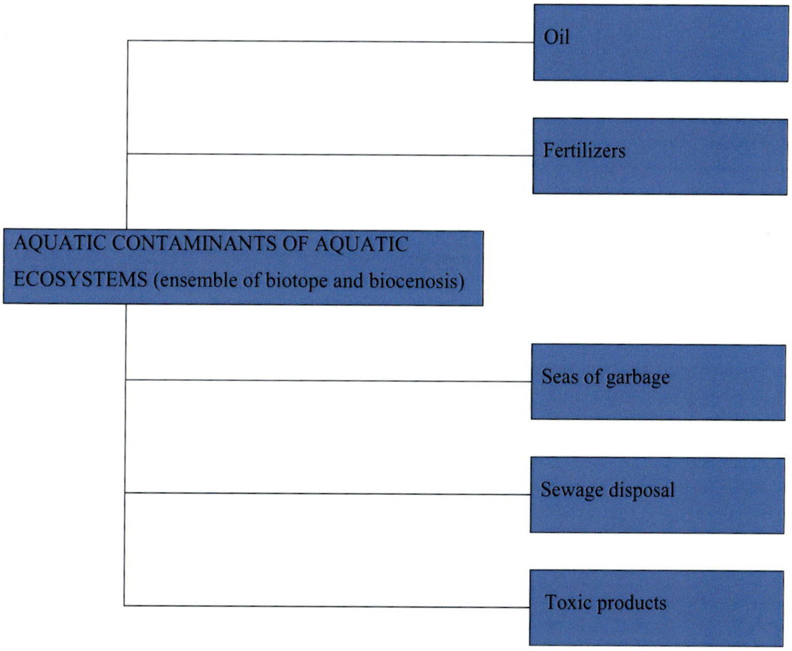

FIGURE 1.2

Aquatic contaminants of marine ecosystems (ensemble of biotope and biocenosis) from land-based activities.

The physicochemical characteristics and their meaningcal and the meaning characteristics

Transparency	Transparency depends on the intensity of the light on the water surface, the angle of incidence of the rays, the materials dissolved in the water and the suspended substances. It is an essential property of water basins, conformable to which the primary productivity is achieved. The most transparent water is in the Sargasso Sea, through which the Secchi disc can be seen at up to 66.5 m depth. In general, the transparency of ocean waters varies between 50–60 m.
Turbidity	Turbidity is due to the presence in the water of gaseous or solid particles in suspension. Sources of turbidity can be sand, mud, fine alluvium, organic substances, and planktonic organisms. Turbidity influences the activity of hydrobionts, either directly by physicochemical action, or indirectly by reducing the quantity of light.
Density	The density of water changes based on the temperature, pressure, salt concentration, and the content of suspended substances. The highest water density is recorded at a temperature of 4°C.
Surface tension	Surface tension is the force acting on the separating surface of a liquid in a gaseous medium.
Viscosity	The viscosity of the aquatic environment is about 60 times higher than the air in identical conditions, disfavoring the movement of animals in the aquatic environment.
Hydrostatic pressure	The hydrostatic pressure depends on the height of the water column and its density. For fresh water, the density is maximum at a temperature of 4.3°C. In principle, the pressure increases by 1 atmosphere every 10.3 m and for water at 9.98 m.
Temperature	Temperature influences water properties (quantity of O_2, water chemistry, etc.; speed of biochemical processes increases approximately twice as the temperature increases by 10°C) and the activity of hydrobionts (locomotion, refabrication, feeding, rate of metabolic processes).
Light Dependence	Depending on the light, biorhythms are formed, some migrations are triggered, certain forms of refabrication are initiated, etc.
pH	Ionic reaction. Most commonly, aquatic life is only possible between pH values 5 -9.
Salinity	Salinity is the total quantity of salts contained in surface waters.
Nitrogen	Nitrogen is one of the components of substances dissolved in water, coming mainly from the excretions of fish.
Phosphorus	Phosphorus is essential for biological processes but in large quantities favors eutrophication circumstances.
Oxygen (O_2) and carbon dioxide (CO_2)	The maximum quantity of O_2 in the water depends on the temperature; the higher it is, the lower the quantity of O_2 in the water (at 25°C there is 18% less O_2 than at 15°C). CO_2 comes from the respiration of fish, plants, bacteria.
The hardness of water	The hardness of water corresponds to the set of substances based on Ca and Mg, which it contains. Many other substances are found in solution in water. Their content is generally low and does not cause problems; some are even beneficial, such as trace elements. Various elements are grouped under this term, including vitamins and metals, which in very small quantities are indispensable to life.

FIGURE 1.3

Physicochemical characteristics of the aquatic environment.

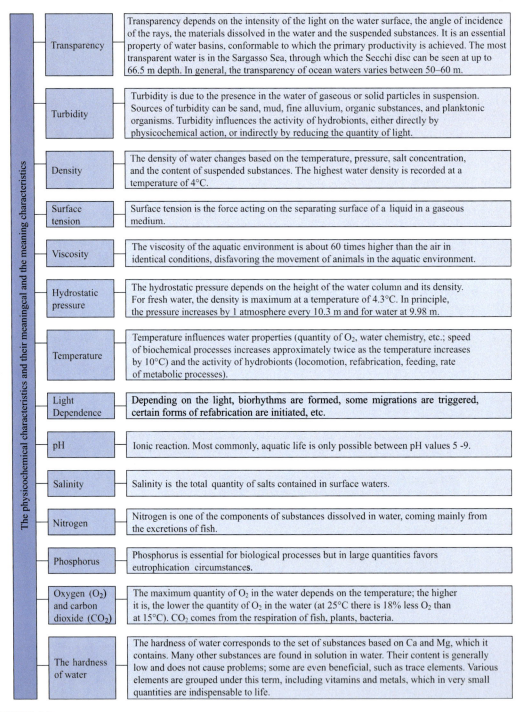

numbers have grown tremendously in the last decade, today totaling over 7000 oil installations. Based on statements by the United Nations Secretariat (UNS), about 27% of the worldwide totals of crude oil and gas are obtained from offshore environments.

Water properties are being regulated by standards, due to their relevance for the safety of life and their important role in economic activities. Water properties are monitored by standardized indicators, and calculations and analyses are performed on both surface and groundwater. The World Health Organization (WHO) and various health and medical investigation councils and environmental protection agencies are promoting a preventive approach to protecting drinking water applied by consumers. The protection of properties of water intended for consumption is a significant objective included in the EC directives, aimed at protecting the public health.

The water, air, and soil, the components of the biosphere, are related environmental elements. In each of them different flora and fauna grow and live, which, in order for the system to endure, must be sustained with a clean and healthy environment. For this to occur, it is mandatory to understand the sources of pollution, that is, the contaminants that could harm the evolution of life and have dire consequences on life and the environment, leading to measures to prevent and decrease contamination of all the environmental elements (Brown & Takada, 2017).

1.2 Sources of contamination in aquatic ecosystems

Most environmental contaminants endanger not only human environments and lives, but also the integrity and function of ecosystems (namely, all living things of a biocenosis). Water pollution is the predominant form of contamination from the moment the contaminants enter seas, rivers, lakes, and wetlands.

Aquatic and coastal ecosystems are characterized by their complexity and sensitivity to different organic and inorganic pollutants. Priority investigations of potential areas of human and environmental hazards have identified and monitored different biomarkers in the hazard assessment process in order to improve hazard integration and assessment.

Biomarkers (a biological characteristic that may be molecular, anatomical, physiological, or biochemical) are identified as functional biological measures that occur due to exposure to contaminants or stressors, which are signaled at the physiological and behavioral level of a suborganism. Attempts are currently being made to identify new responsive biomarkers, the characteristics of which can be measured and evaluated objectively, to act as indicators of a standard or pathological biological process, to be used in imaging in new biological tests to understand the immediate and long-term effects of different compounds on aquatic life in general and on human health in particular.

Water pollution affects life on a global scale. Water is the source of life for organisms in all environments, as without water there can be no life. Water quality has begun to deteriorate more and more as a result of bacteriological, chemical, and physical changes.

Water pollution is the alteration of the biological, chemical, and physical qualities of water, caused directly or indirectly, naturally or anthropically. Polluted water is considered unfit for standard uses. The pollution can take place continuously, as is the case with sewage in cities or waste from manufacturing that is discharged into water, or discontinuously, at irregular or regular time intervals; it can be temporary or accidental, as in the case of unforeseen damage (Kalogianni et al., 2017).

Profound water pollution is caused by agriculture, manufacturing, shipping, and household activities. The manufacturing discharges into natural water include toxic products, organic and inorganic compounds, plant and animal waste, hydrocarbons, solvents, and heat. The substances can be in a solid or liquid state, miscible or immiscible with water, easily or barely volatile, and more or less toxic.

Water plays the most significant role in ecological balance, and its pollution is a current problem with serious repercussions for the population. Currently, there are tons of impurities in the atmosphere that rain brings to the earth, and years of technological advancements in manufacturing and agriculture have intensified the pollution of water sources. The effects of water source pollution are complex and varied, depending on the nature and the concentration of the contaminants.

Contaminants introduced into waters from both natural and artificial sources are numerous, having significant consequences on both surface and groundwater. From the point of view of the source, water pollution can thus be either natural or artificial.

Natural pollution is due to sources of natural contamination, for example, the interaction of water with the atmosphere when gases are dissolved in it, or the pollution that occurs when water passes through soluble rocks (when the water is infused with different salts), or as a result excessive growth of aquatic vegetation and living things, etc.

Artificial pollution is due to sources of wastewater of any kind, whether rainwater, mud, residues, from shipping, etc. Based on the degree of transformation of the water contaminants, we distinguish the deposition of suspended compounds from wastewater discharged into a body of receiving water on its bed, and which begins as soon as the gases have accrued from the fermentation of organic matter in the compounds. Deposited suspensions entrain the rest of the suspensions and bring them to the surface of the water, from where they are then delivered by the water current (Fan et al., 2018).

Water contaminants can be categorized, based on their nature and the damage caused, into the different types described in the following paragraphs.

Organic compounds (artificial or natural sources) are the main water pollutants. Organic compounds of natural sources (vegetable and animal) consume oxygen from the water both for growth and after death. Organic matter consumes oxygen (O_2) from the water during decomposition, to a greater or lesser extent, based on the quantity of organic matter discharged, leading to the destruction of fish stock and in general of all aquatic organisms (diversity of plants, invertebrates, fish, wildlife, etc., is affected).

At the same time, oxygen is also needed for aerobic self-purification processes, in particular for aerobic bacteria that oxidize organic compounds and which, in the end, lead to water self-purification. The normative dissolved oxygen concentration varies between 4 and 6 mg/dm^3, based on the category of use; going below this limit causes the aerobic processes to cease, with very serious repercussions.

The most significant organic compounds from natural sources are carbohydrates, crude oil, lignin, aquatic biotoxins, tannin, etc. Artificial organic compounds come from the manufacturing in refineries (diesel, gasoline, oils, organic solvents, etc.), organic chemical manufacturing, and petrochemical manufacturing (detergents, halogenated hydrocarbons, hydrocarbons, etc.) (Bergmann et al., 2017).

Inorganic, suspended, or dissolved compounds are the most common in industrial wastewater. Of these, the heavy metals (Pb, Cu, Zn, Cr), chlorides, sulfates, and others are the most significant. Inorganic salts lead to increased water salinity, and some of them can increase hardness. Chlorines in large quantities make water unsuitable for drinking as well as for industrial water supplies, irrigation, and other applications.

Due to bioaccumulation, heavy metals have toxic effects on aquatic organisms (diversity of plants, invertebrates, fish, and other wildlife) while inhibiting self-purification processes. Nitrogen and phosphorus salts produce the rapid growth of algae on the water surface. Very hard water leads to deposits on pipes, increasing their roughness and reducing their delivery and heat transfer abilities.

Suspended materials, organic or inorganic, are deposited on the channel bed, forming banks that can impede navigation; the consumption of oxygen from the water if the materials are of organic source causes the formation of foul-smelling gases.

Floating suspended compounds, such as crude oil, foam due to detergents, and petroleum products, damage the effluents. Thus, they give the water an unpleasant taste and smell, prevent oxygen absorption at the water surface and consequently self-purification, are deposited on different installations, clog filters, are toxic to aquatic fauna and flora, and make water unusable for supplying cooling, irrigation, leisure, and so on. Toxic compounds cannot be retained by water treatment plants and some of them can reach the human body, instigating disease. These organic or inorganic materials, sometimes even in very low concentrations, can quickly destroy the flora and fauna of the receptor.

Radioactive compounds, radionuclides, and radioisotopes are some of the most hazardous toxic compounds. The discharge of radioactive wastewater into surface water and groundwater presents special dangers due to the action of radiation on living organisms. The effects of radioactive compounds on organisms depend both on the concentrations of radionuclides and on how they act, from outside or inside the body, with internal sources being the most hazardous (Karydis & Kitsiou, 2013).

Compounds with pronounced acidity or alkalinity, evacuated with wastewater, lead to the destruction of aquatic flora and fauna and the degradation of hydrotechnical construction, vessels, and installations mandatory for navigation. They prevent the use of water for leisure, irrigation, water supplies, etc. For example, the toxicity of sulfuric acid (H_2SO_4) to wildlife depends on the pH value, with fish dying at $pH = 4.5$.

Sodium hydroxide (NaOH), applied in many industrial processes, is very soluble in water and can rapidly raise the pH and alkalinity of water, interfering with different applications of water. Concentrations of more than 25 mg/L NaOH destroy the fish fauna.

Dyes, mainly from textile, paper, tanneries, etc., prevent the absorption of oxygen (O_2) and the standard processes of self-purification and photosynthesis (e.g., interfering with the sunlight energy in the chemical equation for photosynthesis: $6CO_2 + 12H_2O + \textit{Light Energy} \rightarrow C_6H_{12}O_6 + 6O_2 + 6H_2O$ or $[6CO_2 + 6H_2O \rightarrow C_6H_{12}O_6 + 6O_2]$; Where: CO_2- carbon dioxide; H_2O-water; light is required; $C_6H_{12}O_6$-glucose; O_2-oxygen; So, the equation may be stated as: Six carbon dioxide molecules and six water molecules react to produce one glucose molecule and six oxygen molecules. The reaction requires energy in the form of light to overcome the activation energy needed for the reaction to proceed. Carbon dioxide and water don't spontaneously convert into glucose and oxygen).

Caloric energy, a characteristic of hot water from thermal power plants and some industries, causes much damage to supplies of drinking and industrial water and prevents the normal growth of aquatic flora and fauna. Due to the increase in water temperature, the concentration of dissolved oxygen decreases, making the life of aquatic organisms difficult.

Microorganisms of any kind that reach a body of water, whether they develop improperly or not, can disrupt the growth of other microorganisms or other lifeforms. Microorganisms from tanneries, slaughterhouses, and manufacturing plants can be very harmful, instigating infection of the effluents that they make unusable (MacKenzie et al., 1995).

1.3 Causes of aquatic ecosystem pollution by hydrocarbons

Pollution of aquatic ecosystems by hydrocarbons is a worrying phenomenon, which has occurred on an unprecedented scale since the 1960s. The sources and causes of this pollution have grown year by year in proportion to the emergence and proliferation of greater operational risks, especially between 1970 and 1980. Drilling, extraction, delivery, transfer operations, loading/unloading, refining, storage, and other activities represent imminent risks, given the hazardous nature of oil and petroleum products.

Huge quantities of hydrocarbons reach aquatic ecosystems. The average global quantity of oil that pollutes seas and oceans annually is about 6 Mt, of which about 3.3 Mt comes from land sources (refineries in coastal regions, industrial and urban clearances, river inputs, atmospheric sources, etc.); about 2.1 Mt (approximately 35%) from ocean transport/shipping; about 600,000 tons from natural seeps from the ocean floor; and about 8000 tons from drilling and offshore activities.

The analysis of statistical data on the sources and causes of oil pollution of the aquatic ecosystems published by the different international, regional, and national institutions shows that the largest share of the oil polluting the Earth's oceans and seas is from land-based sources (Rajendran et al., 2018). Along with this category, among the most significant sources of oil pollution are cited as marine shipping, natural seeps, oil extraction from offshore deposits, refineries and the oil terminals (Table 1.1). Thus the industrial sources are responsible for most of the quantity of oil that pollutes the aquatic ecosystems globally, with approximately 60.8% of the total.

Table 1.1 Sources of oil contamination in the aquatic ecosystem, based on the United Nations Environment Program (UNEP).

No.	Percentage	Sources of oil and ocean contamination
1.	60.8	Industrial leakages
2.	2.1	Offshore fabrication
3.	1.2	Oil refineries/terminals
4.	10.2	Natural seeps
5.	14.4	Nautical delivery, excluding oil tankers
6.	6.6	Oil tanker incidents
7.	4.7	Operation of oil tankers

Notwithstanding that land-based sources provide most of the oil that pollutes the oceans and seas globally, far exceeding all the other sources taken together, the worst-case accidents were due to shipping. The ecological disasters caused by shipping incidents of oil tankers (e.g., Torrey Canyon, Amoco-Cadiz, Exxon Valdez, Prestige) show that the hazards involved in ocean transport of oil have much greater risks than those of other sources.

About 110,000 tons of oil are discharged annually in the Black Sea basin, which makes the effects of pollution felt in the ecological balance of the entire basin. Continental sources pollute the Black Sea the most. The rivers alone spill an approximate 53,000 tons a year, that is, about 50% of the total quantity of oil spilled annually into the sea. Among the other terrestrial sources, the following are highlighted: approximately 30,000 tons of oil come from domestic wastewater; approximately 15,000 tons come from manufacturing, including oil manufacturing operations; and the remaining approximately 12,000 tons is the input from other sources, which includes marine transport of oil. Pollution due to oil spills as the result of shipping incidents is insignificant, this sea basin having so far avoided profound incidents (Naser, 2013).

The causes of pollution of the aquatic ecosystems with hydrocarbons are analyzed, as a rule, in relation to two main forms of pollution: operational and accidental.

Operational pollution consists of spills that occur unintentionally, in the following situations: during loading-unloading operations of oil tanks; during bunkering operations (heavy and light refills for ship engines); during ship voyages, by evacuating the ballast and bilge water without sufficient purification; while stationed in ports, when there are leaks from washing the oil tankers in preparation to switch to the delivery of other types of oil, etc.

Accidental pollution is caused by aquatic disasters, the most significant of which are collisions, strandings, shipwrecks due to rupture of the hull or water holes, fires, and explosions.

Since 1974, there have been over 10,000 episodes of oil pollution as a result of operational and accidental discharges from oil shipping/transport, but of these, the majority (about 83.5%) are in the category of minor, with discharges of less than 7 tons. The percentage of discharges over 700 tons is only about 3.7%, and aquatic disasters hold a share of 46.6% of the total. During this period, there were 573 collisions, 580 failures, 709 incidents due to failure of the ship's hull (shipwreck, water holes, etc.), 133 explosions, and 2361 incidents from other or unknown causes (Ménesguen & Lacroix, 2018). Aquatic disasters account for about 46.6% of the total of 9351 incidents.

Operational pollution, in terms of the number of episodes, has the largest share and is largely the result of poor management in the prevention of water pollution, both onboard ship and in shipping companies and at the terminals of oil companies, port operators, or bunkering companies. Compared to major accidental spills, operational pollution has not had obvious environmental repercussions, notwithstanding that globally the annual quantity of hydrocarbons from operational sources is about four times higher (Tibbetts, 2015).

However, the real ecological catastrophes were caused by the aquatic disasters involving large oil tankers, as a result of which, not infrequently, the quantity discharged by some ships exceeded about 50,000 tons, reaching even 287,000 tons in the case of the oil tanker Atlantic Empress.

From an analysis of the risk factors that were at the source of the aquatic disasters, we can highlight the major contribution of five types of risk elements:

- Hazards caused by human error.
- Hazards caused by the ship's hidden defects.

- Hazards from acts of war, terrorism, and piracy.
- Hazards due to extreme hydrometeorological circumstances (thunderstorms with very vigorous winds and sea conditions, tsunamis, hurricanes, etc.).
- Hazards due to dangerous physicochemical properties of certain delivered goods, including oil and petroleum products.

In the period 1974–2007, major oil spills of over 700 tons mainly had accidental causes: collisions-573, failures-580, incidents due to damage to the hull-709, fires and explosions-133, other causes or unknown causes-2361. After the 1970s, when most spills in human history were recorded, the 10-year average of spills of more than 700 tons decreased from one decade to the next (Richardson et al., 2017).

1.4 The effects of water pollution

A particularly serious effect of water pollution is the eutrophication of lakes, also called "lake death," due to their increased fertility through the intake of nutrients, especially phosphates and nitrates, which promote the proliferation of phytoplankton and aquatic plants. Little by little, the lake becomes clogged, narrows, and disappears.

The extent and diversity of pollution damage is easy to measure. First of all, human health is at stake. After that, many economic activities are threatened. Finally, the degradation of aquatic life is fraught with repercussions, as it tends to decrease foods from marine sources at a time when their wider use is being examined. The fact that pollution can harm tourism is easy to understand: rare are those who have never encountered a dirty beach.

The fact that pollution can be fatal to oyster crops is self-evident. Likewise, it is easy to understand that our health can be severely affected by many pollutants: it is well-known that certain oils spilled in the sea contain carcinogens (Benedetti et al., 2015).

Following is a list of other contaminants commonly found in water.

1.4.1 Aluminum

Aluminum (Al) is the most common metal on Earth, making up about 8% of the mass of the Earth's core. Different quantities of Al are naturally present in groundwater and surface water (including drinking water sources). It is currently applied in the fabrication of machinery, equipment, construction, and in different types of food packaging (cans, and especially Al foils). Al builds up in the bones and brain and has been found to be linked to the onset of Alzheimer's disease. Due to the hazards of Al consumption, cooking or boiling tea/coffee in Al pots is not recommended (Golding et al., 2015).

1.4.2 Ammonia

Ammonia (NH_3) is a colorless gas with a strong odor that is naturally released by the degrading of nitrogenous materials (such as amino acids). It is produced by our body, being processed by the enzymes of healthy humans. It is produced for commercial purposes by a catalytic reaction between nitrogen and hydrogen and applied in the manufacture of toxic products such as fertilizers, explosives, plastics, and petroleum articles. Ammonium is very soluble in water, but does not pose a hazard to our health. However, it should be reduced due to its strong smell (Cao et al., 2011).

1.4.3 Arsenic

Arsenic (As) is a metal, most commonly found as arsenic sulfide. It is applied in manufacturing in the alloying of transistors, lasers, and semiconductors and in the manufacturing of dyes, glass, paper, textiles, adhesives, and ceramics. It is introduced into water by dissolving minerals and ores and through the atmosphere. In its natural state, it is found dissolved in rocks, hence the content of As in groundwater used as a source of drinking water. It is determined to be carcinogenic to humans, and is very poisonous, causing skin burns and circulatory problems (Lunde, 1977).

1.4.4 Barium

Barium (Benzene or Barium (Ba)) is an element present in volcanic and sedimentary rocks, making up about 0.039% and being the 14th most common element of the Earth's crust. Due to its high reactivity, Ba is not found in nature in a free state, but only in the form of compounds, the most common being barite and the least witerite (Ba carbonate). Ba compounds have a wide range of uses in manufacturing, in pyrotechnics, as rat poison, as raw material for the manufacture of glass, and in oil operations to increase the density of drilling mud. When ingested, it can cause side effects such as high blood pressure (Kanduc et al., 2011).

1.4.5 Benzene

Benzene (Ba) is a hydrocarbon with a colorless, highly flammable, and volatile aromatic core with a distinct odor. It is found in coal and oil and is naturally obtained from volcanoes and forest fires, being present in many other combustion products, including cigarette smoke. It is a solvent widely applied in chemical manufacturing and is a significant precursor to the chemical synthesis of drugs, plastics, synthetic rubber, and dyes. Its compounds are applied in the manufacture of adhesives, detergents, dyes, explosives, nylon, pesticides, plastics, resins, and tires. Benzene is also applied as a fuel additive, but due to the negative effects on health and to reduce the hazard of groundwater contamination with this compound, the maximum allowable emission of about 1% benzene has been imposed. Benzene reaches the water through factory discharges and air contamination. It can be inhaled, ingested, or can penetrate the skin. Once in the body, it concentrates in fats and bone marrow, for which it is toxic, blocking the formation of blood cells. Benzene irritates the eyes, the skin, and the respiratory tract (Soundarajan & Mohan Das, 2019).

Swallowing the fluid may cause it to be aspirated into the lungs, leading to chemical pneumonia. Exposure to a higher level than allowed can lead to loss of consciousness and death. The compound may affect the central nervous system, instigating dizziness.

Cigarette smoke contains a large quantity of this compound. A smoker inhales 10 times more benzene than a nonsmoker, and lifelong passive smoking increases the risk of cancer. Benzene has a boiling point of 80°C. Thus the higher the temperature, the faster it is released and it can reach larger quantities in the body. Benzene poisoning is disastrous because it is carcinogenic and damages the liver. Also, when the temperature increases significantly, the movement of the air is faster, and the dust is suspended; consequently we inhale larger quantities (Kacar et al., 2016).

The limit value specified by legislation in Romania is 0% starting in 2010.

1.4.6 Cadmium

Cadmium (Cd) is a gray-metallic transition metal that belongs to the group of rare metals, being found only in the form of inorganic compounds, especially in zinc ores. The level of Cd found in the atmosphere is from 0.1 to 5.0 ng/m^3; in the Earth's crust it is from 0.1 to 0.5 μg/g; in aquatic sediments it is 1 μg/g; in seawater it is 0.1 mg/L. It is also found in manufacturing as an inevitable by-product of the extraction process of zinc, lead, and copper. It is mainly applied in the electrogalvanizing process and in atomic-nuclear reactors.

At the same time, Cd compounds are applied in the manufacture of pigments and dyes, as stabilizers in plastic materials, in the electrodes of nickel-Cd alkaline batteries, in printing, textile manufacture, photography, lasers, semiconductors, pyrotechnics, dental amalgams, fluorescent lamps, jewelry, engraving, automotive and aircraft manufacture, pesticides, and polymerization catalysts (Bielmyer-Fraser et al., 2018).

About 10% of Cd released comes from secondary sources, mainly from dust coming from iron and steel scrap in recycling. The largest quantity of Cd is released into the environment, about 25,000 tons per year. About half of this quantity of Cd is released into rivers, rocks and air, from forest fires and volcanic eruptions, and the rest is released due to human activities (Cd is found in polyphosphate and pesticide fertilizers, so it reaches the soil in the environment). Environmental exposure to Cd can occur through the consumption of staple foods, especially seeds, cereals, and leafy vegetables, which absorb naturally occurring Cd, or by contaminating soils with sewage sludge, fertilizers, and contaminated groundwater (Chiarelli et al., 2019).

Exposure to Cd occurs mainly in workplaces where Cd products are manufactured, especially in nonferrous metallurgy plants, in battery factories, and during welding.

Humans working in smelters, mining, textile manufacturing, Cd alloy manufacturing, jewelry manufacturing, stained glass manufacturing, waste recovery, and battery manufacturing are at a high risk of exposure to Cd. In the case of smokers, there is an exposure to a significant concentration of Cd.

Cigarette smoke carries Cd to the lungs. Cd is first delivered to the liver through the blood. There, it combines with proteins and forms compounds that are delivered to the kidneys. Cd accumulates in the kidneys, where it affects the filtration mechanism. It causes the excretion of essential proteins and sugars from the human body and kidneys later. Cd accumulated in the kidneys requires a very long time to be eliminated from the body (Le Croizier et al., 2018).

Cd is present in galvanized pipes and solder alloys applied in the installation of water heaters and thus can contaminate water along its distribution route, only in the case of acid water. However, the toxicity of Cd in food is very rare and occurs as a result of environmental contamination or the consumption of foods high in Cd. Studies on its involvement in the lung and prostate cancer have yielded inconclusive results (Marshall et al., 2019).

1.4.7 Calcium

Calcium (Ca) is a gray alkaline earth metal, the fifth element in terms of its presence in the Earth's crust and the diffusion of ions dissolved in seawater, after sodium, chlorine, magnesium, and sulfates. Pure Ca is found in nature only in sedimentary rocks: limestone, dolomite, and gypsum. Ca salts and Ca ion solutions (Ca^{2+}) are colorless regardless of the contribution of Ca. Ca is applied as a reducing agent in the extraction of uranium, zirconium, and thorium metals; as a deoxidant, desulfurizer, or decarbonizer for different alloys; as an alloying agent in the fabrication of alloys based on Al, beryllium, Cu, lead, and magnesium; and in the preparation of cement and mortar applied in construction. The Ca compounds also have a wide range of uses in manufacturing of glass, dental work, insecticides, fertilizers, the manufacture of yeast, rockets, fireworks, paint, and chalk. Calcium hypochloride is applied in pool water disinfection as a bleaching agent, and calcium permanganate is applied as a rocket fuel and as a water sterilizer. In solution, the Ca ion has a variable taste, being declared to have a slightly salty, sour, or mineral taste. Ca plays a very significant role in the human body, being the fifth most common element in the body. Ca is one of those minerals that, when missing from the body, can cause many health problems. Especially during the growth period of children, Ca has an essential role in strengthening bones and teeth, regulating muscle function, ensuring better functioning of the heart muscle, and in blood clotting and regulating the endocrine system. Whether it occurs in childhood or adulthood, Ca deficiency can influence the processes that take place in the nervous system. Ca enters water through the decomposition of rocks, especially limestone, and from the soil through infiltration. Ca in water is inorganic, so it cannot be absorbed by the body (Saari et al., 2017).

1.4.8 Chlorine

Chlorine (Cl) is a yellow-green gas, with a characteristic suffocating and irritating odor, being present in the Earth's crust in a proportion of about 0.02%. In the form of the Cl ion, it is a part of common salts and other compounds, is abundant in nature, and mandatory for many lifeforms, including humans. Cl makes up the majority of salts in ocean water—Cl ions represent about 1.9% of the ocean mass, but are also found in the form of solid deposits in the Earth's crust. Chlorine also has everyday uses as a disinfectant, bleach, in the fabrication of the military gas mustard gas, stationery, antiseptics, dyes, food, insecticides, paints, petroleum products, plastics, medicine, textiles, solvents, as well as in many other consumer products. And its compounds in the form of hypochlorous acid, obtained by hydrolysis of sodium hypochlorite, are applied to eliminate bacteria and other microbes from drinking water and swimming pools.

Infusion solutions, called saline, are 0.9% NaCl solutions. Like sodium, chlorine is usually brought into the body in the form of sodium chloride (table salt), with a daily requirement of 2–3 g, which corresponds to 3–5 g of salt, a chemical made up of approximately 61% chlorine and approximately 39% sodium (Pan et al., 2019).

From a dietary point of view, sea salt and even coarse salt should be preferred over refined salt, which is pure sodium chloride. Chlorine is deadly in large quantities. Because it is heavier than air, it replaces oxygen in the lungs. Excess salt destroys vitamin E (tocopherol) and is toxic to the body. To replace the intestinal flora destroyed by the chlorine in tap water, increased sodium intake is recommended. According to scientists, chlorinated water contains a variety of compounds, called trihalomethanes (THMs), that produce genetic mutations that can cause cancer. These compounds are absorbed by the skin, leading to the growth of disease in the organs, where it is more easily assimilates (absorbs being fat-soluble) (Zeng et al., 2017).

In Romania, the limit imposed for drinking water is 250 mg/L Cl, and the exceptionally allowed value is 400 mg/L of chlorides.

1.4.9 Chromium

Chromium (Cr) is a hard, silvery-white, corrosion-resistant metal; in its native form it is easy to process and is quite common in the Earth's crust, but only in the form of chemical combinations, such as chromite. Its name derives from the Greek word chroma, which means color, due to the fact that most of its compounds are intensely colored. Since the last century, it has been observed that the addition of Cr to steel contributes to the increase of its mechanical characteristics as well as its corrosion resistance. Cr and derivatives are applied in the manufacture of engines, the use in combination as an anticorrosive alloy on metal parts to be subjugated to high temperatures, in the fabrication and burning of bricks, in the green coloring of glass, in the use of chromates as a pigment in the manufacture of dyes, and in the fabrication of acoustic recording magnetic tapes (Alomary & Belhadj, 2007).

Cr is a mineral that the body uses for standard functioning, such as digesting food. Cr exists in many natural foods, such as bread yeast, meat, potatoes (especially if cooked in the peel), cheeses, spices, wholemeal bread, cereals, fresh fruits, and vegetables. Cr deficiency can cause different eye conditions. There is a link between low Cr levels and the risk of glaucoma. It slows down Ca loss, so it can be helpful in preventing bone loss in menopausal women. Consumption of untreated tap water provides a fairly large quantity of Cr to the body and cooking in stainless steel pots increases the Cr content of food. The Cr found in food is not harmful. However, taking large quantities of Cr can cause stomach problems. Too much Cr can affect the kidneys, liver, and nerves, and can cause an irregular heartbeat (Jitar et al., 2015).

In Romania, the allowed concentration in drinking water is 0.05 mg/L Cr.

1.4.10 Copper

Copper (Cu) is a reddish metal, a very good conductor of electricity and heat. Cu with a purity of over 99% is applied in the manufacture of gas and water pipes, roofing materials, utensils, and ornamental objects. Because Cu is a good conductor of heat, it is applied in boilers and other devices that involve heat transfer, and in Cu foil. One of the disadvantages of Cu is the phenomenon of oxidation (greening), which can often be seen on old Cu vessels or coins. Following some analyses, it was found that Cu sources are also found in the thermal waters of Acas, in Satu Mare (a locality in Romania). In the existing thermal waters at Baile Herculane (locality in Romania) you can also find a high Cu content with antiseptic and antiallergen rolea. Cu is common in foods, fruits, and meat: in the liver, such as beef liver (39 mg/kg), oysters (9 mg/kg), nuts, vegetables, most cereals, grapes (the richest source), bananas (1 mg/kg), shellfish (4–10 mg/kg), green vegetables, cow's milk (0.06 mg/kg), seeds, mushrooms, chocolate (36 mg/kg), and cocoa powder.

Despite the fact that it is a metal, Cu is also an indispensable element of life. It is found in all body tissues, but most of the Cu in the body is found in the liver, and smaller quantities are also found in the brain, heart, kidneys, and muscles. As beneficial effects, Cu helps the body to use iron in the blood, reducing the actions of free radicals on tissues (Zhou et al., 2018).

Eating Cu-containing foods can also prevent certain diseases or deficiencies, such as allergies, baldness, AIDS, leukemia, osteoporosis, and stomach ulcers. Despite these beneficial effects, the nonassimilation of Cu in the body can lead to three rare diseases: Wilson's disease, when the body is unable to regulate the absorption of Cu and thus Cu accumulates in the liver; Menkes disease, when the body cannot make Cu reserves, hence its lack; and toxicity through Cu (a disease of unknown source and very rare, especially in children), when Cu accumulates in the liver. The presence of Cu in tap water can be observed through green stains on clothes and sanitary installations, when it is in a concentration higher than 1 mg/L. Cu dissolved in water has a pungent, unpleasant, and long-term effect; it has negative effects on the kidneys and causes gastrointestinal burns (Bao et al., 2018).

In Romania, the limit imposed for drinking water is 0.05 mg/L Cu, and the exceptionally allowed value is 0.1 mg/L Cu.

1.4.11 Magnesium

Magnesium (Mg) is the eighth element and the third metal after aluminum and iron (Al and Fe) in abundance in the solid Earth's crust, forming about 2% of its mass, and the third as a component in salts dissolved in seawater. Mg is an alkaline-earth, silvery-white metal, and is consequently only found in combination with other elements in deposits of magnesite (Mg carbonate), dolomite, and other minerals. Mg compounds, mainly Mg oxide, are applied as a refractory material for furnaces in steel manufacturing (in the fabrication of cast iron and steel), the nonferrous metals, glass, and cement industries.

Mg oxide and other compounds are applied in agriculture (as fertilizers), chemistry, and the construction industry. Mg carbonate powder is applied by athletes (to prevent perspiration of the palms). Mg stearate is a white powder with lubricating properties. In the pharmaceutical industry, it is applied in the manufacturing process of tablets.

Mg-Al alloy is applied mainly in the preservation of beverages, and other Mg alloys are applied as components of the structures of cars and in the construction of aircraft and missiles. Mg is the fourth most abundant mineral element in the body and is an element of great relevance in human biology. The body contains about 21–28 g of Mg; of this about 53% is disposed in the bones, 19% in tissues, and 1% in extracellular fluid.

Mg has a role in the metabolism of vitamin C, Ca, phosphorus, sodium, and potassium; it is essential for the proper functioning of muscles and nerves, and intestinal absorption depends, above all, on the quantity of Mg ingested.

Mg prevents the formation of Ca deposits in kidney and gallstones. It is a growth constituent, a widespread tonic, a cell regenerator, a mental balancer, and a liver drainer.

Mg is an inorganic element present in the blood plasma and helps to maintain a healthy cardiovascular system by preventing heart disease; it has antidepressant action, is also known as an anti-stress mineral, and, in combination with Ca, it acts as a natural tranquilizer. It improves the body's defense reactions and fights against the effects of senescence (Pearce & Mann, 2006).

Specialist studies show that approximately a quarter of the adult population suffers from a Mg deficiency, and this is due to the stressful lives of humans, and also their poor/ with quantities scanty diet of fruits, vegetables, whole grains, oilseeds, and dried vegetables. Mg deficiency it's getting bigger with the consumption of alcohol, caffeine, birth control pills, excess sugar, and diuretics. Mg deficiency in the body is recognized as the cause of a common disease, from which more than 10% of the population suffers, namely spasmophilia.

Mg requirements vary with age and sex. On average, to calculate the optimal requirement for an adult, multiply the weight in kilograms by the number 6, obtaining the dose in mg/day: kg $\times$ 6 = ? mg Mg/day.

In surface waters, Mg occurs both from natural sources and from the discharge of industrial wastewater (mainly from the glass and ceramic industry, refractory bricks, and chemical and metallurgical manufacturing). Mg, together with Ca, is involved in the hardness of water and gives a willow taste, slightly bitter, to the water. The most notable adverse effect of high Mg content in drinking water is the laxative effect. Otherwise, it poses no health hazard (Liu et al., 2016).

In Romania, the limit imposed for drinking water is 50 mg/L Mg, and the exceptionally allowed value is 80 mg/L Mg.

Oil spills have dramatic effects on aquatic ecosystems due to the exposure of organisms and seabirds to chemical compounds. Oil pollution particularly affects large species such as seabirds and aquatic mammals, which live long lives and spend time on the surface of the water. Oil spills also contain contaminants such as polycyclic aromatic hydrocarbons (PAHs) that affect aquatic communities over the long run. Due to the affinity for fatty tissues, these hydrocarbons can strongly affect the components of food chains. Although there are many studies on the serious effects of oil ingestion on avian physiology, there is too little investigation on exposure to oil contamination. A better understanding of the long-term consequences of oil spills gives a strong advantage in managing and ameliorating the effects on the environment (Mearns et al., 2019).

A study of the situations in which oil tanker shipping incidents occurred shows that more than half of the incidents were due to severe hydrometeorological conditions, most occurring in the Gulf of Mexico, the Mediterranean Sea, The North Sea, the English Channel, the western Iberian Peninsula, the Sea of Japan, the northeastern coasts of the United States, and the southern coasts of Africa, during the high-frequency seasons of these circumstances. These elements were the cause of some of the famous historical incidents, as in the cases of Amoco-Cadiz, Castillo de Bellver, Prestige, Torrey Canyon, and others.

More than two-thirds of the incidents examined occurred in the cold season of the year, when thunderstorms at sea are more frequent: in the northern hemisphere between October and March, and in the southern hemisphere from March to October.

The hostile actions that resulted in the sinking of oil tankers during the two World Wars have been extensively investigated; during this period there was a huge and uncontrolled risk for oil tankers carrying oil, more precisely destruction through terrorism (Lan et al., 2015).

The French oil tanker Limburg, with approximately 7000 tons of oil on board, was subjugated to a terrorist offensive in the region of the Gulf of Aden on October 2002. Due to the gas clearances that occur in tanks, petroleum and oil products present a high danger of fires and explosions. Consequently, special measures are needed on board these ships including gas equipment, fire alarms and fire extinguishing systems, and operational plans for the safety of the navigation and the cargo (Raddadi et al., 2017).

The case of the oil tanker Exxon Valdez, which failed in the Alaska Gulf, is the most bizarre example. The ship's captain was accused of driving the ship while intoxicated, instigating an ecological disaster with particularly severe effects. Studies carried out by the IMO, Lloyd's, ITOPF, and other institutions involved in ship management show that the main causes of oil tanker shipping incidents include noncompliance with safety rules and measures, through the easy registration of ships under a flag of compliance, allowing ships older than 15 years to remain in service, thus subjecting ships to additional risks. Many shipping companies register their ships under the flags of states such as Liberia, Singapore, Panama, Bahamas, Malta, Cyprus, and others, not only to pay lower taxes and duties, but also to pass inspections more easily. The most demanding of the classification or registry companies of the respective states, being avoided, for the financial reasons of some shipping companies. As a result, a large proportion of ships under a flag of convenience do not meet all the requirements for the safety of navigation and the safety of nautical delivery, and thus are exposed to shipping incident risks.

According to ITOPF statistics, out of a sample of 100 oil tankers with accidental discharges of more than 1000 tonnes, 66 oil tankers were sailing under a flag of convenience. Regarding the older ships, it was found that most of the lost ships were over 13 years old. In fact, the average age of the worldwide oil fleet is 10 years, but ships over the age of 20 represent about 13.6% of the total. Virtually all major incidents in which oil tanks broke in two and then sank occurred on ships over the age of 13 (Vergeynst et al., 2018).

Another significant cause of oil and ocean pollution by hydrocarbons is the extraction and processing/handling of aquatic oil and natural gas.

The execution of drilling works, the actual handling of hydrocarbons, their storage at sea or near the coast, and the delivery by means of ships or subaquatic pipelines involve a series of hazards regarding aquatic pollution. And in the case of these activities, we can see the two types of pollution, as in the case of navigation: operational and accidental.

Operational contamination is due to compilation of relatively small leaks during standard operations of drilling and operating platforms, transfer facilities, storage and land delivery. However, when constant leaks with significant flows occur, such as in the case of imperfect or damaged connections, pollution can reach hazardous levels.

Accidental pollution during the manipulation of aquatic hydrocarbons is caused by major oil spills as a result of accidents at offshore platforms, storage tanks, or subaquatic crude oil pipelines to oil terminals or to land.

The oil leaves traces that cannot be erased for years. Over time, the ecological catastrophes caused by massive oil spills have been more than numerous. Surprised by the sea, the oil tankers lost their compass or were damaged by accidental collision with other ships at sea, leaving behind huge stains that could not be erased, sometimes for years. The oil, which is insoluble in water and much lighter, could not be removed at times due to vigorous thunderstorms, which did not allow lifeboat access to damaged vessels. Or, in states bordering maritime areas unprepared to deal with disasters, much of the fuel spilled at the time of the accident.

However, the unrecovered crude oil, which remains on the surface of the water, spreads rapidly and forms an oily blanket, quite thick, which floats for a long time at sea, affecting the surrounding fauna and flora (Zabbey & Olsson, 2017). Among the incidents of this type are the following cases:

- Santa Barbara Channel (1969), when about 11,200 tons of crude oil were spilled into the sea off the coast of Santa Barbara, California, as a result of the explosion of a drilling rig in the oil field Montesito;
- Ekofisk (1977), accident at the Bravo oil rig in the Ekofisk oil field in the central North Sea. An instantaneous eruption caused the uninterrupted release of approximately 12,000 tons of oil. Half of this quantity evaporated quickly, as the oil in this deposit had a high temperature (90°C) and is very light, and the rest was deposited on the sea surface forming a huge film, over about 1000 km^2;

- Ixtoc One (1979). In the history of contamination incidents recorded on offshore oil rigs, the worst accidental oil spill was at the Ixtoc One offshore drilling rig in the Gulf of Mexico, after which about 600,000 tons of crude oil were spilled into the sea from June 1979 to February 1980, which means the equivalent of three spills the size of the one caused by Amoco-Cadiz.
- The sinking of the Romanian oil tanker UNIREA (1982), of approximately 88,285 Mt, which took place off the coast of Bulgaria, 40 Mm southeast of Cape Kaliakra, was caused by fire and explosions caused by gas accumulations. The hidden defects of the ship, the equipment, and the installations on board had some of the most serious effects on the oil companies.
- On November (2007), due to a vigorous storm, with wind gusts of over 100 km/h and wave heights exceeding 5 m, in the Black Sea (Kerch Strait) the oil tanker "Volgoneft-139" and sulfur-laden "Volnogorsk," "Nahicevan," and "Kovel" freighters sank. Following the shipwreck, 1300 tons of crude oil and about 7000 tons of sulfur were spilled into the sea.

Oil pollution has destroyed local aquatic ecosystems. Tens of thousands of birds and fish migrating from the Black Sea to the Sea of Azov have died. An inventory of animal species that have disappeared or become very rare near the Romanian coast includes taxa from most groups. Thus the oysters beds and bivalve *Barnea candida* beds disappeared from the bottoms near Constanta (locality in Romania); bivalves like *Irus irus*, *Pholas dactylus*, *Gibbomodiola adriatica*, and *Petricola lithophaga* have become rare; *Solen marginatus* and *Teredo navalis* have disappeared. Also, some species of coelenterates such as *Lucernariopsis campanulata* (a delicate species that disappeared with the domains of brown algae of the genus *Cystoseira*) have become rare; sea worms like *Ophelia bicornis*; gastropods such as *Limapontia capitata*; *Gibbula divaricata*, large crustaceans such as *Upogebia pusilla* (tens of thousands of gulls once lived on the sandy bottoms of Mamaia [locality in Romania], and now the species can only be found north of Cape Midia [locality in Romania]), shrimp *Hippolyte inermis*, *Lysmata seticaudata*, *Processa pontica*, *Macropodia aegyptia*, *Pestarella candida*, *Carcinus maenas* (so common in the past, now rarely north of Agigea [locality in Romania]).

A number of oil tankers that were torn in two, or that endured damage to the steering gear or radar equipment and that resulted in shipwrecks were suspected of hidden defects. Notwithstanding, other causes were invoked, in the incidents involving the oil tankers Prestige, Erika, and others, the shipowners requested expertise to prove that the breaking of the respective ships in two was due to hidden defects (Raddadi et al., 2017).

Human error is often cited as the cause of oil tanker aquatic disasters. It is estimated that more than half of aquatic disasters involve the mistakes of commanders or crews. In some cases, incidents could be avoided by taking measures to prevent certain risks. We must convince ourselves that a fight against pollution cannot be the work of one country or one generation, but everything must be thought of universally. It is a great satisfaction to see Latinos from all over the world being attracted to and sometimes even having an enthusiasm for this battle, meant to protect our environment. Notwithstanding the proportions in general, this spills are far from unprecedented, the studies undertaken and the media coverage of these cases have been far below the level of major oil tanker incidents. Aquatic oil pollution can also be caused by acts of sabotaging against offshore oil installations (Mason et al., 2017).

In February 1991, during the Gulf War, Iraqi troops, retreating from Kuwait, opened the taps of huge oil tanks in the conflict zone and set fire to the wells, triggering the largest oil spill in history. About 60 million barrels (150 times more than in the case of the Exxon Valdez) flooded large regions of land and coastline, with some of the oil reaching the Persian Gulf. Prior to Allied bombing, Iraqi soldiers dumped tens of millions of barrels of oil into Gulf waters, blowing up numerous Kuwaiti oil terminals to prevent a sea invasion. The oil layer formed covered approximately 1600 km of coastline. The east coast of AbouAli Island, once made up of sandy beaches highly sought after by tourists, has become a lifeless place (Thibodaux et al., 2014). The ecological catastrophe caused by the spilled oil has severely affected the entire ecosystem in the northern part of the Persian Gulf.

Oil is the most significant fuel of our society but also the main source for products such as synthetic plastics, lubricating oils, bitumen, etc. Oil is a danger to the aquatic environment due to spills into the oceans during transport, amounting to several million tons/year annually. The oceans have long been established as the easiest way to dilute industrial and municipal waste that has been deposited in them.

The main constituent of oil is hydrocarbons (50%–90%) consisting of n-alkanes, branched alkanes, cycloalkanes, and aromatic hydrocarbons (20%). Some of the aromatic hydrocarbons are from the class of PAHs, of which benzopyrene is predominant. Other constituents are polar compounds containing heteroatoms such as O, N, S. Sulfur compounds are numerous, such as H_2S, R-SH, R-S-R, and thiophenic derivatives. Among the physical properties of oil important for the environment are significant density, lower than that which will cause an oil slick to float on the surface of the water over large areas (the thickness of the oil layer on water is 0.1 mm), solubility, and volatility. Oil forms emulsions with water: oil/water and water/oil.

The taste of petroleum hydrocarbons is most unpleasant for humans. All components of crude oil are biodegradable, but it is only that the biodegrading speeds are very different. Both crude oil and refined products contain a variety of toxic compounds. Low molecular weight compounds are not very hazardous for water because they evaporate very quickly. High molecular weight compounds are also less hazardous. The most hazardous are medium molecular compounds such as diesel. For port waters, the limit for dissolved hydrocarbons shall not exceed 0.03 ppm. From the point of view of the source of pollution, it can be located in the Black Sea (breaking an oil pipeline, washing the tanks of an oil tanker, large naval accident, etc.), near the shore (failed barge, clandestine spillage of debris), at the mouths of rivers or urban sewers, on sea beaches or sunbeds related to sea beaches. Depending on the magnitude, the pollution can take the form of incidents (for the solution of which simple means, usually found in the incorporate of the potential polluting agent, are sufficient), accidents (when the quantity discharged exceeds the possibilities of action of a single pollution control unit properly equipped, requiring the intervention of other similar units), or catastrophe (when all existing means in the area must come into play, including those operated by neighboring riparian countries or various other foreign companies). As for direct damages, the economic ones are significant: product losses, disturbance of river traffic and tourism in the area, compromise of fishing, damage to industrial facilities (ships, port facilities), payment of fines and compensation, as well as short-term environmental damage (affecting individuals or species).

Among the consequences, the most significant are the social ones (public health, the shock of public opinion, the disruption of some activities), the political ones (internal, but especially external), as well as the long-term ecological ones (breaking of food chains, numerical decrease of some species or generations, metabolic changes, etc.).

Hydrometeorological constituents, that is, winds, waves, currents, and atmospheric state, also have a role to play in the methods and means of intervention. If we add the particular situations of each pollution accident (e.g., whether or not it is accompanied by fire, whether it threatens special areas or industrial installations of special importance, etc.), as well as the fact that all of these are variable over time, we will more easily understand why there is this great diversity of methods and means of intervention.

The oil leaves traces that cannot be erased for years. In recent times, the ecological catastrophes caused by massive oil spills have been more than numerous.

1.5 The repercussions of pollution of aquatic ecosystems with hydrocarbons

Microalgae are responsive indicators of environmental change and due to the fact that they are the basis of most freshwater and aquatic ecosystems, they are widely applied in environmental hazard assessment. The toxicity of a particular element or compound can be tested at different levels of biological organization.

For unicellular algae monocultures, the actions of toxic compounds can be tested using one of the properties: growth rate, motility, gamete fabrication, cell division, cytoplasmic flow, protein synthesis, cytoplasmic membrane permeability, blocking regions of absorption for a particular nutrient, fabrication or fixation of CO_2, or assimilation or fabrication of O_2.

Of all forms of pollution, the one having the most severe consequences for aquatic ecosystems is oil pollution. Oil, spilled in large quantities, covers a considerable area and severely affects water quality and aquatic life. The most disastrous repercussions have been from oil spills occurring in coastal regions and especially in locations where the oil slick was concentrated on small or shallow areas: the English Channel, the Bay of Biscay, the Gulf of Alaska, the Persian Gulf, etc. (Patlovich et al., 2005).

Experiments and observations carried out on accidental oil spills have revealed that the toxicity of different petroleum products, such as crude oil or even finished products, has very different effects on aquatic organisms. In general, hazardous toxicity has been observed to be caused by the concentration of volatile compounds in hydrocarbons, including light and aromatic paraffins (gasoline, toluene, xylenes, naphthalene, etc.).

These compounds are more soluble in water than in hydrocarbons, which creates the toxic effect of polluting sources.

Oil pollution in the seas and oceans globally has reached such a magnitude in the last 40 years that, according to many experts, aquatic ecosystems will take years to fully recover, both in terms of the health of biotopes and in terms of rehabilitating all species of flora and fauna that remain from the results of pollution (Centers for Disease Control & Prevention CDC, 2008).

The ecological repercussions of oil spills from large oil tankers, coastal refineries, or offshore oil rigs accompanied by the discharge of massive quantities of oil into the sea have proved to be particularly serious and sometimes even catastrophic for aquatic ecosystems.

Studies of oil spills show that the consequences of oil pollution on benthic and pelagic organisms are considerable. In the North Atlantic, according to some estimates, oil pollution annually causes the death of about 500,000 birds, the most affected families being those of Procellaride and Alcidae.

The causes of death of seabirds are attributed to effects such as loss of feathers, wing disturbances, inability to move, impaired intestinal and glandular functions by which birds get freshwater from seawater, and loss of the water-repellent feature of feathers, which leads to inability to maintain body heat in contact with water. During hatching, the oil that covers the feathers of adult birds contaminates the eggs and causes the death of the embryos.

Observations show that approximately 20 mg of hydrocarbons deposited on the eggs of seabirds are already highly toxic. The effects of oil pollution have decimated populations of diverse aquatic flora and fauna, both as genera and as species: benthic and pelagic fish, oysters and mussels, gastropods, crustaceans, dolphins, seals, sea otters, penguins, phytoplankton and zooplankton, coral colonies, seaweed, etc. Moderate doses of oil have been shown to decrease the photosynthetic activity of algae and phytoplankton. At the same time, fish that live in contaminated regions accumulate hydrocarbons in the muscle tissues, which makes them inedible (Dar, Dar et al., 2016; Dar, Kamili et al., 2016; Rogowska & Namieśnik, 2010).

Some species of aquatic fauna, including bivalves, fish, crustaceans, zooplankton, different microorganisms, and bacteria, can consume or absorb certain quantities of hydrocarbons from polluted regions. It has been shown that the tissues of many aquatic organisms can retain some fractions of spilled oil for a long time. In the bodies of fish and other aquatic organisms, these fractions are transformed into different compounds by metabolic processes.

The concentration of hydrocarbons in the body increases more when these creatures feed on microorganisms contaminated with crude oil, with such cases exhibiting a higher mortality rate.

The effects of substances applied in the fight against pollution, especially detergent-based dispersants, which in some cases were more serious than the effects exhibited by oilcloth, were also negative, even to the death of organisms (the cases of Torrey Canyon, Exxon Valdez, etc.) (Dar, Dar et al., 2016; Dar, Kamili et al., 2016; Dar et al., 2020; Mackey & Hodgkinson, 1996).

The increases and changes over time of hydrocarbons discharged into aquatic ecosystems can decisively influence aquatic ecosystems. The length of the contaminated coast depends very much on the type of hydrocarbons, the hydrometeorological elements (wind, currents, tides, etc.), and the viscosity, density, and flow point of crude oils, which can contribute to the dispersion of the oil slick offshore or ability to drift to shore, sometimes pushing long distances, even hundreds of nautical miles. Also, the efficiency of forces and means of intervention for decontamination can have significant effects on the progression of the oil slick. During an oil spill, a polluting oil slick (also called a "film") forms on the surface of the water.

Hydrocarbons, commonly having a lower density than seawater, remain floating on the sea surface and only heavier fractions, such as fuel oil or some manufactured products such as asphalt, can have a slight immersion, remaining in suspension to the surface.

However, the hydrocarbons that arrive in aquatic ecosystems do not remain unaltered; under the action of the environment they undergo different transformations, which causes, over time, the concentration of hydrocarbons accrued from a spill to gradually decrease. This also explains the fact that, notwithstanding that every year quantities of hydrocarbons on the order of millions of tons are discharged into the oceans, the global concentration does not show a significant increase. As shown in recent studies, the analysis of the hydrocarbon content present in ocean waters showed that in offshore waters the concentration is less than 1 ppb, and in coastal waters is about 1^{-10} ppb (Ling et al., 2018).

A higher concentration was highlighted near oil rigs, with values of 2–20 ppb. Unlike the oceans, in the waters of seas it has been estimated that, commonly, the concentration of hydrocarbons is about 3 ppb offshore, 20–50 ppb in territorial waters, and can reach values of 100–1000 ppb in regions contaminated with oil products.

These values are averages and they differ from one sea to another, based on several elements: the volume of oil spills from different sources, the intensity of nautical traffic (especially oil), the characteristics of the sea (surface, volume of water, degree of opening, concentration of biodegradable elements, etc.), intensity of the effects of hydrometeorological circumstances, etc. (Shahidul Islam & Tanaka, 2004).

One of the cases, perhaps the most representative for the study of the ecological repercussions of hydrocarbon contamination, is that of the oil spill caused by the failure of the Amoco-Cadiz oil tanker. The observations and activities undertaken in an extensive investigation program have led to significant findings and conclusions on the long-term consequences of oil contamination on a valuable aquatic ecosystem.

From the conclusions drawn, it emerged that the contamination caused by the discharge of the approximate 223,000 tons of crude oil affected about 300 km of coastline and led to the destruction of 260,000 tons of aquatic biomass, on a contaminated region of about 250,000 ha.

Based on IFREMER estimates, among the approximate 19,000–37,000 birds and hundreds of millions of fish perished in this environmental accident. The coastal region was contaminated with 50,000–60,000 tons of oil, and in the region affected by the oil spill (sea and coast), 30% of the fauna and 5% of the flora were destroyed (Bilal et al., 2018).

Another famous case was the Exxon Valdez, a tanker that spilled 40,000 tons of oil in Prince William Bay, Alaska. This oil slick had catastrophic ecological repercussions as it has lasted in a responsive marine area populated by dolphins, seals, sea otters, salmon, eagles and a refuge for many migratory birds. The sensitivity of the habitat was also increased by the physical characteristics of the sea basin in which the pollution took place: it was a bay with a relatively narrow opening, with shallow depths and dotted with many small islands. The spilled oil contaminated about 1770 km of coastline, including both the coastline of the Gulf of Alaska and the coasts of nearby islands. The aquatic ecosystem has exhibited dramatic effects: the entire habitat has been severely affected, and pollution has killed about 35,000 birds, 1000 sea otters, 300 seals, 250 eagles, 25 whales, and millions of fish (herring. salmon, and other species) (Reid & Whitehead, 2016).

The local population, which lives mainly from fishing, had to endure a dramatic period, as fishery sources, especially those for salmon and herring, were decimated. Another consequence of the contamination of the surface layer of the seas and oceans with hydrocarbons is the alteration of ocean-atmosphere interactions.

Oil negatively influences, on the one hand, the transfers of matter within the biogeochemical circuits, and on the other hand, the flows of caloric energy at the interface between the two environments. It has been estimated that 1 tonne of oil can cover 12 km^2 of the ocean, which means that at the annual quantity of millions of tons spilled across the aquatic ecosystems, the sea surface polluted by this form of contamination would be tens, maybe even hundreds, of millions of square kilometers (Mearns et al., 2017).

In the Atlantic and Indian Oceans, the region annually contaminated with oil (dispersed by sea currents) is estimated at over 50 million km^2, which represents about 30% of the surface of the two oceans. Due to the current concentration of contaminants at the surface, the ocean—which plays a significant role in the exchange of gases with the atmosphere, absorbing some of them—takes up a decreasing quantity of atmospheric CO_2, so excess CO_2 of anthropogenic sources contributes to amplifying the greenhouse effect. At the same time, ocean surfaces contaminated with hydrocarbons, and also with other types of contaminants (detergents, pesticides, PCBs, etc.), reduce the processes of photosynthesis, evaporation, and formation of oxygen and aquatic aerosols, whose role in atmospheric circulation and in supporting some hydrometeorological circumstances is crucial. Due to the affinity for fatty tissues, these hydrocarbons can drastically affect the components of food chains. Notwithstanding that there are many studies on the serious effects of oil ingestion on avian physiology, there is too little investigation on exposure to oil pollution. A better understanding of the long-term consequences of oil spills brings a great advantage in managing and preventing environmental effects.

1.6 Other sources of water pollution spread

1.6.1 Spray drift

This can become a problem when, during spraying, pesticides are delivered to nearby water sources or to responsive locations by drift and concentrate large quantities of plant protection products in the water. Elements influencing spray drift are: wind and tractor speed, wind direction, distance to water/responsive regions, culture, air humidity, droplet size, air influence, and spray distance to target.

Drift can be minimized by using discount nozzles, especially at the ends and edges of the ground, by adjusting the equipment, by adapting the speed of the tractor, and by planning the work based on the weather conditions (Glinski et al., 2018).

1.6.2 Leakage

After application, pesticides can reach the water from treated soils through soil eroded particles. Polluted water can reach water sources, leading in the short to medium term to relatively high concentrations of pesticide products in the water. Elements that influence runoff to watercourses are precipitation conditions, soil permeability, soil moisture, water flow rate, and distance to water.

Leaks can be avoided by adapting soil work to maximize water infiltration, the use of conservation work, contour plowing and cultivation systems, crop rotation, and the planting of live curtains. After treatment, some pesticides can be delivered underground and then into groundwater by infiltrating the soil. This results in low concentrations in the medium and long term, which can sometimes violate very low limits for drinking water in the EU.

The risk of leaching can be reduced by restricting the use of pesticides in regions identified as vulnerable, such as soils with low organic matter, surface soils, soils with surface groundwater, or sandy soils (Fogg et al., 2004).

1.6.3 Drainage

After application, pesticides can be delivered with water that seeps into the soil in the drainage system. Contaminated water then enters surface waters, leading in the short to medium term to significant concentrations of pesticides in watercourses.

The risks can be reduced by avoiding the application of pesticides during the drainage season and before the onset of torrential rains. If possible, it is recommended to isolate and store drained water in an artificial collection region. Also, pesticides should not be applied on drained, cracked, or most dry soils to reduce the risk of contamination (Macchi et al., 2018).

Protecting water and aquatic sources—and ensuring their environmental status—is one of the pillars of EU environmental policy. The Water Framework Directive (WFD), introduced in 2000, and the Aquatic Strategy Framework Directive (DCSMM) of 2008 set out the framework for the management of all aquatic ecosystems.

Their aim is to achieve good ecological status for our aquatic sources and freshwater through an ecosystem or holistic approach. Pollution can take many forms, ranging from organic compounds and other toxic products to residues from the fabrication of different types of energy. The degree of toxicity of a pollutant to human health and ecosystems depends on its chemical nature, quantity or concentration, as well as its persistence.

The different effects of pollutants depend not only on the environment in which they are located (air, water, or soil), but also on the interaction with coexisting pollutants, as well as the duration of exposure. Some types of pollution, such as certain forms of polluted water, poor air quality, industrial waste, garbage, light, thermal, or noise overload, are easy to observe, but there are more subtle forms of pollution detectable only with different tools.

1.7 Conclusions and recommendations

Accidental introduction into aquatic ecosystems of nonnative (or invasive) plant and animal species poses a threat to the ecosystem and is difficult to control. Aquatic organisms that cause this type of contamination are often found on the outer surface of ships. Untreated ballast water also delivers such organisms into the aquatic ecosystems. Currently, aquatic ecosystems are invaded by some exotic species of crab, jellyfish, ringworm, mollusks, and algae. In addition to navigation, fishing and restricted aquaculture, mining, tourism, recreation and military exercises are also practiced in the aquatic ecosystems. The aquatic ecosystems platform and basin are also under intense pressure due to intensive human activities, including urban growth, manufacturing, hydropower and nuclear use, agriculture, and land use. The main elements that threaten the aquatic ecosystems are:

- Decreased natural sources caused by overmanipulation of aquatic life.
- Physical changes in the seabed, coasts, and rivers, and
- Contamination of the sea (by solid and liquid waste).

Water is an essential element for the existence of life and the evolutions of human communities, the planetary water sources being about 1.37 billion km^3, of which 97.2% are disposed in the seas and oceans and 2.7% in groundwater and surface water. Of freshwater, only 1.44% are liquid, the rest being glaciers. Being a partly renewable resource, water needs to be protected from harmful human activities, because contamination affects its ability to self-purify. Aquatic ecosystems, like freshwater ecosystems, perform several vital functions: flood prevention; maintaining local and global climate balance; protecting biological diversity; and last but not least, water filtration, dilution, and storage. Among the consequences, the more significant are the social ones (the shock of public opinion, public health, the disruption of some activities), the political ones (internal, but especially external), as well as the long-term ecological ones (breaking of food chains, metabolic changes, numerical decrease of some species or generations, etc.). Due to the gas clearances that occur in tanks, oil and petroleum products present a high danger of fires and explosions. Consequently, on board these ships, special measures are needed both in terms of equipment with inert gas installations, fire warning and extinguishing systems, and management for the safety of navigation and operation of the cargo.

References

Alimba, C. G., & Faggio, C. (2019). Microplastics in the marine environment: Current trends in environmental pollution and mechanisms of toxicological profile. *Environmental Toxicology and Pharmacology*, *68*, 61–74. Available from https://doi.org/10.1016/j.etap.2019.03.001.

Alomary, A. A., & Belhadj, S. (2007). Determination of heavy metals (Cd, Cr, Cu, Fe, Ni, Pb, Zn) by ICP-OES and their speciation in Algerian Mediterranean Sea sediments after a five-stage sequential extraction procedure. *Environmental Monitoring and Assessment*, *135*(1–3), 265–280. Available from https://doi.org/10.1007/s10661-007-9648-8.

Bao, V., Ho, K., Lai, K., Mak, Y., Mak, E., Zhou, G. J., Giesy, J. P., & Leung, K. (2018). Water-effect ratio of copper and its application on setting site-specific water quality criteria for protecting marine ecosystems of Hong Kong. *Environmental Science and Pollution Research International*, *25*(4), 3170–3182. Available from https://doi.org/10.1007/s11356-017-9428-0.

Benedetti, M., Giuliani, M. E., & Regoli, F. (2015). Oxidative metabolism of chemical contaminants in marine organisms: Molecular and biochemical biomarkers in environmental toxicology. *Annals of the New York Academy of Sciences*, *1340*, 8–19. Available from https://doi.org/10.1111/nyas.12698.

Bergmann, M., Tekman, M. B., & Gutow, L. (2017). Marine litter: Sea change for plastic pollution. *Nature*, *544*(7650), 297. Available from https://doi.org/10.1038/544297a.

Bielmyer-Fraser, G. K., Harper, B., Picariello, C., & Albritton-Ford, A. (2018). The influence of salinity and water chemistry on acute toxicity of cadmium to two euryhaline fish species. *Comparative Biochemistry and Physiology Part C: Toxicology & Pharmacology*, *214*, 23–27. Available from https://doi.org/10.1016/j.cbpc.2018.08.005.

Bilal, M., Rasheed, T., Sosa-Hernández, J. E., Raza, A., Nabeel, F., & Iqbal, H. (2018). Biosorption: An interplay between marine algae and potentially toxic elements – A review. *Marine Drugs*, *16*(2), 65. Available from https://doi.org/10.3390/md16020065.

Brown, T. M., & Takada, H. (2017). Indicators of marine pollution in the North Pacific Ocean. *Archives of Environmental Contamination and Toxicology*, *73*(2), 171–175. Available from https://doi.org/10.1007/s00244-017-0424-7.

Cao, H., Li, M., Dang, H., & Gu, J. D. (2011). Responses of aerobic and anaerobic ammonia/ammonium-oxidizing microorganisms to anthropogenic pollution in coastal marine environments. *Methods in Enzymology*, *496*, 35–62. Available from https://doi.org/10.1016/B978-0-12-386489-5.00002-6.

Centers for Disease Control and Prevention (CDC). (2008). Fatalities among oil and gas extraction workers – United States, 2003–2006. *Morbidity and Mortality Weekly Report*, *57*(16), 429–431.

Chiarelli, R., Martino, C., & Roccheri, M. C. (2019). Cadmium stress effects indicating marine pollution in different species of sea urchin employed as environmental bioindicators. *Cell Stress & Chaperones*, *24*(4), 675–687. Available from https://doi.org/10.1007/s12192-019-01010-1.

Dar, G. H., Bhat, R. A., Kamili, A. N., Chishti, M. Z., Qadri, H., Dar, R., & Mehmood, M. A. (2020). *Correlation between pollution trends of freshwater bodies and bacterial disease of fish fauna. Fresh Water Pollution Dynamics and Remediation* (pp. 51–67). Singapore: Springer.

Dar, G. H., Dar, S. A., Kamili, A. N., Chishti, M. Z., & Ahmad, F. (2016). Detection and characterization of potentially pathogenic *Aeromonas sobria* isolated from fish *Hypophthalmichthys molitrix* (Cypriniformes: Cyprinidae). *Microbial Pathogenesis*, *91*, 136–140.

Dar, G. H., Kamili, A. N., Chishti, M. Z., Dar, S. A., Tantry, T. A., & Ahmad, F. (2016). Characterization of *Aeromonas sobria* isolated from Fish Rohu (*Labeo rohita*) collected from polluted pond. *Journal of Bacteriology & Parasitology*, *7*(3), 1–5. Available from https://doi.org/10.4172/2155-9597.1000273.

Fan, C., Hsu, C. J., Lin, J. Y., Kuan, Y. K., Yang, C. C., Liu, J. H., & Yeh, J. H. (2018). Taiwan's legal framework for marine pollution control and responses to marine oil spills and its implementation on T.S. Taipei cargo shipwreck salvage. *Marine Pollution Bulletin*, *136*, 84–91. Available from https://doi.org/10.1016/j.marpolbul.2018.09.005.

Fogg, P., Boxall, A. B., Walker, A., & Jukes, A. (2004). Leaching of pesticides from biobeds: Effect of biobed depth and water loading. *Journal of Agricultural and Food Chemistry*, *52*(20), 6217–6227. Available from https://doi.org/10.1021/jf040033o.

Glinski, D. A., Purucker, S. T., Van Meter, R. J., Black, M. C., & Henderson, W. M. (2018). Analysis of pesticides in surface water, stemflow, and throughfall in an agricultural area in South Georgia, USA. *Chemosphere*, *209*, 496–507. Available from https://doi.org/10.1016/j.chemosphere.2018.06.116.

Golding, L. A., Angel, B. M., Batley, G. E., Apte, S. C., Krassoi, R., & Doyle, C. J. (2015). Derivation of a water quality guideline for aluminium in marine waters. *Environmental Toxicology and Chemistry/SETAC*, *34*(1), 141–151. Available from https://doi.org/10.1002/etc.2771.

Jitar, O., Teodosiu, C., Oros, A., Plavan, G., & Nicoara, M. (2015). Bioaccumulation of heavy metals in marine organisms from the Romanian sector of the Black Sea. *New Biotechnology*, *32*(3), 369–378. Available from https://doi.org/10.1016/j.nbt.2014.11.004.

Kacar, A., Pazi, I., Gonul, T., & Kucuksezgin, F. (2016). Marine pollution risk in a coastal city: Use of an eco-genotoxic tool as a stress indicator in mussels from the Eastern Aegean Sea. *Environmental Science and Pollution Research International*, *23*(16), 16067–16078. Available from https://doi.org/10.1007/s11356-016-6783-1.

Kalogianni, E., Vourka, A., Karaouzas, I., Vardakas, L., Laschou, S., & Skoulikidis, N. T. (2017). Combined effects of water stress and pollution on macroinvertebrate and fish assemblages in a Mediterranean intermittent river. *The Science of the Total Environment*, *603-604*, 639–650. Available from https://doi.org/10.1016/j.scitotenv.2017.06.078.

Kanduc, T., Medaković, D., & Hamer, B. (2011). Mytilus galloprovincialis as a bioindicator of environmental conditions: The case of the eastern coast of the Adriatic Sea. *Isotopes in Environmental and Health Studies*, *47*(1), 42–61. Available from https://doi.org/10.1080/10256016.2011.548866.

Karydis, M., & Kitsiou, D. (2013). Marine water quality monitoring: A review. *Marine Pollution Bulletin*, *77*(1-2), 23–36. Available from https://doi.org/10.1016/j.marpolbul.2013.09.012.

Lan, D., Liang, B., Bao, C., Ma, M., Xu, Y., & Yu, C. (2015). Marine oil spill risk mapping for accidental pollution and its application in a coastal city. *Marine Pollution Bulletin*, *96*(1-2), 220–225. Available from https://doi.org/10.1016/j.marpolbul.2015.05.023.

Le Croizier, G., Lacroix, C., Artigaud, S., Le Floch, S., Raffray, J., Penicaud, V., Coquillé, V., Autier, J., Rouget, M. L., Le Bayon, N., Laë, R., & Tito De Morais, L. (2018). Significance of metallothioneins in differential cadmium accumulation kinetics between two marine fish species. *Environmental Pollution (Barking, Essex: 1987)*, *236*, 462–476. Available from https://doi.org/10.1016/j.envpol.2018.01.002.

Ling, S. D., Davey, A., Reeves, S. E., Gaylard, S., Davies, P. L., Stuart-Smith, R. D., & Edgar, G. J. (2018). Pollution signature for temperate reef biodiversity is short and simple. *Marine Pollution Bulletin*, *130*, 159–169. Available from https://doi.org/10.1016/j.marpolbul.2018.02.053.

Liu, H., Chen, X., Su, Y., Kang, I. J., Qiu, X., Shimasaki, Y., Oshima, Y., & Yang, J. (2016). Effects of calcium and magnesium ions on acute copper toxicity to glochidia and early juveniles of the Chinese pond mussel *Anodonta woodiana*. *Bulletin of Environmental Contamination and Toxicology*, *97*(4), 504–509. Available from https://doi.org/10.1007/s00128-016-1890-8.

Lunde, G. (1977). Occurrence and transformation of arsenic in the marine environment. *Environmental Health Perspectives*, *19*, 47–52. Available from https://doi.org/10.1289/ehp.771947.

Macchi, P., Loewy, R. M., Lares, B., Latini, L., Monza, L., Guiñazú, N., & Montagna, C. M. (2018). The impact of pesticides on the macroinvertebrate community in the water channels of the Río Negro and Neuquén Valley, North Patagonia (Argentina). *Environmental Science and Pollution Research International*, *25*(11), 10668–10678. Available from https://doi.org/10.1007/s11356-018-1330-x.

MacKenzie, K., Williams, H. H., Williams, B., McVicar, A. H., & Siddall, R. (1995). Parasites as indicators of water quality and the potential use of helminth transmission in marine pollution studies. *Advances in Parasitology*, *35*, 85–144. Available from https://doi.org/10.1016/s0065-308x(08)60070-6.

Mackey, A. P., & Hodgkinson, M. (1996). Assessment of the impact of naphthalene contamination on mangrove fauna using behavioral bioassays. *Bulletin of Environmental Contamination and Toxicology*, *56*(2), 279–286. Available from https://doi.org/10.1007/s001289900042.

Marshall, T. M., Dardia, G. P., Colvin, K. L., Nevin, R., & Macrellis, J. (2019). Neurotoxicity associated with traumatic brain injury, blast, chemical, heavy metal and quinoline drug exposure. *Alternative Therapies in Health and Medicine*, *25*(1), 28–34.

Mason, K. L., Retzer, K. D., Hill, R., & Lincoln, J. M. (2017). Occupational fatalities resulting from falls in the oil and gas extraction industry, United States, 2005-2014. *Morbidity and Mortality Weekly Report*, *66* (16), 417–421. Available from https://doi.org/10.15585/mmwr.mm6616a2.

Mearns, A. J., Bissell, M., Morrison, A. M., Rempel-Hester, M. A., Arthur, C., & Rutherford, N. (2019). Effects of pollution on marine organisms. *Water Environment Research*, *91*(10), 1229–1252. Available from https://doi.org/10.1002/wer.1218.

Mearns, A. J., Reish, D. J., Oshida, P. S., Morrison, A. M., Rempel-Hester, M. A., Arthur, C., Rutherford, N., & Pryor, R. (2017). Effects of pollution on marine organisms. *Water Environment Research*, *89*(10), 1704–1798. Available from https://doi.org/10.2175/106143017X15023776270647.

Ménesguen, A., & Lacroix, G. (2018). Modelling the marine eutrophication: A review. *The Science of the Total Environment*, *636*, 339–354. Available from https://doi.org/10.1016/j.scitotenv.2018.04.183.

Naser, H. A. (2013). Assessment and management of heavy metal pollution in the marine environment of the Arabian Gulf: A review. *Marine Pollution Bulletin*, *72*(1), 6–13. Available from https://doi.org/10.1016/j.marpolbul.2013.04.030.

Pan, Z., Zhu, Y., Li, L., Shao, Y., Wang, Y., Yu, K., Zhu, H., & Zhang, Y. (2019). Transformation of norfloxacin during the chlorination of marine culture water in the presence of iodide ions. *Environmental Pollution (Barking, Essex: 1987)*, *246*, 717–727. Available from https://doi.org/10.1016/j.envpol.2018.12.058.

Patlovich, S., Emery, R. J., & Whitehead, L. W. (2005). Characterization and geographic location of sources of radioactivity lost downhole in the course of oil and gas exploration and production activities in Texas, 1956 to 2001. *Health Physics*, *89*(Suppl. 5), S69–S77. Available from https://doi.org/10.1097/01.hp.0000178539.00699.4b.

Pearce, N. J., & Mann, V. L. (2006). Trace metal variations in the shells of *Ensis siliqua* record pollution and environmental conditions in the sea to the west of mainland Britain. *Marine Pollution Bulletin*, *52*(7), 739–755. Available from https://doi.org/10.1016/j.marpolbul.2005.11.003.

Raddadi, N., Giacomucci, L., Totaro, G., & Fava, F. (2017). *Marinobacter* sp. from marine sediments produce highly stable surface-active agents for combatting marine oil spills. *Microbial Cell Factories*, *16*(1), 186. Available from https://doi.org/10.1186/s12934-017-0797-3.

Rajendran, V., Nirmaladevi, D., S. Srinivasan, B., Rengaraj, C., & Mariyaselvam, S. (2018). Quality assessment of pollution indicators in marine water at critical locations of the Gulf of Mannar Biosphere Reserve, Tuticorin. *Marine Pollution Bulletin*, *126*, 236–240. Available from https://doi.org/10.1016/j.marpolbul.2017.10.091.

Reid, N. M., & Whitehead, A. (2016). Functional genomics to assess biological responses to marine pollution at physiological and evolutionary timescales: Toward a vision of predictive ecotoxicology. *Briefings in Functional Genomics*, *15*(5), 358–364. Available from https://doi.org/10.1093/bfgp/elv060.

Richardson, K., Haynes, D., Talouli, A., & Donoghue, M. (2017). Marine pollution originating from purse seine and longline fishing vessel operations in the Western and Central Pacific Ocean, 2003–2015. *Ambio*, *46*(2), 190–200. Available from https://doi.org/10.1007/s13280-016-0811-8.

Rogowska, J., & Namieśnik, J. (2010). Environmental implications of oil spills from shipping accidents. *Reviews of Environmental Contamination and Toxicology*, *206*, 95–114. Available from https://doi.org/10.1007/978-1-4419-6260-7_5.

Saari, G. N., Scott, W. C., & Brooks, B. W. (2017). Global scanning assessment of calcium channel blockers in the environment: Review and analysis of occurrence, ecotoxicology and hazards in aquatic systems. *Chemosphere*, *189*, 466–478. Available from https://doi.org/10.1016/j.chemosphere.2017.09.058.

Schnurr, R., Alboiu, V., Chaudhary, M., Corbett, R. A., Quanz, M. E., Sankar, K., Srain, H. S., Thavarajah, V., Xanthos, D., & Walker, T. R. (2018). Reducing marine pollution from single-use plastics (SUPs): A review. *Marine Pollution Bulletin*, *137*, 157–171. Available from https://doi.org/10.1016/j.marpolbul.2018.10.001.

Shahidul Islam, M., & Tanaka, M. (2004). Impacts of pollution on coastal and marine ecosystems including coastal and marine fisheries and approach for management: A review and synthesis. *Marine Pollution Bulletin*, *48*(7-8), 624–649. Available from https://doi.org/10.1016/j.marpolbul.2003.12.004.

Soundarajan, K., & Mohan Das, T. (2019). Sugar-benzohydrazide based phase selective gelators for marine oil spill recovery and removal of dye from contaminated water. *Carbohydrate Research*, *481*, 60–66. Available from https://doi.org/10.1016/j.carres.2019.06.011.

Thibodaux, D. P., Bourgeois, R. M., Loeppke, R. R., Konicki, D. L., Hymel, P. A., & Dreger, M. (2014). Medical evacuations from oil rigs off the Gulf Coast of the United States from 2008 to 2012: Reasons and cost implications. *Journal of Occupational and Environmental Medicine/American College of Occupational and Environmental Medicine*, *56*(7), 681–685. Available from https://doi.org/10.1097/JOM.0000000000000221.

Tibbetts, J. H. (2015). Managing marine plastic pollution: Policy initiatives to address wayward waste. *Environmental Health Perspectives*, *123*(4), A90–A93. Available from https://doi.org/10.1289/ehp.123-A90.

Vergeynst, L., Wegeberg, S., Aamand, J., Lassen, P., Gosewinkel, U., Fritt-Rasmussen, J., Gustavson, K., & Mosbech, A. (2018). Biodegradation of marine oil spills in the Arctic with a Greenland perspective. *The Science of the Total Environment*, *626*, 1243–1258. Available from https://doi.org/10.1016/j.scitotenv.2018.01.173.

Zabbey, N., & Olsson, G. (2017). Conflicts – Oil exploration and water. *Global Challenges*, *1*(5), 1600015. Available from https://doi.org/10.1002/gch2.201600015.

Zeng, L., Lam, J., Chen, H., Du, B., Leung, K., & Lam, P. (2017). Tracking dietary sources of short- and medium-chain chlorinated paraffins in marine mammals through a subtropical marine food web. *Environmental Science & Technology*, *51*(17), 9543–9552. Available from https://doi.org/10.1021/acs.est.7b02210.

Zhou, Y., Wei, F., Zhang, W., Guo, Z., & Zhang, L. (2018). Copper bioaccumulation and biokinetic modeling in marine herbivorous fish *Siganus oramin*. *Aquatic Toxicology (Amsterdam, Netherlands)*, *196*, 61–69. Available from https://doi.org/10.1016/j.aquatox.2018.01.009.

CHAPTER

2 Heavy metals as pollutants in the aquatic Black Sea ecosystem

Monica Butnariu

Banat's University of Agricultural Sciences and Veterinary Medicine "King Michael I of Romania" from Timisoara, Timis, Romania

2.1 Background

Although it does not have high biological diversity, the Black Sea has a complex aquatic ecosystem with unique features in terms of physicochemical and biological characteristics. The environment conducive to life generally is located on the continental aquatic platform, up to a depth of 150–200 m, and is largely influenced by environmental conditions and water dynamics. Below that, the presence of hydrogen sulfide (H_2S) causes 85%–90% of the entire body of water (except anaerobic bacteria) to be completely lifeless.

The aquatic ecosystem consists of the biotope (water, substrate nature, geographical and climatic factors, salinity, minerals) and the biocenosis (all aquatic organisms in the biotope, which belong to various species and are functionally interdependent), which together form an integrated ensemble in permanent interaction. A characteristic of the Black Sea ecosystem is its low salinity, which is explained by the significant supply of fresh water it receives from the large rivers that flow into its basin. The movements of the sea waters have a significant role in the dynamics of aquatic species (Tsikliras et al., 2015). Occasional waves and surface currents are reflected in the transport of alluvium. The thermal variation is quite large between the summer months, when the waters reach 25°C–27°C, and the winter ones, when they can drop below 0°C. Temperature variations, with their reversals between the hot and cold seasons, take place only up to a depth of 75–100 m. Below this depth there is a constant temperature of 7°C which rises to the bottom up to 9°C.

The reaction of the medium is alkaline. Water oxygenation varies with season and depth. In the cold period, the surface layers are oversaturated with oxygen. Oxygen concentration decreases with depth. From 150 to 200 m to the bottom of the sea the dissolved gas is hydrogen sulfide (H_2S).

The structure of the biotope determines the configuration of the aquatic ecosystem. The supralittoral floor consists of shoreline areas covered or accidentally splashed by waves. The area has a high humidity and a generally large, or at least significant, amount of organic matter brought by waves or of local origin. Usually organic matter decomposes to form methane and hydrogen sulfide odor deposits.

Bacterial Fish Diseases. DOI: https://doi.org/10.1016/B978-0-323-85624-9.00003-8

The flora consist mainly of certain forms of algae, or rarely lichens, with resistance to environmental and hydrophilic variations. Angiosperms are also found at a lower frequency, especially in the dry part of the supralittoral floor.

In addition to aerobic and less anaerobic bacteria, the fauna includes numerous crustaceans, insects, and worms. Most of these creatures feed on deposits of organic matter. A smaller part consists of small predators. To these must be added passenger life, especially seabirds (Fontaine et al., 2012).

In general, pollution is the main culprit for declining aquatic biodiversity. However, pollution is only one aspect of the range of man-made factors that have made the Black Sea ecosystem one of the most affected sea basins.

2.2 Heavy metal poisoning

2.2.1 General properties of metals

The descriptive functional terms accepted for the classification of metals in environmental studies are: trace metals, micronutrients, and heavy metals. Trace metals are those metals found in concentrations of less than 0.1% (<1000 mg/kg) in the soil.

The term micronutrient describes those elements that are needed in small quantities by certain organisms (e.g., phytoplankton, crustaceans, fish) to perform their metabolic functions.

The term heavy metal describes elements that have an atomic number greater than 20. Heavy metals have also been defined as that group of metals and metalloids with an atomic density greater than 4 g/cm^3 or five times greater than water density.

Metals circulate naturally in the environment, and metallic elements are found in all aquatic organisms (e.g., phytoplankton, crustaceans, fish), where they play a variety of roles. They can be structural elements, stabilizers of biological structures, components of control mechanisms and enzyme activators, or components of redox systems.

Therefore some metals are essential elements, and their deficiency leads to impaired biological functions. When present in excess, however, these essential metals can also become toxic. Other metals do not have an essential function and can give rise to toxic manifestations even when their intakes are small.

Unlike most organic chemicals, which can be removed from tissues by metabolic degradation, metals are indestructible elements and therefore have the potential to accumulate. Excretion is the main mechanism for removing metals from the tissue. Accumulation in the tissue does not necessarily imply the appearance of a toxic effect, because in the case of certain metals, inactive complexes or deposits are formed (Flora & Pachauri, 2010). Heavy metal poisoning (mercury, lead, cadmium, chromium) occurs when the amount of that metal in the body exceeds a certain value.

Some metals can be absorbed only by ingestion, others by respiration, and finally some even by simple contact with the skin. Their drawback is that they produce intoxication from cumulative effects: once ingested/inhaled they are very difficult to eliminate, over years.

They are deposited, depending on the metal, in certain internal organs, muscles, or even bones and teeth. Fish flesh may contain such metals, especially if the environment in which those fish live is heavily polluted.

Accumulation is more common in predatory fish, precisely due to the cumulative effect and contamination "in the pyramid" (it consumes other potentially contaminated fish). The accumulated metal is stored mainly in the liver and muscles, the deposits being all the more important the older the fish.

Given that we rarely reach the age of Methuselah, not too many problems are posed as long as we do not consume fish from certain affected tanks (downstream of chemical plants, settling tanks, etc.). Beware of pools in tailings areas and thermal power plants: from burning, the tailings become rich in certain heavy metals and radioactive components!

On the other hand, it is a fact that the Danube has a smaller amount of pesticides and heavy metals at its discharge than at its entrance to the country, due to the decline of Romanian industry and agriculture. Basically, our rivers now dilute the heavy metals brought in by the Danube (Varga et al., 2002).

Closely related to heavy metal poisoning is lead poisoning, both acute and chronic. People can get sick from lead poisoning in two ways: by ingestion, the most common (but with the "advantage" that only 10% of ingested lead is absorbed), or by inhalation. The risk is increased in circulating areas (from burning lead tetraethyl from gasoline), especially since inhaled lead is absorbed in a proportion of 90%. In this case, we are obviously interested in how much lead consumed fish can contain.

From my point of view, the problem of poisoning from fishermen using lead weights is enormously exaggerated. More than likely, significant economic interests are at the base (Bonaldo et al., 2017). Theoretically, opponents of the use of lead as fishing weights are somewhat correct, but basically problems occur only with the direct ingestion of lead by fish and other aquatic animals. (Lead poisoning, on the other hand, is relatively common in hunters, because bullets can remain in the tissues for a long time, without causing the animal's death.) In addition, in cold water lead forms a film of lead oxide that is practically insoluble and does not allow water molecules to reach the rest of the metal (it insulates the metal as a kind of varnish). X (Pain et al., 2019).

2.2.2 Fish species with "toxic" flesh

There is great confusion between the toxicity of fish flesh and other diseases induced by it. Usually, the only common species that is said to have toxic flesh per se is the sea rooster, improperly called "sea dog" (confusion with the Black Sea ecosystem shark), a species in the Bleniidae family, and especially *Blennius sanguinolentus*, the fish often denigrated by the *Ponticola cephalargoides* fishermen on the Coast. There has not been found in any ichthyological treatise, regardless of age, a phrase that says "the flesh is toxic"; all state instead "the flesh cannot be eaten."

2.2.3 Toxicity of various organs/tissues of fish

True, there are fish species that contain venom glands. However, they are not the subject of this article, most of them having edible flesh.

Heavy metals are chemical elements that are naturally present in ecological systems, but with exploitation they have become metal with toxic potential. All of the exploitation has led to the appearance in the aquatic environment of the risks from anthropic sources, which have become much larger compared to the inputs from natural sources.

A metal can be characterized by an anthropogenic enrichment factor. This factor is the percentage that is associated only with anthropogenic sources, calculated from the total annual emissions of a metal. Thus, this factor has the following values for various metals: Pb—97%, Cd—89%, Zn—72%, Hg—66%, and Mg—12%. This anthropogenic enrichment factor, together with the potential for metal toxicity, indicates the priority to be given when choosing the metals to be analyzed (Lipowicz et al., 2013).

Heavy metals are of interest from an industrial point of view as well as ecologically and biologically. There are many metals that are of interest because of their toxic properties or because they are essential for the health and survival of aquatic animals and plants. It is important to note that most often importance is given to pollution and toxicity.

Metals that have an important role in the life of aquatic organisms (e.g., phytoplankton, crustaceans, fish) can be in large quantities, like K, Mg, Ca, Na; or only in traces (less than 1 mg/kg tissue), as in Cu, Fe, Mn, Zn, Se, Co, Mb, Cr, Ni, Si, V, and As.

The metal problem involves two special situations: in the first case the emphasis is on excess metals in some compartments of aquatic ecosystems, which leads to disruption of their functionality and damage to human health, and in the second case the problem is related to quantitative deficiency of certain metals in agricultural systems, which leads to a limitation of productivity. Heavy metals are considered to be a very important category of metal that are toxic and stable. Compared to the organic pollutants, heavy metals are not biodegradable; they have a less mobile character, such that they can persist much longer in the soil and in sediments (Rehman et al., 2018).

Heavy metals are not created by biological or chemical processes but are not destroyed either. These processes have the ability to make possible only a transition of the metal from one valence to another or from an organic to an inorganic form.

A main problem that is associated with the persistence of metals is related to their ability to bioaccumulate and their bioamplification, phenomena that can cause an increase of the metal in the aquatic ecosystem, posing possible long-term risks in ecological systems. In order to differentiate the toxicity of heavy metals, their chemical properties and their compounds must be taken into account, but the properties of organisms (e.g., phytoplankton, crustaceans, fish) that are exposed to contamination must be taken into account.

However, a high potential for bioaccumulation does not necessarily imply a high potential for toxicity, the situation differing from one element to another. Heavy metal contamination of aquatic ecosystems is a problem of great importance because these metals enter the food chain and have effects on the functioning of the biocenosis (Chaturvedi et al., 2015).

Heavy metals are considered particularly dangerous because, in the process of food preparation, they do not decompose; on the contrary, their concentration per unit of measurement increases. Metals also have the property of accumulating in the human body (bioaccumulation), so they slow down or even block intracellular biochemical processes. Most metals have mutagenic and carcinogenic properties, being difficult to remove from the human body (Ajima et al., 2015).

Lead is part of the group of metals with high toxicity potential. Because of this, the maximum limit set by official health standards (Codex Alimentarius) is very strict: 0.5 mg/kg for all categories of meat. For carp flesh in Romania, the average varies between 0.45 and 0.62 ppm. However, there is a limit below that set for other categories of food, such as that in cans.

Of course, the first source of contamination is waste discharged or emissions from various industries, especially the (non) ferrous metals and steel industry. Within them, an important place is occupied by the processing of lead and lead alloys (batteries for the automotive industry, the arms industry). Some of these residues reach the Danube through river waters, but also through rainwater resulting from rains and melting snow, which effectively washes the surface of the soil while entraining the impregnated residues.

Lead overgrowth has been revealed over time in fish with permanent habitat in the Delta.

Razelm Lake (Romania) is not found on the Danube stream, having a direct connection with the Black Sea ecosystem, which is why the lead content of the shale in this lake is found in traces. It should be noted that lead in contaminated water is not found as such, but in the form of compounds. Some of them are insoluble, so they are in the form of microparticles.

Lead, like all metals, is not degradable, so its concentration in sediments gradually increases, as long as there are sources of pollution. However, there are some lead products that are water soluble; these, although quantitatively lower, are perhaps the most important source of fish pollution. The fish takes in oxygen from the water, which it transfers to the blood in the gills. The water stream for oxygenation has a permanent circuit on the mouth-gill path. During the fish's breathing, the amount of water that bathes the gills is enormous. At the level of the gills, some of the substances in the water are retained and passed directly into the blood. There are also fat-soluble lead compounds (e.g., lead tetraethyl), which can penetrate directly through the skin, adding to those that reach the body in other ways (Matache et al., 2013).

Cadmium is a metal whose toxicity appears to exceed that of lead (maximum limit set at 0.075 ppm). It is widely used as a coating to protect ferrous surfaces from corrosion. It is rarely found in fish species in our country (Kim & Kim, 2016).

Zinc is a metal with relatively low toxic potential; it is found in fish flesh in disable quantities. However, when the content is high, the consequence of contamination from pollution sources, measures are needed to restrict consumption. The maximum regulated limit for food is 50 ppm. The values found in continuously studied fish are below this limit. Sometimes, higher values have been found (in saltwater fish), but still within normal limits.

Tin has an almost insignificant toxic potential. Therefore, in the official health norms maximum limits are established only for cans in metal boxes. Tin is widely used in the canning industry to protect food stored in cans made of other metals. However, the tin layer on the inner surface has a relatively poor stability. In prolonged contact with small amounts of hydrogen sulfide, which can form in the sterilization process, it forms tin sulfide, which (relatively soluble) passes into the content. In this case, the first indication is the metallic taste imparted to the contents. The tin alloy is made of tin with a small lead alloy (1% or even more). Therefore when stains or large white-bluish areas (marbling) are found on the inside of tin cans, it is evidence of the appearance of tin sulfide (Seto et al., 2013).

Copper is naturally found in most foods, but in very small amounts. When the content is high, it presents a toxicological risk. The maximum limit recommended by the Codex Alimentarius for fish is 3 ppm. This condition is currently being challenged by most researchers.

Numerous laboratory analyses have shown that in the copper circuit in the body, the maximum level of accumulation and concentration is achieved in the liver, where the content exceeds the value of 3 ppm naturally.

In the case of fish species from the Delta, the copper content is in the range of 1.3–1.5 ppm, thus falling within the regulated values. In contrast, in oceanic fish (mackerel and horse mackerel) the value exceeds twice the allowed limit (Freitas et al., 2018).

2.2.4 Bioaccumulation factor

Given the close link between fish fauna and the environment in which they live, after determining the concentration of heavy metals in water, sediment, and fish samples, it was necessary to determine the percentage in which the studied species accumulated metals from water or sediment. The bioaccumulation factor (BAF) can be defined as the ratio between the concentration of the analyzed metal in an organism and the concentration of the same metal in water, during steady state. The absorption of metals can be different in each organism, as if following a passive diffusion mechanism analogous to that of oxygen absorption. In the case of sediments, the BAF is calculated with the same formula, replacing the metal concentration in the body with the metal concentration in the sediment (Jitar et al., 2015).

Some studies characterize the aquatic environment as the ultimate "recipient" of natural or anthropogenic heavy metals. In the lists of potentially toxic substances, metals are considered to have a priority role in monitoring water quality. Some heavy metals, such as mercury, lead, and cadmium, are compounds that cannot be degraded naturally, remaining in the environment for a long time, a context in which their danger is highlighted by the high potential for accumulation in the food chain.

Sewage, both domestic and industrial, exerts a significant pressure on the aquatic environment, due to the loads of organic matter, nutrients, and hazardous substances. In rural areas, 95.9% of the population is not connected to sewage systems, so the management of rural wastewater has been a major problem. Aquatic organisms (e.g., phytoplankton, crustaceans, fish) exposed to these contaminants accumulate these elements and sometimes the amounts of metals compared to the body mass of these organisms increase with the evolution of the food chain. If fishing activities occur in the contaminated aquatic area and subsequently the catches are consumed, the toxic elements are transferred to the human body, where they can cause a number of diseases whose severity depends on the accumulated metal and the amount accumulated (Tang et al., 2020).

The Black Sea is a main source of aquatic catches for riparian countries. In a market study conducted by the Ministry of Agriculture and Rural Development, it was stated that in the period 2005–13 the average share of fish production in Romania was as follows: 68% aquaculture, 27% inland fishing, and 5% marine fishing.

Fish are among the most important groups of wild species, from an ecological but also an economic perspective, representing a valuable food resource for both humans and wildlife. They play a major role in the functioning of aquatic ecosystems. At the same time, fish are recognized bioindicators in assessing the pollution of an aquatic environment with various chemicals. As a result, this is possible by tracking the dynamics of heavy metal concentrations in aquatic organisms (e.g., phytoplankton, crustaceans, fish).

In order to avoid a polluting disaster, such as that in Minamata, Japan, manifested by chronic mercury and cadmium poisoning of the human and animal population following the consumption of contaminated fish, continuous monitoring of environmental pollution is mandatory (Dvorak et al., 2018).

2.3 The role of heavy metals as pollutants

The spread of metals in water, sediments, and the atmosphere results from their presence in the earth's crust. In their natural concentrations, metals play an essential role in many biochemical processes in the body, but any concentration that exceeds the background concentration can become toxic.

As a result of anthropogenic activities, current levels are higher than under natural conditions, posing a threat to aquatic organisms, as many metals are harmful even in moderate concentrations. The toxic potential of metals depends on their bioavailability and their physicochemical properties. These properties depend on the atomic structure of metals, rendered in the periodic table of elements.

Metals are divided into the following categories: alkaline, alkaline-earth, transitional, metalloid.

Examples of metals that are more relevant to the environment in terms of their toxic effects are the following: cadmium (Cd), chromium (Cr), cobalt (Co), copper (Cu), lead (Pb), mercury (Hg), nickel (Ni), tin (Sn), vanadium (V), and zinc (Zn). Arsenic is also considered a dangerous metal, although chemically it is actually a semimetal (metalloid).

The general sources of pollution of the aquatic environment are represented by cities and coastal industries, wastewater and industrial waste, household waste and stormwater; shipping, unloading of waste at sea; wrecks, ammunition lost or thrown away intentionally, offshore drilling rigs, and atmospheric deposits (Gao et al., 2019). The terrestrial sources that generate heavy metals are mainly represented by wastewater treatment plants, manufacturing industries, mining, and agriculture. Metals are transported either in forms dissolved in water or as an integral part of sediments.

2.3.1 Peculiarities of heavy metals found in aquatic ecosystems

In the aquatic environment, the toxicity and bioavailability of metals are influenced by various abiotic factors, such as pH, water hardness, alkalinity, and accumulation of humic substances.

The toxicity of metals in water increases in proportion to alkalinity, pH, salinity, temperature, and conductivity. The two most important water parameters that influence the accumulation of metals in the biota are pH and salinity. There is a strong correlation between these parameters and the accumulation of metals, so that an acidic environment causes an increase in the accumulation of metals in the biota. For many metals, alkalinity is a much more important cofactor than hardness. Heavy metal salts (Mn, Co, Ni, Cr, As, Cd, Pb, Fe, Sn, Sb, Au, Ag, Cu, Hg) are stable and toxic compounds, so they can be forms of severe pollution for surface waters (Bankar et al., 2018).

Inorganic, insoluble, or partially soluble organic complexes are less toxic than simple ions. Analyzing their impact on fish, heavy metals can be classified as essential elements [iron (Fe), zinc (Zn), copper (Cu), magnesium (Mg), selenium (Se), cobalt (Co), vanadium (Vn)] and nonessential elements (potentially toxic trace elements) [aluminum (Al), arsenic (As), mercury (Hg), lead (Pb), cadmium (Cd), bismuth (Bi)].

Some metals have a higher toxic impact on the environment than others, such as Cd, Cr, Cu, Pb, Ni, and Zn. Metals in an aquatic environment can cause the loss of biodiversity by exerting toxic effects on the biota. Heavy metals are of particular importance in water bodies due to their inability to decompose, long persistence, bioaccumulation, and biomagnification in the food chain. The bioaccumulation of heavy metals in aquatic organisms occurs as a result of their very long half-life, but also is due to the fact that the body does not have physiological elimination capabilities. In this way, exposure to heavy metals becomes cumulative, initially asymptomatic. When the level of biostored metal is high enough, signs and symptoms begin to appear progressively, which are proportional to the degree of exposure of the body. The symptoms are very individual, so there may be different symptoms at a similar level of exposure (Hao et al., 2019).

The second element is represented by bioacceleration. This phenomenon consists in the fact that these heavy metals accumulate in the body in proportion to the evolution of the food chain. Excessive pollution and unhealthy nutrition are the main factors affecting the organisms.

Heavy metals are characterized by two important properties: bioaccumulation in the body and bioacceleration, as shown in Fig. 2.1.

Currently, we no longer talk about acute intoxications with heavy metals, except among workers in an at-risk industry or in the case of industrial or environmental accidents. However, today we are concerned about a new pathology associated with heavy metals, which is represented by low-toxicity exposure. This type of particular exposure is characterized by heavy metal levels near normal, but the measurements made to determine the level of exposure in blood, urine, or tissues, among which the most used is hair, do not provide a complete, real, or accurate picture of the biostored amounts. For example, methylmercury is stored in the brain, making it extremely difficult to estimate the amount stored there. It is this characteristic that defines and explains the levels of low toxicity, which normally should not produce symptoms, but which in reality produce negative effects felt by the organisms (e.g., phytoplankton, crustaceans, fish).

Within these low-toxicity exposures, a predominant enzymatic inactivation occurs at the level of selenoenzymes or enzymes possessing thiol groups. In this way, dozens of enzymes essential to the body's homeostasis are inactivated, causing chaos in the functioning of various devices and systems (Yılmaz et al., 2017).

The contamination of the waters of the seas and oceans with mercury has caused a phenomenon of accumulation of this metal in the aquatic things in these areas. The phenomenon of bioacceleration has been described for the first time in fish. As a result of this process, the more evolved the fish in the food chain, the higher the amount of mercury accumulated. Making a comparison between a sardine and a shark, we can conclude that the latter has accumulated an amount of mercury over 300 times higher, knowing that the shark ingested all species of fish contaminated in turn with mercury, below its trophic level (Saher & Siddiqui, 2019).

Once in the aquatic environment, these metals can follow several paths: dissolved in the water column, stored in sediments, volatilized in the atmosphere, taken over by organisms (e.g.,

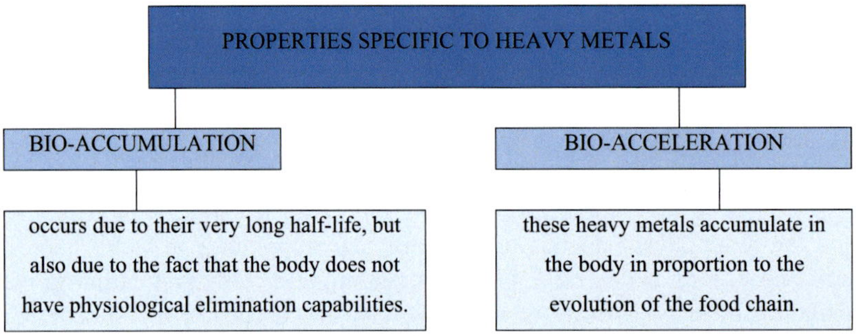

FIGURE 2.1

The properties specific to heavy metals, which are manifested in aquatic organisms.

phytoplankton, crustaceans, fish). Metals are also generated as a result of natural rock erosion processes. This process is intensified following mining activities that expose various ores containing metals. Leaks from tailing dumps and tailing ponds introduce substantial amounts of metals into water resources. It is considered that, in the absence of appropriate measures, mining activities pose a high long-term risk of the release of heavy metals into the environment.

Any activity involving the extraction or processing of metals can be a source of fine metallic particles dispersed in the atmosphere. Rust and other forms of corrosion lead to the spread of metals in the environment during the use or storage of various metal equipment.

The burning of fossil fuels or various categories of waste also causes the release of metals into the atmosphere.

The greatest deposition of metal particles obviously occurs in the vicinity of mines, smelters, or other categories of metal-processing activities, which are the major sources of emissions. But most particles are so small they can be transported over enormous distances by wind. In particular, mercury, which is present in gaseous form in the atmosphere, can be widely dispersed, far from the sources.

Road transport is also responsible for significant lead emissions, following the use of lead-containing fuels as an additive (Ighariemu et al., 2019).

The metals released into the atmosphere are deposited at ground level, where they remain for a long time. Under certain conditions, for example lowering the pH, the metals in the soil, especially mercury and cadmium, are solubilized and reach the water resources.

Understanding the mechanisms of interaction between heavy metals and aquatic organisms involves the following aspects: the bioavailability and how to take over metals; the intervention of possible protective mechanisms; the susceptibility of aquatic organisms (e.g., phytoplankton, crustaceans, fish) to the various effects of exposure (Wang et al., 2018).

2.4 Bioavailability of heavy metals for aquatic organisms

Heavy metals are present in water, sediment, and the bodies of fish. The distribution of heavy metals in the aquatic environment is as follows: distributed in the water column, precipitated in sediments or accumulated in the benthic substrate and related to other organic elements or taken up by organisms (e.g., phytoplankton, crustaceans, fish).

The metallic elements transported in the aquatic systems are subjected to specific processes, manifested under the direct influence of the physicochemical parameters of the water. The phenomenon of bioavailability has been defined as the relative ease of an element to be transferred from the environment to a specific location in an organism of interest and refers to both the water component and the sediment component.

The bioavailability of metals in the water column refers to their water-soluble capacity, the soluble form of the elements being one of the most toxic forms on fish. The purification of the water column is achieved due to important mechanisms such as adsorption and sedimentation processes. Sediments have the ability to immobilize metal ions through specific processes such as adsorption, flocculation, and coprecipitation. Therefore sediments can act as “hot spots” of high concentration of metals. In shallow lakes, metals are more likely to be resuspended and may cause secondary contamination of the aquatic environment (Kim et al., 2016). Thus sediments can act both as a decanter and as a source of metals in the aquatic environment. A process of particular importance is that of bioaccumulation in aquatic organisms. The rate of bioaccumulation is influenced by certain factors such as temperature and the physiological state of the body (sex, age, size).

Heavy metals, when ingested in excessive amounts, can lead to the creation of random bonds with cellular biomolecules, such as enzymes or proteins to form complexes that can compromise their structure and/or function. Heavy metals tend to accumulate in organisms at the top of the food chain (such as fish in the aquatic environment), through biomagnification effects. The biomagnification process is a complex process that involves increasing the concentration of toxins upwards in the food chain.

The bioavailability of metals is defined by the fraction of the total concentration of metals that has the potential to accumulate in the body. The factors that control the bioavailability of metals are the following:

- the biological characteristics of the organism (metal assimilation efficiency, feeding strategies, size/age, reproductive stage);
- metal geochemistry (water–sediment distribution, metal speciation);
- physicochemical factors of the environment, which influence the factors listed earlier (temperature, salinity, pH, ionic strength, concentration of dissolved organic carbon, total solid suspensions) (Speelmans et al., 2010).

The speciation of heavy metals in the aquatic environment is of fundamental importance due to the fact that the bioavailability and toxicity of metals depend on their chemical form in water. Speciation is in turn dependent on the specific physicochemical factors of the aquatic environment.

Heavy metals are present in the aquatic environment in various chemical forms (dissolved, colloidal, or particulate), as a result of the balance between metal ions and inorganic and organic complexes.

The bioavailability of metals is one of the determining factors of their accumulation in aquatic organisms. The uptake of metals occurs directly from seawater through the permeable surfaces of the body, in the case of dissolved forms, as well as through food, in the case of particulate forms.

The uptake of metals from seawater is influenced by metal speciation, the presence of organic or inorganic complexes, pH, temperature, salinity, and redox conditions. Intestinal uptake depends on similar factors, plus feeding rate, intestinal transit time, and digestion efficiency (Carbonaro et al., 2019).

Numerous studies have shown that the hydrated free ion form is the predominant bioavailable form for copper, cadmium, and zinc, although exceptions have been reported. Thus the importance of other chemical forms of dissolved metals, such as complexes formed with low molecular weight organic ligands, should not be neglected.

It has been observed that the presence of organic ligands increases the bioavailability of cadmium in mussels and fish, due to the facilitation of the diffusion of the hydrophobic compound in membrane lipids.

Organic compounds of some metals may be much more bioavailable than ionic forms, the best example being organomercuric compounds, which are soluble lipids and penetrate rapidly into the body, having an increased toxicity to mercury chloride.

Adsorption on suspensions affects the total concentration of metals present in the water column. The association of metals in the particulate phase is also critical for the process of uptake by organisms through food ingestion.

Sediments accumulate insoluble metal compounds, which can under certain conditions be released into the interstitial water, thus adding to the soluble or suspended metals in the water column.

The concentrations of heavy metals in the sediments and suspensions are much higher than in seawater, so a small fraction of them can be an important source for uptake, especially for filter organisms (e.g., phytoplankton, crustaceans, fish) and those buried in sediment. Fine-grained oxidized particles are expected to be the most important source of available metals (Menegário et al., 2017).

Numerous studies have shown that the bioavailability of metals for bivalve molluscs that feed in sediment depends on the type of sedimentary particles. If the particles are coated with bacterial extracellular polymers or fulvic acids, the bioavailability of cadmium, zinc and silver is significantly increased.

Binding to iron oxyhydroxides decreases the bioavailability of metals in the sediment. The nature of different forms of metals in the aquatic environment remains a variable that is not fully understood. Dissolved or particulate forms of metals have different uptake and accumulation pathways and require in-depth studies.

Specific pathways for taking up free ionic forms and those complexed with organic ligands must be identified and characterized. It is not known whether there are specific mechanisms for different valence states or for different types of inorganic ionic complexes (Stenzler et al., 2017).

The transfer of heavy metals along aquatic food chains is of interest in environmental health research for several reasons. Heavy metals are generally heavier than iron in the periodic table. Some of them are necessary for human metabolism, such as organic compounds of iron, cobalt, copper, and zinc. However, not all organic heavy metal compounds are beneficial, and some do not exist in water and can only be found in food.

From water, we can consume only dangerous inorganic compounds, as well as a whole range of other heavy metals that should be avoided by any means. This range is called xenobiotic because it dramatically damages all life forms; lead is an example of a xenobiotic.

First, the accumulation of heavy metals in aquatic organisms may ultimately result in the trophic transfer of metals to humans, leading to a potential risk to public health from the consumption of contaminated aquatic products. The best-known and most tragic example was the appearance of Minamata disease in Japan, following the consumption of aquatic products containing high concentrations of methylmercury.

From the point of view of aquatic ecosystem health, metals can be toxic to aquatic organisms, preventing the functioning of the aquatic ecosystem through a wide range of harmful effects (Matsuyama et al., 2018).

Aquatic organisms play an important role in the biogeochemical cycles of metals in the aquatic environment. The factors that influence the accumulation of metals are the relative amounts of metals present in the environment, as well as their chemical form. However, there is considerable variation in the concentration of heavy metals between species, tissues, and even between individuals collected from the same location.

This is due to the fact that the uptake and elimination of heavy metals are determined by biological parameters, which include the permeability of external surfaces, feeding strategies, quantities and types of internal ligands, efficiency of excretory systems, nutritional status, growth, season, and reproductive stage (Gopalakrishnan et al., 2007).

Aquatic organisms have a certain selectivity in the accumulation of heavy metals, and a distinction must be made between essential and nonessential metals. Essential metals such as copper, zinc, manganese, iron, or cobalt are vital components of many respiratory enzymes and pigments. Consequently, aquatic organisms must provide metal tissues in sufficient quantities for metabolic and respiratory needs. Deficiency of these metals, but also accumulation above certain levels, produce harmful effects.

Nonessential metals (lead, arsenic, mercury, cadmium) are very toxic, even at very low levels, especially if they accumulate at metabolically active sites. The bodies of aquatic organisms are obliged to limit the accumulation of nonessential metals or to change them into nontoxic forms.

Toxic heavy metals interfere with the normal metabolic functions of the essential elements. Binding to protein macromolecules disrupts normal biological function. The catalyzed formation of metals by oxygen free radicals is involved in the production of many pathological changes, including mutagenesis, carcinogenesis, and aging. Thus, although metals are essential components of life, they become harmful when they are present in excess (Zhang et al., 2016).

Rising bioavailable levels in the aquatic environment is a problem for human health and aquatic ecosystems. Bioaccumulation is a common topic recently addressed in the field of research and analysis of environmental risk because it is the body's exposure to various environmental pollutants. The last decade has shown that the bioaccumulation and bioamplification of chemicals, through the food chain, can be a necessary condition for highlighting adverse effects in species and individuals.

The redox potential, together with other physicochemical parameters (pH, different organic compounds, humic substrate, complex particles, presence or absence of other metals, anions, different ionic bonds, temperature, salinity, light intensity, dissolved oxygen) plays a role important in the bioaccumulation of heavy metals. At a low redox potential, the metals bind to the sulfides in the sediment, thus becoming immobile. High salinity influences the formation of metal chlorides, preventing their absorption by plants and other aquatic organisms. The accumulation of metals in aquatic sediments is a risk to the aquatic ecosystem, and the concentrations of heavy metals in them can provide historical information about the pollution of an area. Unlike many organic pollutants, metals do not degrade, but remain in the environment (Wei et al., 2020).

2.5 Effects of heavy metal pollution on aquatic ecosystems

In recent decades, increased contaminant inputs and habitat destruction have produced drastic changes in aquatic ecosystems. In this direction, the scientific interest in the following fields has increased: the accumulation and toxic effects of contaminants on aquatic organisms; taking over and accumulating contaminants in aquatic resources intended for human consumption.

The effects of heavy metal can be detected at several levels of biological organization, from the entire aquatic ecosystem to the subcellular and molecular level. The most relevant ecotoxicological assessments, from an ecological point of view, are those that describe changes in the structure and function of aquatic ecosystems. These measurements are often difficult, lengthy, and do not allow the degree of aquatic ecosystem change to be correlated with a particular level of contamination. At the cellular and molecular level, pathological changes and biochemical markers have been identified that occur as a result of exposure to heavy metal (Kanhai et al., 2014).

Correlations were established between specific heavy metal present in certain concentrations and pathological or biochemical responses. However, correlating the effects at the individual level with the alterations at the level of communities or populations is quite difficult. There are concerns about the relevance of applying physiological and biochemical methods to assess the effects of pollution at the population level.

Interindividual variability in response to heavy metals is of major importance, as it is the key to understanding the selection mechanisms that accompany pollution-induced ecological changes. Heavy metals that do not exert selection pressure are not considered to cause significant biological effects at the aquatic ecosystem level, as they do not result in community restructuring. There is a wide range of methods available for assessing the effects of heavy metals in the aquatic environment, of particular importance being the integrated assessment, using several methods, each pursuing a different level of biological organization (Schertzinger et al., 2018).

The interactions between heavy metal and aquatic organisms involve several aspects. The first stage in aquatic toxicology studies consists of assessing the type of pollutant, its bioavailability, and the methods of uptake by the bodies of aquatic organisms. The distribution of heavy metal in the aquatic environment includes accumulation in the benthic substrate, distribution in the water column, and uptake by aquatic organisms (e.g., phytoplankton, crustaceans, fish).

The fraction present in the water column (bound to colloids, particles, or dissolved) and in food represents the bioavailable fraction. There are two major ways of taking it in: the respiratory system (gills) and the digestive system.

Food as a source of uptake is especially important in benthic aquatic ecosystems, where sediment-associated heavy metal are significant sources for aquatic ecosystems (Dar, Dar et al., 2016; Dar, Kamili et al., 2016; Gholizadeh & Patimar, 2018).

The uptake of contaminants leads to their concentration in the tissues. The rate of bioconcentration depends on many factors, such as temperature, physiological state (sex, season) and the potential for biomagnification along the trophic level.

The first stage of the contaminant-organism impact is represented by the interaction with endogenous molecules. These interactions fall into three main groups, shown in Fig. 2.2. Bioaccumulation signifies the bioconcentration, to which is added the food chain transfer; then the sum (elimination + growth dilution) is subtracted.

The contaminant may be sequestered and then neutralized, and/or may have specific interactions with endogenous molecules (inhibition of some enzymes) and/or may be metabolized by the enzymes of the biotransformation system. All these interactions can lead to:

- long-term storage (neutralized fraction);
- direct or indirect toxic effects (after biotransformation);
- excretion of contaminants or their metabolites.

The toxic effects of heavy metal have repercussions at the cellular, tissue, or body level, thus changing the integrity of the population and ultimately the entire aquatic ecosystem. The response

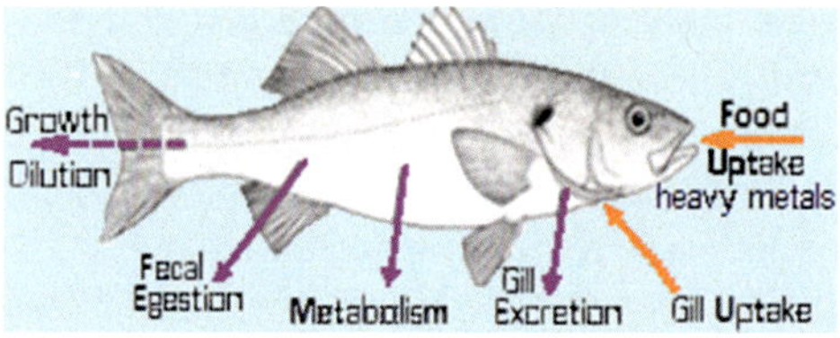

FIGURE 2.2

Schematic presentation of the phenomenon of bioaccumulation of heavy metals in fish [bioconcentration + food chain transfer − (elimination + growth dilution)].

time to the impact with contaminants can vary from hours for the molecular and cellular level, to several years for the population and community (Mohamed et al., 2016).

Specificity, in the sense of identifying the pollutant that causes an effect, can only be achieved at the molecular level. A physiological response, such as the effect on growth, is a nonspecific response to a number of environmental stimuli, providing a measure of the overall impact.

Even if in this case the ecological relevance exists, it is necessary to elucidate the molecular mechanisms. The term biomarker defines biological parameters that change in an organism exposed to environmental contaminants (Lacave et al., 2018). The concept of biomarkers does not only mean biochemical measurements, but also includes cellular pathology, physiological processes and even the behavior of an organism exposed to heavy metal.

This creates the possibility of sequential use of biomarkers, starting with nonspecific (physiological) ones and ending with specific biochemical and cellular biomarkers (e.g., mixed-function oxidase activity, metallothioneins, intracellular granules, tissue damage). The magnitude of the response of biomarkers, together with the determination of tissue concentrations of heavy metal, contributes to the overall assessment of the impact of pollution.

The behavior of metals in aquatic organisms is described by the mechanisms of uptake, storage, excretion, and regulation. General models of metal uptake and accumulation will be discussed in an attempt to understand and explain the variability of metal tissue levels.

First of all, in the process of food preparation the metals do not decompose; on the contrary, their concentration increases. Second, metals have the property of accumulating in the human body, so they slow down or even block intracellular biochemical processes. Third, most metals possess mutagenic and carcinogenic properties. Heavy metals are difficult to remove from the human body. A few words about the most common metals follow (Kim et al., 2018).

2.6 Methods of taking heavy metals from the bodies of organisms

The motivation for numerous studies on heavy metals comes from the need to understand the impact of pollution on aquatic communities. There has been a tendency to investigate mainly the uptake of heavy metals in solution in seawater, although the nutritional pathway is also of great importance in the uptake of heavy metals.

Heavy metals present in the water-soluble form penetrate the cells of aquatic organisms (e.g., phytoplankton, crustaceans, fish) either through passive diffusion processes (invertebrates) or through a "carrier" molecule (fish) (Braz-Mota et al., 2018; Dar, Dar et al., 2016; Dar, Kamili et al., 2016; Dar et al., 2020).

The absorption of heavy metals in solution in seawater occurs both through the general surface of the body and through specialized areas such as the gills or intestinal walls. In most large crustaceans and mollusks, the barrier through which the metals are taken over is represented by the respiratory surfaces, the rest of the body being protected by the calcareous or chitinous exoskeleton. Once heavy metal ions have crossed the barrier into the body, they are rapidly bound by intracellular ligands. The presence in the cell of ligands with high affinity for heavy metals, such as glutathione and metallothioneins, ensures their continuous removal from the inner layer of the cell membrane, thus reducing the concentration of cations in the cell and maintaining a permanent gradient to the external environment.

The uptake of heavy metals bound to suspended matter is mainly related to the activity of the digestive tract and is an important source for organisms (Armant et al., 2017). In aquatic invertebrates (mollusks, crustaceans), metal-bearing suspensions are taken up by endocytosis, an active transport mechanism. At the level of lysosomes the biological material is degraded, and in this way the metal becomes available to the cell or can remain bound to the resulting compounds. Water can be an important source of heavy metals, as a result of discharges, the activity of sewage and pretreatment plants, the discharge of sewage water, and household waste. The hardness of the water and the content of organic compounds can determine its enrichment with heavy metals.

As mentioned, not all heavy metals are harmful to the human body; some of them are essential in the development of metabolic processes (Pereira et al., 2016). For example, magnesium is involved in protein synthesis, blood sugar control, and prevents cardiac arrhythmia; zinc is involved in the proper functioning of the thyroid, immune system, and reproductive system. Some adverse effects are listed in the following paragraphs.

- Arsenic (by definition it is not a heavy metal, but it is an extremely toxic metal) inactivates over 200 enzymes involved in the body's functioning. Many of these enzymes are dedicated to important functions, such as energy pathways at the cellular level, protein synthesis, and nucleic acid repair (DNA). Chronic arsenic exposure is a multisystemic disease that affects the entire body. Like mercury, arsenic is known to be a particle capable of inducing carcinogenesis. Sources of exposure are water contaminated with arsenic, but also samples of rice from polluted areas (Juncos et al., 2019).
- Cadmium binds at the mitochondrial level, blocking cellular respiration. It produces the depletion of some molecules essential in the antitoxic defense, among which is glutathione, but also of some antioxidant enzymes such as catalase and superoxide-dismutase (Le Croizier et al., 2019).
- Lead is a toxic particle that has an affinity for the bone system. In addition, it exerts its toxic effects on hemoglobin synthesis, causing anemia. Accumulation of lead in the body will affect kidney function, with a progressive decrease in glomerular filtration rate. In most aquatic organisms there is a balance of lead, excreting an amount equal to the lead introduced into the body, and the tissue level of the metal remains below the concentrations that cause pathological changes. However, an increase in intake will lead to metal accumulation or a positive lead

balance. Because lead is a chemical metal very similar to calcium, the body assimilates it as calcium. The first route of transport is thus plasma and soft tissue membranes. It is then distributed to the teeth in growing children and in bones at all ages. Ocean fish and refined chocolate are a source of lead and mercury (Lee et al., 2019).

- Manganese toxicity is frequently associated with iron deposits. It blocks calcium channels, and chronic intoxication causes dopamine depletion in the brain (Jiang et al., 2017).
- Mercury/organic mercury reacts with sulfhydryl groups in the aquatic organisms, thus interfering with cellular and subcellular functions. It modifies the synthesis of proteins and the transcription of nucleic acids, especially DNA. These changes underlie current hypotheses that place mercury among potentially carcinogenic toxins. Mercury has a toxic effect at the subcellular level, affecting mitochondria and the integrity of cell membranes, intensifying the production of oxygen free radicals and oxidative stress, and modifying neurotransmission in the nervous system. It can cause damage both to the brain and to the peripheral nervous system. It reduces the immune capacity, with the damage of some cells involved in the immune defense. The main manifestation of mercury exposure is neurotoxicity. Trivial manifestations, such as decreased attention span and concentration and unwarranted nervous fatigue, may be the first signs of mercury poisoning. The presence of peripheral signs of toxicity represented by paresthesias as tingling in the limbs can also complete the clinical picture. Mercury can combine with a methyl group, turning into methylmercury, which is found in a variety of environmental pollutants (heavy metals). The chemical element mercury is less labile, but produces a similar set of toxic manifestations (Korbas et al., 2008).
- Nickel causes lung tissue damage with the slow development of malignancies. Large amounts of nickel used in animal feed can lead to a decrease in nitrogen content and a decrease of growth. The epidemiological investigations of workers related to the production of refined nickel show that nickel and its compounds can cause diseases of the nasal cavity and throat, as well as the lungs. Malignant kidney formations occurred in rats when nickel was introduced into the kidneys. Teratogenic effects, such as exencephaly, brittle ribs, and soft palate decomposition, occur in mammals that have been influenced by various nickel compounds (Palermo et al., 2015).

For the most part, heavy metals bind to oxygen, nitrogen, and sulfhydryl groups in proteins, leading to altered enzymatic activity. This affinity of metals for sulfhydryl groups is also protective in metal homeostasis. Increased synthesis of heavy metal-binding proteins in response to their high levels in the body is the first defense mechanism against poisoning (Rahmani et al., 2018).

For example, metalloproteins are induced by many metals. These molecules are rich in thiol ligands, which allow the high-affinity binding of cadmium, copper, silver, and zinc among other elements.

Other proteins involved in the transport of heavy metals and their excretion through the formation of ligands are ferritin, transferrin, albumin, and hemoglobin. All these mechanisms of action can be diagramed as shown in Fig. 2.3.

The symptoms of heavy metal poisoning are not difficult to recognize, due to the fact that they are usually severe, have rapid onset, and can be associated with a recent event (exposure or ingestion). Excess heavy metals have inhibitory effects on the development of aquatic organisms (Gandar et al., 2017).

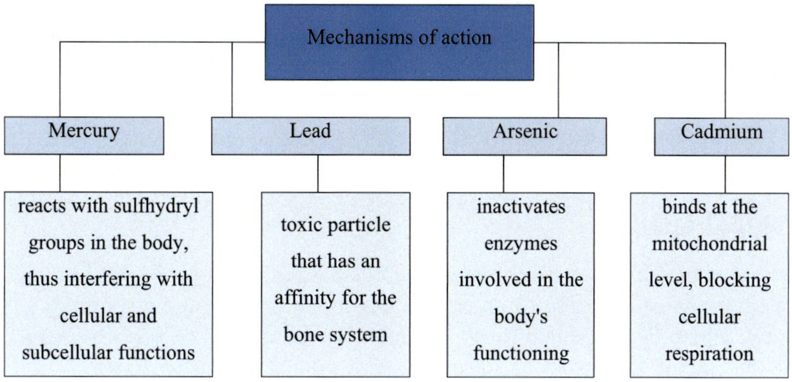

FIGURE 2.3

Mechanisms of action of mercury, lead, arsenic, and cadmium.

2.7 Methods of accumulation and disposal of metals

Bioconcentration is a term used in aquatic toxicology and refers to the accumulation of a chemical (heavy metals) when water is the sole source of the chemical in or on an organism (e.g., phytoplankton, crustaceans, fish).

A bioconcentration factor (BCF) describes the level of chemical (heavy metals) division between the environment and the body. BCF is usually expressed in units of liters per kilogram and is a measure of the milligrams of chemicals per kilogram of body mass, to milligrams of chemicals per liter of water. The factor is the chemical concentration (heavy metals) of the body in relation to the concentration of water (Zhong et al., 2018).

There are different ways to measure and evaluate *bioconcentration* and *bioaccumulation*. Some of the methods used include octanol-water partition coefficient, BCF, biological sediment accumulation factor, BAF, and fugacity-based BCF, among others.

A BCF higher than 1 is an indication that lipophilic or hydrophobic chemicals are present. It is an indication of the probability that a substance will bioaccumulate. These lipophilic substances have a high affinity for lipids, allowing them to accumulate in tissues that have a high amount of fat, unlike aqueous environments such as cytosol. In the prediction of the chemical distribution in the environment, scientific models are used which, in turn, allow the prediction of the biological variety of lipophilic chemicals. One such model is the equilibrium compartmentalization model, which assumes a steady-state scenario, and the fate of a substance in a system would be modeled, leading to a prediction of the phases of concentrations and endpoints. Other scientific models are fleeting models and food web models, among others (Wood et al., 2004).

2.7.1 Bioconcentration: applications in toxicology

BCFs allow the prediction of contamination (heavy metals) levels of aquatic organisms, depending on the chemical concentration of the surrounding water. BCFs do not explicitly take into account metabolism and therefore need to be taken into account in models from other stages by elimination, absorption, or degradation equations for the aquatic organisms (e.g., phytoplankton, crustaceans, fish) concerned.

In addition to prediction, another application of biocentric factors includes body burden, which describes the total amount of chemical (heavy metals) in the aquatic organisms (Brix et al., 2017).

2.7.2 Biological factors

Biological factors are essential in determining the occurrence of bioconcentration and there is competition against the rate of exposure of an aquatic organism through the respiratory surfaces and the rate of excretion, which is a chemical (heavy metals) loss. The ratio of surface area to body volume can determine the rate of absorption of substances from the surrounding water.

The primary factor that affects the values of BCFs is the species of concern, because it determines the biological factors that change the bioconcentration (Antony Jesu Prabhu et al., 2018).

2.7.3 Environmental parameters

Some environmental factors affect bioconcentration. Water quality can affect bioavailability. Pollutants (heavy metals) and the natural content of the particles can bind to the particles found in the water, preventing the absorption and ingestion of contaminated particles by the aquatic organisms. Temperature changes also affect bioenergetics and metabolic transformation.

Change in pH, which is affected by temperature changes, influences the bioavailability of ionic contaminants (Johnston et al., 2010).

2.7.4 The effects of bioconcentration and bioaccumulation on the aquatic ecosystem

The long-term effects of bioaccumulation and bioconcentration are a reduction in the wild population and damage to aquatic organisms. When heavy metals, such as mercury, cadmium, and lead, bind to cell walls, there is a risk of disruption of cell surfaces, which affects the metabolism, leading to death in some cases.

According to a 2001 study on the effects of silver on zooplankton, metal toxicity results in problems in the reproductive system. The study found that there was a decrease in the number of eggs, as well as an interruption in the development of the ovaries. A 2015 study on the bioavailability and toxicity of metals in fresh water showed that fish exposed to cadmium had coarse granules in their cytoplasm. Another observation was the dilation of blood vessels in most fish intestines (Gárriz et al., 2019).

Bioaccumulation, as discussed earlier, is the accumulation of substances (heavy metals), such as heavy metals, in aquatic organisms. Bioaccumulation occurs when the aquatic organisms absorb substances much faster than they are eliminated or consumed by catabolism and excretion (Bonsignore et al., 2018). The key difference between *bioaccumulation* and *biomagnification* is that bioaccumulation refers to the accumulation of a toxic chemical in the body of an aquatic organism, while biomagnification is the increase in the concentration of a toxic chemical going from lower levels to higher levels in a food chain. Pollutants (heavy metals) must have a long life to cause biomagnification. They must also be mobile so that they can easily enter biological systems through food or water. If it is not mobile, a pollutant can remain inside an organism and will not move to the next trophic level. Moreover, if a pollutant is fat soluble, it tends to remain in the aquatic organisms for a longer period (Nasyitah Sobihah et al., 2018).

Bioamplification represents the increase in the concentration of a pollutant (heavy metals) at a certain trophic level of a link between two trophic chains. Certain chemicals are found in the highest ecological concentration in predators at the top of the food chain.

The tragic poisoning during the 1950s in Minimata, Japan is the best-known example of mercury bioamplification. When mercury enters the sea as waste, bacteria turn it into methylmercury and it enters small organisms, such as shrimp. After that, many of these small creatures are eaten by larger fish, which accumulate a large amount of mercury, which means that absorption increases and the amount of toxins accumulates. Eventually, it reaches the peak consumer, humans, who consume the most mercury, which, in the case of Minamata, led to serious illness.

The factory in Minimata used mercury and dumped methylmercury-rich waste into the gulf, where it accumulated in the fish. Many residents of the city depended on fish as a crucial part of their diet and were poisoned. The exact number of deaths is not certain, but more than a thousand people died and many others suffered nerve injuries (Zheng et al., 2019).

Heavy metal poisoning or, more correctly, the toxicity of metals is the harmful effect of some metals in certain forms and doses. Some metals are toxic when they form soluble compounds, but not all have a biological role (they are not essential metals) or are toxic in some form (e.g., lead, mercury and cadmium).

The concept of "heavy metals" is often used synonymously with "toxic metals," but there are also light metals that in certain circumstances can become toxic, some being essential, such as iron, selenium, copper, chromium, and zinc.

Once in the body, the metal will accumulate and/or be excreted. Accumulation can occur as a result of physiological mechanisms in the case of essential metals, which are directed to tissues to perform metabolic functions (Milenkovic et al., 2019). For example, the most important way that humans expose themselves to mercury is through fish and seafood. When aquatic animals absorb mercury, it tends to remain in the body, where it accumulates over time. Large predatory fish usually contain higher concentrations of mercury because they feed on smaller animals that have already ingested mercury. Therefore the consumption of larger predatory fish, such as tuna or swordfish, usually leads to the ingestion of a higher dose of mercury compared to the consumption of smaller fish, which are at a lower level in the food chain.

Exposure to mercury can also occur in the uterus if the mother eats fish or seafood. This can have a significant lifelong impact on the development of the baby's brain and nervous system, affecting memory, language, attention, and other skills. It is estimated that, in Europe alone, more than 1.8 million children are born each year with mercury levels higher than the recommended safety limits (Bosch et al., 2016).

At the tissue level, there may be a need for metal storage in order to ensure a stock of metals that meets metabolic requirements. Excess heavy metal ions that exceed metabolic and storage needs are toxic and must be removed from the vicinity of biologically important molecules. The heavy metals can be removed from the body or biotransformed, before storage in specific tissues, into nontoxic inert forms.

Aquatic organisms have developed various strategies for taking over, storing, or disposing of heavy metals. Either the penetration of heavy metals ions is restricted, accompanied by mechanisms that ensure a low tissue requirement of essential metals, or the penetration of all metal ions accompanied by mechanisms for storage or disposal of excess metals is allowed (Begum et al., 2013).

The heavy metals in the body can be stored initially in the tissues where they were taken in (gills, intestines, skin). They then reach the tissues where detoxification, long-term storage, or elimination takes place (digestive gland, kidneys). Three main mechanisms for removing heavy metals have been described in aquatic organisms, as described in the following paragraphs.

One mechanism, represented by the loss of the surface of the body or gills, characterizes easily mobilizable heavy metals, which are adsorbed on the external mucus or complexed to intracellular or extracellular ligands with low affinity.

Another mechanism is intestinal elimination. In invertebrates, heavy metals are eliminated along with residual bodies resulting from intracellular digestion, which are excused from the digestive gland into the intestine.

Urine excretion is another way of eliminating heavy metals (Ruha et al., 2009).

2.8 Identification and adjustment of concentrations of metals in tissue

The different tissue concentrations of heavy metals are explained on the basis of the specific properties of each tissue: uptake, retention, and excretion.

By combining these processes, some aquatic organisms are potentially able to regulate the concentrations of certain heavy metals in the body. There are many concerns about the regulatory capacity of aquatic organisms in the context of their use as biological indicators of heavy metal pollution (Łuczyńska et al., 2018).

For an organism to be used as an indicator of pollution, there must be a simple correlation between the level of heavy metals in the environment and in its tissues. Aquatic organisms such as phytoplankton, crustaceans, and fish, capable of regulating their heavy metal concentrations, do not meet this criterion.

In some tissues, the levels of some metals are maintained within narrow limits by regulatory mechanisms that do not involve the accumulation of excess metal. In other tissues, heavy metal concentrations may be more variable. This may reflect either a less rigorous regulation of uptake and excretion, or that the tissue has storage capacity, allowing the accumulation of metals in nontoxic, metabolically inert forms (Bar-Ilan et al., 2009).

The heavy metals in these deposits can be gradually eliminated from the body or can continue to accumulate throughout life. In some organisms (e.g., phytoplankton, crustaceans, fish), the entire heavy metals load of the body can be found in a single tissue. Fluctuations in heavy metal concentrations in certain tissues can be masked when analyzing the total content throughout the body.

The conclusion that an aquatic organism has regulatory capabilities only on the basis that the total level shows small variations with increasing exposure concentration may be incorrect if one neglects to observe a significant increase in heavy metal levels in a target tissue or vital organ (Minghetti & Schirmer, 2016).

Aquatic organisms take heavy metals from food or water, transport them, store them, and excrete them, in order to maintain a continuous flow that controls the concentration of free cations in cells and fluids. In the cells of different tissues, heavy metals can reach high concentrations, due to the ability of some cells to accumulate excess heavy metals in a nontoxic form, by binding to soluble compounds or compartmentalization into membrane vesicles and granules. For example, the bioaccumulation of heavy metals and its presumptive cytological outcomes in neurons (more exactly, ciliated olfactory sensory receptor neuron) of mudskipper has been investigated and it was concluded that they cause particular neurodegenerative disorders in fish (Sarkar & De, 2016).

The different biochemical processes involved in heavy metals homeostasis do not show the same degree of activity in all cells of an organism. In different organs of the same aquatic organisms, heavy metals can accumulate differently. The determining factor for heavy metals concentrations in the body is their bioavailability of water and food. The nature of the metal (essential or nonessential, chemical properties) and the physiological state of the body influence the uptake, distribution, and accumulation in tissues and excretion.

The significance of heavy metal levels is discussed in relation to the health of aquatic organisms and their use in the biomonitoring of metal pollution.

2.9 Conclusions

Water is not a commercial product like any other, but a legacy that must be preserved, protected, and treated as such. Water pollution is the process of altering its physical, chemical, or biological quality, which is produced by human activity. A body of water can be polluted not only when it shows visible changes (discoloration, iridescence of petroleum products, unpleasant odors) but also when, although appearing clear and clean, it contains, even in a small amount, potentially toxic substances. Excess heavy metals have inhibitory effects on the development of aquatic organisms (phytoplankton, crustaceans, fish). They can affect the growth of mollusks, oxygen consumption, etc. In fish and crustaceans exposed to high concentrations of heavy metals, histological changes occur such as changes in the appearance of the gills, necrosis, or fatty degeneration of the liver.

Elucidating the effects at the cellular level allows us to understand the ways in which heavy metals can alter the metabolism and physiology of aquatic organisms. The cell membrane is the first target structure when heavy metals penetrate the cell. It has been shown that metals can bind to membrane proteins and phospholipids, altering their structure and functions. Heavy metals can stimulate lipid peroxidation processes, a complex sequence of biochemical reactions, defined as "oxidative damage of polyunsaturated lipids." The whole process results in the production of extremely toxic compounds for the cell, due to the high reactivity towards the other cellular components (soluble and membrane proteins, DNA).

The biological role of essential metals is correlated with their high affinity for active groups of structural enzymes and proteins. The toxic effects of nonessential metals can be produced by their tendency to substitute essential metals and to compete for biological ligands. Toxic heavy metals with high affinity for active groups of proteins can affect the structure and function of these molecules and ultimately the physiology of the cell.

The lethal effects of some heavy metals on crustaceans have been attributed to the inhibition of enzymes involved in cellular respiration. The histological changes observed in fish and crustaceans following chronic exposure to heavy metals are side effects caused by disturbance of nutrition due to inhibition of catabolic pathway enzymes. The effects on growth and development were attributed to inhibition of enzymes involved in protein synthesis and cell division. It is considered that, depending on the heavy metal, the degree of bioaccumulation, and the vulnerability of enzyme systems, it is possible to have a multitude of effects.

References

Ajima, M. N., Nnodi, P. C., Ogo, O. A., Adaka, G. S., Osuigwe, D. I., & Njoku, D. C. (2015). Bioaccumulation of heavy metals in Mbaa River and the impact on aquatic ecosystem. *Environmental Monitoring and Assessment*, *187*(12), 768. Available from https://doi.org/10.1007/s10661-015-4937-0.

Antony Jesu Prabhu, P., Stewart, T., Silva, M., Amlund, H., Ørnsrud, R., Lock, E. J., Waagbo, R., & Hogstrand, C. (2018). Zinc uptake in fish intestinal epithelial model RTgutGC: Impact of media ion composition and methionine chelation. *Journal of Trace Elements in Medicine and Biology*, *50*, 377–383. Available from https://doi.org/10.1016/j.jtemb.2018.07.025.

Armant, O., Gombeau, K., Murat El Houdigui, S., Floriani, M., Camilleri, V., Cavalie, I., & Adam-Guillermin, C. (2017). Zebrafish exposure to environmentally relevant concentration of depleted uranium impairs progeny development at the molecular and histological levels. *PLoS One*, *12*(5), e0177932. Available from https://doi.org/10.1371/journal.pone.0177932.

Bankar, A., Zinjarde, S., Shinde, M., Gopalghare, G., & Ravikumar, A. (2018). Heavy metal tolerance in marine strains of *Yarrowia lipolytica*. *Extremophiles*, *22*(4), 617–628. Available from https://doi.org/10.1007/s00792-018-1022-y.

Bar-Ilan, O., Albrecht, R. M., Fako, V. E., & Furgeson, D. Y. (2009). Toxicity assessments of multisized gold and silver nanoparticles in zebrafish embryos. *Small (Weinheim an der Bergstrasse, Germany)*, *5*(16), 1897–1910. Available from https://doi.org/10.1002/smll.200801716.

Begum, A., Mustafa, A. I., Amin, M. N., Chowdhury, T. R., Quraishi, S. B., & Banu, N. (2013). Levels of heavy metals in tissues of shingi fish (*Heteropneustes fossilis*) from Buriganga River. *Environmental Monitoring and Assessment*, *185*(7), 5461–5469. Available from https://doi.org/10.1007/s10661-012-2959-4.

Bonaldo, R. M., Pires, M. M., Guimarães Jr, P. R., Hoey, A. S., & Hay, M. E. (2017). Small marine protected areas in Fiji provide refuge for reef fish assemblages, feeding groups, and corals. *PLoS One*, *12*(1), e0170638. Available from https://doi.org/10.1371/journal.pone.0170638.

Bonsignore, M., Salvagio Manta, D., Mirto, S., Quinci, E. M., Ape, F., Montalto, V., Gristina, M., Traina, A., & Sprovieri, M. (2018). Bioaccumulation of heavy metals in fish, crustaceans, molluscs and echinoderms from the Tuscany coast. *Ecotoxicology and Environmental Safety*, *162*, 554–562. Available from https://doi.org/10.1016/j.ecoenv.2018.07.044.

Bosch, A. C., O'Neill, B., Sigge, G. O., Kerwath, S. E., & Hoffman, L. C. (2016). Heavy metals in marine fish meat and consumer health: A review. *Journal of the Science of Food and Agriculture*, *96*(1), 32–48. Available from https://doi.org/10.1002/jsfa.7360.

Braz-Mota, S., Campos, D. F., MacCormack, T. J., Duarte, R. M., Val, A. L., & Almeida-Val, V. (2018). Mechanisms of toxic action of copper and copper nanoparticles in two Amazon fish species: Dwarf cichlid (*Apistogramma agassizii*) and cardinal tetra (*Paracheirodon axelrodi*). *The Science of the Total Environment*, *630*, 1168–1180. Available from https://doi.org/10.1016/j.scitotenv.2018.02.216.

Brix, K. V., Tellis, M. S., Crémazy, A., & Wood, C. M. (2017). Characterization of the effects of binary metal mixtures on short-term uptake of Cd, Pb, and Zn by rainbow trout (*Oncorhynchus mykiss*). *Aquatic*

Toxicology (Amsterdam, Netherlands), *193*, 217–227. Available from https://doi.org/10.1016/j.aquatox.2017.10.015.

Carbonaro, R. F., Farley, K. J., Delbeke, K., Baken, S., Arbildua, J. J., Rodriguez, P. H., & Rader, K. J. (2019). Modeling the fate of metal concentrates in surface water. *Environmental Toxicology and Chemistry/SETAC*, *38*(6), 1256–1272. Available from https://doi.org/10.1002/etc.4417.

Chaturvedi, A. D., Pal, D., Penta, S., & Kumar, A. (2015). Ecotoxic heavy metals transformation by bacteria and fungi in aquatic ecosystem. *World Journal of Microbiology and Biotechnology*, *31*(10), 1595–1603. Available from https://doi.org/10.1007/s11274-015-1911-5.

Dar, G. H., Bhat, R. A., Kamili, A. N., Chishti, M. Z., Qadri, H., Dar, R., & Mehmood, M. A. (2020). *Correlation between pollution trends of freshwater bodies and bacterial disease of fish fauna. Fresh water pollution dynamics and remediation* (pp. 51–67). Singapore: Springer.

Dar, G. H., Dar, S. A., Kamili, A. N., Chishti, M. Z., & Ahmad, F. (2016). Detection and characterization of potentially pathogenic *Aeromonas sobria* isolated from fish *Hypophthalmichthys molitrix* (Cypriniformes: Cyprinidae). *Microbial Pathogenesis*, *91*, 136–140.

Dar, G. H., Kamili, A. N., Chishti, M. Z., Dar, S. A., Tantry, T. A., & Ahmad, F. (2016). Characterization of *Aeromonas sobria* isolated from fish rohu (*Labeo rohita*) collected from polluted pond. *Journal of Bacteriology & Parasitology*, *7*(3), 1–5. Available from https://doi.org/10.4172/2155-9597.1000273.

Dvorak, P., Andreji, J., Faltová Leitmanová, I., Petrách, F., & Mraz, J. (2018). Accumulation of selected metals pollution in aquatic ecosystems in the Smeda river (Czech Republic). *Neuro Endocrinology Letters*, *39*(5), 380–384.

Flora, S. J., & Pachauri, V. (2010). Chelation in metal intoxication. *International Journal of Environmental Research and Public Health*, *7*(7), 2745–2788. Available from https://doi.org/10.3390/ijerph7072745.

Fontaine, M. C., Snirc, A., Frantzis, A., Koutrakis, E., Oztürk, B., Oztürk, A. A., & Austerlitz, F. (2012). History of expansion and anthropogenic collapse in a top marine predator of the Black Sea estimated from genetic data. *Proceedings of the National Academy of Sciences of the United States of America*, *109*(38), E2569–E2576. Available from https://doi.org/10.1073/pnas.1201258109.

Freitas, F., Lunardi, S., Souza, L. B., von der Osten, J., Arruda, R., Andrade, R., & Battirola, L. D. (2018). Accumulation of copper by the aquatic macrophyte *Salvinia biloba* Raddi (Salviniaceae). *Brazilian Journal of Biology*, *78*(1), 133–139. Available from https://doi.org/10.1590/1519-6984.166377.

Gandar, A., Laffaille, P., Canlet, C., Tremblay-Franco, M., Gautier, R., Perrault, A., Gress, L., Mormède, P., Tapie, N., Budzinski, H., & Jean, S. (2017). Adaptive response under multiple stress exposure in fish: From the molecular to individual level. *Chemosphere*, *188*, 60–72. Available from https://doi.org/10.1016/j.chemosphere.2017.08.089.

Gao, F., Li, J., Sun, C., Zhang, L., Jiang, F., Cao, W., & Zheng, L. (2019). Study on the capability and characteristics of heavy metals enriched on microplastics in marine environment. *Marine Pollution Bulletin*, *144*, 61–67.

Gárriz, Á., Del Fresno, P. S., Carriquiriborde, P., & Miranda, L. A. (2019). Effects of heavy metals identified in *Chascomús shallow* lake on the endocrine-reproductive axis of pejerrey fish (*Odontesthes bonariensis*). *General and Comparative Endocrinology*, *273*, 152–162. Available from https://doi.org/10.1016/j.ygcen.2018.06.013.

Gholizadeh, M., & Patimar, R. (2018). Ecological risk assessment of heavy metals in surface sediments from the Gorgan Bay, Caspian Sea. *Marine Pollution Bulletin*, *137*, 662–667. Available from https://doi.org/10.1016/j.marpolbul.2018.11.009.

Gopalakrishnan, S., Thilagam, H., & Raja, P. V. (2007). Toxicity of heavy metals on embryogenesis and larvae of the marine sedentary polychaete *Hydroides elegans*. *Archives of Environmental Contamination and Toxicology*, *52*(2), 171–178. Available from https://doi.org/10.1007/s00244-006-0038-y.

Hao, Z., Chen, L., Wang, C., Zou, X., Zheng, F., Feng, W., Zhang, D., & Peng, L. (2019). Heavy metal distribution and bioaccumulation ability in marine organisms from coastal regions of Hainan and Zhoushan, China. *Chemosphere*, *226*, 340–350. Available from https://doi.org/10.1016/j.chemosphere.2019.03.132.

Ighariemu, V., Belonwu, D. C., & Wegwu, M. O. (2019). Levels of some heavy metals and health risks assessment of three different species of catfishes in Ikoli Creek, Bayelsa State, Nigeria. *Biological Trace Element Research*, *189*(2), 567–573. Available from https://doi.org/10.1007/s12011-018-1484-x.

Jiang, W. D., Tang, R. J., Liu, Y., Wu, P., Kuang, S. Y., Jiang, J., Tang, L., Tang, W. N., Zhang, Y. A., Zhou, X. Q., & Feng, L. (2017). Impairment of gill structural integrity by manganese deficiency or excess related to induction of oxidative damage, apoptosis and dysfunction of the physical barrier as regulated by NF-κB, caspase and Nrf2 signaling in fish. *Fish & Shellfish Immunology*, *70*, 280–292. Available from https://doi.org/10.1016/j.fsi.2017.09.022.

Jitar, O., Teodosiu, C., Oros, A., Plavan, G., & Nicoara, M. (2015). Bioaccumulation of heavy metals in marine organisms from the Romanian sector of the Black Sea. *New Biotechnology*, *32*(3), 369–378. Available from https://doi.org/10.1016/j.nbt.2014.11.004.

Johnston, B. D., Scown, T. M., Moger, J., Cumberland, S. A., Baalousha, M., Linge, K., van Aerle, R., Jarvis, K., Lead, J. R., & Tyler, C. R. (2010). Bioavailability of nanoscale metal oxides TiO(2), CeO(2), and ZnO to fish. *Environmental Science & Technology*, *44*(3), 1144–1151. Available from https://doi.org/10.1021/es901971a.

Juncos, R., Arcagni, M., Squadrone, S., Rizzo, A., Arribére, M., Barriga, J. P., Battini, M. A., Campbell, L. M., Brizio, P., Abete, M. C., & Ribeiro Guevara, S. (2019). Interspecific differences in the bioaccumulation of arsenic of three Patagonian top predator fish: Organ distribution and arsenic speciation. *Ecotoxicology and Environmental Safety*, *168*, 431–442. Available from https://doi.org/10.1016/j.ecoenv.2018.10.077.

Kanhai, L. D., Gobin, J. F., Beckles, D. M., Lauckner, B., & Mohammed, A. (2014). Metals in sediments and mangrove oysters (*Crassostrea rhizophorae*) from the Caroni Swamp, Trinidad. *Environmental Monitoring and Assessment*, *186*(3), 1961–1976. Available from https://doi.org/10.1007/s10661-013-3510-y.

Kim, H. T., & Kim, J. G. (2016). Uptake of cadmium, copper, lead, and zinc from sediments by an aquatic macrophyte and by terrestrial arthropods in a freshwater wetland ecosystem. *Archives of Environmental Contamination and Toxicology*, *71*(2), 198–209. Available from https://doi.org/10.1007/s00244-016-0293-5.

Kim, J. M., Baars, O., & Morel, F. (2016). The effect of acidification on the bioavailability and electrochemical lability of zinc in seawater. *Philosophical Transactions. Series A, Mathematical, Physical, and Engineering Sciences*, *374*(2081), 20150296. Available from https://doi.org/10.1098/rsta.2015.0296.

Kim, J. Y., Kim, S. J., Bae, M. A., Kim, J. R., & Cho, K. H. (2018). Cadmium exposure exacerbates severe hyperlipidemia and fatty liver changes in zebrafish via impairment of high-density lipoproteins functionality. *Toxicology In Vitro: An International Journal Published in Association with BIBRA*, *47*, 249–258. Available from https://doi.org/10.1016/j.tiv.2017.11.007.

Korbas, M., Blechinger, S. R., Krone, P. H., Pickering, I. J., & George, G. N. (2008). Localizing organomercury uptake and accumulation in zebrafish larvae at the tissue and cellular level. *Proceedings of the National Academy of Sciences of the United States of America*, *105*(34), 12108–12112. Available from https://doi.org/10.1073/pnas.0803147105.

Lacave, J. M., Vicario-Parés, U., Bilbao, E., Gilliland, D., Mura, F., Dini, L., Cajaraville, M. P., & Orbea, A. (2018). Waterborne exposure of adult zebrafish to silver nanoparticles and to ionic silver results in differential silver accumulation and effects at cellular and molecular levels. *The Science of the Total Environment*, *642*, 1209–1220. Available from https://doi.org/10.1016/j.scitotenv.2018.06.128.

Le Croizier, G., Lacroix, C., Artigaud, S., Le Floch, S., Munaron, J. M., Raffray, J., Penicaud, V., Rouget, M. L., Laë, R., & Tito De Morais, L. (2019). Metal subcellular partitioning determines excretion pathways and sensitivity to cadmium toxicity in two marine fish species. *Chemosphere*, *217*, 754–762. Available from https://doi.org/10.1016/j.chemosphere.2018.10.212.

Lee, J. W., Choi, H., Hwang, U. K., Kang, J. C., Kang, Y. J., Kim, K. I., & Kim, J. H. (2019). Toxic effects of lead exposure on bioaccumulation, oxidative stress, neurotoxicity, and immune responses in fish: A review. *Environmental Toxicology and Pharmacology*, *68*, 101–108. Available from https://doi.org/10.1016/j.etap.2019.03.010.

Lipowicz, B., Hanekop, N., Schmitt, L., & Proksch, P. (2013). An aeroplysinin-1 specific nitrile hydratase isolated from the marine sponge *Aplysina cavernicola*. *Marine Drugs*, *11*(8), 3046–3067. Available from https://doi.org/10.3390/md11083046.

Łuczyńska, J., Paszczyk, B., & Łuczyński, M. J. (2018). Fish as a bioindicator of heavy metals pollution in aquatic ecosystem of Pluszne Lake, Poland, and risk assessment for consumer's health. *Ecotoxicology and Environmental Safety*, *153*, 60–67. Available from https://doi.org/10.1016/j.ecoenv.2018.01.057.

Matache, M. L., Marin, C., Rozylowicz, L., & Tudorache, A. (2013). Plants accumulating heavy metals in the Danube River wetlands. *Journal of Environmental Health Science and Engineering*, *11*(1), 39. Available from https://doi.org/10.1186/2052-336X-11-39.

Matsuyama, A., Yano, S., Taninaka, T., Kindaichi, M., Sonoda, I., Tada, A., & Akagi, H. (2018). Chemical characteristics of dissolved mercury in the pore water of Minamata Bay sediments. *Marine Pollution Bulletin*, *129*(2), 503–511. Available from https://doi.org/10.1016/j.marpolbul.2017.10.021.

Menegário, A. A., Yabuki, L., Luko, K. S., Williams, P. N., & Blackburn, D. M. (2017). Use of diffusive gradient in thin films for in situ measurements: A review on the progress in chemical fractionation, speciation and bioavailability of metals in waters. *Analytica Chimica Acta*, *983*, 54–66. Available from https://doi.org/10.1016/j.aca.2017.06.041.

Milenkovic, B., Stajic, J. M., Stojic, N., Pucarevic, M., & Strbac, S. (2019). Evaluation of heavy metals and radionuclides in fish and seafood products. *Chemosphere*, *229*, 324–331. Available from https://doi.org/10.1016/j.chemosphere.2019.04.189.

Minghetti, M., & Schirmer, K. (2016). Effect of media composition on bioavailability and toxicity of silver and silver nanoparticles in fish intestinal cells (RTgutGC). *Nanotoxicology.*, *10*(10), 1526–1534. Available from https://doi.org/10.1080/17435390.2016.1241908.

Mohamed, S. A., Elshal, M. F., Kumosani, T. A., Mal, A. O., Ahmed, Y. M., Almulaiky, Y. Q., Asseri, A. H., & Zamzami, M. A. (2016). Heavy metal accumulation is associated with molecular and pathological perturbations in liver of *Variola louti* from the Jeddah Coast of Red Sea. *International Journal of Environmental Research and Public Health*, *13*(3), 342. Available from https://doi.org/10.3390/ijerph13030342.

Nasyitah Sobihah, N., Ahmad Zaharin, A., Khairul Nizam, M., Ley Juen, L., & Kyoung-Woong, K. (2018). Bioaccumulation of heavy metals in maricultured fish, *Lates calcarifer* (Barramudi), *Lutjanus campechanus* (red snapper) and *Lutjanus griseus* (grey snapper). *Chemosphere*, *197*, 318–324. Available from https://doi.org/10.1016/j.chemosphere.2017.12.187.

Pain, D. J., Mateo, R., & Green, R. E. (2019). Effects of lead from ammunition on birds and other wildlife: A review and update. *Ambio*, *48*(9), 935–953. Available from https://doi.org/10.1007/s13280-019-01159-0.

Palermo, F. F., Risso, W. E., Simonato, J. D., & Martinez, C. B. (2015). Bioaccumulation of nickel and its biochemical and genotoxic effects on juveniles of the neotropical fish Prochilodus lineatus. *Ecotoxicology and Environmental Safety*, *116*, 19–28. Available from https://doi.org/10.1016/j.ecoenv.2015.02.032.

Pereira, P., Puga, S., Cardoso, V., Pinto-Ribeiro, F., Raimundo, J., Barata, M., Pousão-Ferreira, P., Pacheco, M., & Almeida, A. (2016). Inorganic mercury accumulation in brain following waterborne exposure elicits a deficit on the number of brain cells and impairs swimming behavior in fish (white seabream-Diplodus sargus). *Aquatic Toxicology (Amsterdam, Netherlands)*, *170*, 400–412. Available from https://doi.org/10.1016/j.aquatox.2015.11.031.

Rahmani, J., Fakhri, Y., Shahsavani, A., Bahmani, Z., Urbina, M. A., Chirumbolo, S., Keramati, H., Moradi, B., Bay, A., & Bjørklund, G. (2018). A systematic review and meta-analysis of metal concentrations in canned tuna fish in Iran and human health risk assessment. *Food and Chemical Toxicology: An International Journal Published for the British Industrial Biological Research Association*, *118*, 753–765. Available from https://doi.org/10.1016/j.fct.2018.06.023.

Rehman, K., Fatima, F., Waheed, I., & Akash, M. (2018). Prevalence of exposure of heavy metals and their impact on health consequences. *Journal of Cellular Biochemistry*, *119*(1), 157–184. Available from https://doi.org/10.1002/jcb.26234.

Ruha, A. M., Curry, S. C., Gerkin, R. D., Caldwell, K. L., Osterloh, J. D., & Wax, P. M. (2009). Urine mercury excretion following meso-dimercaptosuccinic acid challenge in fish eaters. *Archives of Pathology & Laboratory Medicine*, *133*(1), 87–92. Available from https://doi.org/10.1043/1543-2165-133.1.87.

Saher, N. U., & Siddiqui, A. S. (2019). Occurrence of heavy metals in sediment and their bioaccumulation in sentinel crab (*Macrophthalmus depressus*) from highly impacted coastal zone. *Chemosphere*, *221*, 89–98.

Sarkar, S. K., & De, S. K. (2016). Electron microscope based X-ray microanalysis on bioaccumulation of heavy metals and neural degeneration in mudskipper (*Pseudapocryptes lanceolatus*). *Journal of Microscopy and Ultrastructure*, *4*(4), 211–221. Available from https://doi.org/10.1016/j.jmau.2016.03.002.

Schertzinger, G., Ruchter, N., & Sures, B. (2018). Metal accumulation in sediments and amphipods downstream of combined sewer overflows. *The Science of the Total Environment*, *616-617*, 1199–1207. Available from https://doi.org/10.1016/j.scitotenv.2017.10.199.

Seto, M., Wada, S., & Suzuki, S. (2013). The effect of zinc on aquatic microbial ecosystems and the degradation of dissolved organic matter. *Chemosphere*, *90*(3), 1091–1102. Available from https://doi.org/10.1016/j.chemosphere.2012.09.014.

Speelmans, M., Lock, K., Vanthuyne, D. R., Hendrickx, F., Du Laing, G., Tack, F. M., & Janssen, C. R. (2010). Hydrological regime and salinity alter the bioavailability of Cu and Zn in wetlands. *Environmental Pollution (Barking, Essex: 1987)*, *158*(5), 1870–1875. Available from https://doi.org/10.1016/j.envpol.2009.10.040.

Stenzler, B., Hinz, A., Ruuskanen, M., & Poulain, A. J. (2017). Ionic strength differentially affects the bioavailability of neutral and negatively charged inorganic Hg complexes. *Environmental Science & Technology*, *51*(17), 9653–9662. Available from https://doi.org/10.1021/acs.est.7b01414.

Tang, C. H., Chen, W. Y., Wu, C. C., Lu, E., Shih, W. Y., Chen, J. W., & Tsai, J. W. (2020). Ecosystem metabolism regulates seasonal bioaccumulation of metals in atyid shrimp (*Neocaridina denticulata*) in a tropical brackish wetland. *Aquatic Toxicology (Amsterdam, Netherlands)*, *225*, 105522. Available from https://doi.org/10.1016/j.aquatox.2020.105522.

Tsikliras, A. C., Dinouli, A., Tsiros, V. Z., & Tsalkou, E. (2015). The Mediterranean and Black Sea fisheries at risk from overexploitation. *PLoS One*, *10*(3), e0121188. Available from https://doi.org/10.1371/journal.pone.0121188.

Varga, C., Kiss, I., & Ember, I. (2002). The lack of environmental justice in Central and Eastern Europe. *Environmental Health Perspectives*, *110*(11), A662–A663. Available from https://doi.org/10.1289/ehp.110-a662.

Wang, M., Tong, Y., Chen, C., Liu, X., Lu, Y., Zhang, W., He, W., Wang, X., Zhao, S., & Lin, Y. (2018). Ecological risk assessment to marine organisms induced by heavy metals in China's coastal waters. *Marine Pollution Bulletin*, *126*, 349–356. Available from https://doi.org/10.1016/j.marpolbul.2017.11.019.

Wei, Y., Zhang, H., Yuan, Y., Zhao, Y., Li, G., & Zhang, F. (2020). Indirect effect of nutrient accumulation intensified toxicity risk of metals in sediments from urban river network. *Environmental Science and Pollution Research International*, *27*(6), 6193–6204. Available from https://doi.org/10.1007/s11356-019-07335-9.

Wood, C. M., McDonald, M. D., Walker, P., Grosell, M., Barimo, J. F., Playle, R. C., & Walsh, P. J. (2004). Bioavailability of silver and its relationship to ionoregulation and silver speciation across a range of salinities in the gulf toadfish (*Opsanus beta*). *Aquatic Toxicology (Amsterdam, Netherlands)*, *70*(2), 137–157. Available from https://doi.org/10.1016/j.aquatox.2004.08.002.

Yılmaz, A. B., Yanar, A., & Alkan, E. N. (2017). Review of heavy metal accumulation on aquatic environment in Northern East Mediterrenean Sea part I: Some essential metals. *Reviews on Environmental Health*, *32*(1-2), 119–163. Available from https://doi.org/10.1515/reveh-2016-0065.

Zhang, Y., Lu, X., Wang, N., Xin, M., Geng, S., Jia, J., & Meng, Q. (2016). Heavy metals in aquatic organisms of different trophic levels and their potential human health risk in Bohai Bay, China. *Environmental Science and Pollution Research International*, *23*(17), 17801–17810. Available from https://doi.org/10.1007/s11356-016-6948-y.

Zheng, N., Wang, S., Dong, W., Hua, X., Li, Y., Song, X., Chu, Q., Hou, S., & Li, Y. (2019). The toxicological effects of mercury exposure in marine fish. *Bulletin of Environmental Contamination and Toxicology*, *102*(5), 714–720. Available from https://doi.org/10.1007/s00128-019-02593-2.

Zhong, W., Zhang, Y., Wu, Z., Yang, R., Chen, X., Yang, J., & Zhu, L. (2018). Health risk assessment of heavy metals in freshwater fish in the central and eastern North China. *Ecotoxicology and Environmental Safety*, *157*, 343–349. Available from https://doi.org/10.1016/j.ecoenv.2018.03.048.

CHAPTER

Effects of heavy metals and pesticides on fish

3

Raksha Rani[1], Preeti Sharma[1], Rajesh Kumar[2] and Younis Ahmad Hajam[1]

[1]Division Zoology, Department of Biosciences, Career Point University, Hamirpur, India [2]Department of Biosciences, Himachal Pradesh University, Shimla, India

3.1 Introduction

Many types of organic and inorganic contaminants, including plastics, pharmaceuticals, pesticides, and metals, are released by humans into the environment, both aquatic and terrestrial. These contaminants have various impacts on aquatic organisms as well as terrestrial ones (Saaristo et al., 2018; Scott & Sloman, 2004; Zala & Penn, 2004). Ecotoxicology research has shown that pollutants have direct effects on the physiology of animals and increase the death rate (Ashauer et al., 2013; Butcher et al., 2006). However, less research has been carried out on behavioral effects of multistress conditions, including environmental contaminants, on wild species (Saaristo et al., 2018; Zala & Penn, 2004). Contaminants from agricultural activities, industrial resources, abandoned military installations, and urban areas are directly or indirectly released into bodies of water, and because of these toxins, aquatic ecosystems are at higher risk of adverse effects (Byrne et al., 2015; Kaur & Dua, 2015; Pinto et al., 2015; Van der Oost et al., 2003). Moreover, due to climate change, abiotic factors like temperature level and precipitation are being altered, which can also affect the functions of organisms in freshwater ecosystems, such as reproduction and feeding. Thus, by identifying the main pollution activities, their sources, and their fate in freshwater ecosystems, we can protect global aquatic species (Liu et al., 2018a, 2018b).

Heavy metals are important pollutants that quickly disperse in the body and are gradually metabolized in and excreted from organisms. Contaminants like lead (Pb), chromium (Cr), zinc (Zn), and mercury (Hg) are the major heavy metal ions released from industrial sources and are water pollutants (Štrbac et al., 2015). Globally, numerous types of metalloids and heavy metals have been released into the environment as a result of a number of increasing human activities (Dhanakumar et al., 2015; Pfeiffer et al., 2015). After their release into the environment, these heavy metals and metalloids enter rivers and lakes. Some metals like nickel (Ni), chromium (Cr), copper (Cu), and zinc (Zn) are damaging in high concentrations, but in trace amounts they are biologically essential for the normal growth and functioning of organisms. On the other hand, lead (Pb), arsenic (As), cadmium (Cd), and mercury (Hg) are all highly toxic even at low levels, and these are nonessential to life.

To understand the complete picture of contamination by heavy metals within an ecosystem, which also represents a threat to human health, it is important to identify the source and extent of

Bacterial Fish Diseases. DOI: https://doi.org/10.1016/B978-0-323-85624-9.00016-6

the pollution through sediment and water sampling, and also to evaluate the magnitude of bioaccumulation in consumed fish species (Perez-Cid et al., 2001). Sediments usually determine the quality of the water of rivers and streams related to heavy metals. Sediments have a deep layer and a surface layer. The surface layer is in contact with water and is rich in organic pollutants and other matter. Sediments have the capability to store and release heavy metals into the water because of the regular contact of the surface layer with water. Contaminants that are insoluble and are released in the riverbed may be detected in sediments (Gibson et al., 2015; Quesada et al., 2014). Moreover, metals are accumulated in benthic invertebrates due to the contamination of sediments, because these organisms have the capability to tolerate the heavy metals (Kalantzi et al., 2014). The tolerance shown by these animals may result in contamination of other trophic levels, as a result of transfer up the food chain. The concentration of heavy metals is found to be higher in sediments as compared to other water levels (Harguinteguy et al., 2014). In the water these elements can also expand and accumulate along the food chain and build up harmful concentrations in higher trophic level species like fish. These pollutants cause various diseases in fish, and if humans regularly consume the fish, then serious health problems can result, including cardiovascular damage, cancers, damage related to the nervous system, various lung diseases, kidney and gastrointestinal problems, or even death (Moreau et al., 2007; Weill et al., 2015). The risk is greater than ever, as people are simultaneously being exposed to high levels of heavy metals in foods, drinking water, and/or their domestic or occupational environments (Tchounwou et al., 2012).

Chlorpyrifos (CPF) is an organophosphorus (OP) pesticide widely used in agriculture that causes toxicity to fish (Schimmel et al., 1983; Yin & Wang, 2018). OP pesticides are a major cause of the inhibition of acetylcholinesterase in nerve cells of fish (Yin & Wang, 2018). They also cause many secondary effects, like oxidative stress, intestinal microbiota dysbiosis, toxicity to the immune system, toxicity to the nervous system, and disruption of the endocrine system (Eddins et al., 2010; Oruç, 2010; Richendrfer et al., 2012; Sharbidre et al., 2011a, 2011b; Wang et al., 2019; Yin & Wang, 2018; Zhang et al., 2017). Olsvik et al. (2015), Olsvik et al. (2017), Sanden et al. (2018), and Olsvik et al. (2019) conducted in vitro and in vivo studies with the methylated forms of chlorpyrifos and pirimiphos (PP) in Atlantic salmon, to study the molecular effects of organophosphorus pesticides found in salmon feeds. It was shown in these studies that CPF and PP interrupted lipid metabolism in fish and induced oxidative stress in their livers. Moreover, polychlorinated biphenyls are estrogenic and antiestrogenic pollutants used as insecticides. These act as endocrine disruptors and affect fish reproduction.

In summary, insecticides may affect biochemical and physiological processes and cause serious problems in fish health status. The toxic manifestations of various contaminants are shown in Fig. 3.1.

3.2 Toxicity due to pesticides in fish

Many types of substances are used to control organisms that harm agricultural fields and crops, including insects, plant diseases, and water weeds. The pesticides used are often highly toxic to nontarget organisms, including fish. These pesticides impair the metabolism of fish and may even cause death. The increase in the human population worldwide has led to a rapid pace of

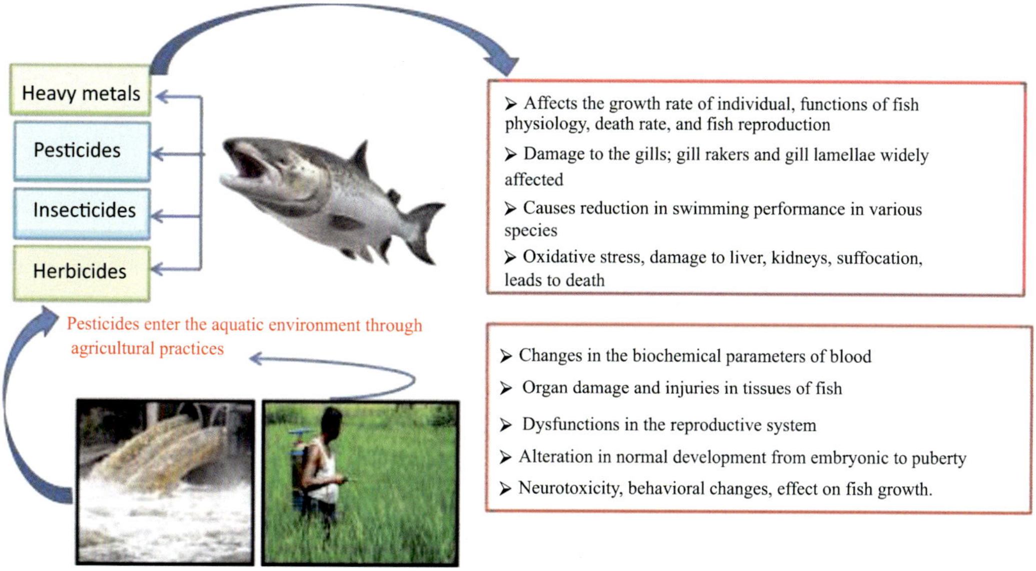

FIGURE 3.1

The effects of various contaminants on fish.

industrialization, and that has induced serious wastewater disposal problems. Every type of waste, whether domestic or industrial, treated or untreated, can contain contaminants, including heavy metals, pesticides, herbicides, and insecticides, released into bodies of water. All these contaminants have contributed to massive numbers of fish deaths in aquatic ecosystems (Murthy et al., 2013). Various research studies have shown that pesticides are toxic to fish and it also has been shown that, at chronic levels, they have adverse effects such as oxidative stress, mutagenesis, developmental changes, histopathological changes, inhibition of ACHE activity, and carcinogenicity. Along with other organophosphate compounds present in the environment, these pesticides can induce lethal or sublethal effects in fish (Murthy et al., 2013). Before the use of the modern synthetic pesticides, naturally occurring pesticides were used for centuries, but currently demand for modern pesticides is increasing day by day. In the United States, over a billion pounds of pesticides are used annually, at a value of one billion dollars per year. There are different categories of pesticides: fungicides (used to kill or prevent the growth of fungi), herbicides (used to kill unwanted plants), insecticides (used to kill insects), rodenticides (used to kill rodents), acaricides (used to control mites and ticks), nematicides (used to control soil, leaf, and stem-dwelling nematodes), molluscicides (used to control snails and slugs), among others. Insecticides, herbicides, and fungicides are the most commonly used pesticides. In the United States, the Federal Insecticide, Fungicide, and Rodenticide Act (FIFRA) stipulates that all pesticides must meet certain criteria to be registered with the Environmental Protection Agency (EPA) before they can be used. The specifications depend on the method used on a particular pest, toxicity to mammals, environmental residue, analysis methods, and other conditions. Consideration is also given to the pesticide ingredients, manufacturing process, physical and chemical properties, and its environmental state: whether it is mobile, volatile, or has accumulation potential in plants and fish. Trade name, usage directions, active ingredients, toxicity rating, and the EPA registration number must be shown on each pesticide label. Toxicity, exposure time, dose rate, and persistence in the environment are the major functions of a pesticide that has the capacity to harm fish and other aquatic organisms. A dose is used as a standard toxicity measurement: it is actually the amount of pesticide needed to cause death in a particular species. This standard is called lethal concentration, or LC50, and it refers to the concentration of a pesticide that kills 50% of a test population within a set period of time, usually 24–96 h.

Bioavailability, bioconcentration, biomagnification, and persistence in the environment are the main factors affecting exposure of fish and other aquatic organisms. Bioavailability means the amount of pesticide in the environment available to fish and wildlife. Different pesticides have different functions and by nature some are diluted in water or volatize into the air and some quickly breakdown after application. Some attach to soil particles suspended in the water column, thereby reducing their availability. Biomagnification means a pesticide builds up at each successive level of the food chain and therefore it passes up from one successive level to another: it is consumed through water by plants that are, in turn, eaten by insects, which are eaten by fish, and so on. The pesticide concentration increases at each step in the food chain. When trout or bass eat contaminated insects or other organisms, then the concentration of contaminant increases in their body and it can pass up to humans through the food chain.

The length of time a pesticide can stay in the environment is called persistence. The persistence of a pesticide depends upon how rapidly it degrades and the degradation is mainly a function of its chemical composition. Persistent pesticides cannot degrade easily and are thus more available to freshwater animals like fish (Helfrich et al., 2009). Photodecomposition, high air or thermal degradation, moisture conditions, biological action (microbial decay), and soil conditions (pH) can all degrade persistent pesticides over time.

3.3 Disadvantages of pesticides

Pesticides become an environmental hazard when they enter the aquatic system. Unintentionally, the fish death rate has increased globally due to these pesticides. Not only fish, but also other aquatic animal populations are decreasing due to pesticides, such as frogs, turtles, water birds, and others. Endangered and rare wildlife species have also been affected by pesticides. Pesticides have the capacity to kill some fish, like salmon, and other aquatic animals after very short exposures. Pesticides can also cause problems through contact over a long period of time. These can change an organism's swimming ability, reduce feeding capability, hinder escape from predators, and reduce the ability to maintain position in a river or pond. Migrating salmon may fail to avoid predators due to coming in contact with pesticides. Some researchers also found an interruption in schooling behavior, as an example of a common sublethal effect of pesticides. Migration of fish to the sea is inhibited by some herbicides and this leads to a disruption of their life cycle. Sometimes there is a delay in spawning of salmon; because salmon avoid polluted areas, they will migrate through such areas, so they have to adjust their migration. Due to this behavior change in salmon, there is disruption in the immune system and in the endocrine system during early stages of development. Pesticides can also act as sex hormone blockers, causing abnormality in sexual development, sex ratios, and mating behaviors; they may also alter the hormonal functions and bone development. Pesticides also have many indirect effects, altering the aquatic habits of fish, disturbing food supplies, reducing growth, and decreasing chances of fish survival. It has been reported that chronic toxicity may cause death, species or individual elimination, and can have many other effects, such as loss of natural defense mechanisms, appetite loss, blindness, subfertility or infertility, hyperexcitability, or other changes or alteration in the bodies of fish (Dutta & Meijer, 2003; Helfrich et al., 2009; Khan & Law, 2005).

3.4 Routes of pesticide exposure in fish

Dermal exposure (absorption of pesticide through skin), inhalation (direct uptake of pesticides during respiration), or oral exposure (drinking pesticide-contaminated water) are the three primary ways through which fish and other aquatic animals can be exposed to pesticides. Secondary causes can also cause toxicity to fish indirectly. Fish and other aquatic animals consume contaminated insects that contain pesticides and they may cause toxicity. Since pesticides first come into contact with the water column and other substances, they may act as secondary causes (Kingsbury & Kreutzweiser, 1980; Schnick et al., 1986). Fish exposure to pesticides has become an extensive problem. Most of these cases are underreported and understudied. There is no accurate estimate of how many fish are killed due to these pesticides, and much research remains to be done in this area.

3.5 Effects of pesticides on fish

Most of the main pollutants or contaminants reside in the water column, including different types of chemicals, heavy metals, insecticides, and pesticides coming from industrial or agricultural activities. Insecticides can be categorized as synthetic and natural groups, and both types cause problems in the environment. Insecticides disperse everywhere in the environment and enter into the aquatic ecosystems through many different pathways. In urban areas these insecticides enter into water ecosystems mainly through rainfall, runoff, and atmospheric deposits. Another pathway is the contamination of water resources through municipal and industrial discharges. Other pesticides eventually find ways to enter into rivers, ponds, and lakes, and these pesticides are very toxic to nontarget organisms (Arjmandi et al., 2010). Contamination by insecticides on surface water is thought to have adverse health effects, affecting survival, reproduction, and growth of organisms. Many types of wastewater have different concentrations of contaminants present and it was found in many studies that these are very toxic to fish species especially (Banaei et al., 2008). Fish are the most sensitive aquatic animal and may be exposed rapidly to these pesticides. Hence, many physiological and biochemical processes are damaged by pollutants such as insecticides. Physiology and health status of fish are impaired by different kinds of insecticides (Banayi et al., 2009).

3.5.1 Residual effects of insecticides

Some insecticides are not very stable in water and can degrade easily. Malathion is one of them: it is very unstable in water and degrades up to 99% in just 7 days. Fenitrothion is somewhat stable; it takes 29 days to degrade up to 97%. Diazinon degrades more than 95% in just 6 days and is considered to be very stable. BPMC is somewhat stable and degrades 80% in 32 days. XMC degrades only 45% in 34 days and it is the most stable insecticide.

Organophosphorus insecticide uptake by fish is generally higher than that of carbamate insecticides. Spinal curvature can be caused by diazinon, fenitrothion and BPM (Kanazawa, 1975). Organochlorine insecticide residues are present in the highest numbers in water and are becoming a part of the earth cycles. They are detected at the highest levels in brain tissues of fish (Bakre et al., 1990).

Diazinon usage has increased in today's world, which is a reason why we are facing these health-related as well as environmental issues. Diazinon is acutely toxic to a wide variety of animals, and can also lead to sublethal toxicity; it causes damage to most of the body organs and tissues, causes reproductive system damage, and also has had various ecological impacts (Sabra & Mehana, 2015).

Compounds from organophosphates and carbamate insecticides are also used widely. These can contaminate surface water and cause toxicity to various aquatic organisms, especially those living at low levels in an aquatic system. Persistence of various pesticides like diazinon, chlorpyrifos, marathon, and carbaryl in water from various sites was detected by Bondarenko et al. (2004). Marathon or carbaryl persistence was found low as compared to diazinon and chlorpyrifos persistence in sea water. Environmental scientists today are very concerned about the extent of pesticide use and the modes of application, especially in agriculture. To detect pesticide concentrations in water and fish, four organophosphorus pesticides—dichlorovos, fenitrothion, chlorpyrifos, and diazinon—and four organochlorine pesticides—4,4-dichlorodiphenyldichloroethane (4,4-DDE), P.P-dichlorodiphenyltrichloroethan (PP-DDT), dichlorophenyldichloroethylene (2,4-DDE), and propiconazol—were investigated. The percentage of pesticides was higher than the permissible limit as detected by Essumang et al. (2009). Doudoroff et al. (1953) found that in goldfish Aldrin caused less toxicity than toxophan, but was more highly toxic than BHCC and DDT and evidently killed all goldfish.

3.5.2 Bioaccumulation of insecticides

Insecticides of different concentrations present in the aquatic ecosystem are accumulated by fish, and after accumulation the insecticides are absorbed in body tissues, gills, and the alimentary canal. These insecticides have the capability to dissolve in liquids, fats, oils. etc., and thus are able to penetrate biological membranes and also to enhance the fish sensitivity to insecticides in the water. In the fish body, these are quickly metabolized in the body tissues. By the process of bioaccumulation, the insecticides increase inside the fish body, which is very harmful to the organisms that consume these contaminated fish and can also cause death of the fish, or long-term damage in different tissues and organs (Ballesteros et al., 2011). Moreover, fish are a relatively lower level in the food chain and the contamination may increase from lower levels to higher levels and spread from their predators to consumers. Human health and survival are affected by this contamination. Thus the bioaccumulation of insecticides in fish and their tissues and biomagnification in humans are thought to be major threats (Ferrari et al., 2004). The bioaccumulation rate of insecticides depends upon a number of factors, including fish species, stages of life in fish, fat reservation in the tissues of fish, diet of fish, chemical and physical properties of the insecticides, and the rate of water pollution.

3.5.3 Biotransformation of insecticides and the toxic mechanisms

Insecticide biotransformation can be subdivided into phases I, phase II, and phase III. These reactions can takes place simultaneously or sequentially. Phase I reactions yields polar, water-soluble metabolite. Most of the products of this phase can become substrates for phase II. It involves oxidation with cytochrome P450, reduction and hydrolysis. Phase II yields a large polar metabolites by adding endogenous hydrophilic groups to form water-soluble inactive compounds that can be excreted by the body. It involves methylation, glucuronidation, acetylation, sulfation, conjugation with glutathione and amino acids. Phase III takes place after phase II, where a chemical substance can undergo further metabolism and excretion. It classifies into the following superfamilies: ATP-binding cassette (ABC) or Solute carrier (SLC) transporters. Cytochrome P450 enzymes are the most important enzymes in phase I reactions, and include CYP1A and CYP3A. Biotransformation of endogenous and exogenous compounds is carried out by cytochrome P450 enzymes to form a water soluble and polar compound. CYP1A, CYP2B, CYP2E1, and CYP3A are the different cytochrome p450 enzymes recognized in the livers of fish, which are very important in the detoxification of different insecticides (Ferrari et al., 2007). The common pathways of biotransformation of different insecticides include three types of reactions: hydroxylation, dealkylation, and epoxidation.

UDP−glucuronosyltransferases (UGTs) are the enzymes that catalyze most phase II reactions. Excretion of some of the metabolites may take place through the genital products, skin, gills, and urine excreted as sulfated metabolites and stool as glutathione-conjugated metabolites (Nebbia, 2001). Some metabolites produced are thought to be much more dangerous than the potential compounds and may cause severe damage in the fish body. Free radical production or reactive oxygen species (ROS) generation during the process of detoxification can cause oxidative stress (Monterio et al., 2009). ROS can react with macromolecules like proteins, lipids, and DNA in living cells and attack randomly, which can induce cytotoxicity and may cause severe disturbances in the physiological processes (Jin et al., 2010). The serious forms of damage, such as cells losing their functions, damage to the DNA, inactivation of enzymes, and oxidation of hormones, are caused by lipid peroxidation and these indicate oxidative stress (Sabra & Mehana, 2015).

3.6 Acute toxicity of insecticides

The toxicity rate due to insecticides in fish is determined after 96 h and it is highly concentration dependent. If the concentration is between 10 and 100 ppm, then it is slightly toxic; if toxicity is between 1 and 10 ppm, then it is moderately toxic; if the toxicity is between 0.1 and 1 ppm then it is highly toxic; and if the toxicity is less than 0.1 ppm, then it is extremely toxic. This is regarded as acute toxicity of insecticides (Banaee et al., 2011a, 2011b). Fish populations exposed to insecticides have a sudden and severe death rate in the case of acute toxicity. Lethargy, extension of fins, change in body color, muscle spasms, abrupt swimming in circles, nervous system disorders, respiratory disorders, etc. are symptoms of insecticide toxicity in fish (Banaee et al., 2011a, 2011b). In order to identify the dose concentration and time at which fish death occurs, an acute toxicity test is used (LC50 in ppm or mg/L). Fish from different species or from the same family show sensitivity differences in high insecticide concentrations. Different factors are responsible for influencing the acute toxicity of different insecticides, for example, the age of fish, fish sex, genetic properties and fish body size, quality of water and physicochemical parameters, and insecticide formulation and purity.

3.6.1 Sublethal toxicity of insecticides

Sublethal toxicity is defined as levels that do not induce mortality in the subject population, but still it can have many deleterious effects. In fish, sublethal toxicity can affect body organs, such as the hepatic portal system, respiratory system, nervous system, cardiovascular system, hematopoietic system, immune system, and reproductive system. Insecticides cause changes in the biochemical parameters of the blood (Kavitha & Rao, 2009). The growth rate of fish is decreased due to insecticides, and many disorders in the reproductive system can appear. These can cause distortion of the spinal cord and changes in the histopathology (Benli & Özkul, 2010), and can also lead to defects in gills, liver, spleen, kidney and renal tubules, along with disorders of the endocrine system, dysfunctioning of the brain, behavior changes, and defects in genes. All these factors indicate fish exposure to insecticides.

3.7 Chronic toxicity of insecticides

Tests available for the detection of chronic toxicity are commonly carried out by measuring the long-term effects of low concentrations of insecticides on various biological systems or characteristics of the fish, including fish survival, growth, and reproduction, as well as aspects related to the fish nervous system and physiology. In chronic toxicity, the type of injury to fish is similar to the damage caused during sublethal toxicity, but the incidence and rate of injury from chronic toxicity may be more, or even sometimes less, than sublethal toxicity damage. Therefore determination of chronic toxicity is crucial in the toxicology of insecticides (Mathur, 1999).

3.7.1 Effects of insecticides on different parameters in fish

A blood biochemistry test is used to determine the various changes in the biochemical parameters of blood. When body tissues are injured, there is a release of specific enzymes into blood plasma and their abnormality is recognized by the test. With severe damage to body tissues, biochemical parameter synthesis may decrease in cells and some of the biochemical factors in the fish blood will be reduced. Following are some examples of fish studied after exposure to insecticides: *Oreochromis niloticus* exposed to carbaryl (Matos et al., 2007); *Cyprinus carpio* exposed to diazinon (Banaei et al., 2008); and *O. niloticus* exposed to bifenthrin (Agrahari et al., 2007; Velisek et al., 2009).

3.7.2 Tissue and organ damage

Organ damage and tissue injuries in fish lead to a reduction in the survival and growth of fish, reduction in reproduction success, or increase in vulnerability to pathological changes. Rate of recurrence and intensity of tissue injury depend upon insecticide concentration and length of the time period since fish were exposed to toxins. Fanta et al. (2003) demonstrated that most of the insecticides cause either definite damage or unclear or nonspecific damage to the body tissues. Velmurugan et al. (2009) reported histopathological lesions in the liver of *Cirrhinus mrigala* and *C. carpio*. Due to exposure to sublethal concentrations of some insecticides like dichlorvos and diazinon, histopathological changes were also noticed by Banaee (2012).

3.8 Reproductive dysfunction

Reproductive success of fish can also be affected by changes in environmental parameters and the physiological condition of the fish. When fish are exposed to various contaminants such as fungicides, insecticides, herbicides, heavy metals, and xenobiotics, various types of disorders occur in the process of fish reproduction. Banaei et al. (2008) found that fish exposed to insecticides had various dysfunctions in their reproductive system. The insecticides caused a decrease in the fecundity or egg-laying capacity of fish, damage to the testis or ovarian histology (Banaei et al., 2008), impairment in the process of vitellogenesis (Haider & Upadhyaya, 1985), steroidogenesis disruption, delay in gonads maturation, changes in parental and reproductive behavior, olfactory response impairment, alteration in reproductive migration, disruption in mating behavior between male and female, and alteration in spawning time (Jaensson et al., 2007). Insecticides can also act as endocrine disrupting agents, which can interrupt the functioning of the endocrine system of fish. Insecticides also have many effects on the hypothalamo-pituitary gonadal axis. This can also lead to reproductive failure in fish. Eggs and milt of fish can also be exposed to insecticides, affecting level of fertilization, rate of egg hatching, and survival ability of larvae (Moore & Waring, 2001).

3.9 Developmental disorders

The study of developmental disorders caused by insecticides highlights the links between the toxin concentrations and alterations in normal development from embryonic to puberty periods. The chances of fish survival are reduced due to the alterations in their development and growth. In viviparous fish, the young, through the yolk or via parental contact, are directly exposed to insecticides, and in other types, their larvae or embryos come in contact with insecticides (Viant et al., 2006). Deformity in the spinal cord occurs, mainly scoliosis or lordosis, and abnormalities in morphology were also found. Xu et al. (2010) reported changes in fish embryo, yolk sac edema, and curved bodies in larvae. Todd and Van Leeuwen (2002) proved that carbaryl insecticide causes abnormal fetal development of fish embryo. Decline in the growth of fish is due to disorders in feeding behavior, reduction in rate of feeding, alterations in metabolic processes, and wasting of energy to overcome the stress caused by exposure to insecticides. Carbohydrate, protein, and lipid metabolism disorders were also reported (Begum, 2004).

3.9.1 Neurotoxicity

Metabolism of neurotransmitters such as γ-aminobutrate (GABA) or functioning of butyryl cholinesterase or ACHE (acetylcholine) activities are inhibited by organophosphate and carbamate insecticides. Neuronal functions were changed by pyrethroid (a type of synthetic insecticide) by interfering with or changing the functions of ion channels present in the nerve cell membrane, changing the concentration of intercellular calcium ions, and inhibiting the receptors of the GABA neurotransmitters. Organochlorine insecticides alter the ion transport across the nerve cell membrane, and change the nerve stimulating ability. To access the exposure of fish to these insecticides, changes in ACHE in brain, plasma, or muscles and other body tissues or activity of the GABA neurotransmitter in fish brains can be determined (Banaee, 2010). Due to the inactivation of the ACHE enzyme, the neurotransmitter acetylcholine accumulates in cholinergic synapses, and this can lead to synaptic blockage and disruption of signal transmission (Ferrari et al., 2007). Furthermore, this can also induce changes in swimming behavior, muscle spasms, and other effects.These can also impair feeding behavior, escaping from predators, and reproduction behavior. Thus fish exposure to organophosphorus and carbamate insecticides is identified by considering ACHE inhibition, as it is thought to be a specific biomarker (Wang et al., 2009).

3.9.2 Behavioral alterations

Fish show various types of behavioral changes when they are exposed to insecticides, including in feeding activities, competition among species, predation, interactions between species-species, and reproduction behavior. When common carp and rainbow trout were exposed to diazinon, common behavioral responses were reported by Banaee et al., 2011a and Banaee et al., 2011b. When insecticides interfere with the nervous system and sensory receptors, certain behavioral changes occur. Fish mobility can also be affected by the inhibition of ACHE activity by certain insecticides (Bretaud et al., 2000).

3.9.3 Genotoxicity

Insecticides that damage genetic information are referred to as genotoxic chemicals, possessing physical and chemical properties that allow interaction with genetic material and that damage or inactivate DNA. Interaction between the chemical and DNA may cause a mutation that is an inheritable change in the genotype of the cell. This defect may be transferred to the next generation, or from generation to generation. Insecticides also have many carcinogenic effects. Tumors can form in different tissues of the fish due to their exposure to these xenobiotics, because xenobiotics have properties to damage genetic information. They can also cause damage to the chromosomes of fish eggs and larvae. By behaving as genes, some insecticides cause reproductive dysfunction, suppression of the immune system, or unusual concentrations of steroid hormones in plasma (Begum, 2004).

3.9.4 Immunosuppression

Functioning of the immune system and immune system depression can be altered by insecticides. Insecticides can cause uncontrolled growth of cells that alters the defense mechanisms of the host, such as innate immunity and acquired immunity against different pathogens. For defense against the different kinds of pathogens, the immune system is crucial. Sublethal levels of insecticides have been identified as stressors that cause suppression of fish immune systems (Werimo et al., 2009). Because of the immune-depressive effect, the exposure to sublethal concentrations of insecticides is what probably makes fish vulnerable to infectious diseases (Zelikoff et al., 2000).

3.9.5 Effect on growth of fish

Toxicity due to insecticides shows alteration in the biosynthesis of nucleic acid. Decrease in the content of RNA is due to disturbances in nucleic acid metabolism. Nucleic acid biosynthesis is also affected by the influence of organophosphate on alkaline phosphatase activity in different body tissues of the fish. Changes in the replication of DNA and chromosomal aberrations are caused by dichlorvos. Due to changes in DNA replication and chromosomal aberrations, uncontrolled cell proliferation and mutations also occur. Enzymes that are involved in the DNA replication are inhibited by these insecticides and mutations can affect the gene expression. These can also reduce the RNA level in the synthesis of protein (Banaee, 2011).

3.9.6 Histopathological alterations due to insecticide toxicity

To study the toxicology and monitoring of water pollution, histopathological investigations on different tissues of exposed fish are useful tools. Information about the nature of a toxicant can be learned by assessing the changes in fish tissue in fish exposed to different concentrations of insecticides. Information about the organ functions and health of fish are thus studied in histopathology. Reduction in survival, growth, and fitness of fish; changes in reproduction behavior; or increased vulnerability to pathological agents are due to damage in tissues and organs. Insecticide detoxification inside liver tissues generates reactive oxygen species (ROS); these free radicals react with lipid, protein, carbohydrates, and nucleic acid and cause oxidative stress to aquatic organisms (Üner et al., 2006). Oxidative stress is a serious metabolic disorder, causing damage to cellular components (Sepici-Dinçel et al., 2009). Moreover, organophosphate insecticides reduce the reconstruction of necrotic tissues through the process of phosphorylation and methylation of cellular proteins (Murray et al., 2003). Pathological changes can result in different body tissues and organs of fish, such as gills, liver, kidney, and spleen, that are exposed to different insecticides, disturbing the body's homeostasis and finally leading to physiological disorders (Abdul-Latif, 2008).

3.9.7 Herbicides

Another type of pesticide, commonly used on agricultural crops, gardens, and lawns, is the herbicide. Herbicides can also be used to control the unusual growth of algae, submerged water grasses, flowering plants in water, and others. Herbicides have serious effects on mortality of fish. Fish kills may result after the application of herbicides, even when the herbicide is not directly applied.

Indirect death of fish may occur, possibly from suffocation because decomposition of herbicides in water weeds reduces the level of oxygen in the water. Copper sulfate, fluridone, Sonar, 2—4-D, glyphosphate, Rodeo, diquat, Weedtrine, Endothall, Aquathol, and Hydrothol are some examples of herbicides. Fish exposed to sublethal doses of glyphosphate experience degeneration of liver and fibrosis if the herbicide resides up to 2 or more days inside the fish body (Topal et al., 2015). Moreover, it also alters the swimming performance, meaning performance declines as compared to the initial performance. Thus systemic and muscular effects are caused due to xenobiotics and lead to reduction in swimming performance.

3.9.8 Fungicides

Fungicides are not as toxic to fish and other aquatic organisms as are herbicides. However, due to their side effects on the environment, some fungicides have been banned. Mercury is present in some fungicides and their use was discontinued for home and other agricultural uses in the United States in 1976. Fungicides containing mercury collected in the environment and contaminated the food chain, causing fish kills. Some fungicides are highly toxic to fish, and some are highly poisonous to beneficial soil invertebrates. Use of fungicides should be avoided, or they should be managed near aquatic systems very carefully (Topal et al., 2015). Recently, Michelle (2009) reported, through extensive sampling, on 50 widely used pesticides, including insecticides, herbicides, and fungicides, in water (present in up to 56% of analytes), in sediment (presence detected was intermittent), and in biota (scarce presence). Benzimidazoles, organophosphorus, and ureas were present at high concentrations in water. However, in sediments and biota, organophosphorus (chlorpyrifos, 93% of sediments) was primarily present. Hence, all these pesticides, including the fungicides, are responsible for the decline in fish fauna in aquatic ecosystems. Fish are very sensitive organisms, and can become contaminated very easily. Thus fish mortality is high due to direct or indirect application of pesticides in agriculture or gardens or near lakes and ponds (Table 3.1).

3.10 Toxicity due to heavy metals in fish

Heavy metals have various adverse effects and these can affect the growth rate of individuals, functions of fish physiology, and death rate of fish. In the fish body, heavy metals enter through the gills, digestive tract, and the surfaces of the body. The major entrance is the gills, with a lesser portion of heavy metals entering through the body surface (Selda & Nurşah, 2012). Food sources also act as a path for heavy metal accumulation, which leads to biomagnification and the extension of contaminants up the food chain. Fish are an excellent source of omega-3 fatty acids, protein, zinc, iron, and calcium (Toth & Brown, 1997), and the sea is a very important source of food, including protein and fatty acids, for human intake (World Health Organization, 1999). Size, stages of development, and salinity are the factors that are crucial to marine organisms with regard to toxicity of heavy metals (Grosell et al., 2007). Organisms affected by heavy metals respond by accumulating the heavy metals inside their bodies or passing them to the next trophic level in the food chain (Shah & Altindağ, 2005). Fish take heavy metals from the surrounding water or by ingesting the

Table 3.1 The effects of different pesticides on fish.

Sr. no.	Different pesticides	Impact on the body of fish	Impact of pesticides	References
1.	Organophosphate and Carbamate	Acetylcholinestrase (ACHE)	Brain ACHE and muscle were inhibited in fish ↓swimming performance and peroxidative damage in brain and gills	Ahmad et al. (2000), Cattaneo et al. (2008)
2.	Dichlorvos	Chromosomal aberrations	Chromosomal abnormalities: centromeric gaps, chromatid gaps, chromatid breaks, subchromatid breaks, reduction, extra-fragments, pycnosis, for example, *Channa punstatus*	Rishi and Grewal (1995)
3.	Oleondrin, Endosulfan	Protein contents	↓ hepatic protein content, muscular, intestinal, gills and blood of fish, for example, *Channa punstatus*, *Cyprinus carpio*	Jenkins et al. (2003), Tiwari and Singh (2004)
4.	Chlorpyrifos, Malathion	Immune system	Disruption in the immune system of fish due to the introduction of low concentrations of pesticides	Satyavardhan (2013)
5.	Dicrotophos, Phosmet	Endocrine system	Pesticides at low concentrations act as blockers of sex hormones, causing abnormal sex development, feminization of males, abnormal sex ratios, and unusual mating behavior	Satyavardhan (2013)
6.	Chlorpyrifos, malathion, acephate, naled, dicrotophos, diazinon, dimethoate, and azinphos-methyl	Juvenile salmonid fish	Long-term exposure to certain pesticides can increase stress in juvenile salmonids, ↓ ability to feed, to avoid predators, to defend territories, and to maintain position in the river system.	Dutta and Meijer (2003), Ewing (1999)
7.	Aldrin, dieldriny, DDT, HBC, and Chlordan	Cyprinid and catfish	Hyperactivity, zigzag movement, loss of buoyancy, elevated cough, loss of schooling behavior, swimming near the upper surface Induced ↓ ovarian weight, retardation growth of the pre-vitellogenic oocytes (e.g., Ham fish) Degeneration of the immature oocytes and rupture of the follicular epithelium and disturbance in the endocrine/hormonal imbalance	Mohan (2000)

food in the surrounding environment (Said et al., 1992). Natural water is being contaminated by human activities that release chemicals or heavy metals into the environment.

3.10.1 Effects of cadmium on fish

Cadmium is one of the most toxic nonessential heavy metals present in the aquatic ecosystem and terrestrial environment. As with most of the heavy metals like lead and mercury, cadmium is also reported to represent a hazard to public health. Cadmium is released into the environment with burning of municipal waste and fossil fuels and also in smelting of zinc, copper, and lead (Sidoumou, 1991). Industrial disposal and household waste are also responsible for the release of cadmium into freshwater. The rate of reproduction of aquatic organisms is greatly affected by these heavy metals, finally leading to a decline in the population of aquatic organisms, such as fish (Sridhara et al., 2010). Exposure to cadmium causes various bodily dysfunctions in fish, including kidney damage, chronic toxicity, changed reproductive capacity, alterations in the functions of kidney, tumors, and hepatic dysfunction. A major part of the human diet is fish and numerous studies have been reported on heavy metal pollution and adverse effects of heavy metals in different edible fish species (Magalhães et al., 2007).

3.10.2 Effects of mercury in fish

Mercury is the most common pollutant among the heavy metals. Mercury enters the fish through the surrounding water and can collect in the fish body. Moving through the food chain, it passes from one trophic level to another. The presence of mercury in freshwater may be from rock weathering or anthropogenic activities, including industrial or domestic wastewater release (Adeyeye et al., 1996). Mercury causes many adverse effects on the growth and structure of fish, and even the survival of fish when present at higher concentrations. Mercury collects in the fish gills and causes depression of respiratory tissues, that is, hypoxia, which causes fish death. It also causes alterations in the physiology of the heart of fish (Olaifa et al., 2004). The hatching ability of fish and their hematology have also been affected by sublethal levels of mercury. It can also cause alterations in fish behavior. Moreover, other disorders have also been observed, including lack of balance, with the fish first swimming restlessly, gulping air, going through a dormancy period, and finally dying (Kori-Siakpere & Ubogu, 2008). Mercury can cause potential toxicity to fish including acid-base disturbances and disturbances in ion regulation, damage in the gill arches, damage in the liver, kidney, blood biochemical parameters, toxicity to the nervous system, and hypoxia (Murugan et al., 2008; Ribeiro et al., 2006).

3.10.3 Effects of lead in fish

Lead (Pb) is considered to be a strong environmental pollutant. Lead toxicity is a major focus because of its high impact on human health (Healey, 2009; Rossi & Jamet, 2008), due to lead entering the aquatic environment through industrial and sewage waste and runoff. More damage to aquatic life is being caused due to the increase in the level of lead in the water; it enters the fish body through gills and body surfaces and accumulates in body tissues or organs of fish, such as bones, muscles, blood, and fat. When accumulation is at a high level, it becomes highly toxic

(Yildirim et al., 2009). This toxicity can alter the functioning of the blood and nervous system of fish and other aquatic organisms (Kalay et al., 1999; McCoy et al., 1995). It is also deposited in the liver, spleen, kidneys, digestive system, and gills of fish. Damage in the gill epithelium and suffocation occur because of acute lead toxicity. Due to the chronic toxicity of lead, changes in gill structure and alterations in defense responses are observed in exposed fish. Chronic lead toxicity can also cause alteration of blood biochemical parameters, causing damage to RBCs, WBCs, and the nervous system (El-Badawi, 2005).

3.10.4 Effects of aluminum in fish

The third most abundant metal on earth is aluminum (Al), after oxygen and silicon (Authman, 2011). It is considered toxic in its ionic form when it is soluble (Walton et al., 2009). It is used in the treatment of water as a coagulating agent and is present in the atmosphere, mostly in the air near industrial areas (Camargo et al., 2009; Silva et al., 2018). Aluminum toxicity to fish depends upon the pH of water and the physical and chemical properties of water. Aluminum causes toxicity in fish respiratory systems and the mechanism of toxicity is related to osmotic balance, which can lead to mucous coagulation on the fish gills and the fusion of gill lamellae and filaments (Abdul-Latif, 2008). Aluminum disrupted the endocrine system of mature *O. niloticus* females (Correia et al., 2010). Aluminum toxicity in fish causes an increase in total RBC count. At low concentrations it caused a decrease in the growth of fish (Bjerknes et al., 2003). Beyond structural gill damage, it can also cause various changes in the fish physiology and disturbances in hematology, respiratory system, ion regulation, cardiovascular system (Laitinen & Valtonen, 1995), reproductive system (Vuorinen et al., 2003), metabolism (Brodeur et al., 2001), and endocrine system (Waring et al., 1996).

3.10.5 Effects of arsenic in fish

Mineral mines, manufacturing industries, smelting procedures, and power plants are the sources by which arsenic (As) reaches aquatic ecosystems. Arsenic is present in pesticides also, and the use of these pesticides in agriculture provides another pathway to the aquatic ecosystems. Arsenic is widely used to kill aquatic plants that are overgrown at fishing areas (Sorensen, 1991). In aquatic organisms and other water sources, it is collected in large amounts. Compounds of arsenic absorb rapidly inside the fish body and cause toxicity to fish. These compounds (arsenates) are capable of causing more toxicity than arsenic alone (Sorensen, 1991). Arsenic accumulates in large quantities inside various fish body organs, such as the liver, as reported in different teleosts such as green sunfish, rainbow trout, Japanese medaka and Mozambique tilapia. Due to the increase in mucous production, suffocation occurs or immediate death results in acute exposures. Chronic exposures of fish can cause metalloid accumulation up to the level of toxicity and can lead to various disease conditions (Hughes, 2002). Gills are the primary part of the fish through which they are continuously exposed to arsenic, as well as intake of As-contaminated food. In *Clarias batrachus* arsenic causes its head kidney cells to swell (Ghosh et al., 2007) and it also causes liver lesions in *C. batrachus* (Datta et al., 2009). Furthermore, in freshwater teleosts it caused alteration in the histology of kidneys in many species, like rainbow trout (Kotsanis & Iliopoulou-Georgudaki, 1999) and *Salvelinus namaycush* (Pedlar et al., 2002). It caused depletion of lymphocytes and melano-macrophage centers in *C. batrachus*

(Datta et al., 2009). Alteration in the liver cells caused necrosis, proliferation in the bile duct, and dying of liver cells. Moreover, some other alterations were caused from exposure to arsenic, including formation of inclusions in internal cells, necrosis body formation, and fibrous body (Sorensen, 1991).

3.10.6 Effects of chromium in fish

Chromium (Cr) is an essential nutrient metal for carbohydrate metabolism (Farag et al., 2006). Effluents released from textile industries, leather tanneries, metal finishing industries, printing and dyeing industries, photographic industries, and pharmaceutical industries are responsible for releasing chromium into the aquatic ecosystems (Abbas & Ali, 2007; Arunkumar et al., 2000). If these effluents are poorly treated, then Cr (VI) is added into the surrounding water bodies, and inside those water bodies it causes damage to fish (Li et al., 2011). On the basis of the physical and chemical properties of chromium, there are two very stable forms of chromium, chromium (III) and chromium (VI), present in the surface water. Chromium (VI) is found to be very toxic and acts as a carcinogenic due to its ability to cross the cell membrane (Eisler, 2000; Lushchak et al., 2009). Cr is assimilated in fish tissues primarily in liver and occurs at much higher concentrations than the external environment (Dar et al., 2016b, 2016a; Eisler, 2000; Lushchak et al., 2009). The impact of chromium on various organs of fish, like kidneys, gills, and liver, affects a number of processes, including metabolic and physiological ones, and alters the fish growth and behavior (Vinodhini & Narayanan, 2008). There are several adverse effects of Cr in the fish body that cause alterations in hematology, morphology, and histology, and reduce the growth and finally lead to the production of reactive oxygen species or dysfunctioning of the immune system (Vera-Candioti et al., 2011). Exposure to sublethal toxicity in *Oreochromis mossambicus* caused changes in the histology of liver, including accumulation of fat, melano-macrophage center increases and necrosis, changes in gill lamellae, alterations in interstitial tissue of ovary, and hypertrophy in testes (Ackermann, 2008). Secretion of excess amounts of mucus, damage to the respiratory epithelium of the gills, and finally death of the fish due to suffocation are adverse effects of chromium.

3.10.7 Effects of copper in Fish

Another essential metal, that is, copper (Cu), is important for the metabolism of cells of various living organisms and is considered as a key constituent metabolic enzyme (Monteiro et al., 2009a, 2009b). However, at high concentration, copper is highly toxic to aquatic animals and may cause intracellular damage (Hernåndez et al., 2006). It occurs as a natural mineral, as it is an abundant element on earth and used widely (Sfakianakis et al., 2015). It enters into aquatic ecosystems through the use of various pesticides, including fungicides, algaecides, and herbicides (Michael, 1984). To control the growth of algae and phytoplankton and control fish-related diseases, an algaecide, copper sulfate (CuSO4), is often used. Copper enters the fish body via direct exposure and diet, taken from the surrounding aquatic environment (Sfakianakis et al., 2015). Copper shows different affinities to accumulate in the liver of fish, even at low concentrations (Jezierska & Witeska, 2006). Copper causes mucous production in very large amounts on the body surface, between the filaments of gills and under the gill covers. Various histological changes are observed in the gills, various sensory receptors such as chemoreceptors and mechanoreceptors, and other body tissues of fish (Sorensen, 1991). Moreover, changes in the histology and morphology in the livers of fish exposed to copper were observed by Varanka et al. (2001). Fish exposed to high doses of copper were affected with visible external lesions, like necrosis, on the livers of *C. carpio*, *Carassiusauratus*, and *Corydoraspaleatus*. Vacuolization of endothelial cells in fish liver was reported by Arellano et al. (1999) after coming in contact with copper. Figueiredo-Fernandes et al. (2007) reported vacuolization of liver cells, necrosis of liver, shrinkage, and an increase in sinusoidal spaces in the liver of copper-exposed fish. Sublethal concentration of copper caused alteration in the gills of Nile tilapia (*O. niloticus*) and necrosis and vacuolization of livers (Figueiredo-Fernandes et al., 2007). Furthermore, changes appeared in testes histopathology, such as testicular hemorrhage necrosis, pyknosis, primary spermatogonia disintegration, and changes in interstitial tissues of testes, of *O. mossambicus* exposed to Cu. There was a decrease in the consumption of oxygen and an increase in the activity of operculum due to the damage in the gills when *Esomus danricus* fish were exposed to copper (Vutukuru et al., 2005). Waterborne copper caused chloride cell dystrophies in *S. senegalensis* fish, reported Arellano et al. (1999). Na, K-ATPase activity of fish gill was more sensitive to chronic exposure of copper in *Oncorhynchus mykiss* (Kamunde & Wood, 2003). In the intestine of the fish *Opsanus beta* Na + /K + -ATPase enzyme activity was elevated when exposed to copper, as reported by Grosell et al. (2007). Copper at high concentrations inhibited the activity of catalase (CAT) enzyme in the liver, gills, and muscle of *C. carpio* (Radi & Matkovics, 1988). Due to the exposure to sublethal concentrations of copper, the process of breakdown of glycogen, that is, glycogenolysis, was stimulated in Labeo *rohita* (Radhakrishnaiah et al., 1992). When significant metal accumulation occurred inside the liver of fish (*Gasterosteus aculeatus*), production of reactive oxygen species occurred and it led to oxidative stress (Sanchez et al., 2005). Moreover, copper-exposed fish also developed alterations in blood biochemical parameters such as increase in hemoglobin and hematocrit in blood, which caused swelling of RBCs (Cyriac et al., 1989). Moreover, copper caused other reproductive effects even at low levels, such as spawning blockage, reduction in egg production, abnormal newly hatched fry, and reduced survival rate in young (Sorensen, 1991). Sublethal concentrations of copper caused a reduction in heart rate and cardiac activity in fish (Gainey & Kenyon, 1990).

3.10.8 Effects of nickel in fish

Nickel (Ni) is present everywhere in the environment, in soil, air, and water, and it enters the environment through both natural and man-made sources. It is released during the conversion of nickel into alloys or nickel mining and is discharged directly in wastewater and from oil-burning power plants; coal-burning power plants are also responsible for the release of nickel into the aquatic ecosystem (Al-Attar, 2007). Nickel allergy, dermatitis, and toxicity to body organs are some adverse health effects caused by exposure to nickel. Toxicity of nickel to aquatic organisms is affected by the physical or chemical properties of water. Nickel causes adverse effects on the gills of fish; the gill lamellae and gill chambers are filled with mucus and become dark red in color (Yang et al., 2007). The freshwater fish *O. niloticus* when exposed to nickel at sublethal concentrations developed various toxic effects including elevation in the level of serum cholesterol, total albumin, protein, ALT, AST, amylase, and decreases in values of sodium and chloride (Al-Attar, 2007). It was also reported that fingerlings of *C. carpio* had decreased blood biochemical parameters such as RBCs, WBCs, and hemoglobin content (Al-Ghanim, 2011; Dar et al., 2016b, 2016a, 2020).

Table 3.2 Effect of different heavy metals on fish.

S. no.	Heavy metals	Source	Effect in fish	References
1.	Copper	Fungicides, algaecides, molluscicides, insecticides	Liver, gill, kidney, hematopoietic tissue, mechanoreceptors, chemoreceptors	Monteiro et al., 2009a, Monteiro et al., 2009b, Sfakianakis et al. (2015), Sorensen (1991)
2.	Nickel	Industry, oil-burning power plants, coal-burning power plants, trash incinerators	Decrease in heart rate and memory impairment	Al-Attar (2007), Al-Ghanim (2011), Yang et al. (2007)
3.	Mercury	Pesticides, batteries, paper industry	Ovaries and the muscles of fish	Dhanakumar et al. (2015), Pfeiffer et al. (2015), Štrbac et al. (2015)
4.	Lead	Agricultural drain water, sewage sludge, fly ash from coal-fired power plants, battery plant, metal ores	Bones, muscles, blood, chronic toxicity, oxidative stress, and neurotoxicity	El-Badawi (2005), Yildirim et al. (2009)
5.	Arsenic	Manufacturing companies, mineral or strip mines, smelting operations, and electric generating stations (power plants)	Suffocation, gill epithelium, kidney, necrotic bodies, and fibrous bodies	Datta et al. (2009), Hughes (2002), Sorensen (1991)

3.10.9 Effects of zinc in fish

Zinc (Zn) is another trace element present abundantly on earth. Zinc is also present in living organisms as a micronutrient and is found in every cell; it is also involved in the synthesis of nucleic acid and is present in many enzymes (Sfakianakis et al., 2015). Zinc and its compounds are used widely in medicine and commerce. Waterborne zinc at high levels may cause toxicity to fish (Niyogi & Wood, 2006). It can affect fish either alone or with other metals like copper (Alabaster & Lloyd, 1982). Zinc causes toxicity to fish gills; the uptake of calcium is disturbed, which can lead to hypocalcemia and ultimately death (Khan et al., 2011). Moreover, toxicity due to zinc varies among freshwater and marine water fish, with various toxic effects on growth, reproduction, survival, and hatching of fish eggs developing (Khan et al., 2011). Zinc was found to cause mortality, retardation in growth, spawning inhibition, and sometimes threatened the survival of fish. Furthermore, it damaged the liver, gills, kidney, and muscles of fish (Sorensen, 1991). Proliferation of gills, mucous cell stimulation, and elevation in the production of mucus were found, along with increases in serum transaminase activity due to zinc exposure (Mallatt, 1985). Chloride cells detached from the epithelium of gills and resulted in gill damage; the epithelial cells detached from the basal lamina and increased the subepithelial space. The fish *O. niloticus* showed congested and pale gills, as reported by Abd El-Gawad (1999). There were alterations in the histopathology in ovarian tissues, liver tissues, and swelling in the body cells of the fish *Tilapia nilotica* due to zinc exposure. New methods for the protection of fish and aquatic ecosystems from heavy metals, along with methods for eliminating contaminants, need to be applied (Table 3.2).

References

Abbas, H. H., & Ali, F. K. (2007). Study the effect of hexavalent chromium on some biochemical, citotoxicological and histopathological aspects of the Orechromis spp. fish. *Pakistan Journal of Biological Sciences*, *10*(22), 3973.

Abd El-Gawad, A. M. (1999). Histopathological studies on the liver and gills of *Tilapia nilotica* (*Oreochromis niloticus*) exposed to different concentrations of lead acetate, and zinc sulphate. *Journal-Egyptian German Society of Zoology*, *30*, 13–22.

Abdul-Latif, H. A. (2008). The influence of calcium and sodium on aluminum toxicity in Nile Tilapia (*Oreochromis niloticus*). *Australian Journal of Basin and Applied Sciences*, *2*, 747–751.

Ackermann, C. (2008). *A quantitative and qualitative histological assessment of selected organs of Oreochromis mossambicus after acute exposure to cadmium, chromium and nickel*. Doctoral dissertation, University of Johannesburg.

Adeyeye, E. I., Akinyugha, N. J., Fesobi, M. E., & Tenabe, V. O. (1996). Determination of some metals in *Clarias gariepinus* (Cuvier and Vallenciennes), *Cyprinus carpio* (L.) and *Oreochromis niloticus* (L.) fishes in a polyculture fresh water pond and their environments. *Aquaculture (Amsterdam, Netherlands)*, *147* (3–4), 205–214.

Agrahari, S., Pandey, K. C., & Gopal, K. (2007). Biochemical alteration induced by monocrotophos in the blood plasma of fish, *Channa punctatus* (Bloch). *Pesticide Biochemistry and Physiology*, *88*(3), 268–272.

Ahmad, S., Scopes, R. K., Rees, G. N., & Patel, B. K. (2000). *Saccharococcus caldoxylosilyticus* sp. nov., an obligately thermophilic, xylose-utilizing, endospore-forming bacterium. *International Journal of Systematic and Evolutionary Microbiology*, *50*(2), 517–523.

Alabaster, J. S., & Lloyd, R. (1982). Finely divided solids. *Water quality criteria for freshwater fish. Second edition. Butterworth, London, UK*, 1–20.

Al-Attar, A. M. (2007). The influences of nickel exposure on selected physiological parameters and gill structure in the teleost fish, *Oreochromis niloticus*. *Journal of Biological Sciences*, *7*, 77–85.

Al-Ghanim, K. A. (2011). Impact of nickel (Ni) on hematological parameters and behavioral changes in *Cyprinus carpio* (common carp). *African Journal of Biotechnology*, *10*(63), 13860–13866.

Arellano, J. M., Storch, V., & Sarasquete, C. (1999). Histological changes and copper accumulation in liver and gills of the Senegales sole, Solea senegalensis. *Ecotoxicology and Environmental Safety*, *44*(1), 62–72.

Arjmandi, R., Tavakol, M., & Shayeghi, M. (2010). Determination of organophosphorus insecticide residues in the rice paddies. *International Journal of Environmental Science & Technology*, *7*(1), 175–182.

Arunkumar, R. I., Rajasekaran, P., & Michael, R. D (2000). Differential effect of chromium compounds on the immune response of the African mouth breeder Oreochromis mossambicus (Peters). *Fish & Shellfish Immunology*, *10*(8), 667–676.

Ashauer, R., Thorbek, P., Warinton, J. S., Wheeler, J. R., & Maund, S. (2013). A method to predict and understand fish survival under dynamic chemical stress using standard ecotoxicity data. *Environmental Toxicology and Chemistry*, *32*(4), 954–965.

Authman, M. M (2011). Environmental and experimental studies of aluminium toxicity on the liver of Oreochromis niloticus (Linnaeus, 1758) fish. *Life Science Journal*, *8*(4), 764–776.

Bakre, P. P., Misra, V., & Bhatnagar, P. (1990). Residues of organochlorine insecticides in fish from Mahala water reservoir, Jaipur, India. *Bulletin of Environmental Contamination and Toxicology*, *45*(3), 394–398.

Ballesteros, M. L., Gonzalez, M., Wunderlin, D. A., Bistoni, M. A., & Miglioranza, K. S. B. (2011). Uptake, tissue distribution and metabolism of the insecticide endosulfan in *Jenynsia multidentata* (Anablepidae, Cyprinodontiformes). *Environmental Pollution*, *159*(6), 1709–1714.

Banaee, M. (2010). *Influence of silymarin in decline of sublethaldiazinon-induced oxidative stress in rainbow trout (Oncorhynchus mykiss)*. Doctoral dissertation, PhD Thesis, Aquaculture and Environmental Department, Natural Resource Faculty, Natural Resource and Agriculture Collage, Tehran University, Iran.

Banaee, M. (2012). Adverse effect of insecticides on various aspects of fish's biology and physiology. *Insecticides—Basic and Other Applications*, *6*, 101–126.

Banaee, M., Mirvaghefi, A. R., Amiri, B. M., Rafiee, G. R., & Nematdost, B. (2011a). Hematological and histopathological effects of diazinon poisoning in common carp (*Cyprinus carpio*). *Journal of Fisheries*, *64* (1), 1–13.

Banaee, M., Sureda, A., Mirvaghefi, A. R., & Ahmadi, K. (2011b). Effects of diazinon on biochemical parameters of blood in rainbow trout (*Oncorhynchus mykiss*). *Pesticide Biochemistry and Physiology*, *99*(1), 1–6.

Banaee, M., Sureda, A., Mirvaghefi, A. R, & Rafei, G. R (2011). Effects of long-term silymarin oral supplementation on the blood biochemical profile of rainbow trout (Oncorhynchus mykiss). *Fish physiology and biochemistry*, *37*(4), 885–896.

Banaei, M., Mir, V. A., Rafei, G. R., & Majazi, A. B. (2008). Effect of sub-lethal diazinon concentrations on blood plasma biochemistry. *International Journal of Environmental Research*, *2*(2), 189–198.

Banayi, M., Mirvaghefi, A., Ahmadi, K., & Ashori, R. (2009). The effect of diazinon on histophatological changes of testis and ovaries of common carp (*Cyprinus carpio*). *Journal of Marine Biology*, *1*(2), 14–25.

Begum, G. (2004). Carbofuran insecticide induced biochemical alterations in liver and muscle tissues of the fish *Clarias batrachus* (linn) and recovery response. *Aquatic Toxicology*, *66*(1), 83–92.

Benli, A. Ç. K., & Özkul, A. (2010). Acute toxicity and histopathological effects of sublethalfenitrothion on Nile tilapia, *Oreochromis niloticus*. *Pesticide Biochemistry and Physiology*, *97*(1), 32–35.

Bjerknes, V., Fyllingen, I., Holtet, L., Teien, H. C., Rosseland, B. O., & Kroglund, F. (2003). Aluminium in acidic river water causes mortality of farmed Atlantic Salmon (*Salmo salar* L.) in Norwegian fjords. *Marine Chemistry*, *83*(3–4), 169–174.

Bondarenko, S., Gan, J., Haver, D. L., & Kabashima, J. N. (2004). Persistence of selected organophosphate and carbamate insecticides in waters from a coastal watershed. *Environmental Toxicology and Chemistry: An International Journal*, *23*(11), 2649–2654.

Bretaud, S., Toutant, J. P., & Saglio, P. (2000). Effects of carbofuran, diuron, and nicosulfuron on acetylcholinesterase activity in goldfish (*Carassius auratus*). *Ecotoxicology and Environmental Safety*, *47*(2), 117–124.

Brodeur, J. C., Økland, F., Finstad, B., Dixon, D. G., & McKinley, R. S. (2001). Effects of subchronic exposure to aluminium in acidic water on bioenergetics of Atlantic salmon (*Salmo salar*). *Ecotoxicology and Environmental Safety*, *49*(3), 226–234.

Butcher, J., Diamond, J., Bearr, J., Latimer, H., Klaine, S. J., Hoang, T., & Bowersox, M. (2006). Toxicity models of pulsed copper exposure to *Pimephales promelas* and *Daphnia magna*. *Environmental Toxicology and Chemistry: An International Journal*, *25*(9), 2541–2550.

Byrne, S., Miller, P., Waghiyi, V., Buck, C. L., von Hippel, F. A., & Carpenter, D. O. (2015). Persistent organochlorine pesticide exposure related to a formerly used defense site on St. Lawrence Island, Alaska: Data from sentinel fish and human sera. *Journal of Toxicology and Environmental Health, Part A*, *78*(15), 976–992.

Camargo, M. M., Fernandes, M. N., & Martinez, C. B. (2009). How aluminium exposure promotes osmoregulatory disturbances in the neotropical freshwater fish *Prochilus lineatus*. *Aquatic Toxicology*, *94*(1), 40–46.

Cattaneo, R., Loro, V. L., Spanevello, R., Silveira, F. A., Luz, L., Miron, D. S., Fonseca, M. B., Moraes, B. S., & Clasen, B. (2008). Metabolic and histological parameters of silver catfish (*Rhamdia quelen*) exposed to commercial formulation of 2, 4-dichlorophenoxiacetic acid (2, 4-D) herbicide. *Pesticide Biochemistry and Physiology*, *92*(3), 133–137.

Correia, T. G., Narcizo, A. D. M., Bianchini, A., & Moreira, R. G. (2010). Aluminum as an endocrine disruptor in female Nile tilapia (*Oreochromis niloticus*). *Comparative Biochemistry and Physiology Part C: Toxicology & Pharmacology*, *151*(4), 461–466.

Cyriac, P. J., Antony, A., & Nambisan, P. N. (1989). Hemoglobin and hematocrit values in the fish *Oreochromis mossambicus* (Peters) after short term exposure to copper and mercury. *Bulletin of Environmental Contamination and Toxicology;(USA)*, *43*(2), 315–320.

Dar, G. H., Bhat, R. A., Kamili, A. N., Chishti, M. Z., Qadri, H., Dar, R., & Mehmood, M. A. (2020). *Correlation between pollution trends of freshwater bodies and bacterial disease of fish fauna. Fresh water pollution dynamics and remediation* (pp. 51–67). Singapore: Springer.

Dar, G. H., Dar, S. A., Kamili, A. N., Chishti, M. Z., & Ahmad, F. (2016a). Detection and characterization of potentially pathogenic *Aeromonas sobria* isolated from fish *Hypophthalmichthys molitrix* (Cypriniformes: Cyprinidae). *Microbial Pathogenesis*, *91*, 136–140.

Dar, G. H., Kamili, A. N., Chishti, M. Z., Dar, S. A., Tantry, T. A., & Ahmad, F. (2016b). Characterization of *Aeromonas sobria* Isolated from Fish Rohu (*Labeo rohita*) collected from polluted pond. *Journal of Bacteriology & Parasitology*, *7*(3), 1–5. Available from https://doi.org/10.4172/2155-9597.1000273.

Datta, S., Ghosh, D., Saha, D. R., Bhattacharaya, S., & Mazumder, S. (2009). Chronic exposure to low concentration of arsenic is immunotoxic to fish: Role of head kidney macrophages as biomarkers of arsenic toxicity to *Clarias batrachus*. *Aquatic Toxicology*, *92*(2), 86–94.

Dhanakumar, S., Solaraj, G., & Mohanraj, R. (2015). Heavy metal partitioning in sediments and bioaccumulation in commercial fish species of three major reservoirs of river Cauvery delta region, India. *Ecotoxicology and Environmental Safety*, *113*, 145–151.

Doudoroff, P., Katz, M., & Tarzwell, C. M. (1953). Toxicity of some organic insecticides to fish. *Sewage and Industrial Wastes*, *25*, 840–844.

Dutta, H. M., & Meijer, H. J. M. (2003). Sublethal effects of diazinon on the structure of the testis of bluegill, *Lepomis macrochirus*: A microscopic analysis. *Environmental Pollution*, *125*(3), 355–360.

Eddins, D., Cerutti, D., Williams, P., Linney, E., & Levin, E. D. (2010). Zebrafish provide a sensitive model of persisting neurobehavioral effects of developmental chlorpyrifos exposure: Comparison with nicotine and pilocarpine effects and relationship to dopamine deficits. *Neurotoxicology and Teratology*, *32*(1), 99–108.

Eisler, R. (2000). *Handbook of chemical risk assessment: Health hazards to humans, plants, and animals, three volume set* (Vol. 1). CRC Press.

El-Badawi, A. A. (2005). *Effect of lead toxicity on some physiological aspects of Nile tilapia fish,* Oreochromis niloticus. *International conferences of the veterinary research division*. Cairo: NRC.

Essumang, D. K., Togoh, G. K., & Chokky, L. (2009). Pesticide residues in the water and fish (lagoon tilapia) samples from lagoons in Ghana. *Bulletin of the Chemical Society of Ethiopia*, *23*(1), 19–27.

Ewing, R. D. (1999). *Diminishing returns: Salmon decline and pesticides. Funded by the Oregon pesticide education network*. Corvallis, OR: Biotech Research and Consulting. Inc. (55 p).

Fanta, E., Rios, F. S. A., Romão, S., Vianna, A. C. C., & Freiberger, S. (2003). Histopathology of the fish *Corydoras paleatus* contaminated with sublethal levels of organophosphorus in water and food. *Ecotoxicology and Environmental Safety*, *54*(2), 119–130.

Farag, A. M., May, T., Marty, G. D., Easton, M., Harper, D. D., Little, E. E., & Cleveland, L. (2006). The effect of chronic chromium exposure on the health of *Chinook salmon* (*Oncorhynchus tshawytscha*). *Aquatic Toxicology*, *76*(3–4), 246–257.

Ferrari, A., Venturino, A., & de D'Angelo, A. M. P. (2004). Time course of brain cholinesterase inhibition and recovery following acute and subacute azinphosmethyl, parathion and carbaryl exposure in the goldfish (*Carassius auratus*). *Ecotoxicology and Environmental Safety*, *57*(3), 420–425.

Ferrari, A., Venturino, A., & de D'Angelo, A. M. P. (2007). Effects of carbaryl and azinphos methyl on juvenile rainbow trout (*Oncorhynchus mykiss*) detoxifying enzymes. *Pesticide Biochemistry and Physiology*, *88*(2), 134–142.

Figueiredo-Fernandes, A., Ferreira-Cardoso, J. V., Garcia-Santos, S., Monteiro, S. M., Carrola, J., Matos, P., & Fontaínhas-Fernandes, A. (2007). Histopathological changes in liver and gill epithelium of Nile tilapia, *Oreochromis niloticus*, exposed to waterborne copper. *Pesquisa Veterinária Brasileira*, *27*(3), 103–109.

Gainey, L. F., Jr, & Kenyon, J. R. (1990). The effects of reserpine on copper induced cardiac inhibition in Mytilusedulis. *Comparative Biochemistry and Physiology Part C: Comparative Pharmacology*, *95*(2), 177–179.

Ghosh, D., Datta, S., Bhattacharya, S., & Mazumder, S. (2007). Long-term exposure to arsenic affects head kidney and impairs humoral immune responses of *Clarias batrachus*. *Aquatic Toxicology*, *81*(1), 79–89.

Gibson, B. D., Ptacek, C. J., Blowes, D. W., & Daugherty, S. D. (2015). Sediment resuspension under variable geochemical conditions and implications for contaminant release. *Journal of Soils and Sediments*, *15*(7), 1644–1656.

Grosell, M., Blanchard, J., Brix, K. V., & Gerdes, R. (2007). Physiology is pivotal for interactions between salinity and acute copper toxicity to fish and invertebrates. *Aquatic Toxicology*, *84*(2), 162–172.

Haider, S., & Upadhyaya, N. (1985). Effect of commercial formulation of four Organophosphorus insecticides on the ovaries of a freshwater teleost, Mystusvittatus (Bloch)-A histological and histochemical study. *Journal of Environmental Science & Health Part B*, *20*(3), 321–340.

Harguinteguy, C. A., Cirelli, A. F., & Pignata, M. L. (2014). Heavy metal accumulation in leaves of aquatic plant Stuckeniafiliformis and its relationship with sediment and water in the Suquía river (Argentina). *Microchemical Journal*, *114*, 111–118.

Healey, N. (2009). Lead toxicity, vulnerable subpopulations and emergency preparedness. *Radiation Protection Dosimetry*, *134*(3–4), 143–151.

Helfrich, L. A., Weigmann, D. L., Hipkins, P. A., & Stinson, E. R. (2009). *Pesticides and aquatic animals: A guide to reducing impacts on aquatic systems*. Virginia Cooperative Extension.

Hernández, P. P., Moreno, V., Olivari, F. A., & Allende, M. L. (2006). Sub-lethal concentrations of waterborne copper are toxic to lateral line neuromasts in zebrafish (*Danio rerio*). *Hearing Research*, *213*(1–2), 1–10.

Hughes, M. F. (2002). Arsenic toxicity and potential mechanisms of action. *Toxicology Letters*, *133*(1), 1–16.

Jaensson, A., Scott, A. P., Moore, A., Kylin, H., & Olsén, K. H. (2007). Effects of a pyrethroid pesticide on endocrine responses to female odours and reproductive behaviour in male parr of brown trout (*Salmo trutta* L.). *Aquatic Toxicology*, *81*(1), 1–9.

Jenkins, F., Smith, J., Rajanna, B., Shameem, U., Umadevi, K., Sandhya, V., & Madhavi, R. (2003). Effect of sub-lethal concentrations of endosulfan on hematological and serum biochemical parameters in the carp *Cyprinus carpio*. *Bulletin of Environmental Contamination and Toxicology*, *70*(5), 0993–0997.

Jezierska, B., & Witeska, M. (2006). *The metal uptake and accumulation in fish living in polluted waters. Soil and water pollution monitoring, protection and remediation* (pp. 107–114). Dordrecht: Springer.

Jin, Y., Chen, R., Liu, W., & Fu, Z. (2010). Effect of endocrine disrupting chemicals on the transcription of genes related to the innate immune system in the early developmental stage of zebrafish (*Danio rerio*). *Fish & Shellfish Immunology*, *28*(5–6), 854–861.

Kalantzi, I., Papageorgiou, N., Sevastou, K., Black, K. D., Pergantis, S. A., & Karakassis, I. (2014). Metals in benthic macrofauna and biogeochemical factors affecting their trophic transfer to wild fish around fish farm cages. *Science of the Total Environment*, *470*, 742–753.

Kalay, M., Ay, Ö., & Canli, M. (1999). Heavy metal concentrations in fish tissues from the Northeast Mediterranean Sea. *Bulletin of Environmental Contamination and Toxicology*, *63*(5), 673–681.

Kamunde, C., & Wood, C. M. (2003). The influence of ration size on copper homeostasis during sublethal dietary copper exposure in juvenile rainbow trout, *Oncorhynchus mykiss*. *Aquatic Toxicology*, *62*(3), 235–254.

Kanazawa, J. (1975). Uptake and excretion of organophosphorus and carbamate insecticides by fresh water fish, motsugo, *Pseudorasbora parva*. *Bulletin of Environmental Contamination and Toxicology*, *14*(3), 346–352.

Kaur, R., & Dua, A. (2015). Scales of freshwater fish *Labeo rohita* as bioindicators of water pollution in Tung Dhab Drain, Amritsar, Punjab, India. *Journal of Toxicology and Environmental Health, Part A*, *78*(6), 388–396.

Kavitha, P., & Rao, J. V. (2009). Sub-lethal effects of profenofos on tissue-specific antioxidative responses in a Euryhyaline fish, *Oreochromis mossambicus*. *Ecotoxicology and Environmental Safety*, *72*(6), 1727–1733.

Khan, F. R., Irving, J. R., Bury, N. R., & Hogstrand, C. (2011). Differential tolerance of two Gammaruspulex populations transplanted from different metallogenic regions to a polymetal gradient. *Aquatic Toxicology*, *102*(1–2), 95–103.

Khan, M. Z., & Law, F. C. (2005). Adverse effects of pesticides and related chemicals on enzyme and hormone systems of fish, amphibians and reptiles: A review. *Proceedings of the Pakistan Academy of Sciences*, *42*(4), 315–323.

Kingsbury, P. D., & Kreutzweiser, D. P. (1980). *Environmental impact assessment of a semi-operational permethrin application*. Canadian Forestry Service, Forest Pest Management Institute, Department of the Environment.

Kori-Siakpere, O., & Ubogu, E. O. (2008). Sublethal haematological effects of zinc on the freshwater fish, Heteroclarias sp.(Osteichthyes: Clariidae). *African Journal of Biotechnology*, *7*(12), 2068–2073.

Kotsanis, N., & Iliopoulou-Georgudaki, J. (1999). Arsenic induced liver hyperplasia and kidney fibrosis in rainbow trout (Oncorhynchus mykiss) by microinjection technique: a sensitive animal bioassay for environmental metal-toxicity. *Bulletin of environmental contamination and toxicology*, *62*(2), 169–178.

Laitinen, M., & Valtonen, T. (1995). Cardiovascular, ventilatory and haematological responses of brown trout (*Salmo trutta* L.), to the combined effects of acidity and aluminium in humic water at winter temperatures. *Aquatic Toxicology*, *31*(2), 99–112.

Li, Z. H., Li, P., & Randak, T. (2011). Evaluating the toxicity of environmental concentrations of waterborne chromium (VI) to a model teleost, *Oncorhynchus mykiss*: A comparative study of in vivo and in vitro. *Comparative Biochemistry and Physiology Part C: Toxicology & Pharmacology*, *153*(4), 402–407.

Liu, J., Lu, G., Zhang, F., Nkoom, M., Yan, Z., & Wu, D. (2018a). Polybrominated diphenyl ethers (PBDEs) in a large, highly polluted freshwater lake, China: Occurrence, fate, and risk assessment. *International Journal of Environmental Research and Public Health*, *15*(7), 1529.

Liu, W., Zeng, Z., Chen, A., Zeng, G., Xiao, R., Guo, Z., Yi, F., Huang, Z., He, K., & Hu, L. (2018b). Toxicity effects of silver nanoparticles on the freshwater bivalve *Corbicula fluminea*. *Journal of Environmental Chemical Engineering*, *6*(4), 4236–4244.

Lushchak, V., Kubrak, O. I., Lozinsky, O. V., Storey, J. M., Storey, K. B., & Lushchak, V. I. (2009). Chromium (III) induces oxidative stress in goldfish liver and kidney. *Aquatic Toxicology*, *93*(1), 45–52.

Magalhães, M. C., Costa, V., Menezes, G. M., Pinho, M. R., Santos, R. S., & Monteiro, L. R. (2007). Intra- and inter-specific variability in total and methylmercury bioaccumulation by eight marine fish species from the Azores. *Marine Pollution Bulletin*, *54*(10), 1654–1662.

Mallatt, J. (1985). Fish gill structural changes induced by toxicants and other irritants: A statistical review. *Canadian Journal of Fisheries and Aquatic Sciences*, *42*(4), 630–648.

Mathur, S. C. (1999). Future of Indian pesticides industry in next millennium. *Pesticide Information*, *22*(4), 9–23.

Matos, P., Fontaı, A., Peixoto, F., Carrola, J., & Rocha, E. (2007). Biochemical and histological hepatic changes of Nile tilapia *Oreochromis niloticus* exposed to carbaryl. *Pesticide Biochemistry and Physiology*, *89*(1), 73–80.

McCoy, C. P., O'Hara, T. M., Bennett, L. W., Boyle, C. R., & Lynn, B. C. (1995). Liver and kidney concentrations of zinc, copper and cadmium in channel catfish (*Ictalurus punctatus*): Variations due to size, season and health status. *Veterinary and Human Toxicology*, *37*(1), 11.

Michael, P. (1984). *Ecological methods for field and laboratory investigations*. Tata McGraw-Hill Publishing Co., Ltd.

Michelle, D. (2009). *Research associated, fisheries and wildlife. Virginia Cooperative Extension Virginia Experiment Station.*

Mohan, M. R. (2000). Malathion induced changes in the ovary of freshwater fish, *Glossogobius giuris* (Ham). *Pollution Research*, *19*(1), 73–75.

Monteiro, S. M., dos Santos, N. M., Calejo, M., Fontainhas-Fernandes, A., & Sousa, M. (2009b). Copper toxicity in gills of the teleost fish, *Oreochromis niloticus*: Effects in apoptosis induction and cell proliferation. *Aquatic Toxicology*, *94*(3), 219–228.

Monteiro, S. M., dos Santos, N. M., Calejo, M., Fontainhas-Fernandes, A., & Sousa, M. (2009). Copper toxicity in gills of the teleost fish, Oreochromis niloticus: effects in apoptosis induction and cell proliferation. *Aquatic toxicology*, *94*(3), 219–228.

Monteiro, D. A., Rantin, F. T., & Kalinin, A. L. (2009a). The effects of selenium on oxidative stress biomarkers in the freshwater characid fish matrinxã, Bryconcephalus (Günther, 1869) exposed to organophosphate insecticide Folisuper 600 BR®(methyl parathion). *Comparative Biochemistry and Physiology Part C: Toxicology & Pharmacology*, *149*(1), 40–49.

Moore, A., & Waring, C. P. (2001). The effects of a synthetic pyrethroid pesticide on some aspects of reproduction in Atlantic salmon (*Salmo salar* L.). *Aquatic Toxicology*, *52*(1), 1–12.

Moreau, R., Thabut, D., Massard, J., Gangloff, A., Carbonell, N., Francoz, C., Nguyen-Khac, E., Duhamel, C., Lebrec, D., & Poynard, T. (2007). Model for end-stage liver disease score and systemic inflammatory response are major prognostic factors in patients with cirrhosis and acute functional renal failure. *Hepatology*, *46*(6), 1872–1882.

Murray, L., Cooper, P. J., Wilson, A., & Romaniuk, H. (2003). Controlled trial of the short-and long-term effect of psychological treatment of post-partum depression: 2. Impact on the mother-child relationship and child outcome. *The British Journal of Psychiatry*, *182*(5), 420–427.

Murthy, K. S., Kiran, B. R., & Venkateshwarlu, M. (2013). A review on toxicity of pesticides in fish. *International Journal of Open Scientific Research*, *1*(1), 15–36.

Murugan, S. S., Karuppasamy, R., Poongodi, K., & Puvaneswari, S. (2008). Bioaccumulation pattern of zinc in freshwater fish *Channa punctatus* (Bloch.) after chronic exposure. *Turkish Journal of Fisheries and Aquatic Sciences*, *8*(1), 55–59.

Nebbia, C. (2001). Biotransformation enzymes as determinants of xenobiotic toxicity in domestic animals. *The Veterinary Journal*, *161*(3), 238–252.

Niyogi, S., & Wood, C. M. (2006). Interaction between dietary calcium supplementation and chronic waterborne zinc exposure in juvenile rainbow trout, *Oncorhynchus mykiss*. *Comparative Biochemistry and Physiology Part C: Toxicology & Pharmacology*, *143*(1), 94–102.

Olaifa, F. E., Olaifa, A. K., & Onwude, T. E. (2004). Lethal and sub-lethal effects of copper to the African catfish (*Clariasgariepinus*) juveniles. *African Journal of Biomedical Research*, *7*(2), 65–70.

Olsvik, P. A., Berntssen, M. H., & Søfteland, L. (2015). Modifying effects of vitamin E on chlorpyrifos toxicity in Atlantic salmon. *PLoS One*, *10*(3), e0119250.

Olsvik, P. A., Berntssen, M. H., & Søfteland, L. (2017). In vitro toxicity of pirimiphos-methyl in Atlantic salmon hepatocytes. *Toxicology in Vitro*, *39*, 1–14.

Olsvik, P. A., Berntssen, M. H. G., Søfteland, L., & Sanden, M. (2019). Transcriptional effects of dietary chlorpyrifos-methyl exposure in Atlantic salmon (*Salmo salar*) brain and liver. *Comparative Biochemistry and Physiology Part D: Genomics and Proteomics*, *29*, 43–54.

Oruç, E. Ö. (2010). Oxidative stress, steroid hormone concentrations and acetylcholinesterase activity in *Oreochromis niloticus* exposed to chlorpyrifos. *Pesticide Biochemistry and Physiology*, *96*(3), 160–166.

Pedlar, R. M., Ptashynski, M. D., Evans, R., & Klaverkamp, J. F. (2002). Toxicological effects of dietary arsenic exposure in lake whitefish (*Coregonus clupeaformis*). *Aquatic Toxicology*, *57*(3), 167–189.

Perez-Cid, B. P., Boia, C., Pombo, L., & Rebelo, E. (2001). Determination of trace metals in fish species of the Ria de Aveiro (Portugal) by electrothermal atomic absorption spectrometry. *Food Chemistry*, *75*(1), 93–100.

Pfeiffer, M., Batbayar, G., Hofmann, J., Siegfried, K., Karthe, D., & Hahn-Tomer, S. (2015). Investigating arsenic (As) occurrence and sources in ground, surface, waste and drinking water in northern Mongolia. *Environmental Earth Sciences*, *73*(2), 649–662.

Pinto, M. F., Louro, H., Costa, P. M., Caeiro, S., & Silva, M. J. (2015). Exploring the potential interference of estuarine sediment contaminants with the DNA repair capacity of human hepatoma cells. *Journal of Toxicology and Environmental Health, Part A*, *78*(9), 559–570.

Quesada, S., Tena, A., Guillén, D., Ginebreda, A., Vericat, D., Martínez, E., Navarro-Ortega, A., Batalla, R. J., & Barceló, D. (2014). Dynamics of suspended sediment borne persistent organic pollutants in a large regulated Mediterranean river (Ebro, NE Spain). *Science of the Total Environment*, *473*, 381–390.

Radhakrishnaiah, K., Venkataramana, P., Suresh, A., & Sivaramakrishna, B. (1992). Effects of lethal and sublethal concentrations of copper on glycolysis in liver and muscle of the freshwater teleost, *Labeo rohita* (Hamilton). *Journal of Environmental Biology*, *13*(1), 63–68.

Radi, A. A., & Matkovics, B. (1988). Effects of metal ions on the antioxidant enzyme activities, protein contents and lipid peroxidation of carp tissues. *Comparative Biochemistry and Physiology Part C, Comparative Pharmacology and Toxicology*, *90*(1), 69–72.

Ribeiro, C. O., Neto, F. F., Mela, M., Silva, P. H., Randi, M. A. F., Rabitto, I. S., Costa, J. A., & Pelletier, E. (2006). Hematological findings in neotropical fish Hoplias malabaricus exposed to subchronic and dietary doses of methylmercury, inorganic lead, and tributyltin chloride. *Environmental Research*, *101*(1), 74–80.

Richendrfer, H., Pelkowski, S. D., Colwill, R. M., & Créton, R. (2012). Developmental sub-chronic exposure to chlorpyrifos reduces anxiety-related behavior in zebrafish larvae. *Neurotoxicology and Teratology*, *34* (4), 458–465.

Rishi, K. K., & Grewal, S. (1995). Chromosome aberration test for the insecticide, dichlorvos, on fish chromosomes. *Mutation Research/Genetic Toxicology*, *344*(1–2), 1–4.

Rossi, N., & Jamet, J. L. (2008). In situ heavy metals (copper, lead and cadmium) in different plankton compartments and suspended particulate matter in two coupled Mediterranean coastal ecosystems (Toulon Bay, France). *Marine Pollution Bulletin*, *56*(11), 1862–1870.

Saaristo, M., Brodin, T., Balshine, S., Bertram, M. G., Brooks, B. W., Ehlman, S. M., McCallum, E. S., Sih, A., Sundin, J., Wong, B. B., & Arnold, K. E. (2018). Direct and indirect effects of chemical contaminants on the behaviour, ecology and evolution of wildlife. *Proceedings of the Royal Society B: Biological Sciences*, *285*(1885), 20181297.

Sabra, F. S., & Mehana, E. S. E. D. (2015). Pesticides toxicity in fish with particular reference to insecticides. *Asian Journal of Agriculture and Food Sciences*, *3*(1).

Said, A. A., Sheik-Bahae, M., Hagan, D. J., Wei, T. H., Wang, J., Young, J., & Van Stryland, E. W. (1992). Determination of bound-electronic and free-carrier nonlinearities in ZnSe, GaAs, CdTe, and ZnTe. *Journal of the Optical Society of America B*, *9*(3), 405–414.

Sanchez, W., Palluel, O., Meunier, L., Coquery, M., Porcher, J. M., & Ait-Aissa, S. (2005). Copper-induced oxidative stress in three-spined stickleback: Relationship with hepatic metal levels. *Environmental Toxicology and Pharmacology*, *19*(1), 177–183.

Sanden, M., Olsvik, P. A., Søfteland, L., Rasinger, J. D., Rosenlund, G., Garlito, B., Ibáñez, M., & Berntssen, M. H. (2018). Dietary pesticide chlorpyrifos-methyl affects arachidonic acid metabolism including phospholipid remodeling in Atlantic salmon (*Salmo salar* L.). *Aquaculture (Amsterdam, Netherlands)*, *484*, 1–12.

Satyavardhan, K. (2013). A comparative toxicity evaluation and behavioral observations of fresh water fishes to Fenvalerate^{TM}. *Middle East Journal of Scientific Research*, *13*(2), 133–136.

Schimmel, S. C., Garnas, R. L., Patrick, J. M., Jr, & Moore, J. C. (1983). Acute toxicity, bioconcentration, and persistence of AC 222,705, benthiocarb, chlorpyrifos, fenvalerate, methyl parathion, and permethrin in the estuarine environment. *Journal of Agricultural and Food Chemistry*, *31*(1), 104–113.

Schnick, R. A., Meyer, F. P., & Gray, D. L. (1986). A guide to approved chemicals in fish production and fishery resource management. *MP-University of Arkansas, Cooperative Extension Service (USA)*.

Scott, G. R., & Sloman, K. A. (2004). The effects of environmental pollutants on complex fish behaviour: Integrating behavioural and physiological indicators of toxicity. *Aquatic Toxicology*, *68*(4), 369–392.

Selda, O. T., & Nurşah, A. (2012). Relationship of heavy metals in water, sediment and tissue with total length, weight and seasons of *Cyprinus carpio* L., 1758 from Isikli Lake (Turkey). *Pakistan Journal of Zoology*, *44*, 1405–1416.

Sepici-Dinçel, A., Benli, A. Ç. K., Selvi, M., Sarıkaya, R., Şahin, D., Özkul, I. A., & Erkoç, F. (2009). Sublethalcyfluthrin toxicity to carp (*Cyprinus carpio* L.) fingerlings: Biochemical, hematological, histopathological alterations. *Ecotoxicology and Environmental Safety*, *72*(5), 1433–1439.

Sfakianakis, D. G., Renieri, E., Kentouri, M., & Tsatsakis, A. M. (2015). Effect of heavy metals on fish larvae deformities: A review. *Environmental Research*, *137*, 246–255.

Shah, S. L., & Altindağ, A. (2005). Alterations in the immunological parameters of Tench (Tincatinca L. 1758) after acute and chronic exposure to lethal and sublethal treatments with mercury, cadmium and lead. *Turkish Journal of Veterinary and Animal Sciences*, *29*(5), 1163–1168.

Sharbidre, A. A., Metkari, V., & kaPatode, P. (2011a). Effect of diazinon on acetylcholinesterase activity and lipid. *Research Journal of Environmental Toxicology*, *5*(2), 152–161.

Sidoumou, Z. (1991). *Water quality of the Mauritanian coast: Study of trace metals in two bivalve molluscs Venus verrucosa and Donaxrugosus*. Doctoral dissertation, Nice.

Sharbidre, A. A., Metkari, V., & Patode, P. (2011b). Effect of methyl parathion and chlorpyrifos on certain biomarkers in various tissues of guppy fish, *Poecilia reticulata*. *Pesticide Biochemistry and Physiology*, *101*(2), 132–141.

Silva, V., Marques, C. R., Campos, I., Vidal, T., Keizer, J. J., Gonçalves, F., & Abrantes, N. (2018). Combined effect of copper sulfate and water temperature on key freshwater trophic levels—approaching potential climatic change scenarios. *Ecotoxicology and Environmental Safety*, *148*, 384–392.

Sorensen, E. M. (1991). *Metal poisoning in fish*. CRC Press.

Sridhara, G., Hill, E., Muppaneni, D., Pollock, L., & Vijay-Shanker, K. (2010). Towards automatically generating summary comments for java methods, September *Proceedings of the IEEE/ACM International Conference on Automated Software Engineering*, 43–52.

Štrbac, S., Kašanin-Grubin, M., Jovančićević, B., & Simonović, P. (2015). Bioaccumulation of heavy metals and microelements in silver bream (*Brama brama* L.), northern pike (*Esox lucius* L.), sterlet (*Acipenser ruthenus* L.), and common carp (*Cyprinus carpio* L.) from Tisza River, Serbia. *Journal of Toxicology and Environmental Health, Part A*, *78*(11), 663–665.

Tchounwou, P. B., Yedjou, C. G., Patlolla, A. K., & Sutton, D. J. (2012). *Heavy metal toxicity and the environment. Molecular, clinical and environmental toxicology* (pp. 133–164). Basel: Springer.

Tiwari, S., & Singh, A. (2004). Toxic and sub-lethal effects of oleandrin on biochemical parameters of fresh water air breathing murrel, *Channa punctatus* (Bloch.). *Indian Journal of Experimental Biology*, *42*(4), 413–418.

Todd, N. E., & Van Leeuwen, M. (2002). Effects of Sevin (carbaryl insecticide) on early life stages of zebrafish (*Danio rerio*). *Ecotoxicology and Environmental Safety*, *53*(2), 267–272.

Topal, A., Atamanalp, M., Uçar, A., Oruç, E., Kocaman, E. M., Sulukan, E., Akdemir, F., Beydemir, Ş., Kılınç, N., Erdoğan, O., & Ceyhun, S. B. (2015). Effects of glyphosate on juvenile rainbow trout (*Oncorhynchus mykiss*): Transcriptional and enzymatic analyses of antioxidant defence system, histopathological liver damage and swimming performance. *Ecotoxicology and Environmental Safety*, *111*, 206–214.

Toth, J. F., Jr, & Brown, R. B. (1997). Racial and gender meanings of why people participate in recreational fishing. *Leisure Sciences*, *19*(2), 129–146.

Üner, N., Oruç, E. Ö., Sevgiler, Y., Şahin, N., Durmaz, H., & Usta, D. (2006). Effects of diazinon on acetylcholinesterase activity and lipid peroxidation in the brain of *Oreochromis niloticus*. *Environmental Toxicology and Pharmacology*, *21*(3), 241–245.

Van der Oost, R., Beyer, J., & Vermeulen, N. P. (2003). Fish bioaccumulation and biomarkers in environmental risk assessment: A review. *Environmental Toxicology and Pharmacology*, *13*(2), 57–149.

Varanka, Z., Rojik, I., Varanka, I., Nemcsók, J., & Ábrahám, M. (2001). Biochemical and morphological changes in carp (*Cyprinus carpio* L.) liver following exposure to copper sulfateand tannic acid. *Comparative Biochemistry and Physiology Part C: Toxicology & Pharmacology*, *128*(3), 467–477.

Velisek, J., Svobodova, Z., & Machova, J. (2009). Effects of bifenthrin on some haematological, biochemical and histopathological parameters of common carp (*Cyprinus carpio* L.). *Fish Physiology and Biochemistry*, *35*(4), 583–590.

Velmurugan, B., Selvanayagam, M., Cengiz, E. I., & Unlu, E. (2009). Histopathological changes in the gill and liver tissues of freshwater fish, *Cirrhinus mrigala* exposed to dichlorvos. *Brazilian Archives of Biology and Technology*, *52*(5), 1291–1296.

Vera-Candioti, J., Soloneski, S., & Larramendy, M. L. (2011). Acute toxicity of chromium on *Cnesterodon decemmaculatus* (pisces: poeciliidae). *Theoria*, *20*, 81–88.

Viant, M. R., Pincetich, C. A., & Tjeerdema, R. S. (2006). Metabolic effects of dinoseb, diazinon and esfenvalerate in eyed eggs and alevins of Chinook salmon (*Oncorhynchus tshawytscha*) determined by 1H NMR metabolomics. *Aquatic Toxicology*, *77*(4), 359–371.

Vinodhini, R., & Narayanan, M. (2008). Bioaccumulation of heavy metals in organs of fresh water fish *Cyprinus carpio* (Common carp). *International Journal of Environmental Science & Technology*, *5*(2), 179–182.

Vuorinen, P. J., Keinänen, M., Peuranen, S., & Tigerstedt, C. (2003). Reproduction, blood and plasma parameters and gill histology of vendace (*Coregonus albula* L.) in long-term exposure to acidity and aluminum. *Ecotoxicology and Environmental Safety*, *54*(3), 255–276.

Vutukuru, S. S., Suma, C. H., Madhavi, K. R., Juveria, J., Pauleena, J. S., Rao, J. V., & Anjaneyulu, Y. (2005). Studies on the development of potential biomarkers for rapid assessment of copper toxicity to freshwater fish using Esomus danricus as model. *International journal of environmental research and public health*, *2*(1), 63–73.

Walton, R. C., McCrohan, C. R., Livens, F. R., & White, K. N. (2009). Tissue accumulation of aluminium is not a predictor of toxicity in the freshwater snail, *Lymnaea stagnalis*. *Environmental Pollution*, *157*(7), 2142–2146.

Wang, C., Lu, G., Cui, J., & Wang, P. (2009). Sublethal effects of pesticide mixtures on selected biomarkers of *Carassius auratus*. *Environmental Toxicology and Pharmacology*, *28*(3), 414–419.

Wang, X., Shen, M., Zhou, J., & Jin, Y. (2019). Chlorpyrifos disturbs hepatic metabolism associated with oxidative stress and gut microbiota dysbiosis in adult zebrafish. *Comparative Biochemistry and Physiology Part C: Toxicology & Pharmacology*, *216*, 19–28.

Waring, C. P., Brown, J. A., Collins, J. E., & Prunet, P. (1996). Plasma prolactin, cortisol, and thyroid responses of the brown trout (*Salmo trutta*) exposed to lethal and sublethal aluminium in acidic soft waters. *General and Comparative Endocrinology*, *102*(3), 377–385.

Weill, D., Benden, C., Corris, P. A., Dark, J. H., Davis, R. D., Keshavjee, S., Lederer, D. J., Mulligan, M. J., Patterson, G. A., Singer, L. G., & Snell, G. I. (2015). A consensus document for the selection of lung transplant candidates: 2014—an update from the Pulmonary Transplantation Council of the International Society for Heart and Lung Transplantation. *The Journal of Heart and Lung Transplantation*, *34*(1), 1–15.

Werimo, K., Bergwerff, A. A., & Seinen, W. (2009). Residue levels of organochlorines and organophosphates in water, fish and sediments from Lake Victoria-Kenyan portion. *Aquatic Ecosystem Health & Management*, *12*(3), 337–341.

World Health Organization. (1999). *The world health report: 1999: Making a difference*. World Health Organization.

Xu, C. T. W., Lou, C. H., & Zaho, M. (2010). Enantioo-selective separation and Zebra fish embryo toxicity of insecticides betacypermethrin. *Journal of Environmental Sciences*, *22*(5), 738–742.

Yang, R., Yao, T., Xu, B., Jiang, G., & Xin, X. (2007). Accumulation features of organochlorine pesticides and heavy metals in fish from high mountain lakes and Lhasa River in the Tibetan Plateau. *Environment International*, *33*(2), 151–156.

Yildirim, Y., Gonulalan, Z., Narin, I., & Soylak, M. (2009). Evaluation of trace heavy metal levels of some fish species sold at retail in Kayseri, Turkey. *Environmental Monitoring and Assessment*, *149*(1–4), 223–228.

Yin, Q., & Wang, W. X. (2018). Uniquely high turnover of nickel in contaminated oysters Crassostreahongkongensis: Biokinetics and subcellular distribution. *Aquatic Toxicology*, *194*, 159–166.

Zala, S. M., & Penn, D. J. (2004). Abnormal behaviours induced by chemical pollution: A review of the evidence and new challenges. *Animal Behaviour*, *68*, 649–664.

Zelikoff, J. T., Raymond, A., Carlson, E., Li, Y., Beaman, J. R., & Anderson, M. (2000). Biomarkers of immunotoxicity in fish: From the lab to the ocean. *Toxicology Letters*, *112*, 325–331.

Zhang, Z., Liu, Q., Cai, J., Yang, J., Shen, Q., & Xu, S. (2017). Chlorpyrifos exposure in common carp (*Cyprinus carpio* L.) leads to oxidative stress and immune responses. *Fish & Shellfish Immunology*, *67*, 604–611.

CHAPTER

Pesticide toxicity and bacterial diseases in fishes

4

Afrozah Hassan, Shabana Gulzar, Hanan Javid and Irshad A. Nawchoo
Plant Reproductive Biology, Genetic Diversity and Phytochemistry Research Laboratory, Department of Botany, University of Kashmir, Srinagar, India

4.1 Introduction

Pollution is the release of harmful contaminants into the environment that have adverse effects on living creatures or the air, water, and land. Pollutants can take many forms, including chemical substances, energy, heat, light, or noise (Özkara et al., 2016), and comprise a range of chemicals, pesticides, metals, and many other organic compounds (Jabeen et al., 2012; Meitei et al., 2004). Pesticides are substances used to exterminate, repel, or avoid pests (Quackenbush et al., 2006). These pollutants affect water quality, which then affects a number of aquatic organisms (Donohue et al., 2005). When water quality is modified drastically, it leads to mortality of aquatic organisms (Sabae et al., 2014; Sarwar et al., 2007). Pesticides consist of compounds categorized as insecticides, rodenticides (anticoagulants), herbicides [paraquat, diquat, 2,4 dichlorophenoxyacetic acid (2,4-D)], fumigants (ethylene dibromide, methyl bromide), and fungicides (captan, dithiocarbamates) (Ellenhorn & Barceloux, 1988).

Contamination of water surfaces by pesticides represents a significant issue on local, national, and multinational levels and has been well documented worldwide (Cerejeira et al., 2003; Huber et al., 2000; Planas et al., 1997; U.S. Geological Survey, 1999). Pesticides reach bodies of water by different routes, including direct spillover, leaching, unregulated disposal of trash, washing of equipment, and others. Pesticides also result from agriculture drainage to surface water (Richards & Baker, 1993). The significant classes of pesticides generally used for agricultural purposes include carbamates, organophosphates, organochlorines, pyrethroids, neonicotinoids, and triazoles (Sabra & Mehana, 2015; Srivastava et al., 2016). In addition to their advantages, these chemicals have countless disadvantages. Pesticides destroy pests and thus have a significant role in food production, but increased use has led to adverse consequences related to human health and natural ecosystems, including aquatic bodies (Aktar et al., 2009; Forget, 1993; Igbedioh, 1991). Bioaccumulation of these pesticides poses a severe threat to fish survival through disturbing the interactions between organisms and biodiversity (Abedi et al., 2013; Morel et al., 1998; Xie et al., 1996).

4.2 Fish: an important resource

In 1992, the Convention of Biological Diversity stated that the preservation of biodiversity, including the biodiversity of water bodies, necessitates the conservation of all significant types of

Bacterial Fish Diseases. DOI: https://doi.org/10.1016/B978-0-323-85624-9.00020-8

ecosystems. This also includes the thoughtful management of ecosystems that are not a part of protected areas. India has vast and unique natural water resources, ranking ninth in freshwater megabiodiversity among the countries of the world, with 2200 fish species (Mittermeier & Mittermeier, 1997). Fishing is a chief economic activity that offers numerous benefits in India and worldwide:

1. Provides a major source of nutrition.
2. Is a direct source of livelihoods for fishers and fish farmers.
3. Indirectly creates employment for those involved in building vessels, manufacturing equipment, and selling/marketing fish products.

Hence, we can say that fishing is an important activity involving several major areas of human concern: (1) ecological (related to populations and aquatic ecosystems); (2) socioeconomic (those working in the sector) and technological (boats, motors, tools, etc.); and (3) political and administrative (Tursi et al., 2015).

Developing countries have 94% of the total freshwater fisheries of the world (FAO, 2007) and fisheries in these countries provide food and employment to millions of people, playing a significant role in overall economic well-being (The World Fish Center, 2002). Around 55.3 million people in the Mekong River basin alone are dependent on freshwater fish for nutrition and supporting their incomes; the estimated average fish intake is approximately 55.6 kg/person/year (Baran et al., 2007). The Food and Agriculture Organization in 2007 estimated that freshwater fish contribute to more than 6% of the world's annual animal protein supplies for humans. In developing countries, fish exports are also a significant source of foreign revenue. Thus fisheries not only add to national income in these countries, but also offer many employment opportunities, therefore helping to eradicate poverty (FAO, 2006).

4.3 Fish as indicators of pollution

Fish play a significant role in monitoring pollution as they are early indicators of anthropogenic stress on natural ecosystems, mainly because of their life-cycle traits. They can act as indicators for a variety of disturbances, including acidification, diseases, and parasites. Fish have the ability to react directly to various environmental stressors such as toxins, climate change, and flow regime (Dudgeon et al., 2006). Fish can bioaccumulate noxious compounds because of their large body size, rapid growth rates, habitat choices, and trophic level (Holmlund & Hammer, 1999). For example, *Apteronotus albifrons*, a tropical fish of South America, emits a wave of electric signals from its electric organ that are steady under ambient conditions, but vary significantly when the water is polluted. It was reported that *A. albifrons* acted as a biological indicator to detect the presence of potassium cyanide in water. This technique could prove significant for tracking the level of pollution in water bodies (Thomas et al., 1996).

4.4 The impact of pesticides on fish

The National Bureau of Fish Genetic Resources in India surveyed freshwater fish for protection and listed some species under the threatened categor (Anonymous, 1992; Lakra & Sarkar, 2007).

However, fish diversity in the country is continuously declining due to toxic chemicals. This might be because of secondary causes that expose fish and other aquatic organisms to pesticides. In the long run, this leads to toxicity, including the consumption of those fish poisoned by pesticides (Kingsbury & Kreutzweiser, 1980; Schnick et al., 1986). The heedless and destructive dumping of excess chemical concentrates, accidental spillages lead to sporadic expulsions of pollutants that find their way into streams, lakes, and rivers. This accounts, primarily or secondarily, for a major fraction of mortality in fishes. For example, in 1967, it was found that organochlorine pesticides killed about 13% of fish in England and Wales and around 25% in Scotland (Anon, 1971). A concentration of DDT surpassing 3 mg/kg in lakes led to high mortality of trout (Burdick et al., 1964). Female brook trout, when fed with DDT-contaminated pellets, produced eggs that experienced a high rate of mortality (Macek, 1968).

The use of pesticides was a major cause of decreases in the practice of rice-fish culture in Central Thailand during the 1970s (Spiller, 1985), Indonesia during 1968–69 (Koesoemadinata, 1980), Malaysia during 1972 (Lim & Ong, 1985), and Vietnam (Vincke, 1979). Concentrations of these pollutants greater than permitted levels have been found to lead to mortality of all organisms present in those aquatic bodies, mostly fish. However, even at lower concentrations, these pollutants can bioaccumulate, passing all the way through the food chain and ultimately to humans (Abedi et al., 2013; Morel et al., 1998; Xie et al., 1996).

Food organisms present in aquatic ecosystems are also fundamentally diminished by the use of pesticides, reducing fish survival, as they are dependent on these food organisms (Helfrich et al., 2009). The migratory behavior of fish changes as a result of pesticides (Nagaraju et al., 2011), showing that their life cycle is disturbed. Pesticides may alter the capability of salmonid fish to transfer from freshwater to seawater. Other important indicators of environmental stress are behavioral modifications that may influence existence (Byrne & O'halloran, 2001), predominantly in nonmigratory species. The triazole fungicide propiconazole is reported to be a major cause of such behavioral changes (Gill et al., 1990; Li et al., 2011; Srivastava et al., 2016).

Under stress conditions, organophosphate induces hyperglycemia in fish and an upsurge in cortisol levels (Borges et al., 2007; Das & Mukherjee, 2003; Jee et al., 2005). *Cyprinus carpio* and *Cnesterodon decemmaculatus* showed a reduced level of acetylcholinesterase (AChE), as they were highly sensitive to pollutants (De la Torre et al., 2002). Since fish are explicitly sensitive to water pollution, certain physiological and biochemical processes may be affected by pollutants such as insecticides. Banaee (2013) studied pesticide toxicity in fish and showed that, at nonlethal levels, pesticides can have specific effects such as oxidative damage, inhibition of AChE activity, histopathological changes, developmental changes, mutagenesis, and carcinogenicity. Thus fish collected from polluted waters do not retain freshness, as compared to fish from nonpolluted waters.

In India, the production of pesticides increased during 1958–98 from 5000 metric tons to 102,240 metric tons. In 1996–97 the demand for pesticides in terms of value was about Rs. 22 billion (US$ 0.5 billion), which on the world market accounted for about 2% of usage (Aktar et al., 2009). As already mentioned, lethal or sublethal effects in fish may be induced by pesticides in the environment (Mathur & Tannan, 1999). Pesticides may show their effects in different ways: they may act directly and cause death, or indirectly cause malnutrition from the destruction of food organisms, or affect growth rate, reproduction, and behavior, with indication of tissue damage. The affected fish are unable to compete with other fish and therefore become vulnerable to predators. Therefore the capability to withstand normal stresses, including temperature variation, temporary malnutrition, and reproduction, is weakened. At younger stages of the life cycle, survival is negligible for most of the species, and at this stage fish are especially vulnerable to some pesticides; therefore it is disastrous if the survival rate is further reduced (Holden, 1973).

The toxicity of a pesticide is determined by the degree to which a toxin produces adverse effects. Brief exposure to a less toxic chemical may have a negligible effect on fish, but if exposure to the same chemical is longer lasting, it may prove detrimental (Arbuckle & Sever, 1998; Barone et al., 2000; Dar et al., 2016a,b). Pesticides in different concentrations are toxic to different fish species. Hence LC50 (lethal concentration required to kill 50% of a population) values of pesticides in fish species also differ (Table 4.1). When different aquatic organisms are exposed to a pesticide, their survival is dependent upon the bioconcentration, biomagnification, bioavailability, and persistence of that pesticide in the environment. Bioavailability can be described as the degree and rate at which a substance (pesticide) is absorbed into a living system.

Some insecticides easily break down after utility. However, some do not break down easily. Rather, they bind firmly to soil debris suspended in the water bodies and therefore become less bioavailable (Maurya et al., 2019). Bioconcentration is the collection and accumulation of a chemical on or in an organism at a level that is more than that present on land or water. The persistence of pesticides in the environment or accumulation of these pesticides drastically affects water body function (Dar et al., 2020; Maurya et al., 2019). Sublethal amounts of pesticides can diminish population abundance and also the rate of adult survival (Gupta, 2004; Hoppin et al., 2002; Kamel & Hoppin, 2004).

Table 4.1 Presence of pesticides in different fish species.

Pesticide	Name of the fish	Reference
Malathion	*Labeo rohita*	Thenmozhi et al. (2011)
Elsan	*Channa punctatus*	Sambasiva Rao et al. (2009)
Endosulfan	*Channa striatus*	Ganeshwade et al. (2012)
Acephate	*Fathead minnow*	Waynon et al. (1980)
Alaclor	*Rainbow trout*	Waynon et al. (1980)
Methyl parathin	*Catla catla*	Ilavazhahan et al. (2010)
DDT	*Rainbow trout*	Waynon et al. (1980)
Dimethoate	*Heteropneustes Fossilis*	Pandey et al. (2009)
Cypermethrin	*Colisa fasciatus*	Singh et al. (2010)
Carbofuran	*Yellow perch*	Waynon et al. (1980)
Biosal	*Cyprinus carpio*	Sial et al. (2009)
Karate	*Cyprinus carpio*	Bibi et al. (2014)
Metasystox	*Nemacheilus botia*	Nikam et al. (2011)
Diazinon	*Anabas testudineus*	Rahman et al. (2002)

4.5 Mitigation of the impact of pesticides

Collection and interpretation of pesticide data are difficult, especially in developing countries, which face difficulties in organic chemical analysis because of improper facilities, contaminated reagents, and lack of money. Therefore novel techniques including immunoassay procedures can be used for the detection of specific pesticides. These techniques are cost-effective and highly reliable. Immunoassay tests can be used for a variety of pesticides, such as paraquat, aldrin triazines, acid amides, carbamates, and 2,4-D/phenoxy acid (Rickert, 1993). Even though many advancements have been made in controlling point-source pollution, still much remains to be done in controlling nonpoint-source pollution. Nonpoint-source pollution is more difficult to control because of multiple origins, seasonality, and inherent variability (Albanis et al., 1998; Pereira & Hostettler, 1993).

Environmental laws should be followed when dealing with pesticides. The US Environmental Protection Agency (EPA), after considering many characteristics, provides guidelines on whether to register a pesticide for commercial use or not. Therefore certain criteria are taken into consideration, including the method used on the pest, method of analysis, filtrates in the environment, and also other conditions, to determine whether pesticides can be registered for commercial use (Shankar et al., 2013). Other methods should be employed for combating pests that reduce the usage of pesticides, including using resistant varieties of plants with the ability to fight pests; use of insect traps; and crop rotation. Instead of using weedicides, mechanical control should be performed for uprooting weeds. Biochemical control can also be practiced by allowing living organisms, such as grass carp, to feed on water weeds (Helfrich et al., 2009). Despite the fact few pesticides have been found prominently in water, it is essential to evaluate the recurrence and intensity of contamination, and carry out assessments to evaluate the risk caused by these pesticides.

During the past several years, growing attention has been focused on plant protection products and their effect on surface and groundwater quality. Pesticides in use should be enrolled according to the US Federal Insecticide, Fungicide and Rodenticide Act (FIFRA). Also, there should be other management options for protecting threatened fish species (Cowx, 2002; Miller & Pister, 1971). Sanctuaries created for the conservation of threatened fish have appeared in nations like the United States (Miller & Campbell, 1994; Pearse, 1998). Developing countries recently have also acknowledged the need for and significance of monitoring biodiversity, especially in protected areas (Danielsen et al., 2000). The evaluation of any biological damage can be carried out through the monitoring of fish and other creatures through experimental testing (Fig. 4.1).

4.6 Bacterial diseases in fishes

Bacteria are the most important pathogens in warm-water cultivated fish species. They are also the most predominant cause of death in wild fish populations, especially in ecologically affected populations. However, little data is available on the origin and evolution of fish diseases due to bacterial pathogens. There are several variables involved in the insertion of a pathogenic bacterium in a prone host, and to spread to a new host, the bacteria must remain viable, whether in the host or outside (Thune et al., 1993). The emergence and development of a fish disease (Table 4.2) is the result of pathogen, host, and environmental interaction. Collaborative studies involving the characteristics

FIGURE 4.1

(A,B) Fishes collected from contaminated water; (C) Fishes collected from fresh water.

of potential pathogenic fish microorganisms, aspects of fish host biology, and a better understanding of the environmental factors would allow effective steps to be taken to prevent and monitor the key diseases limiting the development of marine fish. The clinical signs caused by each pathogen (external and internal) depend on the host species, the age of the fish, and the stage of the disease (acute, chronic, subclinical carrier). In fact, systemic diseases with elevated mortality rates (i.e., pasteurellosis, piscirickettsiosis) cause internal signs in the infected fish, but afflicted fish often have a healthy external appearance. On the other hand, some diseases with comparatively lower mortality rates (i.e., flexibacteriosis, some streptococcosis, "winter ulcer syndrome") cause severe external lesions, including ulcers, necrosis, and exophthalmia, making the fish unmarketable (Toranzo et al., 2005).

4.7 Major bacterial diseases in fish

4.7.1 Bacterial enteritis of flounder

A disease called "Chokan-hakudaku-sho" in Japanese, meaning a disease condition described by an opaque intestine, occurs in 14- to 30- or 40-day-old Japanese larval flounder. The symptoms include opacity of the intestine and darkening of the color of the body. Mortality is often 90% or higher, especially when it occurs in younger fish (Murata, 1987). The bacteria *Vibrio* sp. is the causative bacterium. It induces intestinal necrosis of flounder larvae (INFL) (Masumura et al., 1989).

Table 4.2 Overview of some bacterial fish diseases.

Scientific name	Common name	Causative agent	Reference
Carassius auratus	Goldfish	*Mycobacterium gordonae*	Sakai et al. (2005)
Danio rerio	Zebrafish	*Mycobacterium haemophilum*	Whipps et al. (2007)
Gymnothorax funebris	Moray eel	*Mycobacterium montefiorense*	Levi et al. (2003)
Oncorhynchust schawytscha	Chinook salmon	*Mycobacterium neoaurum*	Backman et al. (1990)
Morone saxatilis	Striped bass	*Mycobacterium shottsii*	Rhodes et al. (2005)

4.7.2 Abdominal swelling of sea bream and studies on intestinal flora

Abdominal swelling occurs in red and black sea bream (Iwata et al., 1978; Kusuda & Isshiki, 1987) and *V. alginolyticus* has been confirmed to be the causative agent or to be associated with the disease.

4.7.3 Gliding bacterial infection

In red and black sea bream juveniles, this disease has existed since the early 1970s. Among juveniles of sea bream ranging from 20 to 40 mm in total length, the disease prevails with the usual signs of eroded greyish-white mouth, frayed fins, and tail rot. The organism was isolated from the diseased fish by Hikida et al. (1979) and Wakabayashi et al. (1986) and the name *Flexibacter maritimus* was given (Wakabayashi et al., 1986). *V. anguillarum* is a ubiquitous marine bacterium in coastal waters and it acts as an opportunistic pathogen for various juvenile marine fishes such as red sea bream, tiger puffer, and Japanese flounder. *V. ordalii* infections have been reported in juvenile rockfish. Immersion vaccination with *V. ordalii* bacterin was demonstrated to be effective against an experimental challenge in rockfish (Nakai et al., 1989) (Table 4.3).

4.8 Control of bacterial fish diseases

Marine ornamental fish have tremendous economic potential. Intensive management of marine ornamental fish and their captive breeding have not yet made much headway in India, as bacterial diseases are responsible for heavy mortality in these fish. The problems in the culture systems are usually tackled by preventing disease outbreaks or by treating the disease with antibiotics or chemicals (Choudhury et al., 2005). The bacterial infections are considered the major cause of mortality in aquaculture. Among the common fish pathogenic bacteria, *Streptococcus agalactiae, Lactococcus garvieae, Enterococcus faecalis* (all Gram positive), *Aeromonas hydrophila* and *Yersinia ruckeri* (both Gram-negative) cause infectious diseases (Pandey et al., 2009). Some of the disease control methods are:

4.8.1 Improving water quality

The addition of Gram-positive bacteria, such as *Bacillus* spp., is beneficial in improving the water quality with the conversion of organic matter into carbon dioxide in comparison to the

Table 4.3 Some bacterial fish diseases and their causative agent.

S. no	Disease	Causative agent	Affected species	Reference
1	Bacterial enteritis	*Vibrio ichthyoenteri*	Japanese flounder	Muroga et al. (1990)
2	Abdominal swelling	*Vibrio* spp.	Red and black sea bream	Yasunobu et al. (1988)
3	Gliding bacterial infection	*Flexibacter maritimus*	Red and black sea bream	Wakabayashi et al. (1986)
4	Vibriosis	*Vibrio* sp. Zoea	Swimming crab	Muroga et al. (1990)
5	Spotting disease	*Flexibacter* sp.	Ezo sea urchin	Tajima et al. (1997)
6	Mycobacteriosis	*Mycobacterium marinum*	Seabass, turbot, Atlantic salmon	Sudheesh et al. (2012)
7	PseudomonadiasisWinter disease	*Pseudomonas anguilliseptica*	Sea bream, eel, turbot, ayu	Sudheesh et al. (2012)
8	Streptococcosis	*Streptococcus iniae*	Yellowtail, flounder, seabass, barramundi	Sudheesh et al. (2012)
9	BKD	*Renibacterium salmoninarum*	Salmonids	Sudheesh et al. (2012)
10	Piscirickettsiosis	*Piscirickettsia salmonis*	Salmonids	Sudheesh et al. (2012)

Gram-negative bacteria, which convert a greater amount of organic matter into bacterial biomass or slime (Balcázar et al., 2006). Additionally, ammonia and nitrite toxicity can be eliminated by the application of nitrifying bacterial cultures into the fish aquaria; moreover, the temperature, pH, dissolved oxygen, NH_3, and H_2S in the rearing water were found to be in permissible limits when probiotics were added. The improvement of water quality or the fish environment is termed "bioremediation" (Mohapatra et al., 2013).

4.8.2 Nanobioencapsulated vaccine

Recent research studies have examined the use of nanoparticles (NPs) as adjuvant and efficient delivery systems in fish vaccine development due to their nanosize. These NPs can be grasped by a cellular endocytosis mechanism that facilitates the cellular uptake of antigens and increases the presentation ability (Vinay et al., 2018).

4.8.3 Quorum sensing

This is defined as the regulation of gene expression in response to a communication between pathogenic bacterial cells. Many bacteria use this system to regulate physiological activities, and so adding probiotics causes a disturbance of quorum sensing, which is considered a potential antiinfective approach in aquaculture (Defoirdt et al., 2004).

4.8.4 Injection vaccination

Vaccination by injection is the delivery method generally resulting in the best protection and is the only choice for adjuvant vaccines (Harikrishnan et al., 2011). The advantages of injecting a vaccine are attaining high protection and the relatively minimal dose requirement, because correct dosage calculation is easy and economical for larger fish, and a multivalent vaccine can be administered (Lillehaug, 2014).

4.8.5 Prebiotics

These are also referred to as food for probiotics. They are resistant to attack by endogenous enzymes and hence can reach the site of action to promote the proliferation of gut microflora. Some of the prebiotics currently used in animal feed are mannan-oligosaccharides, fructo oligosaccharide, and mixed oligo-dextran (Carbone & Faggio, 2016). Probiotics are cultured products or live microbial food supplements, which beneficially affect the host by improving its intestinal microbial balance. These can help to improve the water quality, aid in food digestion, and modulate the host immune responses, increasing production efficiency and reducing disease incidence (Gatlin & Peredo, 2012).

4.8.6 Plant product application

In aquaculture for disease control, the use of plant products is one of the promising alternatives. They stimulate the immune system of fish, prevent stress, and act as antibacterial and antiparasitic agents due to their active chemical ingredients (Reverter et al., 2014). They can be administered by extracting their active component or the whole plant material can be added to the aquarium directly. Depending on the type of plant part used and the season of harvest of the plant material, their active ingredient may vary, so knowledge of the plant and the season of collection is necessary. Medicinal plants can be administered to fish by injection, oral administration, and through immersion or baths (Miriam et al., 2017). Injecting the extracted material is an effective method for large fish (Awad & Awaad, 2017).

4.9 Conclusion

Exposure of organisms to pesticides poses a continuing health risk to the population of both aquatic and terrestrial ecosystems. Thus human beings face health risks if they consume toxic fish species. Researchers have revealed that higher levels of pesticides may cause certain chronic effects (oxidative damage, embryonic and developmental changes, carcinogenicity, and mutagenesis). Therefore precautions should be taken in order to ensure that potentially hazardous chemicals are not used in situations having a high probability of environmental damage. This indicates that, in order to safeguard the fish population and other aquatic fauna, preventive measures should be taken before the application of pesticides.

It is probable that, if molecular biology techniques are used, the process for detecting pesticides (ecological stressors) in aquatic ecosystems will become cheaper, faster, and more reliable. As there are numerous unsafe impacts of pesticides on aquatic organisms, including both plants and animals, it is essential to design thorough standards and guidelines against the self-assertive utilization of pesticides. As pesticides in nature are toxicant mixes, that is, mixes of organophosphates, they may have noxious or deadly impacts on fish species. Consequently, it is important to constantly manage the convergence of pesticide deposits in food material and monitor the impact of pesticides on people. Increasingly, exploratory work needs to be done to enhance the focus on pesticides and bacteria and investigate critical lethal and sublethal impacts on the flora and fauna. Pollution in the aquatic bodies should be monitored, as it is a main cause of the spread of bacterial disease. This would guarantee greater assurance for protection of the environment, in which natural changes are difficult to distinguish.

References

Abedi, Z., Hasantabar, F., Khalesi, M. K., & Babaei, S. (2013). Enzymatic activities in common carp; *Cyprinus carpio* influenced by sublethal concentrations of cadmium, lead, chromium. *World Journal of Fish and Marine Sciences*, *5*, 144–151. Available from http://idosi.org/wjfms/wjfms5(2)13/6.pdf.

Aktar, W., Sengupta, D., & Chowdhury, A. (2009). Impact of pesticides use in agriculture: Their benefits and hazards. *Interdisciplinary Toxicology*, *2*, 1–12. Available from https://doi.org/10.2478/v10102-009-0001-7.

Albanis, T. A., Hela, D. G., Sakellarides, T. M., & Konstantinou, I. K. (1998). Monitoring of pesticide residues and their metabolites in surface and underground waters of Imathia (N. Greece) by means of solid-phase extraction disks and gas chromatography. *Journal of Chromatography*, *823*, 59–71.

Anon. (1971). *Third report of research committee on toxic chemicals*. London: Agricultural Research Council.

Anonymous. (1992). *Annual report*. India: National Bureau of Fish Genetic Resources, Lucknow, Uttar Pradesh.

Arbuckle, T. E., & Sever, L. E. (1998). Pesticide exposures and fetal death: A review of the epidemiologic literature. *Critical Reviews in Toxicology*, *28*, 229–270.

Awad, E., & Awaad, A. (2017). Role of medicinal plants on growth performance and immune status in fish. *Fish & Shellfish Immunology*, *67*, 40–54. Available from https://doi.org/10.1016/j.fsi.2017.05.034.

Backman, S., Ferguson, H. W., Prescott, J. F., & Wilcock, B. P. (1990). Progressive panophthalmitis in a chinook salmon, *Oncorhynchus tshawytscha* (Walbaum): A case report. *Journal of Fish Diseases*, *13*, 345–353.

Balcázar, J. L., De Blas, I., Ruiz-Zarzuela, I., Cunningham, D., Vendrell, D., & Múzquiz, J. L. (2006). The role of probiotics in aquaculture. *Veterinary microbiology*, *114*(3-4), 173–186.

Banaee, M. (2013). *Physiological dysfunction in fish after insecticides exposure. Insecticides–development of safer and more effective technologies* (pp. 103–143). IntechOpen.

Baran, E., Jantunen, T., & Chong, C. K. (2007). *Values of inland fisheries in the Mekong River Basin (No. 1812)*. WorldFish.

Barone, S., Jr, Das, K. P., Lassiter, T. L., & White, L. D. (2000). Vulnerable processes of nervous system development: A review of markers and methods. *Neurotoxicology*, *21*, 15–36.

Bibi, N., Zuberi, A., Naeem, M., Ullah, I., Sarwar, H., & Atika, B. (2014). Evaluation of acute toxicity of karate and its sub-lethal effects on protein and acetylcholinestrase activity in *Cyprinus carpio*. *International Journal of Agriculture and Biology*, *16*, 731–737.

Borges, A., Scotti, L. V., Siqueira, D. R., Zanini, R., do Amaral, F., Jurinitz, D. F., & Wassermann, G. F. (2007). Changes in hematological and serum biochemical values in jundiá Rhamdia quelen due to sublethal toxicity of cypermethrin. *Chemosphere*, *69*, 920–926.

Burdick, G. E., Harris, E. J., Dean, H. J., Walker, T. M., Skea, J., & Colby, D. (1964). The accumulation of DDT in lake trout and the effect on reproduction. *Transactions of the American Fisheries Society*, *93*, 127–136.

Byrne, P. A., & O'halloran, J. (2001). The role of bivalve molluscs as tools in estuarine sediment toxicity testing: A review. *Hydrobiologia*, *465*, 209–217.

Carbone, D., & Faggio, C. (2016). Importance of prebiotics in aquaculture as immune stimulants. Effects on immune system of *Sparus aurata* and *Dicentrarchus labrax*. *Fish & Shellfish Immunology*, *54*, 172–178. Available from https://doi.org/10.1016/j.fsi.2016.04.011.

Cerejeira, M. J., Viana, P., Batista, S., Pereira, T., Silva, E., Valério, M. J., & Silva-Fernandes, A. M. (2003). Pesticides in Portuguese surface and ground waters. *Water Research*, *37*, 1055–1063.

Choudhury, S., Sree, A., Mukherjee, S. C., Pattnaik, P., & Bapuji, M. (2005). In vitro antibacterial activity of extracts of selected marine algae and mangroves against fish pathogens. *Asian fisheries science*, *18*(3/4), 285.

Cowx, I. G. (2002). *Analysis of threats to freshwater fish conservation: Past and present challenges. Conservation of freshwater fishes: Options for the future* (pp. 201–220). UK: Blackwell Scientific Press.

Danielsen, F., Balete, D. S., Poulsen, M. K., Enghoff, M., Nozawa, C. M., & Jensen, A. E. (2000). A simple system for monitoring biodiversity in protected areas of a developing country. *Biodiversity & Conservation*, *9*, 1671–1705.

Dar, G. H., Bhat, R. A., Kamili, A. N., Chishti, M. Z., Qadri, H., Dar, R., & Mehmood, M. A. (2020). *Correlation between pollution trends of freshwater bodies and bacterial disease of fish fauna. Fresh water pollution dynamics and remediation* (pp. 51–67). Singapore: Springer.

Dar, G. H., Dar, S. A., Kamili, A. N., Chishti, M. Z., & Ahmad, F. (2016a). Detection and characterization of potentially pathogenic *Aeromonas sobria* isolated from fish *Hypophthalmichthys molitrix* (Cypriniformes: Cyprinidae). *Microbial Pathogenesis*, *91*, 136–140.

Dar, G. H., Kamili, A. N., Chishti, M. Z., Dar, S. A., Tantry, T. A., & Ahmad, F. (2016b). Characterization of *Aeromonas sobria* isolated from fish rohu (*Labeo rohita*) collected from polluted pond. *Journal of Bacteriology & Parasitology*, *7*(3), 1–5. Available from https://doi.org/10.4172/2155-9597.1000273.

Das, B. K., & Mukherjee, S. C. (2003). Toxicity of cypermethrin in *Labeo rohita* fingerlings: Biochemical, enzymatic and haematological consequences. *Comparative Biochemistry and Physiology Part C: Toxicology & Pharmacology*, *134*, 109–121.

De la Torre, F. R., Ferrari, L., & Salibian, A. (2002). Freshwater pollution biomarker: Response of brain acetyl cholinesterase activity in two fish species. *Comparative Biochemistry and Physiology Part C: Toxicology & Pharmacology*, *131*, 271–280. Available from https://doi.org/10.1016/S1532-0456(02)00014-5.

Defoirdt, T., Boon, N., Bossier, P., & Verstraete, W. (2004). Disruption of bacterial quorum sensing: an unexplored strategy to fight infections in aquaculture. *Aquaculture*, *240*(1-4), 69–88.

Donohue, I., Styles, D., Coxon, C., & Irvine, K. (2005). Importance of spatial and temporal patterns for assessment of risk of diffuse nutrient emissions to surface waters. *Journal of Hydrology*, *304*, 183–192.

Dudgeon, D., Arthington, A. H., Gessner, M. O., Kawabata, Z. I., Knowler, D. J., Lévêque, C., & Sullivan, C. A. (2006). Freshwater biodiversity: Importance, threats, status and conservation challenges. *Biological Reviews*, *81*, 163–182.

Ellenhorn, M. J., & Barceloux, D. G. (1988). *Chapter 38: Pesticides. Medical toxicology* (pp. 1069–1077). New York: Elsevier.

FAO. (2006). *The state of world fisheries and aquaculture 2006*. Rome: Food and Agriculture Organization of the United Nations.

FAO. (2007). *The state of world fisheries and aquaculture 2006* (p. 180) Rome: Food and Agriculture Organization of the United Nations.

Forget, G. (1993). Balancing the need for pesticides with the risk to human health. In Impact of pesticide use on health in developing countries. In: *Proceedings of a Symposium held in Ottawa, Canada*, 17–20 Septemper 1990. IDRC, Ottawa, ON, CA.

Ganeshwade, R. M., Dama, L. B., Deshmukh, D. R., Ghanbahadur, A. G., & Sonawane, S. R. (2012). Toxicity of endosulfan on freshwater fish *Channa striatus*. *Trends in Fisheries Research*, *1*, 29–31.

Gatlin III, D. M., & Peredo, A. M. (2012). *Prebiotics and probiotics: Definitions and applications*. SRAC Publication no. 4711.

Gill, T. S., Pande, J., & Tewari, H. (1990). Hepatopatho toxicity of three pesticides in a freshwater fish, *Puntius conchonius* ham. *Journal of Environmental Science & Health Part A. Environmental Science and Engineering and Toxicology*, *25*, 653–663. Available from https://doi.org/10.1080/10934529009375587.

Gupta, P. K. (2004). Pesticide exposure-Indian scene. *Toxicology*, *198*, 83–90.

Harikrishnan, R., Balasundaram, C., & Heo, M. S. (2011). Fish health aspects in grouper aquaculture. *Aquaculture (Amsterdam, Netherlands)*, *320*, 1–21.

Helfrich, L. A., Weigmann, D. L., Hipkins, P. A., & Stinson, E. R. (2009). *Pesticides and aquatic animals: A guide to reducing impacts on aquatic systems*. Blacksburg, VA: Virginia Department of Game and Inland Fisheries.

Hikida, M., Wakabayashi, H., Egusa, S., & Masumura, K. (1979). *Flexibacter* sp., a gliding bacterium pathogenicto some marine fishes in Japan. *Nippon Suisan Gakkaishi*, *45*, 421–428.

Holden, A. V. (1973). *Effects of pesticides on fish. Environmental pollution by pesticides* (pp. 213–253). Boston, MA: Springer.

Holmlund, C. M., & Hammer, M. (1999). Ecosystem services generated by fish populations. *Ecological Economics*, *29*, 253–268. Available from https://doi.org/10.1016/S0921-8009(99)00015-4.

Hoppin, J. A., Umbach, D. M., London, S. J., Alavanja, M. C., & Sandler, D. P. (2002). Chemical predictors of wheeze among farmer pesticide applicators in the Agricultural Health Study. *American Journal of Respiratory and Critical Care Medicine*, *165*, 683–689.

Huber, A., Bach, M., & Frede, H. G. (2000). Pollution of surface waters with pesticides in Germany: Modeling non-point source inputs. *Agriculture, Ecosystems & Environment*, *80*, 191–204.

Igbedioh, S. O. (1991). Effects of agricultural pesticides on humans, animals, and higher plants in developing countries. *Archives of Environmental Health: An International Journal*, *46*, 218–224.

Ilavazhahan, M., Tamil, S. R., & Jayaraj, S. S. (2010). Determination of LC50 of the bacterial pathogen, pesticide and heavy metal for the fingerling of freshwater fish *Catla catla*. *Global Journal of Environmental Research*, *4*(2), 76–82.

Iwata, K., Yanohara, Y., & Ishibashi, O. (1978). Studies on factors related to mortality of young red seabream (*Pagrus major*) in the artificial seed production. *Fish Pathology*, *13*, 97–102.

Jabeen, G., Javed, M., & Azmat, H. (2012). Assessment of heavy metals in the fish collected from the river Ravi, Pakistan. *Pakistan Veterinary Journal*, *3*, 107–111.

Jee, J. H., Masroor, F., & Kang, J. C. (2005). Responses of cypermethrin-induced stress in haematological parameters of Korean rockfish, Sebastes schlegeli (Hilgendorf). *Aquaculture Research*, *36*, 898–905.

Kamel, F., & Hoppin, J. A. (2004). Association of pesticide exposure with neurologic dysfunction and disease. *Environmental Health Perspectives*, *112*, 950–958.

Kingsbury, P. D., & Kreutzweiser, D. P. (1980). *Environmental impact assessment of a semi-operational permethrin application* [in Canada]. Report Canadian Forestry Service (Canada). no. FPM-X-30.

Koesoemadinata, S. (1980). *Pesticides as a major constraint to integrated agriculture-aquaculture farming systems* [in many Asian countries notably Indonesia]. In ICLARM Conference Proceedings (Philippines).

Kusuda, R., & Isshiki, T. (1987). Susceptibility of yellowtail fingerlings to yellowtail ascites virus ŽYTAV. *Reports of the USA Marine Biological Institute, Kochi University. Kochi*, *9*, 51–57.

Lakra, W. S., & Sarkar, U. K. (2007). *Fish diversity of Central India* (pp. 1–183). NBFGR Publication, Lucknow, Uttar Pradesh.

Levi, M. H., Bartell, J., Gandolfo, L., Smole, S. C., Costa, S. F., Weiss, L. M., Johnson, L. K., Osterhout, G., & Herbst, L. H. (2003). Characterization of *Mycobacterium montefiorense* sp. nov., a novel pathogenic mycobacterium from moray eels that is related to *Mycobacterium triplex*. *Journal of Clinical Microbiology*, *41*, 2147–2152.

Li, Z. H., Velisek, J., Grabic, R., Li, P., Kolarova, J., & Randak, T. (2011). Use of hematological and plasma biochemical parameters to assess the chronic effects of a fungicide propiconazole on a freshwater teleost. *Chemosphere*, *83*, 572–578.

Lillehaug, A. (2014). Vaccination strategies and procedures. In R. Gudding, A. Lillehaug, & O. Evensen (Eds.), *Fish vaccination* (pp. 141–150). John Wiley & Sons.

Lim, C. S., & Ong, S.-H. (1985). *Environmental problems of pesticide usage in the Malaysian ricefield.* Perception and future consideration. Paper presented at the fourth International Meeting, Chiang Mai, Thailand.

Macek, K. J. (1968). Reproduction in brook trout (*Salvelinus fontinalis*) fed sublethal concentrations of DDT. *Journal of the Fisheries Board of Canada*, *25*, 1787–1796.

Masumura, K., Yasunobu, H., Okada, N., & Muroga, K. (1989). Isolation of a *Vibrio* sp., the causative bacterium of intestinal necrosis of Japanese flounder larvae. *Fish Pathology*, *24*, 135–141.

Mathur, S. C., & Tannan, S. K. (1999). Future of Indian pesticides industry in next millennium. *Pesticide Information*, *24*, 9–23.

Maurya, P. K., Malik, D. S., & Sharma, A. (2019). Impacts of pesticide application on aquatic environments and fish diversity. *Contaminants in Agriculture and Environment: Health Risks and Remediation*, *1*, 111.

Meitei, N. S., Bhargava, V., & Patil, P. M. (2004). Water quality of Purna river in Purna town. *Maharastra state. Journal of Aquatic Biology*, *19*, 77–78.

Miller, J. A., & Campbell, C. E. (1994). *A marine geographic information system for the channel Islands National Marine Sanctuary. The Fourth California Islands Symposium: Update on the Status of Resources* (pp. 135–139). Santa Barbara Museum of Natural History.

Miller, R. R., & Pister, E. P. (1971). Management of the Owens Pupfish, *Cyprinodon radiosus*, in Mono County, California. *Transactions of the American Fisheries Society*, *100*, 502–509.

Miriam, R., Tapissier, N., Pierre, S., & Saulnier, D. (2017). Use of medicinal plants in aquaculture. In B. Austin, & A. Newaj-Fyzul (Eds.), *Diagnosis and control of diseases of fish and shellfish* (pp. 223–261). JohnWiley & Sons Ltd.

Mittermeier, R. A., & Mittermeier, C. G. (1997). Megadiversity: Earth's biolocally wealthiest nation. In Global freshwater biodiversity: Striving for the integrity of fresh water ecosystem. *Mexico City: Sea Wind, Cemex*, *11*, 1–140.

Mohapatra, S., Chakraborty, T., Kumar, V., DeBoeck, G., & Mohanta, K. N. (2013). Aquaculture and stress management: A review of probiotic intervention. *Journal of Animal Physiology and Animal Nutrition*, *97*, 405–430.

Morel, F. M., Kraepiel, A. M., & Amyot, M. (1998). The chemical cycle and bioaccumulation of mercury. *Annual Review of Ecology and Systematics*, *29*, 543–566.

Murata, O. (1987). Infectious intestinal necrosis in flounder. *Fish Pathology*, *22*, 59–61.

Muroga, K., Yasunobu, H., Okada, N., & Masumura, K. (1990). Bacterial enteritis of cultured flounder *Paralichthys olivaceus* larvae. *Diseases of Aquatic Organisms*, *9*, 121–125.

Nagaraju, B., Sudhakar, P., Anitha, A., Haribabu, G., & Rathnamma, V. V. (2011). Toxicity evaluation and behavioral studies of fresh water fish Labeorohita exposed to Rimon. *International Journal of Research in Pharmaceutical and Biomedical Sciences*, *2*, 722–727.

Nakai, T., Muroga, K., & Masumura, K. (1989). Immersion vaccination of juvenile rockfish Sebastes schlegeli by *Vibrio ordalii* and *Vibrio anguillarum antigens. Aquaculture. Science (New York, N.Y.)*, *37*, 129–132.

Nikam, S., Shejule, K. B., & Patil, R. B. (2011). Study of acute toxicity of Metasystox on the freshwater fish, Nemacheilus botia, from Kedrai dam in Maharashtra, India. *Biology and Medicine*, *3*, 13–17.

Özkara, A., Akyıl, D., & Konuk, M. (2016). *Pesticides, environmental pollution, and health. Environmental health risk-hazardous factors to living species*. IntechOpen.

Pandey, R. K., Singh, R. N., Singh, S., Singh, N. N., & Das, V. K. (2009). Acute toxicity bioassay of dimethoate on freshwater airbreathing catfish, *Heteropneustes fossilis* (Bloch). *Journal of. Environment Biology*, *30*, 437–440.

Pearse, J. S. (1998). *Biodiversity of the rocky intertidal zone in the Monterey bay National Marine Sanctuary: A 24 year comparison. California Sea Grant: Report of Completed Projects 1994–1997* (pp. 57–62). USA: La Jolla.

Pereira, W. E., & Hostettler, F. D. (1993). Nonpoint source contamination of the Mississippi River and its tributaries by herbicides. *Environmental Science & Technology*, *27*, 1542–1552.

Planas, C., Caixach, J., Santos, F. J., & Rivera, J. (1997). Occurrence of pesticides in Spanish surface waters. Analysis by high resolution gas chromatography coupled to mass spectrometry. *Chemosphere*, *34*, 2393–2406.

Quackenbush, R., Hackley, B., & Dixon, J. (2006). Screening for pesticide exposure: A case study. *Journal of Midwifery & Women's Health*, *51*, 3–11.

Rahman, M. Z., Hossain, Z., Mollah, M. F. A., & Ahmed, G. U. (2002). Effect of Diazinon 60 EC on *Anabas testudineus*, *Channa punctatus* and *Barbodes gonionotus*. *Naga, the ICLARM Quarterly*, *25*, 8–12.

Reverter, M., Bontemps, N., Lecchini, D., Banaigs, B., & Sasal, P. (2014). Use of plant extracts in fish aquaculture as an alternative to chemotherapy: Current status and future perspectives. *Aquaculture (Amsterdam, Netherlands)*, *433*, 50–61.

Rhodes, M. W., Kator, H., McNabb, A., Deshayes, C., Reyrat, J.-M., Brown-Elliott, B. A., Wallace, R., Trott, K. A., Parker, J. M., Lifland, B. D., Osterhout, G., Kaattari, I., Reece, K., Vogelbein, W. K., & Ottinger, C. A. (2005). *Mycobacterium pseudoshottsii* sp. nov., a slowly growing chromogenic species isolated from Chesapeake Bay striped bass (*Morone saxatilis*). *International Journal of Systematic and Evolutionary Microbiology*, *55*, 1139–1147.

Richards, R. P., & Baker, D. B. (1993). Pesticide concentration patterns in agricultural drainage networks in the Lake Erie basin. *Environmental Toxicology and Chemistry: An International Journal*, *12*, 13–26.

Rickert, D. (1993). *Water quality assessment to determine the nature and extent of water pollution by agriculture and related activities. Prevention of water pollution by agriculture and related activities. Water reports (FAO)*. Roma, Italia: FAO.

Sabae, S. Z., El-Sheekh, M. M., Khalil, M. A., Elshouny, W. A. E., & Badr, H. M. (2014). Seasonal and regional variation of physicochemical and bacteriological parameters of surface water in El-Bahr El-Pherony, Menoufia, Egypt. *World Journal of Fish and Marine Sciences*, *6*, 328–335.

Sabra, F. S., & Mehana, E. S. E. D. (2015). Pesticides toxicity in fish with particular reference to insecticides. *Asian Journal of Agriculture and Food Sciences*, *3*(1).

Sakai, M., Kono, T., Tassakka, A. C. M. A. R., Ponpornpisit, A., Areechon, N., Katagiri, T., Yoshida, T., & Endo, M. (2005). Characterization of a Mycobacterium sp. isolated from guppy *Poecilia reticulata*, using 16S ribosomal RNA and its internal transcribed spacer sequences. *Bulletin of the European Association of Fish Pathologists*, *25*, 64–69.

Sambasiva Rao, K. R. S., Surendra Babu, K., & Ramana Rao K. V. (2009). Toxicity of Elsan to the Indian snakehead *Channa punctatus*.

Sarwar, S., Ahmad, F., & Khan, J. (2007). Assessment of the quality of Jehlum river water for irrigation and drinking at district Muzaffarabad Azad Kashmir. *Sarhad Journal of Agriculture*, *23*, 1041.

Schnick, R. A., Meyer, F. P., & Gray, D. L. (1986). *A guide to approved chemicals in fish production and fishery resource management*. MP-University of Arkansas, Cooperative Extension Service, USA.

Shankar, K. M., Kiran, B. R., & Venkateshwarlu, M. (2013). A review on toxicity of pesticides in fish. *International Journal of Open Scientific Research*, *1*, 15–36.

Sial, I. M., Kazmi, M. A., Kazmi, Q. B., & Naqvi, S. N. U. H. (2009). Toxicity of biosal (phytopesticide) and permethrin (pyrethroid) against common carp, *Cyprinus carpio*. *Pakistan Journal of Zoology*, *41*(3), 235–238.

Singh, S. K., Singh, S. K., & Yadav, R. P. (2010). Toxicological and biochemical alterations of cypermethrin (synthetic pyrethroids) against freshwater teleost fish Colisa fasciatus at different season. *World Journal of Zoology*, *5*(1), 25–32.

Spiller, G. (1985). *Rice-cum-fish production in Asia: Aspects of rice and fish production in Asia.* Bangkok, Thailand: FAO Office for Asia and the Pacific.

Srivastava, P., Singh, A., & Pandey, A. K. (2016). Pesticides toxicity in fishes: Biochemical, physiological and genotoxic aspects. *Biochemical and Cellular Archives*, *16*, 199–218.

Sudheesh, P. S., Al-Ghabshi, A., Al-Mazrooei, N., & Al-Habsi, S. (2012). Comparative pathogenomics of bacteria causing infectious diseases in fish. *International Journal of Evolutionary Biology*, *2012*, 457264.

Tajima, K., Hirano, T., Shimizu, M., & Ezura, Y. (1997). Isolation and pathogenicity of the causative bacterium of spotting disease of sea urchin Strongylocentrotus intermedius. *Fisheries Science*, *63*, 249–252. Available from https://doi.org/10.2331/fishsci.63.249.

Thenmozhi, C., Vignesh, V., Thirumurugan, R., & Arun, S. (2011). Impacts of malathion on mortality and biochemical changes of freshwater fish *Labeo rohita*. *Iran Journal of Environment. Health Science and Engineering*, *8*, 87–394.

Thomas, M., Chretien, D., Florion, A., & Terver, D. (1996). Real-time detection of potassium cyanide pollution in surface waters using electric organ discharges wave emitted by the tropical fish, *Apteronotus albifrons*. *Environmental Technology*, *17*, 561–574.

Thune, R. L., Stanley, L. A., & Cooper, R. K. (1993). Pathogenesis of gram-negative bacterial infections in warmwater fish. *Annual Review of Fish Diseases*, *3*, 37–68.

Toranzo, A. E., Magariños, B., & Romalde, J. L. (2005). A review of the main bacterial fish diseases in mariculture systems. *Aquaculture (Amsterdam, Netherlands)*, *246*, 37–61.

Tursi, A., Maiorano, P., Sion, L., & D'Onghia, G. (2015). Fishery resources: Between ecology and economy. *Rendiconti Lincei*, *26*, 73–79.

U.S. Geological Survey. (1999). *The quality of our nation's waters nutrients and pesticides* (p. 82) U.S. Geological Survey.

Vinay, T. N., Bhat, S., Gon Choudhury, T., Paria, A., Jung, M. H., Shivani Kallappa, G., & Jung, S. J. (2018). Recent advances in application of nanoparticles in fish vaccine delivery. *Reviews in Fisheries Science & Aquaculture*, *26*(1), 29–41.

Vincke, M. M. J. (1979). *Aquaculture en riziere: Situation et role future. Advances in aquaculture.* Rome: FAO.

Wakabayashi, H., Hikida, M., & Masumura, K. (1986). *Flexibacter maritimus* sp. nov., a pathogen of marine fishes. *International Journal of Systematic and Evolutionary Microbiology*, *36*, 396–398.

Waynon, W., Johnson., Mack, T., & Finley. (1980). *Hand book of acute toxicity of chemicals to fish and aquatic invertebrates.* Washington, DC: United states Department of the interior fish and wild life service/ Resource Publication 137.

Whipps, C. M., Butler, W. R., Pourahmad, F., Watral, V. G., & Kent, M. L. (2007). Molecular systematics support the revival of *Mycobacterium salmoniphilum* (ex Ross 1960) sp. nov., nom. rev., a species closely related to *Mycobacterium chelonae*. *International Journal of Systematic and Evolutionary Microbiology*, *57*, 2525–2531.

The World Fish Center. (2002). *Fish: An issue for everyone: A concept paper for fish for all.* WorldFish.

Xie, P., Zhuge, Y., Dai, M., & Takamura, N. (1996). Impacts of eutrophication on biodiversity of plankton community. *Acta Hydrobiologica Sinica*, *20*, 30–37.

Yasunobu, H., Muroga, K., & Maruyama, K. (1988). A bacteriological investigation on the mass mortalities of red sea bream Pagrus major larvae with intestinal swelling. *Aquaculture Science*, *36*, 11–20.

CHAPTER

Impact of aquatic pollution on fish fauna

5

Arizo Jan[1], Tasaduq Hussain Shah[1] and Nighat Un Nissa[2]

[1]Division of Fisheries Resource Management, Faculty of Fisheries, Sher-e-Kashmir University of Agricultural Sciences and Technology of Kashmir, Srinagar, India [2]Department of Zoology, University of Kashmir, Srinagar, India

5.1 Introduction

Aquatic ecosystems are a major recipient of pollutants that, over time, can result in adverse consequences for aquatic life. These consequences only become evident when changes occur in the community, ecosystem, or population, and by that time it may be impossible to counteract the change. This has become a matter of great concern over the last few years, as the contamination of water bodies due to the wide range of pollutants not only poses a threat to public water supplies, but also causes serious harm to aquatic biota. The main causes of aquatic pollution are rapid industrialization, discharge of heavy metals from industrial and domestic wastes, agricultural activities, physical and chemical weathering of rocks, soil erosion including sewage disposal, and atmospheric deposition, all of which extensively contaminate the natural aquatic systems (Khatri & Tyagi, 2015). New agricultural techniques have contributed to enhancing crop production, but at the same time these techniques have polluted the aquatic environment to a greater extent. Escalation in river pollution as well as pollution of other water bodies has become a subject of great concern in recent years due to direct disposal of industrial effluents and urban sewage with little or no treatment. Fish, though nontarget organisms, have become victims of aquatic pollution by accumulating toxins. Then the accumulated toxicants adversely affect the human population through consumption of the fish (Ansari, Marr, & Tariq, 2004).

The aquatic pollution may biologically affect biota by altering their biochemical, respiratory, and immune functions, including changes in population structure and development and structural abnormalities. The presence of pollutants either directly or indirectly in aquatic systems causes measurable environmental and economic impacts. The measures of the environmental impacts are the contaminant accumulation in biota, the significant increase in pollution-related reproduction, and developmental anomalies and other major alterations in biochemical or physiological processes in fish and invertebrates (Arukwe, 2001). The economic impacts may include declines in commercial and recreational fishing activities or closure of fisheries due to the accumulation of toxicants in the body composition of commercially important fish, shellfish, and gamefish (Dar et al., 2020; Dar, Dar et al., 2016; Dar, Kamili et al., 2016).

Bacterial Fish Diseases. DOI: https://doi.org/10.1016/B978-0-323-85624-9.00011-7

Fish, a major aquatic lifeform of crucial importance to humans, can display either acute effects or long-term chronic effects of water pollution, in the form of immune suppression, low metabolic rate, and gill and epithelium damages. A number of histocytological changes have been recommended as biomarkers for pollution monitoring of fish (Au, 2004). The changes in hydrochemical and fauna characteristics due to the introduction of pollutants in aquatic systems represent a major threat. The systematic failure to understand the connectivity between the developmental activities and their impact on freshwater ecosystems results in the loss of freshwater biodiversity (Crook et al., 2015). The destructive influence on the aquatic environment in the form of sublethal pollution from humans results in chronic stress, causing ill effects on aquatic organisms. Hence it should be highlighted that mortalities may not necessarily indicate an outbreak of disease in the fish population (Dar, Dar et al., 2016; Dar, Kamili et al., 2016), as the massive fish kills from release of hydrocarbons into water bodies or spillage of pesticides are not due to a disease as defined by Austin (1998).

5.2 Sources of pollution

Sewage from cities is the major source of surface water pollution. Sources of pollution are generally grouped as point sources and nonpoint sources (Fig. 5.1).

1. **Point-source pollution:** This is the source of contaminants that pave the way to the water body at a specific site/point through a separate conveyance, such as a pipe or a ditch. The source of pollution is relatively negligible compared to other pollution sources and is easily recognized. Discharges/effluents from sewage-treatment points, landfill sites, power stations, and fish farms all constitute forms of point-source pollution.
2. **Nonpoint-source pollution:** This type of pollution is derived from many diffuse contaminations, unlike the point source, which results from a single source. The typical example of nonpoint-source (NPS) pollution is the seeping of nitrogenous compounds from agricultural fields into waterways. Other examples include nutrient run-off in storm water from "sheet flow" over a forest or agricultural field. Water washed off from contaminated areas or urban run-off such as from parking lots, roads, and highways are also included in NPS.

5.3 Impacts of heavy metal pollution on fish health

Heavy metals are present in aquatic systems as trace components naturally, but due to agricultural activities, industrialization, and mining activities their quantities have increased to unacceptable levels (Masindi & Muedi, 2018). The local waters of both developed and developing countries have become contaminated by heavy metals due to industrial development. This type of pollution may have negative effects at cellular levels and disturb the ecological balance of aquatic organisms, either fresh or marine water. The ingestion of contaminated marine aquatic products such as seafoods can lead to various health complications in humans and animals (Smith & Gangolli, 2002). The term heavy metal can be defined as any metallic chemical element with relatively high density and that is poisonous or toxic at low concentrations. Heavy

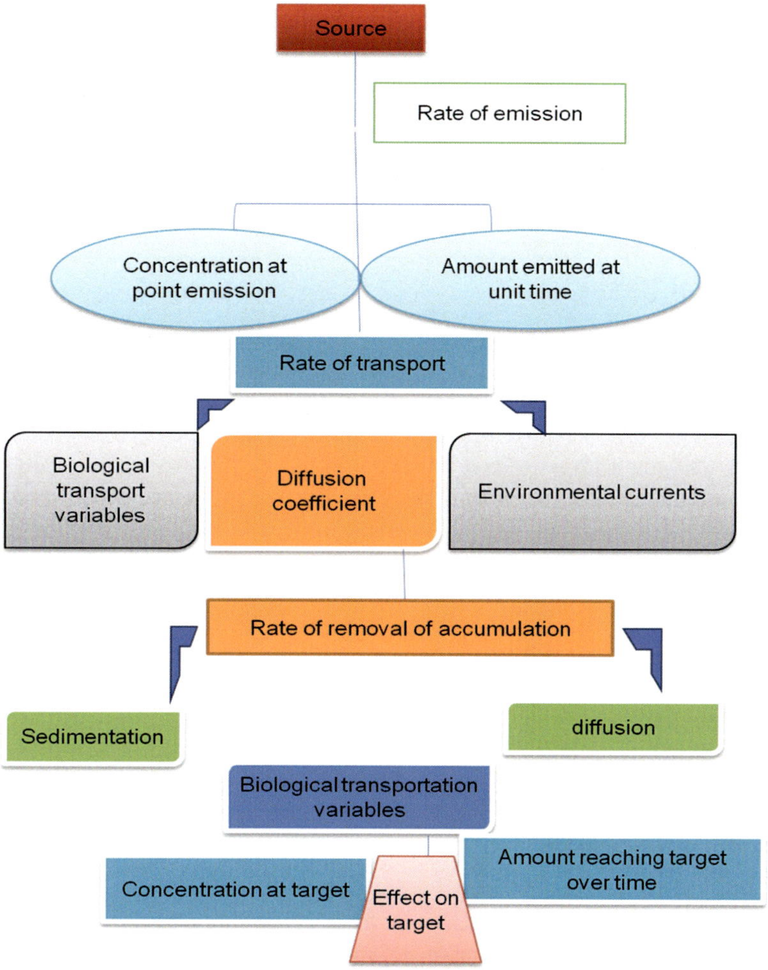

FIGURE 5.1

A generalized pollutant pathway showing point and nonpoint sources (NPS) of pollution. Further, the diagram depicts diffusion, sedimentation, and concentration, thus affecting the target.

metals are mostly defined on density factor, having a specific density of more than 5 g/cm^3: for example, mercury (Hg), cadmium (Cd), arsenic (As), chromium (Cr), thallium (Ti), and lead (Pb).

Heavy metals are toxicants that cause acute disorders in aquatic biota. Many pathological disabilities like renal damage, sporadic fever, cramps in humans, and hypertension may occur due to heavy metal consumption via the food chain in aquatic organisms. Fish, being at the top of the food pyramid, represent an important target for biomagnification of metals and act as possible transfer media to human beings (Zeitoun & Mehana, 2014). Heavy metals tend to accumulate in tissues of organisms that live in polluted water. Accumulation depends on both extrinsic and intrinsic factors. Extrinsic factors include metal concentration, method of metal intake, time of exposure, environmental factors like water temperature, pH, hardness, and salinity; intrinsic factors include fish age and feeding habits, etc. The main sites of metal accumulation are liver, kidney, and gills, with the lowest levels in fish muscles compared to other tissues (Murugan, Karuppasamy, Poongodi, & Puvaneswari, 2008).

5.4 Heavy metal hazards

The industrial pollutants lead (Pb) and cadmium (Cd) have strong negative effects on human and animal health by accumulation in liver and kidney. The lead acts as a substitute for essential elements important for metabolism like calcium, iron, and zinc by acting as a mimetic agent, having both physiological and biochemical effects on fish (Zeitoun & Mehana, 2014). Cd (0.1 mg/g) and Pb (0.4 mg/g) are maximum levels of concentration permitted for sea fish and Cd (2–20 mg/L) is the lethal concentration for different species of fish. The consumption of fish species exposed to high levels of cadmium may result in skeletal damage. This hazard was reported in the 1950s in Japan as Itai-Itai ("Ouch-Ouch") disease, which led to both osteomalacia and osteoporosis. Likewise, acute mercury exposure leads to the well-known Minamata disease, associated with lung damage (Fig. 5.2) (Castro-González & Méndez-Armenta, 2008). With fish, heavy metal toxicity leads to loss of balance or disequilibrium, increased opercular movement, and irregular vertical movements, eventually leading to death. Moreover, cadmium, mercury, lead, and arsenic severely damage the renal and nervous systems of fish as well as gills.

5.5 Fishes as biomarkers

Fish are considered to be a general early warning system for environmental degradation, but they also provide specific measures of the existence of toxic mutagenic and carcinogenic components in biological materials. In fish, the primary target organs of pollution are gills, liver, and kidneys. Fish exposed to heavy metals are reported to have histopathological lesions and increased size. Changes like these may serve as bioindicators for detecting and determining the impact of heavy metals on both organism and ecosystem health. Biomarkers are diagnostic tools to monitor heavy metal contamination through the changes that occur in chemical and physiological behavior on an organism level and to determine the contamination levels directly by observing and measuring these changes in the organism (Sabullah et al., 2015).

5.6 Impact on fish reproduction

Fish species exposed to aquatic pollution, either directly or indirectly, may experience serious reproduction effects (Kime, 1995). This is important for preserving the number of fish species and for the success of fisheries, since reproduction is a basic factor. A wide range of pollutants,

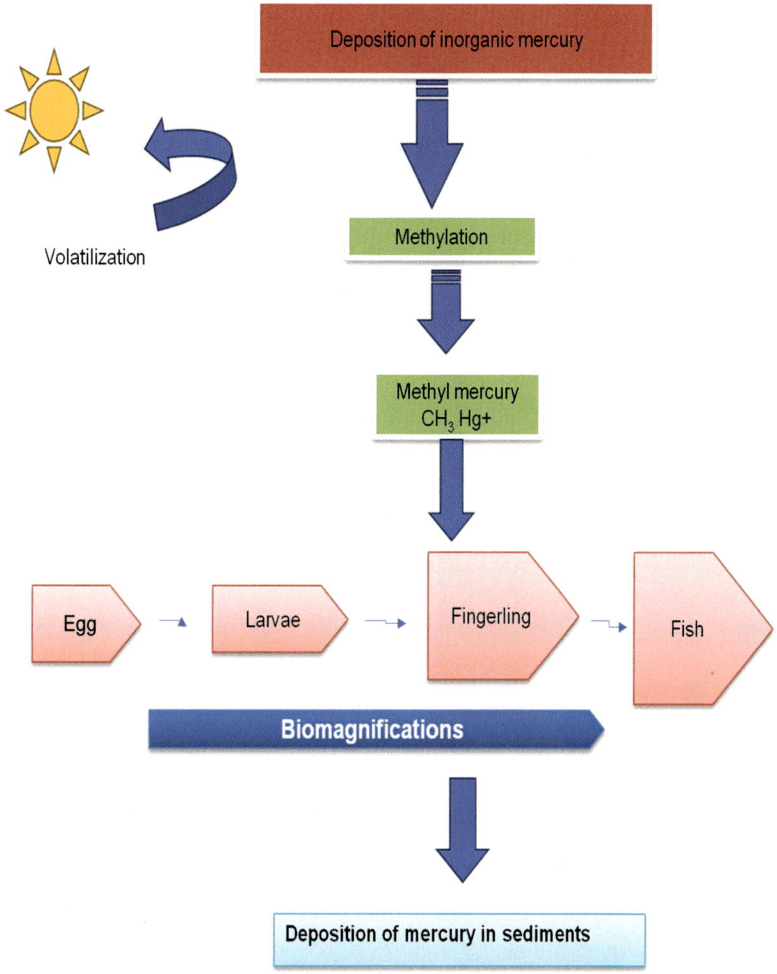

FIGURE 5.2

The pathway of mercury deposition in aquatic systems. Deposition of inorganic mercury into pond from different sources and its route to fish.

including acidification, herbicides, thermal effluents, and others have been found to affect parental care and courtship behavior in several fish families, including Cichlidae, Poecilidae, and Cypridae. The investigations showed changes in normal courtship behavior such as increases or decreases in courtship frequency displays, extended courtship duration, and masculinized females. The effects on parental care included poor nest building activity and offspring defense and major changes in distribution of parental care between the sexes. Aquatic pollutants may also affect the reproductive systems of fish species.

5.6.1 Impacts on male reproductive systems

Unlike other vertebrates, teleost gonads lack medullary tissue and correspond only with the cortex. Most teleosts have paired elongated testes attached to the dorsal wall, while in some they form a single sac. As in mammals, the testes are composed of sperm-producing tubular, lobular compartments and steroid-secreting endocrine interstitial cells. A few studies have looked at the consequences of aquatic pollution on the male reproduction system. Pandey (2000) showed that fish fauna exposure to BHC resulted in vacuolated cells and necrosis in seminiferous tubules (ST), atrophy of interstitial Leydig cells (LC), reduced secretory activities, and gonadosomatic indices and thickening of the sperm duct wall in tilapia. Degeneration of acidophilic granular cells of chromatophores and aggregation of focal leukocytes have been reported from the testis of 11 fish species sampled from the Gulf of Mexico from the sites of petroleum production (Jones & Reynolds, 1997; Stott, McArthur, Tarpley, Sis, & Jacobs, 1980). Abnormal lobular architecture, germinal epithelium dissolution, scattered LC with homogeneous liquified cytoplasm, necrosis, and prominent vacuolization have been reported due to long exposure to carbamide during the preparatory and maturing phase, with no significant effect on spawning and postspawning phases (Pandey, 2000). Dimecron exposure prompted the atrophy of ST, spermatocytes, and sperm mother cells, as well as Leydig cell enlargement, sperm duct thickening, and reduction in the gonadosomatic index (GSI). Exposure to thiourea elicited a lowered value of GSI and an increase in cholesterol level in *Gara mullya*. Dimecron- and carbofuran-treated samples resulted in reduced GSI, necrosis of spermatozoa and spermatogenesis, Leydig cell involution enhancement, and collagenous capsule formation around necrotic germ cells in *Channa punctatus* (Cruz-Landim, Abdalla, & Cruz-Höfling, 2005). Treatment also showed depletion in testicular proteins, RNA, lipids, and ascorbic acid, and it induced an increase in cholesterol and phospholipids in *Channa punctatus*. Moreover, exposure to cythion elicited arrest of spermatogenesis at the spermatid stage and ST were found without sperm, appearing like solid cords, as well as necrosis of spermatocytes and inactive involuted LC with darkly stained nuclei. Emission-treated *Clarias batrachus* resulted in the reduction of 5−3 p-hydroxysteroid, dehydrogenase (HSD) enzyme activity, pyknosis of LC, and lowered GSI, free cholesterol, total lipids, and phospholipids (Cruz-Landim et al., 2005; Jones & Reynolds, 1997). Also, extensive damage to Singhi testis was observed due to chlordecone exposure, including flattening of the ST, degeneration of the germinal epithelium, and atrophied and vacuolated LC (Srivastava & Srivastava, 1994).

5.6.2 Impacts on female reproductive system

Unlike mammals, the reproductive systems of female teleosts have a wide range of reproductive patterns, including vivaparity, and they are highly variable. Most teleosts have paired hollow ovaries; however, in some species the paired organ may combine to form a single solid organ during the early developmental stages. Ovaries are composed of oogonia, oocytes, and their surrounding follicle cells, stroma, and nervous and vascular tissue (Helmstetter et al., 2016; Jones & Reynolds, 1997). The ovaries of 11 species collected from petroleum-induced lesions showed atresia, leukocytic loci, or infiltration of acidophilic and chromatophore cells. Cythion treatment delayed secondary yolk deposition and showed a notable decrease in GSI and cholesterol. Fenitrothion and carbofuran resulted in lower total ovarian protein, lipids, and ascorbic acid; however, elevations in cholesterol and lipids have been observed during different phases of the ovarian cycle. Treatment with mercuric chloride resulted in reduced total ovarian protein and alkaline phosphate and elevated the levels of glycogen and acid phosphatase and cholesterol. Cyanide treatment of rainbow trout resulted in a decrease of GSI and oocyte diameters and a reduction in levels of E_2, T_3, and vitellogenin. The inhibition of ovarian growth and degeneration of vitellogenic oocytes have been observed due to ammonium sulfate treatment (Jones & Reynolds, 1997). It has been observed by a number of workers that treatment of fish species with various pollutants has resulted in underdeveloped ovarian growth as manifested by reduced GSI and fewer maturing oocytes in stage II and III. This pollutant-induced disablement of ovarian function was assumed to mediate through the hypothalamo-hypophysial ovarian axis, since degenerative changes were also observed in the hypothalamus and gonadotropes of treated fish (Harries et al., 1996; Jobling, Sumpter, Sheahan, Osborne, & Matthiessen, 1996). Exposure to different pesticides elicited inhibition of steroidogenesis and mutilation and clumping of yolk in stages II and III of oocyte maturation. Also, reduction in GSI, water, and lipid contents has been observed due to the increasing dose of pesticides in the ovary.

5.7 Effects of pollution on disease outbreak

Aquatic pollution extensively contributes to the occurrence of diseases in fish populations, as it alters the susceptibility of the host to pathogens, which is fundamental in any habitat. The presence of pollutants in the aquatic system may cause this increased susceptibility by causing stress to aquatic animals (Bucke, 1993). Moreover, an aquatic environment contaminated with organic material like fecal debris may lead to an outbreak of diseases through increasing the microbial population in the system. The diseases related to aquatic pollution include fin or tail rot, gill diseases, neoplasia, hepatic damage, and ulceration (Au, 2004; Austin, 1998). For example, fin or tail rot caused by *Amnonas hydrophila* and *Pseudomonas fluorescens* and surface lesions from *Serratia plymuthica* specifically reflect the effects of pollution. Contaminated diets, heavy metals, nitrogenous compounds like ammonia and nitrites, hydrocarbons, pesticides, and other unspecified pollutants have been pointed out by various researchers as triggers for these diseases. Among them, the fin/tail rot is mainly associated with bacterial contamination, as the presence of pollutants weakens the fish by suppressing the immune system, resulting in microbial colonization and thence the occurrence of clinical disease. The presence of carcinogens and the action of viruses may lead to neoplasia. Several hotspots for the occurrence of tumors in fish and shellfish have been described, which certainly correspond to high concentrations of anthropogenic compounds (Austin, 1998; Bucke, 1993).

5.8 Role of heavy metals

Heavy metals play a major role in aggravating microbial diseases in fish fauna. In particular, copper increases susceptibility to vibriosis (i.e., *Vibrio anguillarum*) and infections by *Edwardsiella tarda* by eliciting debilitating effects based on concentration and time of exposure (Afshan et al.,

2014). It has been observed that the exposure of a fish population to copper results in a mucous layer coagulation of the gills, which consequently leads to inhibition of oxygen transport and results in respiratory stress or also a reduction in the process of phagocytosis due to a decrease in the number of lymphocytes and granulocytes (Austin, 1998). Likewise, large-scale surveys have been conducted on dumping sites receiving titanium dioxide acids, and a higher prevalence of diseases like epidermal hyperplasia/papilloma, liver nodules, lymphocystis, and infections with *Glugea spp* was noted as compared to control areas (Ali, Farid, Bharwana, & Ahmad, 2013).

5.9 Role of hydrocarbons and nitrogenous compounds

The long-term exposure of fish populations to hydrocarbons results in impaired mucus, defective immune system development, and increased susceptibility to parasitism; it also can induce liver hypertrophy and hyperplasia and mortalities (Castro-González & Méndez-Armenta, 2008). It has been demonstrated by much experimental evidence in fish exposed to sediments containing a high molecular weight creosote fraction suspended in an aquatic environment that fish may develop fin erosion and epidermal lesions, possibly leading to death. On the other hand, head lesions, that is, around mouth, nares, and opercula, may appear on exposure to low molecular weight creosote fraction (Mézin & Hale, 2000). The deleterious effects of nitrogenous compounds on fish health have been observed by various studies, as in the case of channel catfish (*Ictalurus punctatus*) exposed to nitrite levels of 6 mg/L, which showed increased susceptibility to infection by *Aeromonas hydrophila*.

5.10 Role of pesticides

Numerous diseases have been found in the aquatic environment due to the presence of pesticides like dichloro-diphenyl-trichloroethane and polychlorinated biphenyl, including "cauliflower disease," ulceration, lymphocystis, and liver neoplasia. The long-term surveys on malformation of embryos in the common dab flounder (*Platichthys flesus*), plaice (*Pleuronectes platessa*), and whiting (*Merlangus sp*) have been considered to be linked to pollution with organochorines (Perwaiz, 2020; Sabra & Mehana, 2015).

5.11 Conclusion

The increased load of organic matter, heavy metals, and anthropogenic activities have cumulative negative impacts on fish fauna. There is strong evidence of significant negative effects on fish due to increased pollution from expanding agricultural, domestic, and industrial development. The fish fauna exposed to contaminated aquatic environments are immunosuppressed and more susceptible to diseases as compared to those in less polluted waters. Contamination with heavy metals induces several histopathological changes. There is accumulating evidence that pollutants enter the aquatic system and are found in the tissues of aquatic vertebrate and invertebrate animals and are responsible for the development of disease outbreaks by suppressing the immune system and weakening the fish, which may likely succumb to disease. The key activity in restoration of polluted environments and anticipation of man-made effects on environmental changes is the monitoring of environmental parameters (man-made, natural chemicals, microbiological and biological characteristics). A major challenge exists to monitor aquatic pollution for sustainable fish and fisheries. A prerequisite is to create policy measures and use holistic approaches against aquatic pollution by enforcing these policies for manufacturing industries, factories, sewage disposal plants, and other sources of pollution.

References

Afshan, S., Ali, S., Ameen, U. S., Farid, M., Bharwana, S. A., Hannan, F., & Ahmad, R. (2014). Effect of different heavy metal pollution on fish. *Research Journal of Chemical and Environmental Sciences*, *2*(1), 74–79.

Ali, S., Farid, M., Bharwana, S. A., & Ahmad, R. (2013). Effect of different heavy metal pollution on fish. *Research Journal of Chemical and Environmental Sciences*, *2*(1), 74–79.

Ansari, T., Marr, I., & Tariq, N. (2004). Heavy metals in marine pollution perspective-a mini review. *Journal of Applied Sciences*, *4*(1), 1–20.

Arukwe, A. (2001). Cellular and molecular responses to endocrine-modulators and the impact on fish reproduction. *Marine Pollution Bulletin*, *42*(8), 643–655.

Au, D. (2004). The application of histo-cytopathological biomarkers in marine pollution monitoring: A review. *Marine Pollution Bulletin*, *48*(9–10), 817–834.

Austin, B. (1998). The effects of pollution on fish health. *Journal of Applied Microbiology*, *85*(S1), 234S–242S.

Bucke, D. (1993). Aquatic pollution: Effects on the health of fish and shellfish. *Parasitology*, *106*(S1), S25–S37.

Castro-González, M., & Méndez-Armenta, M. (2008). Heavy metals: Implications associated to fish consumption. *Environmental Toxicology and Pharmacology*, *26*(3), 263–271.

Crook, D. A., Lowe, W. H., Allendorf, F. W., Erős, T., Finn, D. S., Gillanders, B. M., & Hughes, J. M. (2015). Human effects on ecological connectivity in aquatic ecosystems: Integrating scientific approaches to support management and mitigation. *Science of the Total Environment*, *534*, 52–64. Available from https://doi.org/10.1016/j.scitotenv.2015.04.034.

Cruz-Landim, C., Abdalla, F. C., & Cruz-Höfling, M. (2005). Morphological changes of Sertoli cells during the male reproductive cycle of the teleost Piaractus mesopotamicus (Holmberg, 1887). *Brazilian Journal of Biology*, *65*(2), 241–249.

Dar, G. H., Bhat, R. A., Kamili, A. N., Chishti, M. Z., Qadri, H., Dar, R., & Mehmood, M. A. (2020). *Correlation between pollution trends of freshwater bodies and bacterial disease of fish fauna. In Fresh water pollution dynamics and remediation* (pp. 51–67). Singapore: Springer.

Dar, G. H., Dar, S. A., Kamili, A. N., Chishti, M. Z., & Ahmad, F. (2016). Detection and characterization of potentially pathogenic Aeromonas sobria isolated from fish Hypophthalmichthys molitrix (Cypriniformes: Cyprinidae). *Microbial Pathogenesis*, *91*, 136–140.

Dar, G. H., Kamili, A. N., Chishti, M. Z., Dar, S. A., Tantry, T. A., & Ahmad, F. (2016). Characterization of *Aeromonas sobria* Isolated from Fish Rohu (*Labeo rohita*) collected from polluted pond. *Journal of Bacteriology & Parasitology*, *7*(3), 1–5. Available from https://doi.org/10.4172/2155-9597.1000273.

Harries, J. E., Sheahan, D. A., Matthiessen, P., Neall, P., Rycroft, R., Tylor, T., & Sumpter, J. P. (1996). A survey of estrogenic activity in United Kingdom inland waters. *Environmental Toxicology and Chemistry: An International Journal*, *15*(11), 1993–2002.

Helmstetter, A. J., Papadopulos, A. S., Igea, J., Van Dooren, T. J., Leroi, A. M., & Savolainen, V. (2016). Viviparity stimulates diversification in an order of fish. *Nature Communications*, *7*(1), 1–7.

Jobling, S., Sumpter, J. P., Sheahan, D., Osborne, J. A., & Matthiessen, P. (1996). Inhibition of testicular growth in rainbow trout (*Oncorhynchus mykiss*) exposed to estrogenic alkylphenolic chemicals. *Environmental Toxicology and Chemistry: An International Journal*, *15*(2), 194–202.

Jones, J. C., & Reynolds, J. D. (1997). Effects of pollution on reproductive behaviour of fishes. *Reviews in Fish Biology and Fisheries*, *7*(4), 463–491.

Khatri, N., & Tyagi, S. (2015). Influences of natural and anthropogenic factors on surface and groundwater quality in rural and urban areas. *Frontiers in Life Science*, *8*(1), 23–39.

Kime, D. E. (1995). The effects of pollution on reproduction in fish. *Reviews in Fish Biology and Fisheries*, *5* (1), 52–95.

Masindi, V., & Muedi, K. L. (2018). Environmental contamination by heavy metals. *Heavy metals*, *10*, 115–132.

Mézin, L. C., & Hale, R. C. (2000). Effects of contaminated sediment on the epidermis of mummichog, Fundulus heteroclitus. *Environmental Toxicology and Chemistry: An International Journal*, *19*(11), 2779–2787.

Murugan, S. S., Karuppasamy, R., Poongodi, K., & Puvaneswari, S. (2008). Bioaccumulation pattern of zinc in freshwater fish *Channa punctatus* (Bloch.) after chronic exposure. *Turkish Journal of Fisheries and Aquatic Sciences*, *8*(1), 55–59.

Pandey, A. (2000). Aquatic pollution and fish reproduction: A bibliographical review. *Indian Journal of Fisheries*, *47*(3), 231–236.

Perwaiz, A. (2020). Pesticides toxicity in fish with particular reference to insecticides. *International Journal of Research and Analysis in Science and Engineering*, *1*(1).

Sabra, F. S., & Mehana, E. (2015). Pesticides toxicity in fish with particular reference to insecticides. *Asian Journal of Agriculture and Food Sciences*, *3*(01). *(ISSN: 2321–1571)*.

Sabullah, M., Ahmad, S., Shukor, M., Gansau, A., Syed, M., Sulaiman, M., & Shamaan, N. (2015). Heavy metal biomarker: Fish behavior, cellular alteration, enzymatic reaction and proteomics approaches. *International Food Research Journal*, *22*(2), 435–454.

Smith, A., & Gangolli, S. (2002). Organochlorine chemicals in seafood: Occurrence and health concerns. *Food and Chemical Toxicology*, *40*(6), 767–779.

Srivastava, A. K., & Srivastava, A. K. (1994). Effects of chlordecone on the gonads of freshwater catfish, Heteropneustes fossilis. *Bulletin of Environmental Contamination and Toxicology*, *53*(2), 186–191.

Stott, G., McArthur, N., Tarpley, R., Sis, R., & Jacobs, V. (1980). Histopathological survey of male gonads offish from petroleum production and control sites in the Gulf of Mexico. *Journal of Fish Biology*, *17*(5), 593–602.

Zeitoun, M. M., & Mehana, E. (2014). Impact of water pollution with heavy metals on fish health: Overview and updates. *Global Veterinaria*, *12*(2), 219–231.

CHAPTER

6 Bacterial diseases in fish with relation to pollution and their consequences—A Global Scenario

J. Immanuel Suresh, M.S. Sri Janani and R. Sowndharya
Department of Microbiology, The American College, Madurai, India

6.1 Introduction

Fish comprise approximately 35,500 diverse species of vertebrates of the phylum Chordata; they originate in both fresh and salt waters. The study of fish (ichthyology) has broad significance, as they are the oldest vertebrates and can live in nearly any location with enough water, from the highest mountain streams down to the deepest ocean abysses (Warren, 1991). Nearly all are cold-blooded (with the exception of one species, the opah, *Lampris guttatus*, which is warm-blooded). Cold-blooded fish can change the coldest, most uninviting waters into a realm of high productivity (Leahy & Colwell, 1990). Fish have been consumed by humans from ancient times (Leung & Bates, 2013). The global marine fish catch for human food is around 80 million tonnes per year, with 13 million tonnes of invertebrates being caught, such as shrimp, squid, mussels, etc. Fish diseases are current major challenges that disrupt the stable provision of fish globally (Rico et al., 2012). Fish do not normally carry microorganisms like *Enterobacteria* or *Campylobacter* that are found in land animal bodies and are the causes of foremost foodborne illnesses (though fish might acquire such microorganisms from contaminated water).

Most microorganism pathogens are considered to be part of the conventional small flora of the marine environment and are usually thought to be beneficial microbes. Illnesses within a host or its bodily processes are most likely to arise due to various changes in inputs (Arkoosh et al., 1998; Dar, Dar et al., 2016; Dar, Kamili et al., 2016). Some ethnic practices in preparation of fish for consumption place a small number of individuals at high risk from food poisoning; thus when examining various types of diseases related to microorganisms in fish species, one should be careful to perceive the link between the microorganism, host, and environment. However, this is not often a major hazard for most shoppers for fish. Besides microorganism infections in fish, marine pollution has direct and serious impacts on both marine and terrestrial environments. Thus there is a need to investigate the relationships between marine life, human populations, and the atmosphere (Austin, 1999).

6.2 The major sources of marine pollution

To grasp the impacts of marine and coastal pollutants, knowledge of the different major types of sources and pollutants is required, as shown in Fig. 6.1 (Dar et al., 2020).

Bacterial Fish Diseases. DOI: https://doi.org/10.1016/B978-0-323-85624-9.00022-1

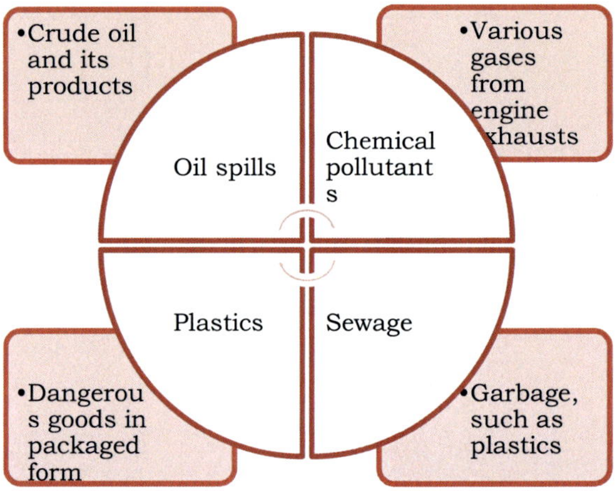

FIGURE 6.1

Sources of marine pollution.

Table 6.1 Marine pollutants of global concern.

S. no	Pollutants	Description
1.	Fossil oil	Hydrocarbon fossil oil and a number of its refined products
2.	Sewage	Wastewater usually from domestic and industrial processes, but might contain a number of cyanogenic substances
3.	Halogenated hydrocarbons	Includes compounds like the PCBs and their degradation and combustion products (e.g., polychlorinated dioxins and dibenzofurans)
4.	Other organic compounds	Endocrine-disrupting chemicals, also called estrogenic chemicals
5.	Trace metals	Notably mercury and other metallic elements (cadmium and lead)
6.	Radionuclides	Particularly cesium-137, strontium-90, and seven plutonium isotopes
7.	Litter (solid waste)	Particularly persistent plastics and lost fishing nets

These pollutants, depending on their concentrations or magnitudes, have negative impacts on marine life, as summarized in Table 6.1.

Several atmosphere contaminants enter into the coastal environment, posing a serious threat to marine organisms and motility and creating a public health risk.

6.3 Bacterial pathologic processes in fish fauna

Bacteria are found everywhere within the marine environment and a variety of diseases can be activated by these microorganisms, which could result in significant mortality in farmed fish

(Cahill, 1990; Dar, Dar et al., 2016; Dar, Kamili et al., 2016). The range of microorganisms in fish isolated from skin, eggs, intestines, and gills had been delineated for a restricted variety of fish species, and also a variety of microorganism genera, isolated from totally diverse fishes, was found to be associated with the aquatic environment of the fish and varied with features like the salinity of the water.

The microorganisms recovered from the gills and skin could also be ephemeral instead of inhabitant on the fishes (Peeler & Taylor, 2011). Microbiological studies carried out on fish from watercourses in African nations (Olayemi et al., 1990) stated that the microbe flora of lungs, skin, and canal of 65 fish contained 16 microorganism genera, including *Escherichia coli*, *Enterobacter aerogenes*, *Klebsiella pneumonia*, *Edwardsiella tarda*, *Aeromonas hydrophila*, *Acinetobacter*, *Staphylococci* and *Micrococcus*. Many marine fish are normally affected or isolated by *Eubacteria* spp., whether or not water animals are largely affected or isolates including *Aeromonas* and *Pseudomonas* spp.

A number of microorganisms can lead to an identical syndrome, generically called pasteurellosis and characterized by external reddening and hemorrhage within the serosa, body wall, and entrails (Nematollahi et al., 2003). Morbidity and mortality are extremely variable, depending on predisposing conditions like concentrations of elements, different water quality issues, stress, or trauma. Lesions are common as the disease progresses, and mortality is high if the stress isn't controlled; fish thus suffer from pollutants due to human activities (Duraisamy & Latha, 2011).

When an organism consumes several other organisms that contain pollutants, the pollutant accumulates more rapidly within it. As the pollutant travels up the food chain, its concentration increases (Austin, 1999). Later, consumers that eat the affected organism accumulate even higher levels of pollutants. This leads to greater chemical contamination of the environment, thus increasing the potential stresses on marine organisms in many of their exposed habitats.

6.4 Impacts of pollution and act

National governments have an obligation to shield our marine resources, according to the Indian Judiciary, which, when coping with cases regarding the environment thought of the right to water and a clean environment as basic to life and upheld it as a basic right. The Judiciary has played an important role in interpreting Article 21 of the Indian Constitution, which guarantees the Right to Life.

The scope of Article 21 of the Constitution has been significantly expanded by the Indian Supreme Court, which has understood the Right to Life to include the right to unpolluted water and additionally the right of access to safe drinking water (Bhavesh & Bharad, 2019). After the Stockholm Conference in 1976, to support environmental conservation, the Indian Constitution was amended by the 42nd Amendment Act. By virtue of this amendment, Article 48-A placed the environment as the responsibility of the government beneath the Directive Principles of State Policy. Article 51-A (g) made environmental protection and conservation a basic duty of all the citizens of India.

The many negative impacts of human activities on the environment should be measured and evaluated, and relevant policies should be developed to conserve the environment (Long et al., 1995). These policies, rules, and laws additionally must be adequately enforced, not just at the international level, but also at the local and national levels (Speare & Ferguson, 1989). Also, numerous international agreements and conventions for environmental protection must be legitimized and legalized by the individual countries of the planet, to make them a part of each country's system. This active participation of all the countries of the planet is vital, as the harmful effects on the environment don't affect just one nation or person in isolation, but instead have a profound impact on the whole planet and every living thing inhabiting the planet (Bhavesh & Bharad, 2019).

A clean and healthy environment is basic to the survival of human civilization; however, ironically, assorted human activities are mainly responsible for inflicting large-scale pollution on the land and in the oceans and air. The abuse and excessive exploitation of natural resources have led to depletion of those resources, which are crucial to human survival.

6.5 Contaminants in the marine environment

There is a great deal of evidence indicating the existence of specific contaminants or pollutants in marine environments, which include the following substances. Hydrocarbon contaminants result from intentional leakage during periods of war or terrorism (Newton & McKenzie, 1995), and also from accidental discharge from tankers (Bourgoin, 1990). Heavy metals such as arsenic (As), copper (Cu), zinc (Zn), cadmium (Cd), lead (Pb), and mercury (Hg) are found in industrial wastes; tin, tributyltin, and triphenyltin are found in fouling resistant dyes used on the undersurfaces of ships to prevent bioattachment and biofouling. Plastics may also cause pollution in marine environments. Inorganic material such as ammonia (NH_3), nitrite (NO_2-), and nitrate (NO_3-) may be obtained from agricultural runoff and found in the marine environment (Thomlinson et al.,1980). Organic material together with fecal debris enter the marine environment from drainage systems. Pesticides, including dioxin, 2,3,7,8-tetrachlorodibenzo-p-dioxin (TCDD), 1, 1, 1-trichloro-2,2-bisi (p-chlorophenyl) ethane (DDT), organochlorines, and pulp mill effluents are major marine environment contaminants.

6.6 Symbiotic microflora in fish

Microflora are a vitally necessary part of the digestive tract in many animals. The physiology of digestion in mammals and fish are similar in some respects (Arkoosh et al., 1998).

The following microorganism taxa are found in fish: the genera *Aeromonas*, *Pseudomonas*, and *Flavobacterium*. True bacteria families are *Enterobacteriaceae* and *Coryneforms* in freshwater fish and the genera *Vibrio* and *Pseudomonas* in marine fishes. The viscera of freshwater fish house *Acinetobacter*, *Enterobacter*, enteric bacteria, *Klebsiella*, *Proteus*, *Serratia*, *Aeromonas (A. caviae, A. hydrophila, A. jandaei, A. sobria*, and *A. veronii)*, *Alkaligenes*, *Eikenella*, genus *Bacteroides*, *Listeria*, *Hafnia alvei*, *Flexibacter*, *Citrobacter freundii*, *Propionibacterium*, *Bacillus*, *Moraxella*, *Pseudomonas*, and *Staphylococcus*.

In marine fish, *Aeromonas*, *Alkaligenes*, *Alteromonas*, *Carnobacterium*, *Flavobacterium*, *Micrococcus*, *Photobacterium*, *Pseudomonas*, *Staphylococcus*, and *Vibrio (V. iliopiscarius)* have been established within the intestines (Bentzon-Tilia et al., 2016). These are the main microflora related to fish. The consumption of infected fish and their products may result in acute or chronic sickness.

Seafood could be an important transmission factor in many microorganism infections. Estuaries and coastal water areas are the main foundations of food globally and are usually contaminated by the activities of the adjacent population and partially treated or unprocessed waste products are released into these water bodies. Industrial products are recognized as a serious carrier of foodborne pathogens like *Enteric* sp., *Staphylococcus aureus*, *Vibrio cholerae*, *Vibrio parahaemolyticus*, *Yersinia enterocilitica*, listeria, *Campylobacter jejuni*, and *E. coli*.

Various characteristics relating to contagious diseases from pathogens in aquatic fauna:

1. A comparatively small number of infective microorganisms are responsible for important economic losses in cultured fish globally.
2. Numerous conventional infectious organisms thought of as typical of freshwater aquaculture are today major problems in marine culture.
3. The external and internal clinical signs elicited by a microorganism depend upon the host, species of fish, and the stage of the disease.
4. Sometimes there is no relationship between external clinical signs of disease and its actual severity; also, mortality from the same pathogens is greater in cultivated fish than in wild fish (Kubečka et al., 2016), probably due to less stress on the wild fish than the levels typically occurring within fish farms.

In India, the role of the food sector in providing economic and biological process security is a major one (Ponnerassery & Sudheesh, 2012). The rising demand for food nationwide and internationally ends up in the production of unscrupulous, under processed and unhealthful products that can harbor numerous species of microorganisms and infective foodborne pathogens.

The presence of these foodborne pathogens causes huge financial losses to fishermen and also to exporters (Heo et al., 1990). Usually, foodborne outbreaks are not properly documented in developing countries, as is done in their Western counterparts; therefore, a fewer number of reports and studies are available in these countries.

In general, marine microorganisms represent unexploited sources of effective remedial and novel drugs (Dadar et al., 2016). The appearance of multidrug-resistant microorganisms constitutes a serious health hazard today, which forces the development of successive effective drugs and vaccines to replace the conventional drugs that are no longer effective.

6.7 The marine environment and its issues in India

The most important predictable sources of marine pollution in India are described in this paragraph. Discarding hazardous substances into the oceans leads to harmful effects on marine ecosystems worldwide. Pollution in the environment, such as extremely contaminated air, plastic bags; acid rain, and lethal gases emitted from industries, along with the weakness of the ozone layer, also has damaging effects on organisms on the water surface and life beneath the surface. In addition, deep seafloor mining causes disturbances to the environments of aquatic lifeforms (Mishra et al., 2017).

Pollution from land-based activities like agriculture, construction, mining, and other commercial activities affects the quality of water and the habitat of the species inhabiting that specific ecosystem. Pollution emitted from ocean-going ships also disrupts life under water.

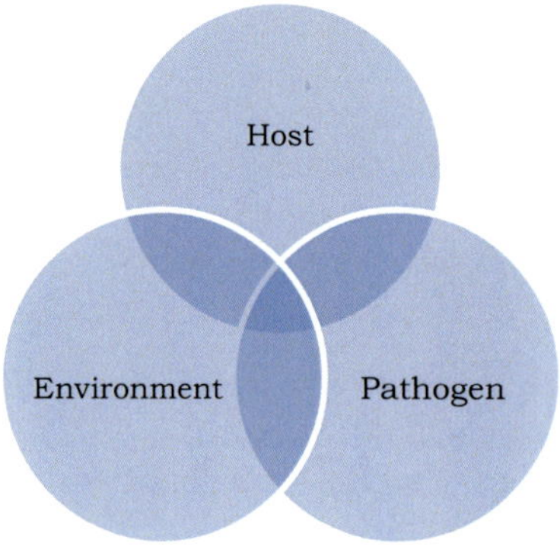

FIGURE 6.2

Interactions between host, environment, and pathogen.

In recent times, the increase in water contamination and vigorous growth in marine fish culture have led to an assortment of fish diseases, such as gill infections and several common diseases of the skin, that have increased in aquaculture fish and that cause serious problems for the aquaculture industry annually.

Oil spills, heavy metal contaminants, pesticide pollution, the effects of agriculture, and domestic wastes are main sources of pollution that contaminate the marine environment. This environmental pollution is occurring in the Bay of Bengal, the Arabian Sea, and the Indian Ocean. Nevertheless, there is only narrow evidence that pollutants are essentially responsible for the development of bacterial fish diseases (Fig. 6.2).

6.8 The outcomes of pollution and bacterial infection in fish fauna

The influence of toxic wastes on fish can lead to the activating of disease conditions. Unambiguously, man-made pollution contributes to the occurrence of infection in fish. It increases the vulnerability of the host to microbes that are ever-present in the environment (Arkoosh et al., 1998). Exposure of marine animals to high concentrations of pollutants may rapidly lead to their death. External lesions on the fins and the opercula, particularly in skin and gills, were reported following exposure to an extensive assortment of chemicals and compounds, including cadmium, creosote, copper, phenol, ammonia, and effluents containing chloride.

6.9 Bacterial pathogens causing diseases in fish due to the effect of marine pollution

As stated earlier, even though highly infectious species from the mainstream of microorganism taxa are involved in bacterial fish diseases, only a comparatively small number of microbes are responsible for the main diseases that lead to high mortalities and economic losses worldwide.

6.9.1 Vibrios

The genus *Vibrio* comprises gram-negative bacteria that are principally infectious to marine and estuarine fish. Vibrios is a leading organism that affects fish and crustaceans, predominantly infectious species like *Vibrio anguillarum, Vibrio harveyii (V. carchariae), V. ordalii*, and *Aliivibrio salmonicida* (previously *Vibrio salmonicida*). The mass of genes that encrypt cell functions and infectious factors are situated within the large fishes. Fishes with tiny body typically encloses genes for ecological acclimatization (Arkoosh et al., 1998).

V. anguillarum is generally considered an etiological agent of vibriosis. The most virulent strains of *V. anguillarum*, such as the O_1 and O_2 serotypes, are often isolated from diseased fish. The O_1 serotype strains cause disease in soft-finned fish, whereas O_2 β strains are typically isolated from cod and other nonsalmonids (Frans et al., 2011).

Vibrio ordalii, formerly called *V. anguillarum* genotype II, may be a very close relation of *V. anguillarum*. Vibriosis caused by these two species has well-supported microscopic anatomy confirmations. *V. anguillarum* has a unique affinity for blood and soft tissues, while *V. ordalii* predominantly presents as aggregates in internal organ muscles and skeletal muscles. *V. ordalii* can cause hemorrhagic septicemia, like *V. anguillarum*, but bacteremia develops later than in *V. anguillarum* infections.

A. salmonicida (previously *V. salmonicida*) is a gram-negative organism that causes coldwater vibriosis in fish; the disease causes deterioration of tissues, rupture of red blood cells, and generalized septicemia in the blood. Order analysis clearly states that A. salmonicida has a rigid system of its order, leaving behind huge target genes and novel genes and has become host-limited, allowing the infectious agent to adapt to novel functions.

6.9.2 Aeromonas

A. hydrophila and other motile aeromonads are common bacteria widely distributed in aquatic environments, found in intake water, chlorinated or well water, sewage, and profoundly contaminated waters, and are often correlated with severe disease among both warm- and cold-blooded vertebrates (Aberoum & Jooyandeh, 2010). The etiology of infectious aeromonad diseases has been investigated using organic chemistry, genetics, and matter nonuniformity of elements of this cluster.

A. hydrophila causes a septic disease with a number of names: "motile aeromonas septicemia," "hemorrhagic blood poisoning," "ulcer disease," or "red-sore disease" in fish (Kim et al., 2017). The sickness induced by the bacteria chiefly affects species of freshwater fish and many species of tropical or ornamental fish. *A.* salmonicida is a motivative factor of microorganism pasteurellosis and is becoming an important pathogen in freshwater fish with major economic impact ,(Ajmal & Hobbs, 1967).

6.9.3 Flavobacterium

Flavobacteria are rod-shaped bacteria omnipresent within soil, freshwater, and seawater environments and are noted for their novel gliding motility. Some species have potential for bioremediation of chemical compounds and organic matter like hydrocarbons. However, a few species are pathogenic to fish. *F. psychrophilum* is the etiological agent of bacterial coldwater disease and rainbow trout fry syndrome (Starliper, 2011). It is a significant factor, having large economic impact and making fish rearing difficult in cultivation.

F. branchiophilum is the main causal agent of bacterial gill disease in many areas worldwide (Madetoja et al., 2002). This disease is distinguished by dangerous morbidity and death rates due to massive bacterial colonization of gill lamellar surfaces and cumulative respiratory organ pathology stemming from high rates of lamellar epithelial necrosis.

Typically, throughout epidemics, the morbidity and death rates increased due to bacterial gill disease (BGD) (Touchon et al., 2011). BGD is a faster moving disease that kills more fish faster than other types of fish infections. In addition, in contrast to the process of necrosis in bacterial gill disease, fish with columnaris infection (*F. columnare*) can have harsh necrosis of all components of the gill because the microorganism infects inside.

6.9.4 Shigella flexneri

Shigella flexneri is a gram-negative bacterium that damages the epithelium layer in fish. The disease caused by this pathogen is known as shigellosis (Gudding, 2014). Up to 165 million cases of shigellosis are estimated per annum, resulting in up to half a million fish deaths. Shigellosis in fish is influenced by pollutants. When fish are exposed to oil spills in marine waters, the mature fish might experience diminished growth, changes in heart and respiration rates, enlarged livers, fin erosion, and reproduction interference (Arkoosh et al., 1998).

6.9.5 Enterobacter amnigenus

Enterobacteriaceae can cause common waterborne fish bacterial infections whenever fish are exposed to stress, such as high temperature and poor water quality. These bacteria commonly reside in the tissues and the gastrointestinal tracts of healthy fish. The stress conditions lead to dangerous outbreaks that cause increased fish deaths (Arkoosh et al., 1998). *Enterobacter amnigenus* can also produce diseases as a primary pathogenic as well as an opportunistic infection (Table 6.2).

Table 6.2 Bacterial disease causing pathogens in relation to marine pollution in fishes.

S. no	Organism	Disease in fishes
1.	*Vibrio harveyi*	Vibriosis
2.	*Shigella flexneri*	Shigellosis
3.	*Enterobacter amnigenus*	Blood infections
4.	*Salmonella typhimurium*	Salmonellosis
5.	*Pseudomonas aeruginosa*	Respiratory tract infections
6.	*Flavobacterium psychrophilum*	Bacterial coldwater disease
7.	*Flavobacterium branchiophilum*	Bacterial gill disease
8.	*Aeromonas hydrophila*	Ulcer disease

6.10 Immune responses in fish

The immune systems of fish incorporate both specific immune responses and nonspecific immune responses. Fish have a substantial barrier that serves as an initial line of defense against antigens (Immanuel Suresh & Stella Mary, 2016). The skin and mucus contain immune reactive molecules that are nonspecific in nature. Both immune responses play a vital role in all stages of infections.

6.11 Fish diseases and their consequences

Many fish diseases, especially those caused by bacteria, can lead to serious economic losses. Although modern human civilization and scientific developments are impressive, our environment is facing numerous setbacks (Sorensen & Larsen, 1986). Consumption of plastic, hydrocarbons, heavy metals, and pesticides, along with their waste, causes pollution, and marine ecosystems are damaged.

These pollutants enter the aquatic environment rapidly and affect the circadian system, leading to immunosuppression as well as damage to gills and the epithelial layer in fish; they may also be increasing fish sensitivity to various bacterial diseases (Frerichs, 1989). The bacterial population differs widely with the pollution conditions. With exposure to a large level of pollution at the water interface, the bacterial population range may increase from a few hundred to a million per milliliter. Most marine bacteria are classified under psychrophiles and halophiles.

Bacterial agents like *V. harveyi*, *S. flexneri*, *E. amnigenus*, *Salmonella typhimurium*, *Flavobacterium* spp., *Aeromonas* spp., and *Pseudomonas aeruginosa* are the source of countless bacterial illnesses in fish.

6.12 Pathogenomics

Genome sequencing of microorganisms, particularly microbial pathogens, has greatly increased our knowledge and understanding of the evolutionary relationships between pathogenic and nonpathogenic species. Also, it has shown how each has developed specialized adaptations benefiting their particular infectious lifestyle (Ponnerassery & Sudheesh, 2012). Over the long term, an understanding of their genomes and biology can enable scientists to design ways of disrupting their infectious behavior.

The genomes of a bacterium are made up of circular or linear chromosomes, extrachromosomal linear or circular plasmids, along with different combinations of those molecules (Rothberg & Leamon, 2008). Functionally associated genes are gathered closely together, and those genes situated on the "core" of the chromosome show a comparatively uniform G + C content along with a particular codon usage (Pallen & Wren, 2007). Bacteria that are closely related usually have similar genomes (Ponnerassery & Sudheesh, 2012).

The size of bacterial genomes is principally the result of two processes that counteract each other: the acquisition of new genes by gene replication or by horizontal gene transfer; and the removal of unnecessary genes. Genome sequencing of bacterial fish pathogens of fish has assisted with greater understanding of their genetics and biology. The genomic flux generated by these gains and losses of genetic information considerably modifies gene content. This method leads to divergence of bacterial species and ultimately their adjustment into new biological functions and ecological niches.

Bacterial pathogens are a significant cause of contagious infections and death in both wild fish and fish nurtured in constrained conditions. At the host-pathogen level of interaction, there is pressure on the bacteria to acclimatize to the harsh host environment, but also to adapt to the ever-changing external environment. With the increase of fish cultivation, greater utilization of water bodies, contamination, globalization, and transboundary movement of aquatic organisms, the list of recent infectious bacteria isolated from fish has been steadily escalating (Romero & Navarrete, 2012). Researchers are searching for better vaccines and drugs to use against harmful bacteria.

6.13 Plastic pollution cause adverse effects in the marine environment

Plastics, which are synthetic or semisynthetic polymerized products, are the most widespread material in use. They are compound substances encompassing thousands of "monomers," that is, a molecule repeated in a long sequence to create a polymer. They are derived from petroleum, from fossil feedstock.

Sandstones and limestone are the main locations where oil is found. Plankton and algae in a chemical reaction in an anoxic environment form kerogen (Tripathi, 2019). Kerogen, once buried deep with high temperature and pressure, forms oil, which rises, being lighter than water, until it is trapped in geological structures fashioned by tectonic processes. Once the crude oil is obtained by drilling, various processes are carried out to produce ethylene.

The monomer ethylene is then polymerized, using different catalysts, to form plastics of various varieties. Typically elements like nitrogen, oxygen, sulfur, chlorine, etc. are added in different quantities to produce a wide range of plastic resins. Thus plastics come from oil and the assumption is that if the oil supply runs out, then there would be no more plastics.

The plastics are flexible, long-lasting, cheap, lightweight, corrosion resistant, and water resistant. Plastics are found in every part of human life; they completely changed the existing designs when they emerged. The advent of the plastics industry 50 years ago was what is now called a "disruptive technology."

On the one hand, plastics are seen as a solution for climate change. Substances like glass fibers or carbon fibers can be added to plastics to create composite materials. These composites could serve as substitutes for wood, steel, or glass. There is energy savings when substances like wood or ore used to make steel are replaced and the carbon dioxide footprint turns out to be smaller and greenhouse gas emissions are reduced (Malti Goel et al., 2019).

Plastics comprise the polymers shown in Fig. 6.3. Generally PE and PP are less dense than ocean water, and have a tendency to float, whereas PS, PA, and PET are denser and they tend to sink.

Plastics are principally of two categories: thermoplastics and thermosets. Thermoplastics have long linear chemical compound series that are weakly bonded. Once a thermoplastic article is heated, these bonds are easily broken, and so they can be melted and remade into alternative items,

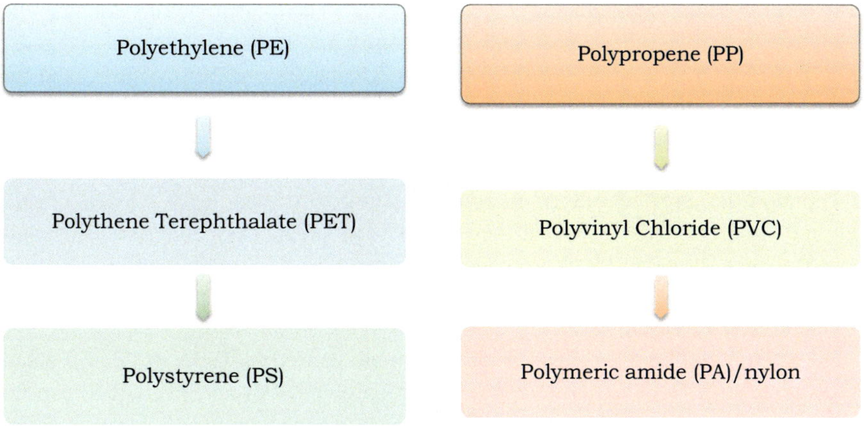

FIGURE 6.3

Plastics as organic polymers of molecular groups encompass primarily the constituents shown.

that is, they are recyclable. Thermosets are created by linear chains that are cross-linked, strongly bonded, and don't melt (Tripathi, 2019). They are hard to recycle, although there are new methods for doing so, such as crushing thermoset articles into fine powder, that can then be used as a filler in reinforced thermosets (Malti Goel et al., 2019).

Bioplastics are biodegradable plastics—microorganisms can break them down into carbon dioxide and water—obtained from renewable biomass sources, like vegetable fats and oils, corn starch, or microbiota. They are made up of agricultural byproducts and also from used plastic bottles and other containers treated by microorganisms. Bioplastics may be the best way for producing single-use plastics.

6.14 The impact of plastic pollution in urban India

Plastic pollution means that deposits of plastic merchandise within the environment harmfully affect life, water areas, the environment, and individuals. Based on their constituent chemical compounds, plastics have features of contaminant, absorption, and adsorption. Degradation of the polymers takes a long time. It is calculable that a styrofoam plastic cup takes 50 years to degrade; a plastic food holder takes 400 years; a disposable nappy takes 450 years, and plastic fishing line takes 600–1000 years to degrade (Tripathi, 2019).

Many birds, mammals, and water lifeforms like fish and turtles have died from being tangled in plastic debris. Plastics can also cause cancer or produce endocrine issues. Chemicals employed in plastic fabrication can cause dermatitis upon contact with human skin (Malti Goel et al., 2019).

Chlorinated plastics release injurious substances into the soil, that then run into groundwater and the ecosystem. Bacteria such as nylon-eating bacterium and *Flavobacteria* disintegrate nylon via the action of the nylonase accelerator. Disintegration of perishable plastics releases methane, a strong greenhouse gas that contributes to global warming (Actis et al., 2011). Harmful chemicals such as bisphenol A and vinylbenzene seep into water. The plastic toxins include diethylhexyl phthalate, which is a toxic carcinogen, cadmium, mercury and lead.

Plastic wastes are classified into meso (from 5 to 20 mm), micro (up to 5 mm) and macro (over 20 mm). Microplastics pollute water bodies and gaseous atmosphere. Municipal waste management has no system for separation of plastics, except for rag pickers. The plastic usually ends up in sewers, streams, and open areas.

Plastic commodities created out of thermocol, styrofoam, clear and colored plastics tend to break down into smaller elements. When they reach water areas, they either stay suspended or settle into the sediment. They can block transfer of oxygen and water percolation in the soil. Due to their long life, they can enter the food chain (Rubio & Rauret, 1996), as microplastics in marine debris can be consumed by fauna.

Ignition of plastic waste, and fires on landfills that are caused by leaks of landfill gas due to disintegration of waste, can lead to incomplete combustion of PP, PS, and PE, which can then lead to high concentrations of carbon monoxide (CO) and harmful emissions. PVC burning produces dioxins, carbon black, and aromatics like pyrene and chrysene. Incineration of brominated and chlorinated plastic results in releases of furans and dioxins, resulting in emanations of oxide, carbon dioxide, and sulfur oxide. It also can generate explosive organic components, smoke, and polychlorinated dibenzofurans.

There are other means of managing plastic waste, like pyrolysis, which involves heating plastic substances at high temperatures in the absence of oxygen; however, this results in pollutant emissions such as hydrogen sulfide, hydrogen chloride, ammonia, and hydrogen cyanide (Table 6.3).

Exposures to single-use plastics take place as these plastics are cut up, but not crushed, and are typically reprocessed or end up in landfills (Hjeltnes & Roberts, 1993). Plastics are the major component of marine pollution. They transfer the organic phenomenon by transporting pollutants.

Three-quarters of all diseases in India are caused by waterborne pathogens (Seshadri et al., 2006). The main potential source is sewage-fed fisheries, and organic management in pools and drains. These strategies supply proceeds additionally to help as an effluent-treatment method.

Table 6.3 Several plastic toxins and their impacts on health.

S. no	Plastic toxins	Impacts on health
1.	Phthalates	Sex hormone compound, disrupts endocrine function and reproduction systems, acute toxicant to aquatic microbes, algae, fish, invertebrate
2.	BFRs (Brominated Flame Retardants)	Thought of as the most unsafe; impact by toxins that alter hormonal function
3.	BPA (Bisphenol-A)	Endocrine disruptor, neuro-, reproductive, and organic process toxicity, cancer risk (breast, prostate, etc.)
4.	Polyvinyl resin and vinyl chloride	Malignant neoplastic disease, mutagenic
5.	Solvents, initiators, catalysts	Toxics are often combustible, cause metabolism and skin issues
6.	Heavy metals—Lead, cadmium	Mutation causes severe damage to genes

6.15 Substitutive uses

Energy recovery is a vital approach to utilizing plastics. Processes like burning and biotransformation are utilized in waste-into-energy plants and they permit changing plastic wastes into oil, and plastic into declined fuel is utilized in cement and brick ovens (Zhang et al., 2011). Plastic waste, when combined with hydrocarbon, is additionally utilized in building highways.

6.16 The Gulf of Mannar

The Gulf of Mannar (GoM) is often referred to as Biologists' Paradise due to its rich marine ecosystem, with virtually 3650 species of existing flora and fauna, besides the seasonally migrating marine mammals such as whales, dolphins, and turtles. Ganesapandian et al. (2011) reported that it is unique because of the occurrence of coral reefs, seaweed beds, and mangroves that serve as egg laying and nourishing floors, as shelter for several types of vital shellfish and finfish. It spreads from Rameswaram to Kanyakumari within the south geographical region of India. This space is held to be among the richest gulfs in the world in terms of ecological diversity. However, marine waste has become a challenging issues in aquatic settings, and not only esthetically, but also as a health risk to humans and marine species.

Studies have been done and measures undertaken in an attempt to save lives of vulnerable aquatic animals and to enhance the social and economic standing of fisherfolk within the GoM region located in India, and conjointly to provide a foundation for marine waste monitoring, to survey and assess the composition, quantity, and distribution of sources of marine litter in the GoM region.

6.17 Impact on marine environment: pollution in the Bay of Bengal

A large number of fertilizer and chemical industries has conventionally been located near the Bay of Bengal. The Bay of Bengal is one of the latest plastic hotspots in Asia. Annually about 2 lakh tons of plastics come into the Bay of Bengal (Sharif et al., 1993). There are many industries, specifically cement, textiles, chemicals, paper, foodstuff processing, medications, petroleum, heavy metals, and others, located around the banks of the Bay of Bengal. Discharge of waste materials as pollutants from this variety of industries has heavily polluted the marine environment near the Bay of Bengal, and this may adversely affect marine organisms. These pollutants cause severe diseases in marine fishes, stated by Rashid (2014).

Due to these pollutants, bacterial pathogens can easily come in contact with marine organisms and they can dramatically affect their growth. This condition leads to high mortality rate in marine fish. Predominantly the adverse effects occur in species at the pinnacle of the food chain. Undesirable economic losses have occurred due to the death of fish by bacterial diseases associated with marine pollution. Fisheries, with their assisting activities, offer a living for countless individuals within the geographical Bay area region; it's conjointly an expansive supply of interchange for several nations within the states (Holmgren, 1994).

Unfortunately, the coastline of the Bay in this geographical region is worsening. The sources for these conditions are sedimentation, contamination, and unrestrained seaside expansion. A few elements of the coastlines are significantly unhealthy, plagued by factory and urban wastes, as well as by unsystematic progress of brackish water systems. This constitutes a threat to the population of wild finfish as well as to mariculture fish. Sirajul Hoque et al. (2015) stated that the harm done is commonly unintentional, being a consequence of unhealthy systems, insufficient information, and little systematization between firms and jurisdictions. The brunt of atmospheric degradation on fish farming within the Bay of Bengal is somewhat moderate. Biodegradable waste is of explicit concern to all nations around the Bay. Saha et al. (2001) Wastes, with no treatment, are directly released into the densely inhabited coastal regions. Rivers, lakes, ponds, bays, etc., are anoxic for shorter or elongated intervals throughout the year, inflicting fish deaths. Additionally, severe health issues are linked with toxic pollution, which affects the living beings. Only the coastline regions, lagoons, and estuaries in some parts of the areas are affected (Chakraborty, 2003). Algal blooms are uncommon and there are few outbursts of diarrheal shellfish poisoning (DSP) or other diseases. Even wherever more concentrations of metals and serious pesticides are found in the water and in the sediments, the deposits in fish and different aquatic species are still within suggested health limits (Rashid, 2014).

6.18 Interaction between pathogens and aquatic environment

Microorganisms play a crucial role in the marine biosphere, but the interconnection between the aquatic biosphere and microbes can result in diseases. Fish are particularly vulnerable to microbial infections when in high-density circumstances (Mira et al., 2002). Disease epidemics reduce productivity and increase the death rate, leading to increased economic loss.

Diseased fish may also cause infections in humans, including vibriosis, shigellosis, bacterial gill diseases, ulcer disease, and salmonellosis, as reported by Braz and Biol (2019). Aquaculture is a fast-growing sector of food production globally, as it fills a worldwide need for protein. Obviously, due to marine pollution and the growing aquaculture industry, a variety of bacterial fish diseases can appear, such as gill infections and other general infections like skin and inflammatory reactions that are more common in farmed fish and that represent a growing problem each year in the aquaculture industry.

6.19 Bacterial fish diseases and their control

The American Fisheries Society has been charged with the task to clarify environmental problems related to fisheries, including inclination of marine fishes to a variety of bacteria such as *V. harveyi*, *S. flexneri*, *E. amnigenus*, *S. typhimurium*, *Flavobacterium* spp., *Aeromonas* sp., and *P. aeruginosa*. Most common are flavobacterial infections in fish, with flavobacterial diseases being increasingly caused by different bacterial species in the family Flavobacteriaceae in farmed fish stocks and in the wild globally (Loch & Faisal, 2015). Fish are being affected by these species due to marine pollution, which results in increased mortality rates in fish as well as economic losses in the aquaculture industry. Increases in bacterial infections are dependent on the susceptibility of the host, condition of the environment, and virulence of the pathogens (Anderson et al., 1974).

Fryer, D. F. reported that to control the disease incidence, a major factor is the avoidance of exposure to bacterial pathogens. From earlier years, cold-blooded vertebrates were vaccinated, but only a slight improvement resulted because the response was insignificant at low temperatures and the necessity of handling each animal made it impractical and difficult on a large scale (Larsen et al., 1994). A commercial vaccine against enteric redmouth disease (ERM) for immunization is also available. Anderson et al. (1974) reported that a vaccine against *V. anguillarum* had been licensed (Meyer, 1991). Other bacterial vaccines are under development and antiviral vaccines will also most likely be available.

6.20 Conclusion

It has been convincingly demonstrated that pollutants enter the marine environment and are found in the tissues of invertebrates and aquatic vertebrates. Nevertheless, there is inadequate evidence that pollutants are essentially accountable for the occurrence of bacterial disease. It is crucial that aquacultural directors have a better comprehension of the epidemiology, microbiology, immunology, physiological therapy, and ecology and are also capable of creating appropriate evaluations of the disease problem and the application of control measures. The present reviewed study concluded that the marine environment is contaminated with pollutants and it causes adverse effects in marine living organisms. Likewise, good practices are necessary in shipping, fishing, transporting goods, and disposing of industrial wastes, without disturbing the marine ecosystem.

References

Aberoum, A., & Jooyandeh, H. (2010). A review on occurrence and characterization of the *Aeromonas* species from marine fishes. *World Journal of Fish and Marine Sciences*, *2*(6), 519–523.

Actis, L. A., Tolmasky, M. E., & Crosa, J. H. (2011). Vibriosis. In (2nd ed.). P. T. K. Woo, & D. W. Bruno (Eds.), *Fish diseases and disorders, viral, bacterial, and fungal infections* (Vol. 3). Oxfordshire, UK: CABI International.

Ajmal, M, & Hobbs, B. C (1967). Species of Corynebacterium and Pasteurella isolated from diseased salmon, trout and rudd. *Nature*, *215*, 142–143, In press. Available from https://doi.org/10.1038/215142a0, PMID: 6069150.

Anderson, D. P., Snieszko, S. F., & Axelrod, H. R. (1974). *Fish immunology – Diseases of fishes* (p. 239) Neptune City, NJ: TFH Publications.

Anonymous. (2019). *Brazilian Journal of Biology*, *80*, And print version ISSN 1519-6984 (pp. 3–4).

Arkoosh, M. R., Casillas, E., Clemons, E., Kagley, A. N., Olson, R., Reno, P., & Stein, J. E. (1998). Effect of pollution on fish disease: Potential impacts on Salmonid populations. *Journal of Aquatic Animal Health*, *10*, 182–190.

Austin, B. (1999). The effects of pollution on fish health. *Journal of Applied Microbiology*, *85*, 234S–242S.

Bentzon-Tilia, M., Sonnenschein, E. C., & Gram, L. (2016). Monitoring and managing microbes in aquaculture – Towards a sustainable industry. *Microbial Biotechnology*, *9*(5), 576–584.

Bourgoin, B. P. (1990). *Mytilus edulis* shell as a bioindicator of lead pollution: Considerations on bioavailability and variability. *Marine Ecology Progress Series*, *61*, 253–262.

Cahill, M. M. (1990). Bacterial flora of fishes: A review. *Microbial Ecology*, *19*, 21–41.

Chakraborty, R. (2003). Accumulation of heavy metals in some freshwater fishes from eastern India and its possible impact on human health. *Pollution Research*, *22*, 353–358.

Dadar, M., Dhama, K., & Vakharia, V. N. (2016). Advances in aquaculture vaccines against fish pathogens: Global status and current trends. *Reviews in Fisheries Science & Aquaculture*, *25*(3), 184–217.

Dar, G. H., Bhat, R. A., Kamili, A. N., et al. (2020). *Correlation between pollution trends of freshwater bodies and bacterial disease of Fish Fauna. Fresh water pollution dynamics and remediation* (pp. 51–67). Singapore: Springer. Available from http://doi.org/10.1007/978-981-13-8277-2_4.

Dar, G. H., Dar, S. A., Kamili, A. N., Chishti, M. Z., & Ahmad, F. (2016). Detection and characterization of potentially pathogenic *Aeromonas sobria* isolated from fish *Hypophthalmichthys molitrix* (Cypriniformes: Cyprinidae). *Microbial Pathogenesis*, *91*, 136–140.

Dar, G. H., Kamili, A. N., Chishti, M. Z., Dar, S. A., Tantry, T. A., & Ahmad, F. (2016). Characterization of *Aeromonas sobria* isolated from Fish Rohu (*Labeo rohita*) collected from Polluted Pond. *Journal of Bacteriology & Parasitology*, *7*(3), 1–5. Available from https://doi.org/10.4172/2155-9597.1000273.

Duraisamy, A., & Latha, S. (2011). Impact of pollution on marine environment – A case study of coastal Chennai. *Indian Journal of Science and Technology*, *4*(3), 259–262, ISSN: 0974-6846.

Frans, I., Michiels, C. W., Bossier, P., Willems, K. A., Lievens, B., & Zediers, H. (2011). *Vibrio anguillarum* as a fish pathogen: Virulence factors, diagnosis and prevention. *Journal of Fish Diseases*, *34*(9), 643–661.

Frerichs, G. N. (1989). Bacterial diseases of marine fish. *The Veterinary Record*, *125*(12), 315–318. Available from https://doi.org/10.1136/vr.125.12.315.

Ganesapandian, S., Manikandan, S., & Kumaraguru, A. K. (2011). Marine litter in the northern part of Gulf of Mannar, southeast coast of India. *Research Journal of Environmental Sciences*, *5*, 471–478.

Gudding, R. (2014). Vaccination as a preventive measure. In R. Gudding, A. Lillehaug, & O. Evensen (Eds.), *Fish vaccination* (1st ed., pp. 12–21)). Oxford, UK: John Wiley & Sons, Inc.

Heo, G. J., Kasai, K., & Wakabayashi, H. (1990). Occurrence of *Flavobacterium branchiophila* associated with bacterial gill disease at a trout hatchery. *Fish Pathology*, *25*, 99–105.

Hjeltnes, B., & Roberts, R. J. (1993). Vibriosis. In R. J. Roberts, N. R. Bromag, & V. Inglis (Eds.), *Bacterial diseases of fish* (pp. 109–121)). Oxford, UK: Blackwell Scientific.

Holmgren, S. (1994). An environmental assessment of the Bay of Bengal region, Bay of Bengal programme BOBP/REP/67. Swedmer Publications.

Immanuel Suresh, J., & Stella Mary, J. (2016). *Plants as immunostimulants in fish, zebrafish – A model organism for regeneration studies* (pp. 123–140). Tamil Nadu, India: Darshan Publishers.

Kim, A., Nguyen, T. L., & Kim, D.-H. (2017). *Modern methods of diagnosis. Diagnosis and control of diseases of fish and shellfish* (pp. 109–145). John Wiley & Sons, Ltd.

Kubečka, J., Boukal, D. S., Cech, M., et al. (2016). Ecology and ecological quality of fish in lakes and reservoirs. *Fisheries Research*, *173*, 1–3.

Larsen, J. L., Pedersen, K., & Dalsgaard, I. (1994). *Vibrio anguillarum* serovars associated with vibriosis in fish. *Journal of Fish Diseases*, *17*(3)), 259–267.

Leahy, J. G., & Colwell, R. R. (1990). Microbial degradation of hydrocarbons in the environment. *Microbiological Reviews*, *54*(3)), 305–315.

Leung, T. L. F., & Bates, A. E. (2013). More rapid and severe disease outbreaks for aquaculture at the tropics: Implications for food security. *Journal of Applied Ecology*, *50*(1), 215–222.

Loch, T. P., & Faisal, M. (2015). Emerging flavobacterial infections in fish: A review. *Journal of Advanced Research*, *6*, 283–300.

Long, E. R., Smith, S. L., & Calder, F. D. (1995). Incidence of adverse biological effects within ranges of chemical concentrations in marine and estuarine sediments. *Environmental Management*, *19*, 81–97. Available from https://doi.org/10.1007/BF02472006.

Madetoja, J., Dalsgaard, I., & Wiklund, T. (2002). Occurrence of *Flavobacterium psychrophilum* in fish-farming environments. *Diseases of Aquatic Organisms*, *52*(2), 109–118.

Meyer, F. P. (1991). Aquaculture disease and health management. *Journal of Animal Science*, *69*(10), 4201–4208.

Mira, A., Klasson, L., & Andersson, S. G. E. (2002). Microbial genome evolution: Sources of variability. *Current Opinion in Microbiology*, *5*(5), 506–512.

Mishra, S. S., Das, R., Choudhary, P., Debbarma, J., & Sahoo, S. N. (2017). Present status of fisheries and impact of emerging diseases of fish and shellfish in Indian aquaculture. *Journal of Aquatic Research and Marine Sciences*, *2017*, 5–26.

Nematollahi, A., Decostere, A., Pasmans, F., & Haesebrouck, F. (2003). *Flavobacterium psychrophilum* infections in salmonid fish. *Journal of Fish Diseases*, *26*(10), 563–574.

Pallen, M. J., & Wren, B. W. (2007). Bacterial pathogenomics. *Nature*, *449*(7164), 835–842.

Pandya, F. A., & Bharad, B. H. (2019). A study on legal aspects of marine environmental protection in India. ISSN: 2581–5830, Volume: II, Issue: III.

Peeler, E. J., & Taylor, N. G. (2011). The application of epidemiology in aquatic animal health – opportunities and challenges. *Veterinary Research*, *42*(1), article no. 94.

Ponnerassery, S., & Sudheesh, A. A.-G. (2012). Comparative pathogenomics of bacteria causing infectious diseases in fish. *International Journal of Evolutionary Biology*, *2012*, 16, Article ID 457264.

Rashid, T. (2014). Pollution in the Bay of Bengal: Impact on marine ecosystem. *Open Journal of Marine Science*, 55–63, Scientific Research Publishing.

Rico, A., Satapornvanit, K., & Haque, M. M. (2012). Use of chemicals and biological products in Asian aquaculture and their potential environmental risks: A critical review. *Reviews in Aquaculture*, *4*(2), 75–93.

Romero, F. C. G., & Navarrete, P. (2012). *Antibiotics in aquaculture – Use, abuse and alternatives. Health and environment in aquaculture* (pp. 160–198). IntechOpen.

Rothberg, J. M., & Leamon, J. H. (2008). The development and impact of 454 sequencing. *Nature Biotechnology*, *26*(10), 1117–1124.

Rubio, R., & Rauret, G. (1996). Validation of the methods for heavy metal speciation in soils and sediments. *Journal of Radioanalytical and Nuclear Chemistry*, *208*, 529–540.

Saha, S. B., Mitra, A., Bhattacharyya, S. B., & Choudhury, A. (2001). Status of sediment with special reference to heavy metal pollution of a brackishwater tidal ecosystem in Northern Sundarbans in West Bengal. *Tropical Ecology*, *42*, 127–132.

Seshadri, R., Joseph, S. W., & Chopra, A. K. (2006). Genome sequence of *Aeromonas hydrophila* ATCC 7966T: Jack of all trades. *Journal of Bacteriology*, *188*(23), 8272–8282.

Sharif, A. K. M., Bilkis, A. B., Roy, S., Silder, M. D. H., Aliidriss, K. M., & Safiullah, S. (1993). *Concentration of radionuclides in coastal sediments from the Bay of Bengal*. Savar, Bangladesh: An Institute of Nuclear Science and Technology, Atomic Energy Research Establishment.

Sorensen, U. B. S., & Larsen, J. L. (1986). Serotyping of *Vibrio anguillarum*. *Applied and Environmental Microbiology*, *51*(3), 593–597.

Speare, D. J., & Ferguson, H. W. (1989). Clinical and pathological features of common gill diseases of cultured salmonids in Ontario. *Canadian Veterinary Journal*, *30*, 882–887.

Starliper, C. E. (2011). Bacterial coldwater disease of fishes caused by *Flavobacterium psychrophilum*. *Journal of Advanced Research*, *2*, 97–108.

Thomlinson, D. L., Wilson, J. G., Harries, C. R., & Jeffney, D. W. (1980). Problems in the assessment of heavy metal levels in estuaries and the formation of a pollution index. *Helgolander Wissenschaftliche Meeresuntersuchungen*, *33*, 566572.

Touchon, M. P., Barbier, P., & Bernardet, J. F. (2011). Complete genome sequence of the fish pathogen *Flavobacterium branchiophilum. Applied and Environmentl Microbiology*, *77*(21), 7656–7662.

Tripathi, N. G. (2019). *Strategies for controlling plastic pollution in India policy.* Climate Change Research Institute.

Warren, J. W. (1991). *Diseases of hatchery fish* (6th ed.). Portland, Oregon: U.S. Fish andWildlife Service, Pacific Region.

Zhang, J., Chiodini, R., Badr, A., & Zhang, G. (2011). The impact of next-generation sequencing on genomics. *Journal of Genetics and Genomics*, *38*(3), 95–109.

Further reading

Abowei, J. F. N., & Briyai, O. F. (2011). A review of some bacteria diseases in Africa culture fisheries. *Asian Journal of Medical Sciences*, *3*(5), 206–217.

Andrews, C. (1988). *The manual of fish health.* Stillwater, MN: Voyageur Press, ISBN 978-1-56465-160-0.

Assefa, A. (2018). Maintenance of fish health in aquaculture: Review of epidemiological approaches for prevention and control of infectious disease of fish. *Veterinary Medicine International*, *18*, Article ID 5432497.

Austin, B., & Austin, D. A. (1993). *Bacterial fish pathogens, disease of farmed and wild fish* (2nd ed.). Chichester: Ellis Horwood.

Axelrod, H. R., & Untergasser, D. (1989). *Handbook of fish diseases.* Neptune, NJ: TFH Publications, ISBN 978-0-86622-703-2.

Decostere, A., Haesebrouck, F., Turnbull, J. F., & Charlier, G. (1999). Influence of water quality and temperature on adhesion of high and low virulence *Flavobacterium columnare* strains to isolated Gill arches. *Journal of Fish Diseases*, *22*, 1–11.

Good, C., Davidson, J., Wiens, J. D., Welch, T. J., & Summerfelt, S. (2015). *Flavobacterium branchiophilum* and *F. succinicans* associated with bacterial gill disease in rainbow trout Oncorhynchus mykiss (Walbaum) in water recirculation aquaculture systems. *Journal of Fish Diseases*, *38*, 409–413.

Harikrishnan, R., & Balasundaram, C. (2015). Modern trends in *Aeromonas hydrophila* disease management with fish. *Reviews in Fisheries Science*, *13*, 281–320.

Harikrishnan, R., Balasundaram, C., & Heo, M. S. (2011). Fish health aspects in grouper aquaculture. *Aquaculture (Amsterdam, Netherlands)*, *320*(1–2), 1–21.

Hellou, J., Warren, W. G., Payne, J. F., Belkhode, S., & Lobel, P. (1992). Heavy metals and other metals in three tissues of cod *Gadus morhua* from the northwest Atlantic. *Marine Pollution Bulletin*, *24*, 452–458.

Holt, R. A. (1987). *Cytophaga psychrophila, the causative agent of bacterial cold-water disease in salmonid fish*, PhD Thesis, Oregon State University, Corvallis, Ore.

Kozinska, A., & Pekala, A. (2012). Characteristics of disease spectrum in relation to species, serogroups, and adhesion ability of motile aeromonads in fish. *The Scientific World Journal*, *2012*, 949358. Available from https://doi.org/10.1100/2012/949358.

Lindesjoo, E., & Thulin, J. (1994). Histopathology of skin and gills of fish in pulp mill effluents. *Diseases of Aquatic Organisms*, *18*, 81–93.

Majumdar, T., Ghosh, S., Pal, J., & Mazumder, S. (2006). Possible role of a plasmid in the pathogenesis of a fish disease caused by *Aeromonas hydrophila. Aquaculture (Amsterdam, Netherlands)*, *256*, 95–104.

Moran, N. A. (2002). Microbial minimalism: Genome reduction in bacterial pathogens. *Cell*, *108*(5), 583–586.

Moyle, P. B., & Cech, J. J. (2004). *Fishes, an introduction to ichthyology* (5th ed.). Benjamin Cummings, ISBN 978-0-13-100847-2.

Noga, E. J. (1996). *Fish disease: Diagnosis and treatment* (pp. 31–35). St. Louis, Missouri.: Mosby.

Trust, T. J., & Bartlett, K. H. (1974). Occurrence of potential fish pathogens in water containing ornamental fishes. *Applied Microbiology*, *28*(1), 35–40.

Vethaak, A. D., & Rheinallt, T. (1992). Fish disease as a monitor for marine pollution: The case of the North Sea. *Fish Biology and Fisheries*, *2*, 1–32.

Vos, J. G., Van Loveren, H., Wester, P. W., & Vethaak, A. D. (1989). Toxic effects of environmental chemicals on the immune system. *Trends in Pharmacological Science*, *10*, 289–293.

CHAPTER

7 Common bacterial infections affecting freshwater fish fauna and impact of pollution and water quality characteristics on bacterial pathogenicity

Zarka Zaheen[1], Aadil Farooq War[2], Shafat Ali[3], Ali Mohd Yatoo[1], Md. Niamat Ali[3], Sheikh Bilal Ahmad[4], Muneeb U. Rehman[5] and Bilal Ahmad Paray[6]

[1]*Centre of Research for Development/Department of Environmental Science, University of Kashmir, Srinagar, India* [2]*Department of Botany, University of Kashmir, Srinagar, India* [3]*Centre of Research for Development, University of Kashmir, Srinagar, India* [4]*Division of Veterinary Biochemistry, Faculty of Veterinary Science & Animal Hisbandry, SKUAST-Kashmir, Srinagar, India* [5]*Department of Clinical Pharmacy, College of Pharmacy, King Saud University, Riyadh, Saudi Arabia* [6]*Department of Zoology, College of Science, King Saud University, Riyadh, Saudi Arabia*

7.1 Introduction

In addition to fish fauna, many other organisms reside in water, including several species of saprophytic bacteria inhabiting phytoplankton and zooplankton as well as sediments and plants. A number of these bacteria live as commensals by colonizing the digestive tract, gills, and skin of fish, which not only supports digestion but also has a useful effect on their immune system. However, these bacteria are denoted as conditionally pathogenic as they can also affect fish health. Many studies conducted worldwide focus on the subject of interactions between bacteria, fish fauna, and the diseases associated with them, thereby indicating their significant part in fish pathology. The development of a disease is a complex process, which depends on the capability of bacteria to cause health illnesses and also on the virulence of the disease-causing agent, immune status of fish, and environmental conditions. Consequently, for any kind of disease development, including the emerging ones as well, the changes that occur in freshwater ecosystems seems to be essential (Johnson & Paull, 2011). The climatic conditions prevailing in a particular zone, region, or country have a significant role in the development of a particular fish disease. This implies that the fish cultured in the Mediterranean Sea, in Northern European countries, and in continental Europe shows different health disorders. For instance, in Central Europe, the most common infections resulting in bacterial diseases in fish are caused by *Aeromonas* sp., resulting in motile aeromonas septicemia (MAS), furunculosis, and motile aeromonas infection (MAI). Meanwhile, amebic gill disease (AGD) and the invasion of copepods reflect the main health problems related to fish fauna in Scandinavian countries. Infections related to *Flavobacterium* sp. are also often witnessed

Bacterial Fish Diseases. DOI: https://doi.org/10.1016/B978-0-323-85624-9.00006-3

(Olesen & Nicolajsen, 2012). The epizootic situation with respect to bacterial fish diseases in Poland does not considerably vary from the situation reflected by other continental European countries. However, in the case of bacterial freshwater fish disease pathology, dynamic changes have been witnessed during recent times. Gram-negative bacteria have been recognized to be as pathogenic to fish as *Aeromonas* species (Pękala-Safińska, 2018).

A number of characteristics of the usual microbial flora that are linked with fish fauna have been observed. These characteristics comprise the alterations that occur in microbial flora in the course of storage processes (Shewan, 1961); the impact of handling and catching approaches on microbial flora, which may result in deterioration (Colwell, 1962; Gillespie & Macrae, 1975); the association between fish microflora and environmental microflora (Horsley, 1973; Sugita et al., 1983; Yoshimizu et al., 1976, 1980); and the formation of basic requirements in order to monitor the alterations in fish farms (Allen et al., 1983; Austin, 1982, 1983; Niemi & Taipalinen, 1982). Numerous researchers have come across different problems associated with taxonomy, while trying to describe the bacterial flora generally linked with various fish species. Shewan (1971), while studying the bacteriology of spoiling and fresh fish, observed that it was quite difficult to recognize various bacterial isolates even to the genus level. He states that most of the articles that were written before 1980 represented the isolated bacteria up to genera level only, but in recent times, documentation to the species level and the recognition of new phena have been established by the determination of G + C ratios and using numerical taxonomy. The existing literature regarding fish microflora reveals very well the concern with respect to fish bacteria, as it reflects their significance as spoilage as well as pathogenic bacteria for human consumption. Every bacterial fish pathogen can live outside the fish in the aquatic environment as they are ubiquitous. The causes for emergence of a fish disease are not entirely recognized, but they might include: introduction of a new fish species that might be sensitive towards a local microbe, variations in the environmental conditions that might favor the potential pathogen, development of various diagnostic tools that might provide an opportunity to define new pathogens, introduction of a novel fish species that might carry a microorganism infectious to native fish species, and decrease of resistance towards infection in the host species (Colorni, 2004). The bacteria that are most commonly identified as fish pathogens include *Edwardsiella* (Blazer et al., 1985; Park et al., 2012), *Aeromonas* (Verner-Jeffreys et al., 2009), *Pseudomonas* (Bartlett, 1974; Berthe et al., 1995), *Mycobacterium* (Zanoni et al., 2008), *Shewanella* (Korun et al., 2009), *Flavobacterium* (Bartlett, 1974), and *Streptococcus* (Bartlett, 1974) (Fig. 7.1). Globally, aquaculture practice has been developed as one of the cost-effective and safest protein sources for human consumption. Since 1995, the fish food production worldwide showed an increase at an average rate of 6.6% annually (Food and Agriculture Organisation Aquaculture (FAO), 2017) and thus in 2016 extended up to 80 million tons (Food and Agriculture Organisation Aquaculture (FAO), 2018). There has been a substantial growth in annual per capita consumption from 1.5 kg in 1961 to 7.8 kg in 2015 in the production of salmon, Nile tilapia, and other freshwater species (Food and Agriculture Organisation Aquaculture (FAO), 2018). In the case of farmed fish in intensive aquaculture, fish resistance may be reduced, when the fish gets affected by several infectious diseases that are aggravated by various stress factors. In order to avoid substantial economic losses and decreased growth performance due to bacterial disease, antibiotic medicines may be required (Romero et al., 2012). Generally, the freshwater fish is infected by the *Aeromonas* genus and these represent the common natives of aquatic ecosystems such as freshwater, sediments, marine waters, and estuarine waters and aquatic animals (shellfish

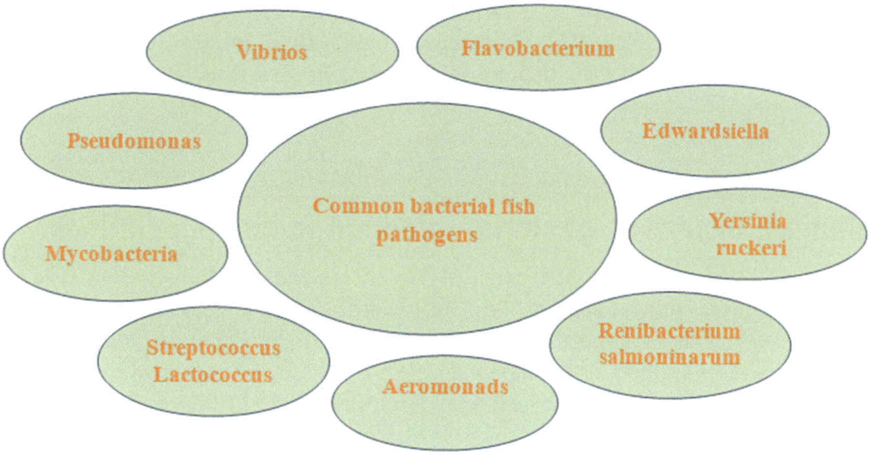

FIGURE 7.1

Common bacterial fish pathogens.

and fish) (Swann & White, 1991). Significant deaths in marine, wild, and farmed freshwater fish species are caused by hemorrhagic, septicemic, and ulcerative diseases in fish due to *Aeromonas* infections, which result in severe economic losses to the aquaculture sector (Beaz-Hidalgo & Figueras, 2013). *Aeromonas hydrophila* and *A. salmonicida* are the two most regularly encountered species in fish farms. Furunculosis is caused by *salmonicida*, which primarily affects salmonids and is a subspecies of *A. salmonicida*. Furunculosis disease causes fish decreases due to the development of muscle boils in chronic form and hemorrhagic septicemia in acute form, which can cause major economic losses (Austin & Austin, 2012). *A. hydrophila*, which may also cause hemorrhagic septicemia, is an omnipresent bacterium usually obtained from freshwater ponds and is also a common native of the gastrointestinal tract of aquatic animals (White & Swann, 1991). *A. hydrophila* also poses a public health risk, as it is a zoonotic pathogen that infects human beings through aquaculture facilities and foodborne infections (Okocha et al., 2018). With regard to the antibiotic resistance and genotypic studies of two important species of *Aeromonas*, that is, *A. hydrophila* and *A. salmonicida*, the subspecies *Salmonicida* showed an incidence of multidrug-resistant plasmids with a high level of interspecies transmission, which includes human bacteria as well (del Castillo et al., 2013; Dar, Dar et al., 2016; Dar, Kamili et al., 2016; Dar et al., 2020; Vincent et al., 2014). By attaching to biofilms on living or nonliving surfaces, the *Aeromonas* may persist, and the dissemination and exchange of antimicrobial resistant genes promotes the existence of *Aeromonas* with *Escherichia coli* in polybacterial mixed biofilms (Talagrand-Reboul et al., 2017). With the use of effective management practices, which includes management of the environment, water, soil, stock, and nutrition, the disease problems in a system can be avoided in the best possible way. Efforts to address the disease problem were made by applying a number of other approaches including chemotherapy, disinfection, and sanitary prophylaxis, with specific importance given to the usage of antibiotics. Moreover, the usage of antibiotics and other chemicals in the case of pond culture was found to be quite undesirable and expensive, and thus may result in development of antibiotic resistance. The application of antimicrobial drugs in Norway has reduced from almost 50 metric tonnes (1987) to 746.5 kg (1997) (Verschuere et al., 2000). In 2007, the application was still at a lower level as in 1997 and vaccinations against specific fish diseases have also been used in recent times (Norwegian Scientific Committee for Food Safety, 2009).

7.2 Pollution: impact on bacterial infection in fish populations

The anthropogenic aspects contribute extensively to the incidence of disease in fish populations because they change the vulnerability of the host towards the pathogens that are omnipresent and form an integral part of any habitat. This is clearly stated by the effect of pollution on fish fauna and initiation of diseases in them (Arkoosh et al., 1998). A study on the impact of temperature and water quality on the adhesion of low as well as high virulence of *Flavobacterium columnare* strains to isolated gill arches recommended that the low virulence strain adhered less as compared to the high virulence strain that adhered more promptly (Decostere et al., 1999). Numerous aspects, comprising elevated temperatures, dipping of gill in bivalent ion rich water, and the occurrence of organic matter or nitrite, enhanced the high virulence strain adhesion. While monitoring bacterial fish pathogens in *Oncorhynchus mykiss* in freshwater farms, 361 bacterial isolates were isolated and diseases like furunculosis, rainbow trout fry syndrome/coldwater disease, and enteric redmouth disease were observed (Dalsgaard & Madsen, 2000).

7.3 Water quality attributes: impact on bacterial pathogenicity in fish populations

While carrying out laboratory examinations on the impact of pH and temperature on the disabling of several fish bacterial and viral pathogens at pH 4.0, pH 12.0 and 60°C, it was found that *Lactococcus garvieae* was the most resistant bacterium to pH 4.0, pH 12.0, and to 60°C as well (Dixon et al., 2012). From an aquaculture farm polluted by agricultural wastewater, various *Vibrio* sp., that is, *V. cholera*, *V. harveyi*, *V. fischeri*, and *V. parahaemolyticus* in diseased *Channa punctatus* were established. It was found *V. cholera*, *V. parahaemolyticus*, and *V. harveyi* most often infect *C. punctatus* (Sankar et al., 2012). In Egypt, a research study was carried out on *A. hydrophila* in the case of cultured *Oreochromis niloticus*, which were collected haphazardly from the fish farm with an incidence rate of 25%. From around 40 cultured *O. niloticus*, nearly 10 isolates of *A. hydrophila* strains were obtained. The collected fish showed clinical symptoms of bilateral exophthalmia, loss of escape reflex, ulcers, and skin darkness, and the postmortem investigations revealed enlargement in internal organs, hemorrhage, and congestion (El Deen et al., 2014). Likewise, another study carried out in Minna Metropolis, Nigeria, on the separation and identification of bacteria in relation to fresh and smoked fish (Clarias gariepinus) under in-vitro assay conditions, showed that the samples were contaminated with several bacterial pathogens (Ibrahim et al., 2014). In water recirculation aquaculture systems, on examining the impact of *F. succinicans* and *F. branchiophilum* in relation to bacterial gill disease in the case of *O. mykiss walbaum* (rainbow trout), it was revealed that *F. branchiophilum* was dominant and was found on the gills of rainbow trout that were affected by regular and environmentally induced bacterial gill disease (Good et al., 2015). Another investigation of *Hypophthalmichthys molitrix* (Silver carp) acquired from a pond at District Poonch of Jammu and Kashmir, India was carried out. The recognition and characterization of possibly pathogenic *A. sobria* from the fish was taken for examination and almost 33 colonies were isolated. The histopathological study of gills displayed epithelial hyperplasia and hemorrhagic gill epithelia (Dar, Dar et al., 2016; Dar, Kamili et al., 2016).

7.4 Common bacteria causing infections in freshwater fish

The common bacterial pathogens of freshwater fish include those described in the following paragraphs (Table 7.1).

7.4.1 Vibrios

The *Vibrio* genus includes the bacteria that are primarily pathogenic to brackish and marine water fish fauna. However, these bacteria are sometimes described in freshwater fish species as well (Hjeltnes & Roberts, 1993; Lightner & Redman, 1998). The vibriosis disease causes huge economic loss to the aquaculture industry, as it is distributed throughout the world (Bergh et al., 2001). The main pathogenic species causing vibriosis includes *V. ordalii*, *V. harveyii (Syn. V. carchariae)*, *V. anguillarum*, and *Aliivibrio salmonicida* (formerly *V. salmonicida*) and they represent a key bacterial disease that affects fish, crustaceans, and bivalves (Austin & Austin, 1999; Austin & Zhang, 2006). So far, more pathogenic species have been recognized and further reported (Actis et al., 2011). Species such as *Moritella viscosa* (Benediktsdóttir et al., 2000) and *V. vulnificus* (Amaro et al., 1992; Biosca et al., 1993) have been found to be associated with fish diseases such as winter ulcer and septicemia, respectively. The genome sequencing of four important pathogenic vibrios, *V. ordalii*, *V. anguillarum*, *V. vulnificus*, and *Aliivibrio salmonicida*, has been achieved and described in the literature (Gulig et al., 2010; Hjerde et al., 2008; Naka et al., 2011). In general, these bacteria possess two chromosomes, one bigger and the other smaller. The maximum genes that code for pathogenic aspects and cell functions are found in the bigger chromosome. The genes for environmental adaptation are generally located in the smaller one. The most considered etiological agent of vibriosis is *V. anguillarum* (Larsen et al., 1994). *V. anguillarum* is the causative agent of hemorrhagic septicemia. The virulent strains commonly obtained from infected fish fauna are the O1 and O2 serotypes (Sørensen & Larsen, 1986; Toranzo & Barja, 1990). *V. ordalii*, formerly known as *V. anguillarum* biotype 2, is closely associated with *V. anguillarum* (Schiewe & Crosa, 1981). On the basis of histological confirmations, it was found that the vibriosis caused by these two species is extremely different (Ransom et al., 1984). *V. ordalii* is typically found in the form of groups in cardiac and skeletal muscles, while *V. anguillarum* has an exceptional affinity for blood and loose connective tissue. In the case of *V. ordalii*, its affinity for blood is small and the bacteremia develops only at the later stages of disease. *Aliivibrio salmonicida* (previously known as *V. salmonicida*) is the causative agent of coldwater vibriosis in marine fish like captive Atlantic cod (*Gadus morhua*), sea-farmed rainbow trout (*O. mykiss*), and farmed Atlantic salmon (*Salmo salar*)

Table 7.1 Common bacterial diseases and their causative agents affecting fish fauna.

S. No	Disease	Causative agent	Host
01.	Vibriosis	*Vibrio ordalii*	Salmonids
02.	Winter ulcer	*Moritella viscosa* (formerly Vibrio viscosus)	*Atlantic salmon*
03.	Vibriosis (Septicemia)	*Vibrio vulnificus*	Eels, tilapia
04.	Vibriosis (Hemorrhagic septicemia)	*Vibrio anguillarum*	Salmonids, turbot, sea bass, striped bass, eel, ayu, cod, and red sea bream
05.	Coldwater Vibriosis	*Aliivibrio salmonicida*	Atlantic cod (*Gadus morhua*), sea-farmed rainbow trout (*Oncorhynchus mykiss*), and farmed Atlantic salmon (*Salmo salar*)
06.	Motile aeromonas septicemia (MAS), red-sore disease, hemorrhagic septicemia, or ulcer disease	*Aeromonas hydrophila*	Catfish, several ornamental or tropical fish species, and a number of bass species
07.	Furunculosis	*Aeromonas salmonicida*	Nonsalmonids and salmonids
08.	Bacterial coldwater disease (BCWD)	*Flavobacterium psychrophilum*	*Cyprinus carpio*, *Anguilla anguilla*, *A. japonica*, *Carassius carassius*, *Plecoglossus altivelis*, *Tinca tinca*, *Rutilus rutilus* and *Perca fluviatilis*
09.	Columnaris disease	*Flavobacterium columnare*	Salmonids and other fish species
10.	Edwardsiellosis (hemorrhagic septicemia)	*Edwardsiella tarda*	Salmon, carp, tilapia, catfish, striped bass, flounder, and yellowtail
11.	Enteric septicemia of catfish (ESC)	*Edwardsiella ictaluri*	Catfish
12.	Enteric redmouth (ERM)	*Yersinia ruckeri*	Salmo gairdneri, eel, sturgeon
13.	Bacterial kidney disease (BKD)	*Renibacterium salmoninarum*	Salmonids
14.	Streptococcosis (meningoencephalitis and septicemia)	*Streptococcus iniae*	Eels, rainbow trout, and other salmonids (Oncorhynchus family), and fishes that belong to the Cichlidae (tilapia) and Ictaluridae (catfish) families
15.	Streptococcosis	*Streptococcus parauberis*	Olive flounder (*Paralichthys olivaceus*) and cultured turbot (*Scophthalmus maximus*)
16.	Lactococcosis (hemorrhagic and hyperacute septicemia)	*Lactococcus garvieae*	Eels, rainbow trout, and other salmonids (*Oncorhynchus* family), and fishes that belong to the Cichlidae (tilapia) and Ictaluridae (catfish) families
17.	Mycobacteriosis	*Mycobacterium fortuitum, M. marinum, M. avium,* and *M. chelonae*	Sea bass, turbot, and Atlantic salmon
18.	Strawberry disease	*Pseudomonas* spp.	Tench (*Tinca tinca*) and rainbow trout (*Oncorhynchus mykiss*)

(Schrøder et al., 1992). This gram-negative bacterium results in sepsis, hemolysis, and tissue degradation in vivo (Hjerde et al., 2008).

7.4.2 Aeromonads

Throughout the world, *A. hydrophila* and other motile aeromonads represent the most common bacteria found in several aquatic environments, which include well water, bottled water, heavily polluted water, sewage, and chlorinated water. These are the causative agents of severe disease in reptiles, amphibians, and birds as well as cultured and wild fishes (Martin-Carnahan & Joseph, 2005). Aeromonads are also recognized as severe developing pathogens of human beings as well (Figueras, 2005). *Aeromonas* sp. are generally found in several environments, which includes water as well, hence fish fauna are at continuous exposure to these bacteria. Under various stress conditions, including unfavorable environmental circumstances and human involvement as in transporting, sorting, and catching fish, the contact with these bacteria becomes dangerous. *Aeromonas* species also form a part of the physiological microflora of the intestine of fish (Austin & Austin, 2016). The biochemical, antigenic, and genetic heterogeneity of the members of this group has made it complex to determine the cause of the diseases that involve aeromonad infections. The *Aeromonas* genus has been categorized into two groups, one being the group of nonmotile psychrophilic species, most importantly characterized by *A. salmonicida* (obligate fish pathogen), and the second one, being typically pathogenic to humans, is the group of motile and mesophilic species, characterized by *A. hydrophila*. The taxonomic confusion between *A. salmonicida* subsp. *salmonicida A449*, *A. hydrophila ATCC 7966T*, *A. caviae*, and *A. veronii* strain B565 has been resolved by carrying out their genome sequencing (Beatson et al., 2011; Li et al., 2011; Reith et al., 2008; Seshadri et al., 2006), which has also provided novel perceptions about the adaptation of these bacteria to countless ecological niches, their virulence mechanisms, and their host adaptive evolution. *A. salmonicida* is a nonmotile gram-negative bacterium that causes furunculosis in nonsalmonids and salmonids. In cultured and wild fish, furunculosis is a significant disease that causes hefty losses to the aquaculture industry globally (Bernoth et al., 1997; Hiney et al., 1999). *A. hydrophila* causes septicemic disease in fish, also known as hemorrhagic septicemia, motile aeromonas septicemia (MAS), red-sore disease, or ulcer disease (Paniagua et al., 1990). This disease mainly affects freshwater fish including catfish, several ornamental or tropical fish species, and a number of bass species. Another very important economic problem in the aquaculture industry is the bacterial hemorrhagic septicemia of fish caused by *A. veronii* (Austin & Austin, 1999).

7.4.3 Flavobacterium

Flavobacterium sp. are naturally occurring bacteria that are found in the aquatic environment and that also form a part of the healthy fish gill microflora. The sources of these bacteria for being a prospective cause of fish disease is yet to be recognized (Austin & Austin, 2016). The *Flavobacterium* genus consists of about 30 species, from which *F. branchiophilum*, *F. columnare*, and *F. psychrophilum* are the significant disease-causing agents in the case of catfish, salmonids, and other cultivated fishes (Madetoja et al., 2002; Nematollahi et al., 2003). Flavobacteria are quite essential, as they are found in marine water and freshwater as well as in soil, and are well-known for their capability of degrading polymeric organic matter like hydrocarbons and also for their unique gliding motility (Leahy & Colwell, 1990). *F. psychrophilum* causes BCWD (bacterial cold-water disease) and is a severe fish pathogen that causes rearing problems in the case of both conservation and commercial aquaculture, and also results in considerable financial losses. *F. psychrophilum* infections are identified worldwide, yet the most vulnerable to BCWD are coho salmon and juvenile rainbow trout. The *F. psychrophilum* infections so far have been described in an extensive range of hosts, which includes *Cyprinus carpio*, *Anguilla anguilla*, *A. japonica*, *Carassius carassius*, *Plecoglossus altivelis*, *Tinca tinca*, *Rutilus rutilus*, and *Perca fluviatilis* (Lehmann et al., 1991; Madetoja et al., 2002). Fingerlings and fry infected with BCWD can show mortalities up to 70% and usually have skin ulcerations at the anus, on the peduncle, on the lower jaw, or on the anterior fin to the dorsal fin (Holt, 1987). In several parts of the world it was observed that *F. branchiophilum* causes bacterial gill disease (BGD) (Heo et al., 1990). This disease displays certain characteristics like high morbidity and mortality rates, which are attributed to huge bacterial colonization of gill lamellar surfaces and progressive branchial pathology resulting from extreme degree of lamellar epithelial necrosis (Ostland et al., 1994). Throughout the world, with respect to the wild fish populations, the ornamental fish industry, and commercial aquaculture, *F. columnare* (previously known as *Flexibacter columnaris*; *Cytophaga columnaris*) causes columnaris disease in the case of salmonids and other fish species (Bernardet, 1989). Typically, during its outburst, the mortality and morbidity rates increase more progressively as compared to BGD. Furthermore, fish infected with columnaris show severe necrosis in the whole gill, as the bacterium occupies more inwardly, contrasting with the necrosis pattern shown in BGD (Speare & Ferguson, 1989). In the cases of *F. psychrophilum* and *F. branchiophilum*, the complete genome sequences are reported (Duchaud et al., 2007; Touchon et al., 2011).

7.4.4 Edwardsiella

A group of gram-negative enteric bacteria, pathogenic to various species, includes the genus *Edwardsiella*. It belongs to the subgroup 3 of γproteobacteria (Abbott & Janda, 2006). *Edwardsiella ictaluri* and *E. tarda* are two essential fish pathogens that are closely linked to each other. Both the species are gram-negative motile rods that ferment glucose to produce gas and acid, and are also cytochrome oxidase negative (Abbott & Janda, 2006). In fish a disease called systemic hemorrhagic septicemia is caused by *E. tarda*, resulting in ulceration and necrosis in internal organs like spleen, kidney, liver, and muscles as well as swelling skin lesions (Meyer & Bullock, 1973). *E. tarda* has the capability to occupy and increase in macrophages and epithelial cells, causing destabilization of the host immune system, resulting in its survival inside the fish (Ling et al., 2000). Another very important disease that affects the catfish industry is ESC (enteric septicemia of catfish) caused by *E. ictaluri*. This disease may appear in chronic form characterized by meningoencephalitis and in an acute form characterized by septicemia and hemorrhagic enteritis (Newton et al., 1989). Hemorrhages on the body, particularly around the fins and mouth, represent the gross external symptoms. The other symptoms comprise exophthalmia, minor ulcerations on the body, and pale gills (Rogers, 1983). Recently, the entire genome sequencing of these two essential fish pathogens has been accomplished and reported, which allows their comparative genomic investigation as well (Wang et al., 2009; Williams et al., 2012). The detection of insertion sequence IS Saen1 of *Salmonella enterica serovar Enteritidis* is a remarkable characteristic (Partridge & Hall, 2003).

7.4.4.1 *Yersinia ruckeri*

Yersinia ruckeri causes another fish disease known as enteric redmouth (ERM) disease. ERM is a bacterial infection of fish fauna; however, it is primarily recognized for its incidence in rainbow trout (*S. gairdneri*) (Furones et al., 1993). In the 1950s, *Y. ruckeri* was originally isolated in the Hagerman Valley (United States) from rainbow trout (Ross et al., 1966) and is currently extensively found in fish fauna all over Europe, Australia, South Africa, and North America (Tobback et al., 2007). Outbursts of ERM generally initiate with low mortalities and show escalation with time, and thus may cause high fatalities. In the case of exposure of chronically infected fish to stressful conditions like poor water quality and high stocking densities, this problem may increase on a large scale (Horne & Barnes, 1999). *Y. ruckeri* does not have a capsule and is a nonspore-forming bacterium, often having a flagellum (Ross et al., 1966). Generally, in the case of biochemical reactions, *Y. ruckeri* is equally homogeneous. However, based on serotype, biotype, and outer membrane protein (OMP) profiles, *Y. ruckeri* strains nowadays are being categorized into clonal types (Davies, 1991). The greatest epizootic outbursts with regard to cultured salmonids are caused by strains of serovars II and I (Stevenson & Airdrie, 1984), equal to serotypes O2b and O1a, respectively (Romalde et al., 1993); moreover, the most predominant is serovar I in rainbow trout (Stevenson, 1997). Recently, uncertainties have been raised about the taxonomic position of *Y. ruckeri* as the multilocus sequence typing showed different phylogenetic divergence of *Y. ruckeri* as compared to the other members of *Yersinia* genus (Kotetishvili et al., 2005). This observation achieved reliability after the genome sequencing of *Y. ruckeri* was carried out, as it showed significantly decreased total genome size of 3.58–3.89 Mb when compared to other members of the genus, which usually displayed a genome size of 4.6–4.8 Mb (Chen et al., 2010).

7.4.5 Renibacterium salmoninarum

A small gram-positive diplobacillus called *Renibacterium salmoninarum* causes bacterial kidney disease (BKD) in fish fauna. BKD is a gradually progressive, systemic infection in salmonids with an insidious nature and a prolonged course (Sanders & Fryer, 1980). *R. salmoninarum* can be transferred through eggs from adults to their progeny (Bullock & Herman, 1988) or from one fish to another (Mitchum & Sherman, 1981). Fish infected with this pathogen may perhaps take months to display the various signs linked with the disease. When it comes to treatment, BKD represents one of the most complex bacterial diseases of fish (Wolf & Dunbar, 1959), primarily because of its capability to avoid phagocytosis and its invasion and persistence inside host cells (Gutenberger et al., 1997; Sudheesh et al., 2007). *R. salmoninarum* is an extremely slow-growing bacterium and the study of its gene functions by the application of various genetic manipulation techniques becomes very difficult. Phylogenetically, *R. salmoninarum* shows a close relation with the nonpathogenic environmental *Arthrobacter* species although being an obligate intracellular pathogen of fish (Wiens et al., 2008). The *R. salmoninarum* has been incorporated in the subdivision "actinomycetes" on the basis of 16S rRNA phylogenetic investigation, and was found associated to a subgroup that includes *Micrococcus*, *Arthrobacter*, *Jonesia*, *Cellulomonas*, *Brevibacterium*, *Stomatococcus*, and *Promicromonospora* (Stackebrandt et al., 1988). Furthermore, *Arthrobacter davidanieli* can be responsible for substantial cross-protection in Atlantic salmon, although not in the case of Pacific salmon, as it is commercially utilized as a vaccine (identified as Renogen) (Salonius et al., 2005). The genome sequencing of the TC1 and FB24 (*Arthrobacter* strains) and ATCC 33209 (*R. salmoninarum* strain) has shown many fascinating characteristics regarding the evolution of this obligate fish pathogen from members of the nonpathogenic genus *Arthrobacter*, by means of horizontal gene acquisition and genomic reduction (Mongodin et al., 2006; Wiens et al., 2008).

7.4.6 Streptococcus and Lactococcus

Gram-positive cocci that belong to *Lactococcus* and *Streptococcus* genera are progressively being identified as vital fish pathogens worldwide (Kitao et al., 1993). There are numerous species of gram-positive cocci that are found linked with infectious diseases of cold as well as warm-water fishes, which includes *L. garvieae* (*syn. Enterococcus seriolicida*) (Vendrell et al., 2006); *Streptococcus iniae*, *S. parauberis*, *S. phocae*, *S. agalactiae* (*syn. Streptococcus difficilis*) (Domeénech et al., 1996; Romalde et al., 2008); *Vagococcus salmoninarum* and *Carnobacterium piscicola* (Wallbanks et al., 1990) and *L. piscium* (Collins et al., 1983; Domenech et al., 1993; Eldar et al., 1996; Williams et al., 1990). *S. iniae* in a number of warm-water fishes causes meningoencephalitis and septicemia, and is a β-hemolytic gram-positive coccus (Perera et al., 1994); however, *S. parauberis* is primarily pathogenic to olive flounder (*Paralichthys olivaceus*) and cultured turbot (*Scophthalmus maximus*) and is an α-hemolytic gram-positive coccus. *L. garvieae* predominantly during summer causes a hemorrhagic and hyperacute septicemia in fishes. The common pathological symptoms of lactococcosis and streptococcosis in fishes include congestion, hemorrhage, dark pigmentation, lethargy, exophthalmos with clouding of the cornea, and erratic swimming (Domeénech et al., 1996; Kusuda et al., 1991). Complete genome sequences of some significant pathogenic species of different strains of *L. garvieae* and *S. parauberis* have been published; these were isolated from humans as well as fish (Aguado-Urda et al., 2011; Morita et al., 2011; Nho et al., 2011; Reimundo et al., 2011). *Streptococcus iniae* and *L. garviae* are of specific significance in the bacterial fish disease pathology, among several species of gram-positive microorganisms. Both these bacterial species cause severe health conditions in diverse species of marine as well as freshwater fishes, including eels, rainbow trout, and other salmonids (Oncorhynchus family), and fishes that belong to the Cichlidae (tilapia) and Ictaluridae (catfish) families (Austin & Austin, 2016).

7.4.7 Mycobacteria

Several species of mycobacteria causing chronic infections in fish have been well identified (Frerichs, 1993; Noga, 1996). From different parts of the world, isolation of a number of fast- as well as slow-growing mycobacteria species has been carried out. These include *Mycobacterium fortuitum*, *M. marinum*, *M. avium*, and *M. chelonae*, isolated from cultured and wild fish suffering from mycobacteriosis (Sanders & Swaim, 2001; Thoen & Schliesser, 1984; Wayne et al., 1986). *M. marinum* is the most significant fish pathogen among these species, commonly isolated from different fish species with granulomas (Stoskopf, 1993). *M. marinum* is transferred to humans while they handle fish in aquaculture tanks and aquariums; therefore it is also called a zoonotic pathogen. It produces aquarium tank or fish tank granuloma (self-limiting and superficial lesions), which involves knees, hands, elbows, and forearms, that is, the cooler parts of the body (Huminer et al., 1986; Smithand & Willett, 1980). The intraspecies sequence homogeneity is significant between various *M. marinum* strains (Yip et al., 2007), even though strain variation has also been described (Ucko et al., 2002). However, it is assumed that only a few *M. marinum* strains might be having a zoonotic prospective (Ucko & Colorni, 2005). *M. marinum* is highly interrelated to *M. ulcerans* followed by *M. tuberculosis* as revealed by phylogenetic studies (Yip et al., 2007). Besides this,

M. tuberculosis and *M. marinum* show response to similar antibiotics and also share significant pathological features and numerous virulence factors as well. Therefore, in order to study the pathogenesis of tuberculosis, *M. marinum* represents an essential model organism (Swaim et al., 2006; Volkman et al., 2004).

7.4.8 Pseudomonas

Pseudomonas sp. represent a large group of microorganisms and are therefore widely spread throughout the environment. These psychrophilic bacteria usually grow at low temperatures and are also characterized as the dominant microorganisms. The *Pseudomonas* sp. are rapidly substituted by mesophilic microorganisms, together with bacteria of the genus *Aeromonas*, at higher temperatures almost above 10°C. *P. fluorescens* is often seen related to fin and skin disease of fish and is considered to be the most significant species in fish pathology. In certain trout farms, *P. fluorescens* infections can result in abrupt mortality throughout the year, irrespective of the temperature of water in the farm. Such a prompt disease development may be because of the mutations, which leads to the emergence of a new ability, permitting the bacteria to acclimatize to new environmental conditions. Some other species of *Pseudomonas* mostly act as an accompanying microflora; these include *P. luteola* or *P. putida*, which are frequently isolated from internal organs of fish (Kozinska, 1999). The *Pseudomonas* sp. in tench (*Tinca tinca*) and rainbow trout (*O. mykiss*) can cause strawberry disease. In the case of silver carp (*Carassius gibelio*) and crucian carp (*Carassius carassius*), comparable systemic infections showing the usual symptoms of septicemia were also observed (Ahne et al., 1982; Csaba et al., 1981; Kozinska, 1999).

7.4.8.1 Emerging prospective pathogens of freshwater fish

Over the last few years, new bacterial infections of serious concern related to freshwater fish have been witnessed and can be called emerging diseases. For any disease to be termed as emerging, the appearance of the disease has to be in a population for the first time, or it may have previously existed, and its prevalence or geographic range must increase rapidly, or it establishes itself in an innovative way (Okamura & Feist, 2011). Shewanellosis could provide us an example of such an emerging disease, as in 2004 it appeared for the first time and caused severe health disorders in rainbow trout and carp (Kozinska & Pekala, 2004). Furthermore, it spread quickly to various fish species in different countries (Pękala et al., 2015; Qin et al., 2012; Rusev et al., 2016).

7.4.9 Plesiomonas shigelloides

Plesiomonas shigelloides belongs to the Enterobacteriaceae family and is found to be both human and animal prospective pathogenic agents. As far as isolation of *P. shigelloides* is considered, there are only a few observations of health disorders in fish that have been described so far. Some of them inflicted high mortalities, reaching up to 40% of the stock in the case of salmonids (Cruz et al., 1986; Vladik & Vitovec, 1974). The clinical symptoms showed redness of the anus and cachexia of fish. Postmortem investigations revealed the occurrence of exudative fluid in the body cavity and the occurrence of punctate petechiae on the peritoneum. The growth of *P. shigelloides* in monoculture or in association with *A. hydrophila* and *Flavobacterium* sp. were observed while carrying out the bacterial analysis of various samples collected from diseased fish (Cruz et al., 1986; Vladik & Vitovec, 1974). Isolation of *P.*

shigelloides was also carried out from sturgeon (*Acipenser sturio*), eels (*Anguilla anguilla*), and African catfish (*Heterobranchus bidorsalis*) (Klein et al., 1993).

7.4.10 Stenotrophomonas maltophilia

Stenotrophomonas maltophilia is one more bacterial species presently often isolated from freshwater fish. This microorganism was isolated from freshwater, salt water, and bottom sediments and was thus considered an omnipresent environmental bacterium (Dungan et al., 2003). This bacterial species was also obtained from agricultural and industrial soils (Sturz et al., 2001) and plant tissues (Taghavi et al., 2009), in the case of terrestrial environments. *S. maltophilia* plays a significant part in processes involving biological purification, as it is involved in the degradation of several xenobiotics (Dubey & Fulekar, 2012). This bacterial species is utilized for promoting plant growth and also for combating plant pathogens (when used as a biological agent) due to release of phytohormones (Peralta et al., 2012). *S. maltophilia* shows an internal resistance to an extensive series of antibacterial agents and is therefore presently categorized as a multidrug-resistant microorganism. It is generally found to be associated with human respiratory diseases (Brooke, 2012). With respect to fish infections caused by *S. maltophilia*, only a few publications are available. Among these publications, the most essential are associated with description of disease outbreak witnessed in channel catfish (*Ictalurus punctatus*) and African catfish (*Heterobranchus bidorsalis*) (Abraham et al., 2016; Geng et al., 2010). In both cases, the clinical investigation of the fish revealed petechiae, depigmentation of the skin, lethargy, edema in the body cavity, and focal hemorrhages. Postmortem examination of fishes displayed congestion of internal organs, intestines filled with gases, and petechiae on their surface. These health disorders and syndromes were called infectious intussusception syndrome (IIS), with mortality rate of 20% of the stock (Geng et al., 2010).

7.4.11 Kocuria rhizophila

Kocuria rhizophila is an outstanding example of an emerging fish pathogen. With recognized pathogenic prospective in the case of salmonids, this represents an entirely new bacterium within the community (Pękala et al., 2018). *Kocuria* sp. are isolated from different environments like freshwater, food, marine sediment, or chicken meat and are well accepted skin commensals in mammals (Becker et al., 2008; Kim et al., 2004). These bacteria composed the physiological microflora of the trout gut from which these were isolated (Kim et al., 2007). *Kocuria rhizophila* infections result in abnormal mortality (nearly 50% of the stock), together with pathological changes in internal organs and external tissues. Clinically, in the case of moribund fish, swollen abdomen, focal lesions, increased skin melanization, exophthalmia, and skin petechiae were often witnessed. Postmortem investigations revealed hemorrhages in the tail muscles primarily in the caudal part, liver congestion, and inflammation of the intestines (Pękala et al., 2018).

7.5 Conclusion

There has been a noticeable rise in the number of bacterial species involved in triggering fish diseases. The common bacterial species that are pathogenic to fish belong to the genera *Aeromonas*,

Vibrio, *Edwardsiella*, *Yersinia*, *Flavobacterium*, *Lactococcus*, *Streptococcus*, *Mycobacterium*, and *Renibacterium*. However, there is an increase in indication among various fish pathogens that the pathogenic species range and the host and geographical variety are broadening, which leads to the appearance of new fish pathogens. The fish diseases pose exceptional and daunting challenges unlike those of animal and human medicine. In addition to this, the existing treatments for fish diseases are usually less effective and the genetics and biology of most of the pathogens are poorly understood, which restricts the utilization of several contemporary science-based pathogen involvement approaches. Prompt development in the genome sequencing of animal and human pathogens has assisted in battling several major diseases and has also led to a better understanding of pathogen evolution, their host adaptation approaches, and their biology. Unfortunately, in the case of fish pathogenic bacteria this kind of growth and advancement with respect to genomics and functional genomics has been very slow. However, the pace of sequencing related to more fish pathogenic bacteria has been improved by the availability of the latest cost-beneficial highly efficient sequencing technologies. With correlated bacteria, comparative pathogenomics has led to an increase in knowledge about their variance in virulence and their capability to acclimatize in different ecological niches.

References

Abbott, S. L., & Janda, J. M. (2006). The genus Edwardsiella. *Prokaryotes*, *6*, 72–89.

Abraham, T. J., Paul, P., Adikesavalu, H., Patra, A., & Banerjee, S. (2016). *Stenotrophomonas maltophilia* as an opportunistic pathogen in cultured African catfish *Clarias gariepinus* (Burchell, 1822). *Aquaculture (Amsterdam, Netherlands)*, *450*, 168–172.

Actis, L. A., Tolmasky, M. E., & Crosa, J. H. (2011). Vibriosis. In P. T. K. Woo, & D. W. Bruno (Eds.), *Fish diseases and disorders, viral, bacterial, and fungal infections* (Vol. 3, pp. 570–605). CABI Publishing.

Aguado-Urda, M., López-Campos, G. H., Blanco, M. M., Fernández-Garayzábal, J. F., Cutuli, M. T., Aspiroz, C., & Gibello, A. (2011). Genome sequence of *Lactococcus garvieae* 21881, isolated in a case of human septicemia. *Journal of Bacteriology*, *193*, 4033–4034.

Ahne, W., Popp, W., & Hoffmann, R. (1982). *Pseudomonas fluorescens* as a pathogen for tench. *Bulletin of the European Association of Fish Pathologists*, *2*(4), 56–57.

Allen, D. A., Austin, B., & Colwell, R. R. (1983). Numerical taxonomy of bacterial isolates associated with a freshwater fishery. *Microbiology (Reading, England)*, *129*(7), 2043–2062.

Amaro, C., Biosca, E. G., Esteve, C., Fouz, B., & Toranzo, A. E. (1992). Comparative study of phenotypic and virulence properties in *Vibrio vulnificus* biotypes 1 and 2 obtained from a European eel farm experiencing mortalities. *Diseases of Aquatic Organisms*, *13*(1), 29–35.

Arkoosh, M. R., Casillas, E., Clemons, E., Kagley, A. N., Olson, R., Reno, P., & Stein, J. E. (1998). Effect of pollution on fish diseases: Potential impacts on salmonid populations. *Journal of Aquatic Animal Health*, *10*(2), 182–190.

Austin, B. (1982). Taxonomy of bacteria isolated from a coastal, marine fish-rearing unit. *Journal of Applied Bacteriology*, *53*(2), 253–268.

Austin, B. (1983). Bacterial microflora associated with a coastal, marine fish-rearing unit. *Journal of the Marine Biological Association of the United Kingdom*, *63*(3), 585–592.

Austin, B., & Austin, D. A. (1999). *Bacterial fish pathogens: Disease in farmed and wild fish* (3rd ed.). New York, NY: Springer.

Austin, B., & Austin, D. A. (2012). *Bacterial fish pathogens* (pp. 481–482). Dordrecht, The Netherlands: Springer.

Austin, B., & Austin, D. A. (2016). *Bacterial fish pathogens: Diseases of farmed and wild fish* (pp. 34–67). The Netherlands: Springer.

Austin, B., & Zhang, X. H. (2006). *Vibrio harveyi*: A significant pathogen of marine vertebrates and invertebrates. *Letters in Applied Microbiology*, *43*(2), 119–124.

Bartlett, K. H. (1974). Occurrence of potential pathogens in water containing ornamental fishes. *Applied Microbiology*, *28*(1), 35–40.

Beatson, S. A., de Luna, M. D. G., Bachmann, N. L., Alikhan, N. F., Hanks, K. R., Sullivan, M. J., & Squire, D. J. (2011). Genome sequence of the emerging pathogen *Aeromonas caviae*. *Journal of Bacteriology*, *193*(5), 1286–1287.

Beaz-Hidalgo, R., & Figueras, M. J. (2013). *Aeromonas* spp. whole genomes and virulence factors implicated in fish disease. *Journal of Fish Diseases*, *36*(4), 371–388.

Becker, K., Rutsch, F., Uekötter, A., Kipp, F., König, J., Marquardt, T., & von Eiff, C. (2008). *Kocuria rhizophila* adds to the emerging spectrum of micrococcal species involved in human infections. *Journal of Clinical Microbiology*, *46*(10), 3537–3539.

Benediktsdóttir, E., Verdonck, L., Spröer, C., Helgason, S., & Swings, J. (2000). Characterization of *Vibrio viscosus* and *Vibrio wodanis* isolated at different geographical locations: A proposal for reclassification of *Vibrio viscosus* as *Moritella viscosa* comb. nov. *International Journal of Systematic and Evolutionary Microbiology*, *50*(2), 479–488.

Bergh, Ø., Nilsen, F., & Samuelsen, O. B. (2001). Diseases, prophylaxis and treatment of the Atlantic halibut *Hippoglossus hippoglossus*: A review. *Diseases of Aquatic Organisms*, *48*(1), 57–74.

Bernardet, J. F. (1989). *Flexibacter columnaris*: First description in France and comparison with bacterial strains from other origins. *Diseases of Aquatic Organisms*, *6*, 37–44.

Bernoth, E. M., Ellis, A. E., Midtlyng, P. J., Olivier, G., Smith, P., Bernoth, E. M., Ellis, A. E., Midtlyng, P. J., Olivier, G., & Smith, P. (Eds.), (1997). *Furunculosis: Multidisciplinary fish disease research*. London, UK: Academic Press.

Berthe, F. C., Michel, C., & Bernardet, J. F. (1995). Identification of *Pseudomonas anguilliseptica* isolated from several fish species in France. *Diseases of Aquatic Organisms*, *21*(2), 151–155.

Biosca, E. G., Llorens, H., Garay, E., & Amaro, C. (1993). Presence of a capsule in *Vibrio vulnificus* biotype 2 and its relationship to virulence for eels. *Infection and Immunity*, *61*(5), 1611–1618.

Blazer, V. S., Shotts, E. B., & Waltman, W. D. (1985). Pathology associated with *Edwardsiella ictaluri* in catfish, *Ictalurus punctatus* rafinesque, and *Danio devario* (Hamilton-Buchanan, 1822). *Journal of Fish Biology*, *27*(2), 167–175.

Brooke, J. S. (2012). *Stenotrophomonas maltophilia*: An emerging global opportunistic pathogen. *Clinical Microbiology Reviews*, *25*(1), 2–41.

Bullock, G. L., & Herman, R. L. (1988). *Bacterial kidney disease of salmonid fishes caused by Renibacterium salmoninarum* (Vol. 78, pp. 0–10). US Fish and Wildlife Service.

Chen, P. E., Cook, C., Stewart, A. C., Nagarajan, N., Sommer, D. D., Pop, M., & Sozhamannan, S. (2010). Genomic characterization of the *Yersinia genus*. *Genome Biology*, *11*(1), R1.

Collins, M. D., Farrow, J. A. E., Phillips, B. A., & Kandler, O. (1983). *Streptococcus garvieae* sp. nov. and *Streptococcus plantarum* sp. nov. *Microbiology (Reading, England)*, *129*(11), 3427–3431.

Colorni, A. (2004). Diseases of Mediterranean fish species: Problems, research and prospects. *Bulletin-European Association of Fish Pathologists*, *24*(1), 22–32.

Colwell, R. R. (1962). The bacterial flora of Puget Sound fish. *Journal of Applied Bacteriology*, *25*(2), 147–158.

Cruz, J. M., Saraiva, A., Eiras, J. C., Branco, R., & Sousa, J. C. (1986). An outbreak of *Plesiomonas shigelloides* in farmed rainbow trout, *Salmo gairdneri* Richardson, in Portugal. *Bulletin of the European Association of Fish Pathologists (Denmark)*, *6*, 20–22.

Csaba, G. Y., Prigli, M., Békési, L., Kováacs-Gayer, E., Bajmócy, E., & Fazekas, B. (1981). Septicemia in silver carp (*Hypophthalmichthys molitrix*, Val.) and bighead (*Aristichthys nobilis* Rich.) caused by *Pseudomonas fluorescens*. In J. Oláh, K. Molnár, & S. Jeney (Eds.), *Fish pathogens and environment in European polyculture* (pp. 111–123). Szarvas: Fisheries Research Institute.

Dalsgaard, I., & Madsen, L. (2000). Bacterial pathogens in rainbow trout, *Oncorhynchus mykiss* (Walbaum), reared at Danish freshwater farms. *Journal of Fish Diseases*, *23*(3), 199–209.

Dar, G. H., Bhat, R. A., Kamili, A. N., Chishti, M. Z., Qadri, H., Dar, R., & Mehmood, M. A. (2020). *Correlation between pollution trends of freshwater bodies and bacterial disease of fish fauna. Fresh Water Pollution Dynamics and Remediation* (pp. 51–67). Singapore: Springer.

Dar, G. H., Dar, S. A., Kamili, A. N., Chishti, M. Z., & Ahmad, F. (2016). Detection and characterization of potentially pathogenic *Aeromonas sobria* isolated from fish *Hypophthalmichthys molitrix* (Cypriniformes: Cyprinidae). *Microbial Pathogenesis*, *91*, 136–140.

Dar, G. H., Kamili, A. N., Chishti, M. Z., Dar, S. A., Tantry, T. A., & Ahmad, F. (2016). Characterization of *Aeromonas sobria* Isolated from fish rohu (*Labeo rohita*) collected from polluted pond. *Journal of Bacteriology & Parasitology*, *7*(3), 1–5. Available from https://doi.org/10.4172/2155-9597.1000273.

Davies, R. L. (1991). Virulence and serum-resistance in different clonal groups and serotypes of *Yersinia ruckeri*. *Veterinary Microbiology*, *29*(3–4), 289–297.

Decostere, A., Haesebrouck, F., Turnbull, J. F., & Charlier, G. (1999). Influence of water quality and temperature on adhesion of high and low virulence *Flavobacterium columnare* strains to isolated gill arches. *Journal of Fish Diseases*, *22*(1), 1–11.

del Castillo, C. S., Hikima, J. I., Jang, H. B., Nho, S. W., Jung, T. S., Wongtavatchai, J., & Aoki, T. (2013). Comparative sequence analysis of a multidrug-resistant plasmid from *Aeromonas hydrophila*. *Antimicrobial Agents and Chemotherapy*, *57*(1), 120–129.

Dixon, P. F., Smail, D. A., Algoët, M., Hastings, T. S., Bayley, A., Byrne, H., & Verner-Jeffreys, D. (2012). Studies on the effect of temperature and pH on the inactivation of fish viral and bacterial pathogens. *Journal of Fish Diseases*, *35*(1), 51–64.

Domeénech, A., Derenaáandez-Garayzábal, J. F., Pascual, C., Garcia, J. A., Cutuli, M. T., Moreno, M. A., & Dominguez, L. (1996). Streptococcosis in cultured turbot, *Scopthalmus maximus* (L.), associated with *Streptococcus parauberis*. *Journal of Fish Diseases*, *19*(1), 33–38.

Domenech, A., Prieta, J., Fernandez-Garayzabal, J. F., Collins, M. D., Jones, D., & Dominguez, L. (1993). Phenotypic and phylogenetic evidence for a close relationship between *Lactococcus garviae* and *Enterococcus seriolicida*. *Microbiologia*, *9*, 63.

Dubey, K. K., & Fulekar, M. H. (2012). Chlorpyrifos bioremediation in *Pennisetum rhizosphere* by a novel potential degrader *Stenotrophomonas maltophilia* MHF ENV20. *World Journal of Microbiology and Biotechnology*, *28*(4), 1715–1725.

Duchaud, E., Boussaha, M., Loux, V., Bernardet, J. F., Michel, C., Kerouault, B., & Bessieres, P. (2007). Complete genome sequence of the fish pathogen *Flavobacterium psychrophilum*. *Nature Biotechnology*, *25* (7), 763–769.

Dungan, R. S., Yates, S. R., & Frankenberger, W. T., Jr (2003). Transformations of selenate and selenite by *Stenotrophomonas maltophilia* isolated from a seleniferous agricultural drainage pond sediment. *Environmental Microbiology*, *5*(4), 287–295.

El Deen, A. N., Dorgham-Sohad, M., Hassan-Azza, H. M., & Hakim, A. S. (2014). Studies on *Aeromonas hydrophila* in cultured *Oreochromis niloticus* at Kafr El Sheikh Governorate, Egypt with reference to histopathological alterations in some vital organs. *World Journal of Fish and Marine Sciences*, *6*(3), 233–240.

Eldar, A., Ghittino, C., Asanta, L., Bozzetta, E., Goria, M., Prearo, M., & Bercovier, H. (1996). *Enterococcus seriolicida* is a junior synonym of *Lactococcus garvieae*, a causative agent of septicemia and meningoencephalitis in fish. *Current Microbiology*, *32*(2), 85–88.

Figueras, M. J. (2005). Clinical relevance of Aeromonas sM503. *Reviews in Medical Microbiology*, *16*(4), 145–153.

Food and Agriculture Organisation Aquaculture (FAO). (2017). *Newsletter. 56*. Retrieved from: http://www.fao.org/3/a-i7171e.pdf.

Food and Agriculture Organisation Aquaculture (FAO). (2018). *The state of world fisheries and aquaculture*. Retrieved from: http://www.fao.org/3/i9540en/I9540EN.pdf.

Frerichs, G. N. (1993). Acid-fast fish pathogens. In V. Inglis, R. J. Roberts, & N. R. Bromage (Eds.), *Bacterial diseases of fish* (pp. 217–233). Oxford, UK: Blackwell Scientific.

Furones, M. D., Rodgers, C. J., & Munn, C. B. (1993). *Yersinia ruckeri*, the causal agent of enteric redmouth disease (ERM) in fish. *Annual Review of Fish Diseases, 3*, 105–125.

Geng, Y., Wang, K., Chen, D., Huang, X., He, M., & Yin, Z. (2010). *Stenotrophomonas maltophilia*, an emerging opportunist pathogen for cultured channel catfish, *Ictalurus punctatus*, in China. *Aquaculture (Amsterdam, Netherlands), 308*(3–4), 132–135.

Gillespie, N. C., & Macrae, I. C. (1975). The bacterial flora of some Queensland fish and its ability to cause spoilage. *Journal of Applied Bacteriology, 39*(2), 91–100.

Good, C., Davidson, J., Wiens, G. D., Welch, T. J., & Summerfelt, S. (2015). *Flavobacterium branchiophilum* and *F. succinicans* associated with bacterial gill disease in rainbow trout *Oncorhynchus mykiss* (Walbaum) in water recirculation aquaculture systems. *Journal of Fish Diseases, 38*(4), 409–413.

Gulig, P. A., de Crécy-Lagard, V., Wright, A. C., Walts, B., Telonis-Scott, M., & McIntyre, L. M. (2010). SOLiD sequencing of four *Vibrio vulnificus* genomes enables comparative genomic analysis and identification of candidate clade-specific virulence genes. *BMC Genomics, 11*(1), 512.

Gutenberger, S. K., Duimstra, J. R., Rohovec, J. S., & Fryer, J. L. (1997). Intracellular survival of *Renibacterium salmoninarum* in trout mononuclear phagocytes. *Diseases of Aquatic Organisms, 28*(2), 93–106.

Heo, G. J., Kasai, K., & Wakabayashi, H. (1990). Occurrence of *Flavobacterium branchiophila* associated with bacterial gill disease at a trout hatchery. *Fish Pathology, 25*(2), 99–105.

Hiney, M., & Olivier, G. (1999). Furunculosis (*Aeromonas salmonicida*). In P. T. K. Woo, & D. W. Bruno (Eds.), *Fish diseases and disorders III: Viral, bacterial and fungal infections* (pp. 341–425). Oxford, UK: CAB Publishing.

Hjeltnes, B., & Roberts, R. J. (1993). Vibriosis. In R. J. Roberts, N. R. Bromag, & V. Inglis (Eds.), *Bacterial diseases of fish* (pp. 109–121). Oxford, UK: Blackwell Scientific.

Hjerde, E., Lorentzen, M. S., Holden, M. T., Seeger, K., Paulsen, S., Bason, N., & Sanders, S. (2008). The genome sequence of the fish pathogen *Aliivibrio salmonicida* strain LFI1238 shows extensive evidence of gene decay. *BMC Genomics, 9*(1), 1–14.

Holt, R.A. (1987). *Cytophaga psychrophila*, the causative agent of bacterial cold-water disease in salmonid fish. Dissertation Abstracts International, B (Sciences and Engineering).

Horne, M. T., & Barnes, A. C. (1999). Enteric redmouth disease (*Y. ruckeri*). In P. T. K. Woo, & D. W. Bruno (Eds.), *Fish diseases and disorders: Viral, bacterial and fungal infections* (Vol. 3, pp. 455–477). CABI Publishing.

Horsley, R. W. (1973). The bacterial flora of the Atlantic salmon (*Salmo salar* L.) in relation to its environment. *Journal of Applied Bacteriology, 36*(3), 377–386.

Huminer, D., Pitlik, S. D., Block, C., Kaufman, L., Amit, S., & Rosenfeld, J. B. (1986). Aquarium-borne Mycobacterium marinum skin infection: Report of a case and review of the literature. *Archives of Dermatology, 122*(6), 698–703.

Ibrahim, B. U., Baba, J., & Sheshi, M. S. (2014). Isolation and identification of bacteria associated with fresh and smoked fish (*Clarias gariepinus*) in Minna Metropolis, Niger State. Nigeria. *Journal of Applied and Environmental Microbiology, 2*(3), 81–85.

Johnson, P. T., & Paull, S. H. (2011). The ecology and emergence of diseases in fresh waters. *Freshwater Biology, 56*(4), 638–657.

Kim, D. H., Brunt, J., & Austin, B. (2007). Microbial diversity of intestinal contents and mucus in rainbow trout (*Oncorhynchus mykiss*). *Journal of Applied Microbiology*, *102*(6), 1654–1664.

Kim, S. B., Nedashkovskaya, O. I., Mikhailov, V. V., Han, S. K., Kim, K. O., Rhee, M. S., & Bae, K. S. (2004). *Kocuria marina* sp. nov., a novel actinobacterium isolated from marine sediment. *International Journal of Systematic and Evolutionary Microbiology*, *54*(5), 1617–1620.

Kitao, T., Inglis, V., Roberts, R. J., & Bromage, N. R. (1993). *Bacterial diseases of fish* (pp. 196–210). Oxford, UK: Blackwell Scientific.

Klein, B., Kleingeld, D., & Bohm, K. (1993). From samples of cultured fish in Germany. *Bulletin of the European. Association of Fish Pathologists*, *13*(2), 70–72.

Korun, J., Akgun-Dar, K., & Yazici, M. (2009). Isolation of *Shewanella putrefaciens* from cultured European sea bass, (*Dicentrarchus labrax*) in Turkey. *Revue de Médecine Vétérinaire*, *160*, 532–536.

Kotetishvili, M., Kreger, A., Wauters, G., Morris, J. G., Sulakvelidze, A., & Stine, O. C. (2005). Multilocus sequence typing for studying genetic relationships among Yersinia species. *Journal of Clinical Microbiology*, *43*(6), 2674–2684.

Kozinska, A. (1999). A typical cases of disorders in cyprinid wintering caused by *Pseudomonas fluorescens* infection. *Bulletin of the European Association of Fish Pathologists*, *19*, 216–220.

Kozinska, A., & Pekala, A. (2004). First isolation of *Shewanella putrefaciens* from freshwater fish-a potential new pathogen of fish. *Bulletin-European Association of Fish Pathologists*, *24*(4), 189–193.

Kusuda, R., Kawai, K., Salati, F., Banner, C. R., & Fryer, J. L. (1991). *Enterococcus seriolicida* sp. nov., a fish pathogen. *International Journal of Systematic and Evolutionary Microbiology*, *41*(3), 406–409.

Larsen, J. L., Pedersen, K., & Dalsgaard, I. (1994). *Vibrio anguillarum* serovars associated with vibriosis in fish. *Journal of Fish Diseases*, *17*(3), 259–267.

Leahy, J. G., & Colwell, R. R. (1990). Microbial degradation of hydrocarbons in the environment. *Microbiology and Molecular Biology Reviews*, *54*(3), 305–315.

Lehmann, J. D. F. J., Mock, D., Stürenberg, F. J., & Bernardet, J. F. (1991). First isolation of *Cytophaga psychrophila* from a systemic disease in eel and cyprinids. *Diseases of Aquatic Organisms*, *10*(3), 217–220.

Li, Y., Liu, Y., Zhou, Z., Huang, H., Ren, Y., Zhang, Y., & Wang, L. (2011). Complete genome sequence of *Aeromonas veronii* strain B565. *Journal of Bacteriology*, *193*, 3389–3390.

Lightner, D. V., & Redman, R. M. (1998). Shrimp diseases and current diagnostic methods. *Aquaculture (Amsterdam, Netherlands)*, *164*(1–4), 201–220.

Ling, S. H. M., Wang, X. H., Xie, L., Lim, T. M., & Leung, K. Y. (2000). Use of green fluorescent protein (GFP) to study the invasion pathways of *Edwardsiella tarda* in in vivo and in vitro fish models. *Microbiology (Reading, England)*, *146*(1), 7–19.

Madetoja, J., Dalsgaard, I., & Wiklund, T. (2002). Occurrence of *Flavobacterium psychrophilum* in fish-farming environments. *Diseases of Aquatic Organisms*, *52*(2), 109–118.

Martin-Carnahan, A., & Joseph, S. W. (2005). Aeromonadaceae. In G. M. Garrity (Ed.), *Bergey's manual of systematic bacteriology* (Vol. 2, pp. 556–580). Baltimore, MD: Williams & Wilkins.

Meyer, F. P., & Bullock, G. L. (1973). *Edwardsiella tarda*, a new pathogen of channel catfish (*Ictalurus punctatus*). *Applied Microbiology*, *25*(1), 155–156.

Mitchum, D. L., & Sherman, L. E. (1981). Transmission of bacterial kidney disease from wild to stocked hatchery trout. *Canadian Journal of Fisheries and Aquatic Sciences*, *38*(5), 547–551.

Mongodin, E. F., Shapir, N., Daugherty, S. C., DeBoy, R. T., Emerson, J. B., Shvartzbeyn, A., & Khouri, H. (2006). Secrets of soil survival revealed by the genome sequence of *Arthrobacter aurescens* TC1. *PLoS Genetics*, *2*(12), e214.

Morita, H., Toh, H., Oshima, K., Yoshizaki, M., Kawanishi, M., Nakaya, K., & Murakami, M. (2011). Complete genome sequence and comparative analysis of the fish pathogen *Lactococcus garvieae*. *PLoS One*, *6*(8), e23184.

Naka, H., Dias, G. M., Thompson, C. C., Dubay, C., Thompson, F. L., & Crosa, J. H. (2011). Complete genome sequence of the marine fish pathogen *Vibrio anguillarum* harboring the pJM1 virulence plasmid and genomic comparison with other virulent strains of *V. anguillarum* and *V. ordalii*. *Infection and Immunity*, *79*(7), 2889–2900.

Nematollahi, A., Decostere, A., Pasmans, F., & Haesebrouck, F. (2003). *Flavobacterium psychrophilum* infections in salmonid fish. *Journal of Fish Diseases*, *26*(10), 563–574.

Newton, J. C., Wolfe, L. G., Grizzle, J. M., & Plumb, J. A. (1989). Pathology of experimental enteric septicaemia in channel catfish, *Ictalurus punctatus* (rafinesque), following immersion-exposure to *Edwardsiella ictaluri*. *Journal of Fish Diseases*, *12*(4), 335–347.

Nho, S. W., Hikima, J. I., Cha, I. S., Park, S. B., Jang, H. B., del Castillo, C. S., & Jung, T. S. (2011). Complete genome sequence and immunoproteomic analyses of the bacterial fish pathogen *Streptococcus parauberis*. *Journal of Bacteriology*, *193*(13), 3356–3366.

Niemi, M., & Taipalinen, I. (1982). *Faecal indicator bacteria at fish farms. Lakes and water management* (pp. 171–175). Dordrecht: Springer.

Noga, E. J. (1996). *Fish disease: Diagnosis and treatment* (pp. 88–93). St. Louis, MO: Mosby.

Norwegian Scientific Committee for Food Safety. (2009). *Criteria for safe use of plant ingredients in diets for aquacultured fish*. Norwegian Scientific Committee for Food Safety, ISBN 978-82-8082-299-4.

Okamura, B., & Feist, S. W. (2011). Emerging diseases in freshwater systems. *Freshwater Biology*, *56*(4), 627–637.

Okocha, R. C., Olatoye, I. O., & Adedeji, O. B. (2018). Food safety impacts of antimicrobial use and their residues in aquaculture. *Public Health Reviews*, *39*(1), 1–22.

Olesen, N. J., & Nicolajsen, N. (2012). Overview of the disease situation and surveillance in Europe in 2011. In: *16th Annual Meeting of the National Reference Laboratories for Fish Diseases*. National Veterinary Institute, Technical University of Denmark, Copenhagen, Denmark, pp. 13–15.

Ostland, V. E., Lumsden, J. S., MacPhee, D. D., & Ferguson, H. W. (1994). Characteristics of *Flavobacterium branchiophilum*, the cause of salmonid bacterial gill disease in Ontario. *Journal of Aquatic Animal Health*, *6*(1), 13–26.

Paniagua, C. A. R. M. E. N., Rivero, O. C. T. A. V. I. O., Anguita, J. U. A. N., & Naharro, G. E. R. M. A. N. (1990). Pathogenicity factors and virulence for rainbow trout (*Salmo gairdneri*) of motile *Aeromonas* spp. isolated from a river. *Journal of Clinical Microbiology*, *28*(2), 350–355.

Park, S. B., Aoki, T., & Jung, T. S. (2012). Pathogenesis of and strategies for preventing *Edwardsiella tarda* infection in fish. *Veterinary Research*, *43*(1), 67.

Partridge, S. R., & Hall, R. M. (2003). The IS1111 family members IS4321 and IS5075 have subterminal inverted repeats and target the terminal inverted repeats of Tn21 family transposons. *Journal of Bacteriology*, *185*(21), 6371–6384.

Pękala, A., Kozińska, A., Paździor, E., & Głowacka, H. (2015). Phenotypical and genotypical characterization of *S. hewanella* putrefaciens strains isolated from diseased freshwater fish. *Journal of Fish Diseases*, *38*(3), 283–293.

Pękala, A., Paździor, E., Antychowicz, J., Bernad, A., Głowacka, H., Więcek, B., & Niemczuk, W. (2018). *Kocuria rhizophila* and *Micrococcus luteus* as emerging opportunist pathogens in brown trout (*Salmo trutta* Linnaeus, 1758) and rainbow trout (*Oncorhynchus mykiss* Walbaum, 1792). *Aquaculture (Amsterdam, Netherlands)*, *486*, 285–289.

Pękala-Safińska, A. (2018). Contemporary threats of bacterial infections in freshwater fish. *Journal of Veterinary Research*, *62*(3), 261–267.

Peralta, K. D., Araya, T., Valenzuela, S., Sossa, K., Martínez, M., Peña-Cortés, H., & Sanfuentes, E. (2012). Production of phytohormones, siderophores and population fluctuation of two root-promoting rhizobacteria in Eucalyptus globulus cuttings. *World Journal of Microbiology and Biotechnology*, *28*(5), 2003–2014.

Perera, R. P., Johnson, S. K., Collins, M. D., & Lewis, D. H. (1994). *Streptococcus iniae* associated with mortality of *Tilapia nilotica* × *T. aurea* hybrids. *Journal of Aquatic Animal Health*, *6*(4), 335–340.

Qin, L., Zhang, X., & Bi, K. (2012). A new pathogen of gibel carp *Carassius auratus* gibelio-*Shewanella putrefaciens*. *Wei Sheng Wu Xue Bao = Acta microbiologica Sinica*, *52*(5), 558–565.

Ransom, D. P., Lannan, C. N., Rohovec, J. S., & Fryer, J. L. (1984). Comparison of histopathology caused by *Vibrio anguillarum* and *Vibrio ordalii* in three species of Pacific salmon. *Journal of Fish Diseases*, *7*(2), 107–115.

Reimundo, P., Pignatelli, M., Alcaraz, L. D., D'Auria, G., Moya, A., & Guijarro, J. A. (2011). Genome sequence of *Lactococcus garvieae* UNIUD074, isolated in Italy from a lactococcosis outbreak. *Journal of Bacteriology*, *193*, 3684–3685.

Reith, M. E., Singh, R. K., Curtis, B., Boyd, J. M., Bouevitch, A., Kimball, J., & Nash, J. H. (2008). The genome of *Aeromonas salmonicida* subsp. salmonicida A449: Insights into the evolution of a fish pathogen. *BMC Genomics*, *9*(1), 427.

Rogers, W. A. (1983). Edwardsiellosis in fishes. In D. P. Anderson, M. Dorson, & Ph Dubourget (Eds.), *Antigens of fish pathogens. Les antigenes des microorganisms pathogenes des poissons* (pp. 153–159). Lyon, France: Fondation Marcel Merieux.

Romalde, J. L., MagariÑos, B., Barja, J. L., & Toranzo, A. E. (1993). Antigenic and molecular characterization of *Yersinia ruckeri* proposal for a new intraspecies classification. *Systematic and Applied Microbiology*, *16* (3), 411–419.

Romalde, J. L., Ravelo, C., Valdés, I., Magariños, B., de la Fuente, E., San Martín, C., & Toranzo, A. E. (2008). *Streptococcus phocae*, an emerging pathogen for salmonid culture. *Veterinary Microbiology*, *130* (1–2), 198–207.

Romero, J., Feijoó, C. G., & Navarrete, P. (2012). Antibiotics in aquaculture – Use, abuse and alternatives. *Health and Environment in Aquaculture*, *159*, 159–184.

Ross, A. J., Rucker, R. R., & Ewing, W. H. (1966). Description of a bacterium associated with redmouth disease of rainbow trout (*Salmo gairdneri*). *Canadian Journal of Microbiology*, *12*(4), 763–770.

Rusev, V., Rusenova, N., Simeonov, R., & Stratev, D. (2016). *Staphylococcus warneri* and *Shewanella putrefaciens* coinfection in Siberian sturgeon (*Acipenser baerii*) and Hybrid sturgeon (*Huso huso* x *Acipenser baerii*). *Journal of Microbiology & Experimentation*, *3*(1), 00078.

Salonius, K., Siderakis, C., MacKinnon, A. M., & Griffiths, S. G. (2005). Use of *Arthrobacter davidanieliasalive* vaccine against *Renibacterium salmoninarum* and *Piscirickettsia salmonis* in salmonids. *Developments in Biologicals: Journal of the International Association of Biological Standardization*, *121*, 189–197.

Sanders, G. E., & Swaim, L. E. (2001). Atypical piscine mycobacteriosis in Japanese medaka (*Oryzias latipes*). *Comparative Medicine*, *51*(2), 171–175.

Sanders, J. E., & Fryer, J. L. (1980). *Renibacterium salmoninarum* gen. nov., sp. nov., the causative agent of bacterial kidney disease in salmonid fishes. *International Journal of Systematic and Evolutionary Microbiology*, *30*(2), 496–502.

Sankar, G., Saravanan, J., Krishnamurthy, P., Chandrakala, N., & Rajendran, K. (2012). Isolation and identification of *Vibrio* spp. in diseased *Channa punctatus* from aquaculture fish farm. *Indian Journal of Geo-Marine Sciences*, *41*(2), 159–163.

Schiewe, M. H., & Crosa, J. H. (1981). *Vibrio ordalii* sp. nov.: A causative agent of vibriosis in fish. *Current Microbiology*, *6*(6), 343–348.

Schrøder, M. B., Espelid, S., & Jørgensen, T. Ø. (1992). Two serotypes of *Vibrio salmonicida* isolated from diseased cod (*Gadus morhua* L.); virulence, immunological studies and vaccination experiments. *Fish & Shellfish Immunology*, *2*(3), 211–221.

Seshadri, R., Joseph, S. W., Chopra, A. K., Sha, J., Shaw, J., Graf, J., & Madupu, R. (2006). Genome sequence of *Aeromonas hydrophila* ATCC 7966 T: Jack of all trades. *Journal of Bacteriology*, *188*(23), 8272–8282.

Shewan, J. M. (1961). The microbiology of sea-water fish. *Fish as Food*, *1*, 487–560.

Shewan, J. M. (1971). The microbiology of fish and fishery products—A progress report. *Journal of Applied Bacteriology*, *34*(2), 299–315.

Smithand, D. T., & Willett, H. P. (1980). Other Mycobacterium species. In K. Joklik, H. P. Willet, & D. B. Amos (Eds.), *Zinsser microbiology* (pp. 674–698). Appleton & Lange.

Sørensen, U. B., & Larsen, J. L. (1986). Serotyping of *Vibrio anguillarum*. *Applied and Environmental Microbiology*, *51*(3), 593–597.

Speare, D. J., & Ferguson, H. W. (1989). Clinical and pathological features of common gill diseases of cultured salmonids in Ontario. *The Canadian Veterinary Journal*, *30*(11), 882–887.

Stackebrandt, E., Wehmeyer, U., Nader, H., & Fiedler, F. (1988). Phylogenetic relationship of the fish pathogenic *Renibacterium salmoninarum* to Arthrobacter, Micrococcus and related taxa. *FEMS Microbiology Letters*, *50*(2–3), 117–120.

Stevenson, R. M. (1997). Immunization with bacterial antigens: Yersiniosis. *Developments in Biological Standardization*, *90*, 117–124.

Stevenson, R. M. W., & Airdrie, D. W. (1984). Serological variation among *Yersinia ruckeri* strains. *Journal of Fish Diseases*, *7*(4), 247–254.

Stoskopf, M. K. (1993). *Nutrition and nutritional diseases of salmonids. Fish medicine* (pp. 354–356). Philadelphia, Pennsylvania: WB Saunders, Co.

Sturz, A. V., Matheson, B. G., Arsenault, W., Kimpinski, J., & Christie, B. R. (2001). Weeds as a source of plant growth promoting rhizobacteria in agricultural soils. *Canadian Journal of Microbiology*, *47*(11), 1013–1024.

Sudheesh, P. S., Crane, S., Cain, K. D., & Strom, M. S. (2007). Sortase inhibitor phenyl vinyl sulfone inhibits *Renibacterium salmoninarum* adherence and invasion of host cells. *Diseases of Aquatic Organisms*, *78*(2), 115–127.

Sugita, H., Oshima, K., Tamura, M., & Deguchi, Y. (1983). Bacterial flora in the gastrointestine of freshwater fishes in the river. *Bulletin of the Japanese Society of Scientific Fisherie*, *49*(9), 1387–1395.

Swaim, L. E., Connolly, L. E., Volkman, H. E., Humbert, O., Born, D. E., & Ramakrishnan, L. (2006). Mycobacterium marinum infection of adult zebrafish causes caseating granulomatous tuberculosis and is moderated by adaptive immunity. *Infection and Immunity*, *74*(11), 6108–6117.

Swann, L., & White, M. R. (1991). *Diagnosis and treatment of* "Aeromonas hydrophila" *infection of fish. AS-cooperative extension service* (pp. 91–92). USA: Purdue University.

Taghavi, S., Garafola, C., Monchy, S., Newman, L., Hoffman, A., Weyens, N., & van der Lelie, D. (2009). Genome survey and characterization of endophytic bacteria exhibiting a beneficial effect on growth and development of poplar trees. *Applied and Environmental Microbiology*, *75*(3), 748–757.

Talagrand-Reboul, E., Jumas-Bilak, E., & Lamy, B. (2017). The social life of Aeromonas through biofilm and quorum sensing systems. *Frontiers in Microbiology*, *8*, 37.

Thoen, C. O., & Schliesser, T. A. (1984). *Mycobacterial infections in cold-blooded animals. The mycobacteria: A source-book*. New York, NY: Dekker.

Tobback, E., Decostere, A., Hermans, K., Haesebrouck, F., & Chiers, K. (2007). *Yersinia ruckeri* infections in salmonid fish. *Journal of Fish Diseases*, *30*(5), 257–268.

Toranzo, A. E., & Barja, J. L. (1990). A review of the taxonomy and seroepizootiology of *Vibrio anguillarum*, with special reference to aquaculture in the Northwest of Spain. *Diseases of Aquatic Organisms*, *9*(1), 73–82.

Touchon, M., Barbier, P., Bernardet, J. F., Loux, V., Vacherie, B., Barbe, V., & Duchaud, E. (2011). Complete genome sequence of the fish pathogen *Flavobacterium branchiophilum*. *Applied and Environmental Microbiology*, *77*(21), 7656–7662.

Ucko, M., & Colorni, A. (2005). Mycobacterium marinum infections in fish and humans in Israel. *Journal of Clinical Microbiology*, *43*(2), 892–895.

Ucko, M., Colorni, A., Kvitt, H., Diamant, A., Zlotkin, A., & Knibb, W. R. (2002). Strain variation in *Mycobacterium marinum* fish isolates. *Applied and Environmental Microbiology*, *68*(11), 5281–5287.

Vendrell, D., Balcázar, J. L., Ruiz-Zarzuela, I., De Blas, I., Gironés, O., & Múzquiz, J. L. (2006). Lactococcus garvieae in fish: A review. *Comparative Immunology, Microbiology and Infectious Diseases*, *29*(4), 177–198.

Verner-Jeffreys, D. W., Welch, T. J., Schwarz, T., Pond, M. J., Woodward, M. J., Haig, S. J., & Baker-Austin, C. (2009). High prevalence of multidrug-tolerant bacteria and associated antimicrobial resistance genes isolated from ornamental fish and their carriage water. *PLoS One*, *4*(12), e8388.

Verschuere, L., Rombaut, G., Sorgeloos, P., & Verstraete, W. (2000). Probiotic bacteria as biological control agents in aquaculture. *Microbiology and Molecular Biology Reviews*, *64*(4), 655–671.

Vincent, A. T., Trudel, M. V., Paquet, V. E., Boyle, B., Tanaka, K. H., Dallaire-Dufresne, S., & Charette, S. J. (2014). Detection of variants of the pRAS3, pAB5S9, and pSN254 plasmids in *Aeromonas salmonicida* subsp. salmonicida: Multidrug resistance, interspecies exchanges, and plasmid reshaping. *Antimicrobial Agents and Chemotherapy*, *58*(12), 7367–7374.

Vladik, P., & Vitovec, J. (1974). *Plesiomonas shigelloides* in rainbow troup septicemia. *Veterinarni Medicina*, *19*(5), 297–301.

Volkman, H. E., Clay, H., Beery, D., Chang, J. C., Sherman, D. R., & Ramakrishnan, L. (2004). *Tuberculous granuloma* formation is enhanced by a mycobacterium virulence determinant. *PLoS Biology*, *2*(11), e367.

Wallbanks, S., Martinez-Murcia, A. J., Fryer, J. L., Phillips, B. A., & Collins, M. D. (1990). 16S rRNA sequence determination for members of the genus Carnobacterium and related lactic acid bacteria and description of *Vagococcus salmoninarum* sp. nov. *International Journal of Systematic and Evolutionary Microbiology*, *40*(3), 224–230.

Wang, Q., Yang, M., Xiao, J., Wu, H., Wang, X., Lv, Y., & Liu, Q. (2009). Genome sequence of the versatile fish pathogen *Edwardsiella tarda* provides insights into its adaptation to broad host ranges and intracellular niches. *PLoS One*, *4*(10), e7646.

Wayne, L. G., Kubica, G. P., et al. (1986). Genus mycobacterium lehmann and neumann, 363AL. In P. H. A. Sneath, N. S. Mair, M. E. Sharpe, & J. G. Holt (Eds.), *Bergey's manual of systematic bacteriology* (Vol. 2, pp. 1436–1457). Baltimore, MD: Williams & Wilkins.

White, M. R., & Swann, L. (1991). Diagnosis and treatment of *Aeromonas hydrophila* infection of fish. *Aquaculture Extension.*, *6*, 91–92.

Wiens, G. D., Rockey, D. D., Wu, Z., Chang, J., Levy, R., Crane, S., & Schipma, M. J. (2008). Genome sequence of the fish pathogen *Renibacterium salmoninarum* suggests reductive evolution away from an environmental *Arthrobacter ancestor*. *Journal of Bacteriology*, *190*(21), 6970–6982.

Williams, A. M., Fryer, J. L., & Collins, M. D. (1990). *Lactococcus piscium* sp. nov. a new Lactococcus species from salmonid fish. *FEMS Microbiology Letters*, *68*(1–2), 109–113.

Williams, M. L., Gillaspy, A. F., Dyer, D. W., Thune, R. L., Waldbieser, G. C., Schuster, S. C., & Lawrence, M. L. (2012). Genome sequence of *Edwardsiella ictaluri* 93-146, a strain associated with a natural channel catfish outbreak of enteric septicemia of catfish. *Journal of Bacteriology*, *194*.

Wolf, K., & Dunbar, C. E. (1959). Test of 34 therapeutic agents for control of kidney disease in trout. *Transactions of the American Fisheries Society*, *88*(2), 117–124.

Yip, M. J., Porter, J. L., Fyfe, J. A., Lavender, C. J., Portaels, F., Rhodes, M., & Stinear, T. (2007). Evolution of *Mycobacterium ulcerans* and other mycolactone-producing mycobacteria from a common *Mycobacterium marinum* progenitor. *Journal of Bacteriology*, *189*(5), 2021–2029.

Yoshimizu, M. (1980). Microflora of the embryo and the fry of salmonids. *Bulletin of the Japanese Society for the Science of Fish*, *46*, 967–975.

Yoshimizu, M., Kamiyama, K., Kimura, T., & Sakai, M. (1976). Studies on the intestinal microflora of salmonids—IV. The intestinal microflora of fresh water salmon. *Bulletin of the Japanese Society for the Science of Fish*, *42*, 1281—1290.

Zanoni, R. G., Florio, D., Fioravanti, M. L., Rossi, M., & Prearo, M. (2008). Occurrence of *Mycobacterium* spp. in ornamental fish in Italy. *Journal of Fish Diseases*, *31*(6), 433—441.

CHAPTER

Global status of bacterial fish diseases in relation to aquatic pollution

8

Rohit Kumar Verma[1], Mahipal Singh Sankhla[2], Swapnali Jadhav[3], Kapil Parihar[4], Shefali Gulliya[5], Rajeev Kumar[6] and Swaroop S. Sonone[3]

[1]*Dr APJ Abdul Kalam Institute of Forensic Science & Criminology, Bundelkhand University, Jhansi, India* [2]*Department of Forensic Science, Vivekananda Global University, Jaipur, India* [3]*Government Institute of Forensic Science, Aurangabad, India* [4]*State Forensic Science Laboratory, Jaipur, Rajasthan, India* [5]*Department of Zoology, DPG Degree College, Gurugram, India* [6]*Department of Forensic Science, School of Basic and Applied Sciences, Galgotias University, Greater Noida, India*

8.1 Introduction

Commercial aquaculture production is being affected by the rising severity of bacterial diseases in fish (Bondad-Reantaso et al., 2005). The fish production industry is a global food source of growing importance in satisfying requirements for aquatic foods (FAO, 2006). Universally, consumption of aquatic foods is growing rapidly, and the foods consumed are more diverse. Enhancing production efficiency for profitability is one of the constant aims related to global water resources. This is especially the case in Asia, with close to a 90% share of the world aquaculture production. Among the Asian nations, China has about 70% of global whole fish farming, making it the world's biggest producer of farmed seafood, with a 490% increase since 1978 (Ellis, 2009). Diseases in fisheries pose risks not only to the fisheries industry but also to human life and health. In addition to zoonoses, the use of some substances and antibiotics in fish can threaten the environment, individual health, and food safety (Sapkota et al., 2008).

Columnaris (cottonmouth) is a bacterial disease that affects numerous cool- and warm-water fishes, usually in warm waters up to 20°C–25°C and higher; however, it's not uncommon to detect columnaris bacterial infection in fish, particularly trout varieties, in water as cool as 12°C–14°C. A number of cultured and free-range fishes are thought to be at risk of potentially infectious diseases. Columnaris infection impacts the aquafarming fish species, particularly catfish species and some aquarium species. Flavobacterium columnare can be cultured from the external fish, with systematic infections of wounds, skin or mucous membrane, and gills, as well as inner tissues, mainly the kidneys of fishes. Cultures can be grown on (Anacker & Ordal, 1959) Cytophaga agar or selective media (Hawke & Thune, 1992). The colony above the initial media plates has many features: light yellow, rhizoid, and tight adhesion (adhesive) on the media. With some simple diagnostic tests, colonies can be confirmed as subcultured (Griffin, 1992).

Bacterial Fish Diseases. DOI: https://doi.org/10.1016/B978-0-323-85624-9.00017-8

Bacterial gill disease, induced by *F. branchiophilum* (Kimura et al., 1992; Von Graevenitz,1990; Wakabayashi et al., 1989), was originally a salmonids disease in hatcheries; this problem is not thought to be present in the wild fish population (Bullock, 1972, 1990; Daoust & Ferguson, 1983; Farkas, 1985; Schachte, 1983). Particularly in endemic regions, bacterial gill disease occurs in aquaculture, usually along with extended host stressors. Experimental production of bacterial gill disease in normal fishes was reported by Ferguson et al. (1991), many researchers have reported that the disease is usually due to a combination of potential factors, such as overpopulation, decreased dissolved oxygen level, increased levels of ammonia gas, or particulate substances present in the water (Bullock, 1972, 1990; Schachte, 1983). In the face of financial issues in the fisheries industry, the incidence of diseases and parasites has increased due to more intense fish cultivation.

Microbial contamination can produce septicemia (Dar et al., 2020; Dar, Dar et al., 2016; Dar, Kamili et al., 2016). To handle the problem, large amounts of antibiotics are used in aquaculture to increase productivity and protect the socioeconomic status of fish producers in developing nations. Environmental contamination, particularly water pollution, by antibiotics has become a serious problem. It also causes the growth of antibiotic-resistant bacteria, which affects control of bacterial diseases in aquatic life. Increased fatality of *Penaeus monodon* larvae due to disease from antibiotic-resistant *Vibrio harveyi* has been reported (Karunasagar et al., 1994).

Transmissible infections within aquatic life are affected by microorganisms, viruses, molds, and parasites. For protection of fish and to better control the aquaculture environment, vaccine administration is an efficient and much-used method. Developing fish vaccines by means of plant biotechnology is a very cost effective and safer method, when compared with live virus vaccines, and could be a significant technology for the aquaculture industry globally, particularly for small fish producers in developing countries, as infection in fishes poses a significant threat to the aquaculture industry and the individual livelihoods that depend upon it (Sapkota et al., 2008).

Control of the health status of aquatic life is a considerable problem that affects food safety and concerns the protection of millions of people, the aquaculture industry, and the environment. Economic growth in a developing country is normally accompanied by a downturn in that country's agricultural sector. Development in supply chains, global trade agreements, and technology have led to a notable increase in world food trade with changes in consumption patterns (Anderson, 2010). The approach utilized to regulate infectious disease complications in fisheries is similar to that applied in the livestock industry. Hence, similar antibiotics to those used in veterinary medicine are employed to regulate transmissible diseases in fish, but there are few antibiotics specifically produced for fish (Santos & Ramos, 2016).

Few experts thought that narrow-based studies concentrating on small numbers of bacterial groups would be more successful than those that are broader-based and try to take into account all the bacteria from fish (Austin, 2006); more recently and in line with other studies of microbial biodiversity, emphasis is being put on molecular-based, culture-independent techniques. Utilizing numerical and conventional techniques could allow recognizing the bacterial cultures of most concern to fish. Identification would be helpful in recognizing potentially pathogenic bacteria, choosing the right strategy for treatment, and also in identifying useful bacteria that can improve fish status, like probiotics. China (62% of world production) is the biggest generator of cultivated food fishes, followed by India (6.5%), Indonesia (5.4%), Vietnam (4.6%), Bangladesh (2.6%), and Norway (1.8%). In the total production, 44 million tonnes of food fishes were cultivated in inland aquaculture, which represented 63.7% of the global aquaculture food fish cultivation in 2013; 24 million tonnes of carp fish were cultivated in 2013, followed by salmon and trout at 2.9 million tonnes, 3.4 million tonnes of Nile tilapia, and 1.1 million tonnes of catfish (FAO, 2016).

The aquatic ecosystem is more highly conducive to pathogenic bacteria individually than the terrestrial environment, and as a result, pathogens can reach higher densities surrounding aquatic animals, which then consume the pathogens when feeding or when drinking. As a result, in many cases the breeding of numerous classes of aquatic animals has resulted in extremely low survival rates (particularly in the larval stage) due to bacterial infectious disease (Verschuere et al., 2000). Bacterial pathogens cause contagious disease and fatality in both wild fish and fish reared in restricted circumstances. The biggest single cause of financial loss in fisheries is infectious disease problems (Meyer, 1991). With the fast growth of aquaculture and its intensity, the increasing usage of resources, contamination, globalization, and transboundary migration concerning water fauna, the number of pathogenic bacteria varieties separated from slowly rising fish (Harvell et al., 1999). An aquatic ecosystem can be the origin of unbound bacteria that can infect humans and impair treatment due to their immune properties. Resistant diseases spread directly from the aquatic environment to many humans, such as *Vibrio cholerae*, *V. parahaemolyticus*, *V. vulnificus*, *Shigella* spp., and *Salmonella* spp. or opportunistic pathogens like *Aeromonas hydrophila*, *Plesiomonas shigelloides*, *Edwardsiella tarda*, *Streptococcus iniae*, and *E. coli*. Resistant bacteria spreading in individuals may occur due to close association with the water source or water animals, by drinking water, or by consumption of seafood. Nearly 80% of antimicrobials utilized in aquaculture access the environment, where they choose bacteria that are resistant to mutations or, importantly, through mobile genetic components that contain many immune determinants that are communicable to other bacteria. Such selection changes biodiversity in aquatic environments and the regular flora of fish and shellfish. The terrestrial bacteria in aquatic ecosystems present with enduring antimicrobials, biofilms, and a large number of bacteriophages with pathogens of individual and animal sources, which enable the interchange of hereditary substances in aquatic and terrestrial bacteria. Some newly found hereditary factors and resistance determinants for quinolones, tetracyclines, and b-lactamases are distributed within aquatic bacteria, fish pathogens, and individual pathogens that are found to have originated in aquatic bacteria (Aly & Albutti, 2014).

Up to 25% of losses in the fisheries industry can be attributed to microbial growth and activity (Wiernasz et al., 2020). Through the applications of community ecology principles to microbiota studies, microbiota ontogeny in fish was evidenced to be influenced by both neutral and nonneutral evolutionary forces (Cheaib et al., 2020; Heys et al., 2020).

8.2 Acidification of bacteria in water

Only a few researchers have studied bacterial processes in acidified ponds, communicating details and opinions of the crucial roles that microorganisms play in decay and mineral reuse. In both Scandinavia and the Adirondacks, segregation of twigs and leaves (governed by mass lost in litter bag) happened gradually at condensed pH over a pH range of about 5.0–7.0 (Francis et al., 1984; Hendrey, 1982) and 4.0–6.0 (Laake, 1976; Traaen, 1980), respectively. Additional experiments at the Experimental Lakes Area suggested that decrease of nitrate and sulfate in anoxic sediments is unaffected by acidification and that these decreasing processes can increase loading ability, in the short term (Kelly et al., 1982; Schindler et al., 1980). The anthropogenic acidification of ponds and rivers can have numerous adverse impacts on main and subordinate manufacturers. Biotic impoverishment, principally regarding invertebrate animals like leeches, gastropods, crustaceans, and fishes such as salmonids, has certainly been noticed in several climatically acidified ponds and rivers of Europe and North America during the previous 40 years (Allan, 1995; Cummins, 1994; Fjellheim & Raddum, 1990; Huckabee et al., 1989; Krause-Dellin & Steinberg, 1986; Meriläinen & Hynynen, 1990; Morris et al., 1989; Ormerod et al., 1990; Schindler et al., 1989). A pH range of 5.5–6.0 appears to be an important threshold below which damage to aquatic life will remain a local and regional ecological issue (Doka et al., 2003). The ecological issue of water acidification impacts animals of all life forms. Prior researchers concentrated on the declines of aquaculture populations, specifically salmonid. Recent examinations revealed that zooplanktons, insect larva, benthic invertebrates, specifically crayfishes, snails, and freshwater mussels, are delicate and are generally decreased or nonexisting in acidified ponds and rivers (Havas, 1986). Hardness of water signified by the concentration of calcium pointedly impacted the number of varieties and their capability to exist in acidified situations. Past research that failed to state the hardness of water or that hard water was utilized for bioassays often stated significantly greater resistance to acidic environments than is noticed in wildlife. Aluminum percolated with acid rain from the washbasin complicates the situation significantly, as different types of aluminum vary in their toxicity. Aluminum and hydrogen ions interrelate both antagonistically and synergistically, depending on the situation (Havas, 1985; Herrmann, 1993; Muniz & Leivestad, 1980; Rosseland & Staurnes, 1994) and in the company of organic compounds in the environment with acidic properties, aluminum toxicity can be counteracted. The significance of certain water quality variables arose, as experts over two landmasses struggled to understand how certain ponds are without life while other, more acidic ponds are teeming with life. Currently, rapidly increasing manufacturing processes in Asia have led to a continuous increase in release of acidic contaminants due to rain at acidic ranges of pH 4 advancing through Japan (Japan Environmental Agency, 1997). In an attempt to describe possible discrepancies, the characteristics of water quality, specifically pH, Al, Ca, and natural acids, are better known. Examinations have shown the release of acidified water into estuaries from acid sulfate soil (ASS) have resulted in fishes becoming more susceptible to fungal infections (Callinan et al., 2005; Virgona, 1992) and bacterial infections (Bromage & Owens, 2009). Principal bacterial procedures for mineral use and ecological operations can also be either reserved or changed as a result of reduced pH ranges. $NH4+$ nitrification resulted in pH ranges less than 5.6 in empirically acidified Canadian ponds, with aquatic denitrification processes being recommended (Rudd et al., 1988, 1990). Extra carbon dioxide reacts with water to form carbonic acid and hydrogen ions, which increases the acidity of the water. Increasing the hydrogen ion bonds with carbonate ions to produce more bicarbonate reduces the concentration of carbonate ion (Fabry et al., 2008; Orr et al., 2005).

8.3 Impact on fish from pollution

Ecological pollutants like pesticides, metals, and many organics pose grave risks to various aquatic creatures. Hence, many studies have categorized the damaging physical processes in creatures exposed to these contaminants. However, the effects of contaminants on fish behavior are less often examined. Since behavior connects physiological functioning with environmental systems, behavioral signs regarding the toxic nature of various substances are important to evaluate the impacts from water pollutants on fish species. We recognize that numerous toxicants can change fish behavior, like predator escape, reproduction, and social functioning. Toxicant exposure usually reduces the appearance of responses that are indispensable to health and also endurance in the normal ecological system, generally after exposures of minor extent rather than those inducing notable mortality (Scott & Sloman, 2004). Contamination effects on the spreading of common diseases in host groups of salmon are currently not known. It is assumed that the outbreak of infection depends upon the interplay of the host and the environment, including the pathogens. Nonexistence of pathogens would improve unfavorable surroundings, which affects disruptions from disease (Arkoosh et al., 1998).

Because of the lack of protein sources for populations in the Third World nations, demand for fish cultivation has increased, as it represents a good source of protein. However, in some regions, fish farmers utilize reprocessed water from farming, industry effluent, or even sewage. These origins of reprocessed water pose a health threat to the fishes, the ecosystem, and the individuals who eat the fish. Our water systems represent the most essential resource that exists on the planet, necessary for survival and also for the improvement of advanced technologies. Rapid industrial expansion is a main reason behind water pollution. Released water has been utilized in various areas of the globe for fish farming (Ashraf, 2005; Gad, 2009; Wong et al., 2001). Fish are usually regarded to be the common organism for pollution monitoring in water sources. Fish species may be seen throughout the water ecosystem; also fish play a significant environmental part in water food chains, due to their roles from low to higher trophic level (Linde-Arias et al., 2008; Van der Oost et al., 2003). Fish species subsist in a water environment, are completely reliant upon it, and are affected by changes within it. Hence, fish varieties may be used as an alarm system to indicate the presence of pollutants in the water (Nussey, et al., 1995). Increasing manufacturing, farming, and industrial substances released into the water systems could cause severe health problems and also threaten the water ecosystem (Mayon, et al., 2006). The heavy metal quantities in fish organs indicate earlier consumption through water or food, and the amount can display the current condition of the organism before toxicity influences the environmental balance of cultures in the water (Birungi, et al., 2007).

8.4 Impact on fish populations from bacterial diseases

Disease factors in fish populations are thought to have vital impacts on the hosts by epizootic or enzootic disease. Enzootic disease can affect the host by lasting effects on physical processes, damaging development and reproduction, as well as existence, whereas epizootic diseases (analogous to an epidemic in humans) characteristically influence population dynamics by decreasing populations due to a temporary large disease outbreak, which if large enough might affect stochastic procedures, producing extinction (Gulland, 1995). Variation of any of these constituents can modify the dynamics of the interaction among pathogen and host, that may change the vulnerability of the fish population to illness. Contamination that causes a modification in the climate can affect the physical well-being of young salmon by increasing their vulnerability to disease, that may in turn possibly adjust the population characteristics (Arkoosh et al., 1998). The degree to which disease-bearing pathogens typically negatively affect the community dynamics of fish species is still mostly unidentified, as are the three-dimensional and sequential ranges of these actions.

The presence of disease in fish populations supposedly relies on the interaction of three factors: the host, the environment, and the pathogen (Snieszko, 1973). Additionally, it is hypothesized that variable fishing pressure directly increases variability in exploited populations (Botsford et al., 1997). Variation concerns about exploited fishes can be affected by ecological forces and mortality from fishing (Jacobson et al., 2001; McFarlane et al., 2002). As we concentrate on the damaged bet-hedging as the chief influencing constituent for increased inconsistency, a fresh examination using a sensitive nonlinear study suggested increased temporal variability of California Current employed fish stocks arises from increased instability in dynamics, which intensifies nonlinear behavior, chiefly the increase in the inherent community development rate (Anderson et al., 2008). Age-truncated or juvenescent populations have increasingly unstable population dynamics. Increases in resident progress measures is due to prematurations (juvenescence) of subjugated people affected by dimensions-selective fishing. As distinct mass reduced including entire biomass persisted mathematically same to the harmed species in an area of Southern California, the number of fledgling fishes has increased. The larger population of short-lived fishes needs an advanced intrinsic measure of development; the population must reproduce major living descendants capita/year to compensate for the shortened lifetime. As a result, exploited populations could be directed to a chaotic or unstabilized state, and therefore show greater differences. However, in adding to the growth age, one must take note that the internal community growth rates also rely on the mortality. Equilibrium among mortality from fishing and adaptive or evolutionary response of animals perhaps regulates the details of population subtleties.

8.5 Consequences of bacterial diseases in fish

Unintentional results and trade-offs are related to providing a fish passageway at barriers, either through fishway building or dam elimination. Good policymaking needs detailed thought about both the costs and benefits related to the choices accessible to managers. Studies have exposed that, while profits of fish passageways and barrier elimination have been interconnected efficiently and are being extensively recognized, the unintentional impacts of such choices and the trade-offs and doubts they create are stated to be less healthy and are frequently ignored or unappreciated. Production is not proposed to be an area of disagreement, in contrast with offering fish passageways or eliminating barriers. These passageway choices should be considered extensively and followed where suitable and conceivable. However, studies also suggest that costs and benefits of such choices can differ from one location to another, making fish passageway choices situation reliant and annoying to investor collections and directors unacquainted with doubts relating to choices to offer fishes passageway or eliminate a barrier. A fusion of unintentional results and trade-offs related to fish passageways might raise consciousness surrounding such questions between piscary executives and experts and help them interconnect such issues to the wider community (McLaughlin et al., 2013). So, the usage of antimicrobial factors in aquaculture results in a wide-ranging ecological application that influences an extensive diversity of microbes (World Health Organization (WHO), 1999). About 40 thousand culture-established cases are presented to the Center for Disease Control every year; actual antimicrobic mediators are serious and might be resurrection for a minimum of 2400 individuals per year. The choice of antimicrobial factor for action against dangerous infections has developed progressively due to growing disinfectant confrontations among *Salmonella* isolates. Earlier, chloramphenicol, ampicillin, and trimethoprim-sulfamethoxazole have been the "action options" for *Salmonella* infections (Lee et al., 1994; McDonald et al., 1987; Riley et al., 1984). However, among 1272 erratically nominated *Salmonella* strains from persons tested at the Center for Disease Control in the National Antimicrobial Resistance Monitoring System in 1996, 10% were resistant to chloramphenicol, 4% to trimethoprim-sulfamethoxazole, and 21% to ampicillin. Contrast to this that almost all of the *Salmonella* strains tested at the CDC have been vulnerable to third-generation cephalosporins and fluoroquinolones (Centers for Disease Control & Prevention, 1996; Herikstad et al., 1997). For such a cause, and because of medical response and favorable pharmacodynamic characteristics, fluoroquinolones and third-generation cephalosporins are the present drug of choice for therapy against offensive *Salmonella* infections in grown fish and offspring, respectively. Should *Salmonella* develop antimicrobial resistance to these two antimicrobial factors, appropriate substitute antimicrobial mediators are not presently accessible and serious individual health consequences are likely. Therefore, these resistance factors can be conveyed by flat gene-transfers to microorganisms of terrestrial environments, including human and animal pathogens, as has been stated for *Vibrio cholera* and *Salmonella enterica* serotype Typhimurium.

While it may seem self-evident that defrayal preferences would be adaptive, little evidence shows that they are. Table 8.1 shows effects of settlement preferences on development and existence of approximately pomacentrid reef fishes; few complete studies exist on other coral-reef fish taxa. The suggestions that do exist are often ambiguous and inconsistent among the studies. While strong settlement preferences have often been expected adaptive, (Tolimieri, 1995) originate no benefit in relations of either convert development or survivorship; however, *S. planifrons* settlers presented robust inclinations for coral species at defrayal.

8.6 Global status of bacterial disease in fishes

Worldwide aquaculture has grown intensely over the last five decades to about 68.3 million tonnes including aquatic plants, that is, 52.5 million tonnes in 2008 at a value of US$98.5 billion, US$106 billion consisting of water flora and amounting to approximately 50% of the global food fish market. Asia controls manufacture, accounting for 89% by volume and 79% by value, with China by far the major manufacturer, with 32.7 million tonnes in 2008. Rapid development in this area is related to a diversity of issues, including already present aquaculture usage, financial development, Controlling the population has become more lenient, and increasing export opportunities. Aquaculture growth in North America and Europe was rapid during the 1980s and 1990s, but has since deteriorated, perhaps owing to controlling limits on places and other inexpensive features, though as marketplaces for seafood and fish they have sustained to raise (Bostock et al., 2010). Financial, ecological, and organism health assistance have emerged as a consequence of the

Table 8.1 Consequences of settlement preferences for coral-reef fishes.

Process	Possible consequences	Taxa	References
Physical variation in visual system, olfactory barbels	Food finding, predator detection	Mullidae (*Upeneus tragula*)	McCormick (1993)
Variation in pigmentation, scale amount, activities	Camouflage, protection, predator avoidance	Labridae, Pomacentridae	Robertson et al. (1988), Victor (1983)
(a) Behavior Defrayal preferences	Patchy spatial and temporal distributions of recruits; enhanced food finding, decreased predation risk	Poinacentridae, Labridae	Booth (1992), Doherty (1982), Eckert (1985), Jones (1987)
(b) Interfaces durable predation	Condensed survivorship, more patchy distributions	Chromis sp., Pomaeentrus spp., Stegastes spp.	Carr and Hixon (1995), Doherty (1982), Shulman (1984), Wellington (1992)
Strong rivalry with residents	Reduced recruit density and enhanced growth rate	Dascyllus spp.	Fjellheim and Raddum (1990), Schachte (1983)
Migration	Augmented mortality or migration redistribution of human resources	Dascyllus spp.	Booth (1991), Booth (unpublished Data), H. Sweatman (personal communication)

extensive usage of medications in the aquaculture industry. The probability that vaccines are responsible for lower death rates, enhanced growth efficacy, and increased production is now a serious issue in disease administration programs in aquaculture. Currently, the most commonly used fish inoculations are chiefly bacterial antigens or inactivated virus, subunit vaccines, and vaccines produced through recombinant DNA technology with antibody-mediated immune response, or live, weakened, and DNA inoculations for cytotoxic-T cells (CTLs) response (Plant & LaPatra, 2011). The extensive and recurrent use of antibiotics in aquaculture in prior times has caused antibiotic resistance development in aquaculture pathogens. Thus, at present, antibiotics are no longer used in treatment of microbial illnesses in some cases. Karunasagar and colleagues state huge deaths in *Penaeus monodon* larva due to *Vibrio harveyi* strains with varied resistance to chloramphenicol, cotrimoxazole, streptomycin, and erythromycin (Karunasagar et al., 1994). The management of short chained fatty acids will reduce abdominal pH, resulting in more helpful bacteriological systems in the GI tract, since neutrophilic bacteria will have major difficulty when in contact with acidophilic bacteria. A main drawback of using short chained fatty acids in aquaculture foods is that they can leach into the water culture and subsequently high amounts would be required to maintain high quality. The effectiveness of short chained fatty acids can be increased by the manufacture of compounds at sites where they need to perform, such as in the gastrointestinal tract. This might be attained by the management of acid-producing probiotic microorganisms, like lactic-acid microbes (Ringø & Gatesoupe, 1998). Presently, methods to protect aquaculture fish from pathogenic microbes without using antibiotics are being commercialized and verified. An all-around method comprising host, pathogen, and environment will possibly be the most applicable in the long term. In this view on biocontrol, steps that prevent or restrict diseases are better for health management

(Subasinghe et al., 2001). However, it would be impractical to imagine that infections can be prevented in all forms and therefore new methods to regulate pathogenic microbes are required to make the aquaculture industry more sustainable. Also, the application of drugs against target microbes may pose risks to aquaculture industry systems and in the following foodstuffs intended for human consumption, which could affect customer health and be the source of unfavorable environmental and financial effects (Brunton et al., 2019; Lulijwa et al., 2020).

8.7 Toxic bacteria in fishes and their occurrence

Reports and reviews on the appearance of pharmaceuticals in the environment (primarily surface waters) suggest that therapeutic composites are omnipresent (Daughton & Ternes, 1999). Examination of blood factors is a significant instrument used to measure the toxic special effects of xenobiotics on fish and biochemical indices (Bojarski & Witeska, 2021; Burgos-Aceves et al., 2019; Hoseini & Yousefi, 2019). Yeast fungi examination as a part of fish microbiota does not require a particular media culture; various broths and agars are acceptable. However, there are dedicated culture media whose choice can be enhanced by the addition of an antibiotic. It appears that yeast may be a vital part of fish intestines. In some cases, the yeast may be more abundant than other bacterias, such as in the deep-sea eel *Synaphobranchus kaupi* (Ohwada et al., 1980), or freshwater adult masu salmon (Yoshimizu et al., 1976) (Fig. 8.1).

8.7.1 Aeromonas

Aeromonas hydrophila causes diverse pathologic illness, which involves severe and immediate impact, impact over an extended period, and possible infection. The condition severity is affected by some factors, such as the virulence of particular types of bacteria and level of pressure exercised on a fish community, and also the stability and physiology of the host. Pathological illness associated with a member of the *A. hydrophila* complex includes dermal ulceration, hemorrhagic septicemia, red sore condition, red rot infection, and scale protrusion illness. The severe and immediate impact of such conditions is indicated by quickly lethal septicemia with some signs of disease. If this occurs, some symptoms are important: exophthalmia, skin reddening, and fluid accumulates in

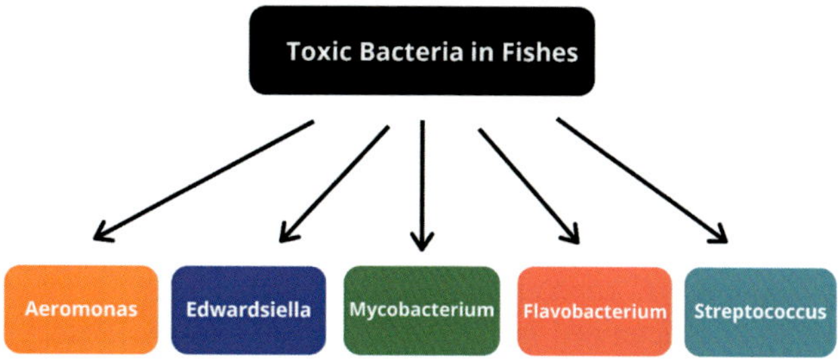

FIGURE 8.1

Toxic bacteria in fish.

the scale (Faktorovich, 1969). *A. hydrophila* is the general cause of the illness, which occurs in warm-water fishes such as channel catfish (*Ictalurus punctatus*), tilapia (Amin et al., 1985), ayu (*Plecoglossus altivelis*) (Miyazaki & Jo, 1985) and also in cold-water fish species and other vertebrates (Janda & Abbott, 1998). Bacteriology study has revealed the appearance of *Aeromonas hydrophila*, *A. sobria*, *Vibrio* spp., and Enterobacteriaceae in the inner part of affected tilapia fingerlings; immersion in 50 ppm solution of formalin regulated the mortality rate (Sugumar et al., 2002). *Aeromonas* infection is also common in warm-water and in a more temperate class of species than in cold-water fish. Infections can occur in fish of any age, but loss normally occurs in fry and little fingerlings. Spreading is seasonal, being high in the spring to the early summer season and then dropping when the water temperature is 65°F–85°F. Spring outbreaks may be associated with reduced infection resistance in fish that are in a poor state from overwintering or after spawning. Reduced resistance can also be induced by greater handling and transportation of fresh fish in the fall.

Aeromonas infections do not adhere to a strict heat range and are described in every month of the year. It is unusually observed that aeromonads could be isolated in the winter from wounds associated with wintertime kill (winter fungi) (Camus et al., 1998). *Aeromonas hydrophila* and motile aeromonads are found in general freshwater localities globally and are often connected with serious illness in both cultured and feral fishes. Determining the etiology of illness includes aeromonad diseases obscured by hereditary, biochemical, and antigenic heterogeneity of association. Motile aeromonads are associated with composite diseases of organisms that are correlated with bacterial hemorrhagic septicemia in fishes. In the early scientific information on fish diseases (Otte, 1963), it was stated that septicemic diseases in fish induced by motile aeromonads were well-known in Europe in the Middle Ages. The bacterial etiology of these early reports was usually indefinite, but the pathology represented was similar to that of red leg infection in frogs, in which *A. hydrophila* was also associated as the causal organism.

8.7.2 Edwardsiella

Edwardsiella tarda (*E. tarda*) is a gram-negative bacterium and inducing factor of dangerous bacterial disease in freshwater and marine water fish species and has a large population globally. The bacterium belongs to the Enterobacteriaceae family (Ling et al., 2000). The bacteria also have a wide host scale and induce illness in reptiles, birds, and individuals as well (Ling et al., 2000). The *E. tarda* bacteria can avoid both serum- and phagocyte-mediated killing, and penetrate cells to infect the host and generate dermatoxins and hemolysin, which are spread by septicemic conditions (Srinivasa Rao et al., 2003). Since there is no vaccine available to regulate edwardsiellosis, chemotherapy is the only efficient mode to control the disease in fish. There is an important concern about the effects on food safety of the chemotherapy method.

After initial examinations, the closely related species *E. piscicida* has recently been declared. In addition to hereditary studies of *E. tarda* isolation from older reports, it has been determined that there are numerous earlier characterized isolates of *E. tarda* that associate to *E. piscicida* species (Buján et al., 2018b; Reichley et al., 2017). Current recategorization and examinations propose that *E. piscicida* is challenging finfish aquaculture more than *E. tarda*. Geographic dispersion of host varieties has influenced the diagnostic methods and implied administration approaches to regulate disease. The presence of edwardsiellosis in eel has so far been restricted to Japanese and Taiwanese aquaculture production (Chang & Liu, 2002; Wakabayashi & Egusa, 1973).

8.7.3 Mycobacterium

Mycobacterium marinum is a pathogen of poikilothermic hosts such as toads and fish that are hereditarily associated with *M. tuberculosis* (Tonjum et al., 1998). The *M. marinum* pathogen induces similar disease in the hosts, consisting of dormancy and granuloma production, and it is responsive to laboratory context as it is an hereditarily tractable and comparatively rapidly increasing mycobacterium; it poses minimal hazard to lab employees (Chapman, 1977; Linell, 1954). Fish mycobacteriosis is one of the constantly recurring illnesses provoked by omnipresent acid-fast bacilli bacteria, also known as nontuberculous mycobacteria (Novotny et al., 2004). This pathogen was initially described by Von Betegh in marine fish (Von Betegh, 1910) and separated by Aronson (1926). The *M. marinum* pathogen belongs to the Actinomycetales order, suborder Corynebacterineae, and family Mycobacteriaceae. These cause fish reservoir granuloma in individuals, as stated in 1951, as a result of *Mycobacterium balnei* (Norden & Linell, 1951; Linell & Norden, 1954); the disease was identified after an examination of 80 individuals who utilized a common swimming pool and were affected by granulomatous lesions on skin. The causative agent of piscine mycobacteriosis was later proved to be *M. marinum*. According to other studies, the hematological parameters showed a slight change in the erythrocyte guides of the Nile tilapia vaccine with *M. marinum*, with distinct cases of microcytic hypochromic anemia, and broad differences in platelet numbers only at 30°C for 8 days. Neutrophilia and slight lymphocytosis resulted in leucocytosis in fish 1 day after illness, which is a specific point of acute disease; 3 days after infection, chronic signs appeared in the form of neutropenia and lymphocytosis, though monocities occurred and spread after days 14 and 35 and by the day 45. Eosinophilia appeared only in 1 fish on day 14 after contagion. Other morphological deformities including young cells, cytoplasmic vacuolization in monocytes, poisonous cytoplasmic granulation in neutrophils, and intense cytoplasmic basophilia were reported at different times in infected fish (Ranzani-Paiva et al., 2004). The researcher stated that sterilization is not useful in regulating fish mycobacteria, due to normal bactericidal confrontation of mycobacteria (Jacobs et al., 2004). Mycobacterium can be efficiently cultivated on liquid media and agar surfaces (Rhodes et al., 2004); as an example, mycobactin is a siderophore utilized to increase the development of gradually increasing *Mycobacterium* on normal media. Nonselective media are not effective for moderate cultivators but suitable in rapid cultivators. There is a need to distinguish among *Mycobacteria* obtained from fish species and the fish-pathogenic gram-positive *Mycobacterium*. *Mycobacteria* sterile separation from the inner tissues ensures that there is no contaminant in the strain (Rhodes et al., 2004). Separations of these bacteria from feces, outer surfaces, and all viscera are ambiguous because of the appearance of outside contaminants (Rhodes et al., 2004). Fish skin models are collected and inoculated on suitable media cultures for growing, such as Middlebrook 7H10 or Lowenstein–Jensen media; to improve the development of *Mycobacterium* spp. and because of the gradually growing organism, it is important to regulate the growing plate for a time varying between 2 and 3 months before recognizing *Mycobacterium* spp. as negative. Identification of *Mycobacterium* spp. from skin or gills of fish with culture media is complicated by the presence of microbiota that compete for the production of *Mycobacterium* on standard culture media. Due to the loss of water, *Mycobacterium* is extremely resilient to the action of both alkaline and acidic chemicals and other composites like benzalkonium chlorides and hypochlorites (Brooks et al., 1984; Rhodes et al., 2004).

8.7.4 Flavobacterium

Flavobacterium psychrophilum was originally isolated from kidney and outside wounds of infected young coho salmon *Oncorhynchus kisutch* (Walbaum) in the state of Washington, and was originally designated as *Cytophaga psychrophila* (Borg, 1960). *Flavobacterium psychrophilum* is a ubiquitous bacterium in the aquatic environment, particularly in freshwater (Groff & LaPatra, 2001). As the etiological agent of bacterial cold-water disease, it is a serious fish pathogen causing substantial economic losses and rearing difficulties in both commercial and conservation aquaculture (Antaya, 2008). Such bacteria seemed plentiful on the exteriors of gills of young trout inside 18–24 h after waterborne contact of fishes with a bacteriological interruption of cells in fish tanks. Infections produced irritation to gills and reduced the fish respiration (Wakabayashi & Iwado, 1985). *Flavobacterium psychrophilum* is a gram-negative bacteria rod, belonging to the Bacteroidetes, that can induce illness in nonsalmonid fish species. This Bacteroidetes agent correlated with illness in eel, *Anguilla anguilla* (L.), and three classes of cyprinids, that is, common carp (*Cyprinus carpio* L.), crucian carp (*Carassius carassius* (L.)), and tench (*Tinca tinca* (L.)), in Europe (Austin & Austin, 1999; Lehmann et al., 1991). The permanent appearance of *Fl. psychrophilum* in the aquatic environment has not yet been described due to lack of appropriate analysis techniques. *Fl. psychrophilum* can be isolated by use of agar media, but in several instances it is inappropriate because of the delayed growth or excess growth of cells or interference by quickly increasing water bacteria. Studies on the existence of *Fl. psychrophilum* microsomes in sterilized laboratories have shown that cells in fresh and undrinkable water have been able to survive in significant amounts for many months (Wiklund et al., 1997).

8.7.5 Streptococcus

The genus *Streptococcus* incorporates several species that induce severe conditions in different hosts. The significant characteristics of genus *Streptococcus* are gram-positive bacteria that, when stained using gram-staining methods, are observed with purple/blue coloration. Most of the fish disease-causing bacteria are gram-negative bacteria that appear with pink coloration with the procedure of gram-staining. This is essential in the handling of bacteria and in their treatment. Streptococcal (strep) illnesses are rare and when they appear in fish can lead to mortality; under some conditions a few aquatic *Streptococcus* species may induce infection in individuals. However, certain latter species do not generally affect strong individuals. Some other bacteria can induce the same diseases, consisting of *Lactococcus*, *Enterococcus*, and *Vagococcus*. All of these bacteria, and the diseases themselves, are known as "strep." Streptococcal disease is a bacterial disease in fish; it was first described in 1957, and cultivated rainbow trout species in Japan were affected (Dar et al., 2020; Dar, Dar et al., 2016; Dar, Kamili et al., 2016; Hoshina et al., 1958). In 1986, fish illness occurred due to *Streptococcus* in Israel and it quickly had outbreaks globally, causing significant financial loss. Two important fish species cultivated in Israel, namely St. Peter's fish, a common freshwater fish, and rainbow trout were affected with this disease. The diseased tilapias (Peter's fish) were sluggish and displayed signs related to the involvement of the central nervous system, such as irregular swimming activity and dorsal stiffness; the clinical sign in trout was septicemia correlated with cerebral impairment. A separate strain was found related to two different phenotypical assemblies concerning streptococci from fish with brain illness, and they were distinct from other specific fish pathogens such as *Enterococcus seriolicida* (Kusuda et al., 1991). From a clinical viewpoint, streptococcosis of fish is a general disease concept utilized to indicate similar, but different, illnesses from any of six distinct varieties of gram-positive cocci, including vagococci, streptococci, and lactococci (Bercovier et al., 1997; Muzquiz et al., 1999). The temperature of the water supposedly plays a decisive role in the beginning of infection induced by certain pathogens. Hence, spreading of those pathogens is connected to diseases from *V. salmoninarum*, *C. piscicola*, and *L. piscium* that normally arise at under 15°C water temperature, referred to as cool-water streptococcosis; infections that occur at over 15°C water temperature are called warm-water streptococcosis, and is from *S. iniae*, *L. garvieae*, *S. difficilis*, and *S. parauberis* (Muzquiz et al., 1999).

8.8 Adverse effects on human health caused by bacterial fish diseases

In 2001, global aquatic production was approximately 37.9 million tons, which represents 41% of production achieved within the limits of human consumption (FAO, 2003). The occurrence and expansion of fish diseases are the effects of the interplay between the pathogens, hosts, and the environment. Hence, multidisciplinary studies include the properties of potentially pathogenic microorganisms for fishes, biological perspectives of fish hosts, and more knowledge of the environmental agents changing such culture, in order to enable appropriate methods of restricting and regulating the major diseases, to reduce the decreases in seawater fish species (Toranzo et al., 2005). However, various sublethal effects were observed at low, environmentally realistic concentrations of antibiotics (Almeida et al., 2019). Fish provide various health advantages; on the other hand, occurrence of toxic elements can have adverse impacts on an individual's health if consumed in large amounts by the individual (Sonone et al., 2020). Fish can be contaminated by various means and this has become a major issue, not only because of the danger to fish but also because of the threat to humans who consume fish as a food source (Fuentes-Gandara et al., 2018) (Fig. 8.2).

8.8.1 Gastrointestinal tract

Improvements have occurred in recent years in our knowledge of the bacteriological bond at both molecular and hereditary stages of endothermic animals, and the electron microscopy technique has contributed greatly to the knowledge increase (Knutton, 1995). Though numerous publications have detailed pathogenesis in fishes, some research has utilized TEM, or transmission electron microscopy analysis, and SEM, or scanning electron microscopy, which are techniques to assess the impact of bacteriological disease on geomorphology in the gastrointestinal (GI) tract of fish. Fishes reside in aquatic environments, and thus their outer surfaces are continually uncovered to probable pathogens, and liquid taken into the GI tract throughout feeding could transfer the pathogens to the mucosal surface of the tract. Additionally, in the nonfeeding initial phases of growth of fish larvae, consumption of water is needed for osmotic regulation (Tytler & Blaxter, 1988) and this gives a

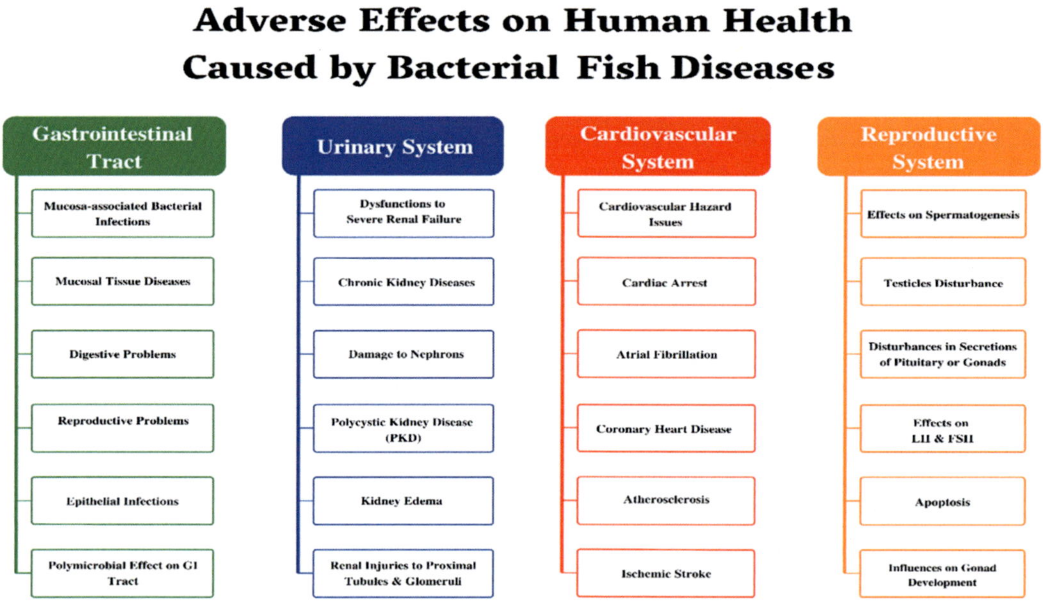

FIGURE 8.2

Adverse effects on human health caused by bacterial fish diseases.

quick entrance into the GI tract for bacteria. It is complicated to analyze the mucosa-associated bacterial populations in the uppermost gastrointestinal (GI) tract and large intestine in healthy individuals; lately there have been restricted examinations on their formation, arrangement, and functioning. Biopsy samples and medical substances are normally taken from diseased individuals, sufferers who have undergone antibiotic treatment, and those individuals whose large bowel has been washed out before examination of the digestive tract or large intestine. As a result, the microbial community associated with these tissues cannot provide the true signs of healthy mucosal diversities. There is difficulty with specimen collection; individual digestive tracts have a body of evidence showing the presence of different mucosa-associated communities (Croucher et al., 1983; Lee et al., 1971; Macfarlane et al., 2004). The digestive systems of fish continuously interact with the outer environment. It is difficult, as polymicrobial ecology associates with the inner and external environments, and their effects on wellness and disease of the fishes. The gut is a multifunctioning organ that digests and absorbs food; it is important in water and electrolyte equilibrium, endocrine control of absorption, metabolism, and the immune system. The gastrointestinal tract microorganisms of fish are distinguished by large community frequency, large variety, and complexity of the interplays. When the whole main community of microorganisms is expressed, bacteria predominate. These are the major components of the stomach microbiota in fishes (Pond et al., 2006; Spanggaard et al., 2000). The colon is the most abundant site colonized by microbes in the gastrointestinal (GI) tract. Colonic microorganisms are generally considered to be uniform objects; this is a simplification, as the bacteria survive in many places under different microhabitat and metabolic conditions inside the mucous layer lining the gastrointestinal tract, the mucous membrane, and on the surface of digestive deposits inside the opening of the intestine (intestinal lumen). The microorganisms are continuously changing, according to nutrient consumption or as distinct sources become accessible. Some data reveal that bacteria in the large bowel can arise separately as unique cells, but they can live in microcolonies or other different communities, including other species, on the exteriors of minute particles of substances. In some conditions, microbial biofilms may have a separate class, such as diseases of the heart valve, catheters, and medical prosthetic devices, but biofilm belonging to distinctive areas of the digestive tract is normally a multispecies consortia whose growth occurs according to climate, nutritional determinants, chemical constituents of the substratum, and host protective mechanisms associated with natural and acquired immunity (Macfarlane & Dillon, 2007). The epithelium of the digestive tract is one of the fastest redeveloping mammal muscles. Epithelial cells of the minor bowel are entirely rehabilitated in a projected 3–4 days in rats; in individuals, this period is much longer. The present assumption is that intestinal stem cells, which stretch to increase to four main cell lineages in the intestinal epithelium, regulate the rapid regeneration (Bach et al., 2000; Cheng & Leblond, 1974).

8.8.2 Cardiovascular system

Ingestion of fish or fish-oil favorably affects numerous cardiovascular risk issues (Mozaffarian, 2009). Variations in the maximum risk effects are usually obvious over a calendar week of variations in intake, and this might result from changed cell-membrane variability and receptor response and subsequent combination of $n-3$ polyunsaturated fatty acids into cell-membrane phospholipids as well as direct association of $n-3$ polyunsaturated fatty acids with cytosolic receptors that control gene transcription (Clandinin et al., 1991; Feller & Gawrisch, 2005; Vanden Heuvel, 2004). Time sequences of benefits differ; for example, some of these specific impacts (such as triglyceride lowering) may require years or months of intake before any effects or medical results are obvious. The reports of benefits for mid-level risk issues are convincing, but implications for specific health impacts on long-lasting cardiovascular disease require confirmation through evidence of real disease termination in patients (Mozaffarian, 2009). The constancy of evidence and extent of the advantages in improvement in heart problems in humans due to fish intake are impressive. The risk decrease is established by reliable indications from research studies and random trials examining fish or its oil consumption on heart risk; randomized controlled trials assessing fish intake/impartial biomarkers of $n-3$ polyunsaturated fatty acid consumption and risk of heart disease (He et al., 2004); studies of characteristic fish consumption and heart disease results that have followed millions of persons for a number of decades through a variety of countries worldwide; and randomized controlled trials of fish or its oil intake, which have joined hundreds of topics and established discoveries in scientific proceedings (Mozaffarian & Rimm, 2006). Fish consumption might also decrease the danger of further coronary and noncoronary results, including but not limited to nonfatal cardiac arrests, atrial fibrillation, ischemic stroke, depression, and cognitive decline; however, the evidence for such measures is still not as strong as for coronary heart disease (CHD) mortality (Lin & Su, 2007; Morris et al., 2005; Mozaffarian et al., 2004; Scott & Sloman, 2004; Sun et al., 2008). Deaths from CHD—that is, recognized or deadly heart attacks and unexpected demise—are clinically distinct situations habitually associated with the final event of ventricular arrhythmia, frequently ischemia-induced ventricular fibrillation. Heterogeneity of the special impacts of fish or fish-oil consumption on cardiovascular outcomes is probable, related to variable dose and time effects and impacts on risk issues (Mozaffarian & Rimm, 2006). Combining all studies, this result was obvious. At consumptions up to 250 mg/day, the relational risk of coronary heart disease death was 14.6% lower (95% confidence interval [CI], 8%–21%) per each 100 mg/day of EPA and DHA, for a total risk reduction of 36% (95% CI, 20%–50%). At higher consumptions, slight extra risk decrease was evident (0.0% change per each 100 mg/day; 95% CI, −0.9% to 0.8%) (Mozaffarian & Rimm, 2006). Though the lipid statement is carefully related to the growth of atherosclerosis, the remark that Inuit infrequently industrialized coronary heart diseases notwithstanding a high consumption of fat encouraged inquiries about protective constituents in their food. Great importance has been placed on polyunsaturated fatty acids, that is, omega 3 ($n-3$) and omega 6 ($n-6$), for protecting against atherosclerosis. Three randomized controlled trials have found benefits from consumption of polyunsaturated fatty acids for patients with previous coronary heart disease. The greatest undoubtedly was the GISSI-Prevenzione trial, in which taking omega 3 fatty acids showed a 20% decrease in risk of overall death and a 45% decrease in risk of sudden cardiac death (Burr et al., 1989; GISSI-Prevenzione Investigators, 1999; Singh et al., 1997).

8.8.3 Kidney

The kidney is a multifaceted body organ containing distinct parts that function in an extremely synchronized manner. A great number of drugs, chemicals, and heavy metals have been shown to affect its construction and function. The amount of renal injury from heavy metals relies on the nature of the metal, dose, route, and period of contact. Together, severe and long-lasting contamination have been established to cause nephropathies with different levels of severity, from tube-shaped dysfunctions to severe renal damage (Barbier et al., 2005). A diversity of defensive events against uranyl nitrate-induced (UN-induced) nephropathy and oxidative injury have been studied. These comprise certain complexing and chelating agents, which process nonionized ring developments with inorganic ions that are consequently removed by the kidney (Domingo et al., 1997; Durakoviae, 1999). In addition, other treatments like melatonin (Belles et al., 2007) were also used to decrease UN nephrotoxicity, but were not originally appropriate for scientific use.

Owing to demographic alterations, kidney diseases have become a major health problem and the number of persons suffering from severe or long-lasting kidney disease is growing. Renal failure leads to decrease in the glomerular filtration rate and the damage of nephrons, the practical components of kidney. Nephron assembly is preserved among vertebrates and has three main sections: glomerulus, distal tubule, and proximal tubule. After renal damage, mammals are able to partially repair structures such as proximal tubules and glomeruli, but are unable to regenerate nephrons, as fish can do (Humphreys et al., 2008). A pool of progenitor cells in fish are triggered after initiation of kidney damage, and begin to form a nephron, first noticeable as basophilic clusters of cells (Diep et al., 2011). One of the miRNAs, miR-21, is upregulated after kidney damage and acts as a pro-proliferative and antiapoptotic factor in kidney regeneration in fish (Hoppe et al., 2015). Receptors have been identified and thought to be liganded by iodomelatonin in various tissues and organs through in vitro autoradiography and conservative binding examination. High-attraction 2-iodomelatonin binding locations have been observed at primary sites in the CNS and presently afterwards into numerous peripheral mammalian and avian tissues, such as lymphocyte, caudal artery, Hadrian gland, spleen, adrenal gland, intestine, kidney, and gonads (Stankov et al., 1993). Histopathic variations into kidney were studied at day 8 and sustained until the end of the research study. The renal-tubule developed as extremely elongated; its epithelial coating was definitely disconnected from tubular cells. An harm to cellular reliability particularly the renal tubules. Edema, dilation, and enlarged nuclei of renal tubules were noted too. Glomeruli demonstrated vacuolization and unsystematic blood capillaries. Pyknotic nuclei and necrosis can be experimental in mesenchymal tissue. Injury developed more markedly by day 15 and was extracritical in groups 3 and 4.

Pathological variations have previously been described within kidneys of fish that are unprotected from numerous contaminants (Anitha Kumari & Sree Ram Kumar, 1997; Banerjee & Bhattacharya, 1994). Higher zinc absorption into kidneys of preserved fishes is perhaps a consequence of the kidney being one of the main organs for decontamination and removal of metal contaminants (Dallinger et al., 1987). The dose level of Zn in current trials seemed to disturb the detoxification cation apparatus in kidney, thus delaying metal removal and increasing its accumulation.

In the realm of kidney biology, zebrafish have been identified as an excellent complement to the study of nephron proliferation and pathogenesis of physiologically changing disease. Kidney studies using the zebrafish model were established from investigations on mutations isolated through zebrafish-forward genetic screens that led to a rich collection of mutations affecting nephron form and function, whose further studies led to information that augmented understanding of the sympathetic mechanisms of ciliogenesis, polycystic kidney disease, and others. Detailed molecular classification of the renal anatomy of zebrafish subsequently revealed the overall conservation of nephron segment pattern and cellular composition in both embryonic and adult kidney structures, connected with other vertebrates, including humans (Poureetezadi & Wingert, 2016).

8.8.4 Reproductive system

The existence of industrial pollutants in the water environment has been of growing concern for the health of the natural world and humans. Mercury (Hg) is a specific worry, due to its bioaccumulation and strong neurotoxicity of some of its forms. Large releases of mercury are the result of anthropogenic actions such as metal production and coalfired energy production, along with normal natural processes such as forest fires and volcanic eruptions. Established from sediments of ponds, records from midcontinent North America show stratigraphic evidence of mercury deposition having tripled in the last 140 years. In the atmosphere, Hg can exist for up to a year, leading to long-range distribution through atmospheric transport. Mercury contamination is a universal problem, with climatic residuals polluting waters in areas distant from anthropogenic or normal point sources (Fitzgerald et al., 1998; Swain et al., 1992).

The sequence of spermatogenesis in stem cell division in established spermatozoa is measured by FSH and LH (Dziewulska & Domagala, 2003; Gomez et al., 1999). Throughout the cycle, the testes grow, which can be calculated by an increase in gonadosomatic index and frequently is largest in the breeding season. Disturbances in testicle function can be categorized as mostly cytotoxicological or endocrine in source (Kime, 1998). Specific cytotoxic impacts are categorized by harm to cellular reliability or changes in cell function, while specific endocrine impacts include disturbance of particular cells as a consequence of changes in secretions of pituitary or gonads. Reproductive damage might be produced through variations in one of the effects involved in changing the hypothalamic pituitary gonadal axis. First, variations in neurotransmitter stages or further limitations of neurotransmission can disturb the release of GnRH or LH from the pituitary and hypothalamus. Next, restriction of GnRH or LH production and synthesis will produce a cascade effect, resulting in diminished gonadal hormone production and consequent reductions in vitellogenesis, gonadal development, and GSI. Third, variations in gonadal steroidogenesis will disturb gamete generation and pituitary hormone production and discharge via positive feedback mechanisms. Lastly, altered lipid balance might be central to disruption of normal development, including steroidogenesis and vitellogenesis. Cellular procedures have also been suggested, which might result in overall noxiousness or specific damage of endocrine processes.

The exoskeleton is a target for methyl-mercury, and interference with cytoskeleton constituents can damage ordinary mitotic division, which occurs throughout gamete development. Apoptosis has been suggested as a process through which methyl-mercury harms the steroidogenic capability of gonads. The reproduction axis is an integrated system; however, numerous current studies examined only one parameter of the axis. Also, data proving the direct participation in reproductive damage or impairment are still absent. The studies offer proof of reproductive inhibition; nonetheless they don't elucidate whether the primary site of action is the gonad aforementioned, or whether it is done through limiting of secretions of the pituitary or altered neurotransmission in parts of the brain. Moreover, changes at the transcriptome or proteome level related to reproductive dysfunction still remain to be determined.

Currently, there seems to be enough evidence accumulating from lab studies to connect contact with mercury with reproductive damage in numerous fishes. It is now essential to control the main site of action and the mechanisms involved in endocrine disturbances at ecologically important contacts with methyl-mercury. Connecting reproductive damage with environmentally applicable effects on fish populations is an additional difficulty. Fish are a source of food for humans and wild animals, and specific effects of mercury contact in fish, as defined in the current appraisal, might likewise occur in these consumers of fish. Consequently, mercury contamination is a concern for both fish populations and those who consume fish, since mercury can act both indirectly, and decrease the food supply, and directly, by concentrating in the food chain and disturbing the top predators, namely, human consumers of fish (Crump & Trudeau, 2009; Sankhla et al., 2018).

8.9 Conclusion

We conclude that bacterial diseases affect fish populations. Several biotechnological innovations in aquaculture along with other developments in technologies lead to optimism toward aquaculture assets. The natural specific consequences of marine acidification are based on related technical information, and the long-term effects of altering seawater chemistry on marine ecosystems can only be assumed. Clinicians should be aware of the clinical signs of bacterial diseases, because sickness and death as a consequence of such infections increases considerably if there are delays in diagnosis and cure. We anticipate that the number of known bacterial diseases will probably increase in the future and affect fish populations.

References

Allan, J. D. (1995). *Stream ecology: Structure and function of running waters*. London: Chapman and Hall.

Almeida, A. R., Tacao, M., Machado, A. L., Golovko, O., Zlabek, V., Domingues, I., & Henriques, I. (2019). Long-term effects of oxytetracycline exposure in zebrafish: A multi-level perspective. *Chemosphere*, *222*, 333–344.

Aly, S. M., & Albutti, A. (2014). Antimicrobials use in aquaculture and their public health impact. *Journal of Aquaculture Research & Development*, *5*(4), 1.

Amin, N. E., Abdallah, I. S., Elallawy, T., & Ahmed, S. M. (1985). Motile Aeromonas septicemia among *Tilapia nilotica* (*Sarotherodonniloticus*) in Upper Egypt. *Fish Pathology*, *20*, 93–97.

Anacker, R. L., & Ordal, E. J. (1959). Studies on the myxobacterium *Chondrococcus columnaris* 1,2 I. Serological typing. *Journal of Bacteriology*, *78*(1), 25–32.

Anderson, C. N. K., Hsieh, C. H., Sandin, S. A., Hewitt, R., Hollowed, A., Beddington, J., May, R. M., & Sugihara, G. (2008). Why fishing magnifies fluctuations in fish abundance. *Nature*, *452*, 835–839.

Anderson, K. (2010). Globalization's effects on world agricultural trade, 1960–2050. *Philosophical Transactions of the Royal Society B*, *365*, 3007–3021.

Anitha Kumari, S., & Sree Ram Kumar, N. (1997). Histopathological alterations induced by aquatic pollutants in *Channa punctatus* from Hussain Sagar lake (A.P.). *Journal of Environmental Biology / Academy of Environmental Biology, India*, *18*(1), 11–16.

Antaya, C. (2008). Current eco-economical impacts of *Flavobacterium psychrophilum. MMG 445 Basic Biotechnology eJournal*, *4*, 7–12.

Arkoosh, M. R., Casillas, E., Clemons, E., Kagley, A. N., Olson, R., Reno, P., & Stein, J. E. (1998). Effect of pollution on fish diseases: Potential impacts on salmonid populations. *Journal of Aquatic Animal Health*, *10*(2), 182–190.

Aronson, J. D. (1926). Spontaneous tuberculosis in salt water fish. *The Journal of Infectious Diseases*, *39*, 315–320.

Ashraf, W. (2005). Accumulation of heavy metals in kidney and heart tissues of *Epinephelusmicrodon* fish from the Arabian Gulf. *Environmental Monitoring Assessment*, *101*, 311–316.

Austin, B. (2006). The bacterial microflora of fish. *The Scientific World Journal*, *6*, 931–945.

Austin, B., & Austin, D. A. (1999). *Bacterial fish pathogens, disease in farmed and wild fish* (3rd ed.). Chichester: Praxis Publishing.

Bach, S. P., Renehan, A. G., & Potten, C. S. (2000). Stem cells: The intestinal stem cell as a paradigm. *Carcinogenesis*, *21*, 469–476.

Banerjee, S., & Bhattacharya, S. (1994). Histopathology of kidney of *Channa punctatus* exposed to chronic non-lethal levels of elsan, mercury and ammonia. *Ecotoxicology and Environmental Safety*, *29*(3), 265–275.

Barbier, O., Jacquillet, G., Tauc, M., Cougnan, M., & Poujeol, P. (2005). Effect of heavy metals on, and handling by kidney. *Nephron Physiology*, *99*, 105–110.

Belles, M., Linares, V., Luisa Albina, M., Sirvent, J., Sanchez, D. J., & Domingo, J. L. (2007). Melatonin reduces uranium-induced nephrotoxicity in rats. *Journal of Pineal Research*, *43*, 87–95.

Bercovier, H., Ghittino, C., & Eldar, A. (1997). Immunization with bacterial antigens: Infections with streptococci and related organisms. *Developments in Biological Standardization*, *90*, 153–160.

Birungi, Z., Masola, B., Zaranyika, M. F., Naigaga, I., & Marshall, B. (2007). Active biomonitoring of trace heavy metals using fish (*Oreochromis niloticus*) as bioindicator species. The case of Nakivubo wetland along Lakealong Lake along Lake Victoria. *Physics and Chemistry of the Earth*, *32*, 1350–1358.

Bojarski, B., & Witeska, M. (2020). Blood biomarkers of herbicide, insecticide, and fungicide toxicity to fish—A review. *Environmental Science and Pollution Research*, *27*, 19236–19250.

Bondad-Reantaso, M. G., Subasinghe, R. P., Arthur, J. R., Ogawa, K., Chinabut, S., Adlard, R., Tan, Z., & Shariff, M. (2005). Disease and health management in Asian aquaculture. *Veterinary Parasitology*, *132*, 249–272.

Booth, D. J. (1991). *Larval settlement and juvenile group dynamics in the domino damselfish (Dascyllus albisella)*. PhD Thesis, Oregon State University.

Booth, D. J. (1992). Larval settlement patterns and preferences hy domino damselfish *Dascyllus alhisella. Journal of Experimental Marine Biology and Ecology*, *155*, 85–10.4.
Borg, A. F. (1960). *Studies on myxobacteria associated with diseases in salmonid fishes*. Washington, DC: American Association for the Advancement of Science, Wildlife Disease No. 8.
Bostock, J., McAndrew, B., Richards, R., Jauncey, K., Telfer, T., Lorenzen, K., & Corner, R. (2010). Aquaculture: Global status and trends. *Philosophical Transactions of the Royal Society B: Biological Sciences*, *365*(1554), 2897–2912.
Botsford, L. W., Castilla, J. C., & Peterson, C. H. (1997). The management of fisheries and marine ecosystems. *Science (New York, N.Y.)*, *277*, 509–515.
Bromage, E., & Owens, L. (2009). Environmental factors affecting the susceptibility of barramundi to Streptococcus iniae. *Aquaculture (Amsterdam, Netherlands)*, *290*, 224–228.
Brooks, R. W., George, K. L., Parker, B. C., & Falkinham, J. O., III (1984). Recovery and survival of nontuberculous mycobacteria under various growth and decontamination conditions. *Canadian Journal of Microbiology*, *30*, 1112–1117.
Brunton, L. A., Desbois, A. P., Garza, M., Wieland, B., Mohan, C. V., Häsler, B., Tam, C. C., Thien Le, P. N., Phuong, N. T., Van, P. T., Nguyen-Viet, H., Eltholth, M. M., Pham, D. K., Duc, P. P., Linh, N. T., Rich, K. M., Mateus, A. L. P., Hoque, Md. A., & Guitian, J. (2019). Identifying hotspots for antibiotic resistance emergence and selection, and elucidating pathways to human exposure: Application of a systems-thinking approach to aquaculture systems. *Science of the Total Environment*, *687*, 1344–1356.
Bullock, G. L. (1972). *Studies on selected myxobacteria pathogenic for fishes and on bacterial gill disease in hatchery-reared salmonids 1972*. Technical paper 60. Washington, DC, US: Fish and Wildlife Service.
Bullock, G. L. (1990). *Bacterial gill disease of freshwater fishes*. Fish disease leaflet 84. Washington, DC: Fish and Wildlife Service.
Burgos-Aceves, M. A., Lionetti, L., & Faggio, C. (2019). Multidisciplinary haematology as prognostic device in environmental and xenobiotic stress-induced response in fish. *The Science of the Total Environment*, *670*, 1170–1183.
Burr, M. L., Fehily, A. M., Gilbert, J. F., Rogers, S., Holliday, R. M., Sweetnam, P. M., Elwood, P. C., & Deadman, N. M. (1989). Effects of changes in fat, fish and fiber intakes on death and myocardial infarction: Diet and reinfarction trial (DART). *Lancet*, *2*, 757–761.
Callinan, R. B., Sammut, J., & Fraser, G. C. (2005). Dermatitis, branchitis and mortality in empire gudgeonHypseleotriscompressa exposed naturally to runoff from acid sulfate soils. *Diseases of Aquatic Organisms*, *63*, 247–253.
Camus, A. C., Durborow, R. M., Hemstreet, W. G., Thune, R. L., & Hawke, J. P. (1998). *Aeromonas bacterial infections-motile aeromonad septicemia* (Vol. 478). Southern Regional Aquaculture Center.
Carr, M. H., & Hixon, M. A. (1995). Predation effects on early post-settlement survivorship of coral-reef fishes. *Marine Ecology Progress Series*, *124*, 31–42.
Centers for Disease Control and Prevention. (1996). *CDC/FDA7- USDA National Antimicrobial Monitoring System 1996 Annual Report*. Atlanta, GA.
Chang, C. I., & Liu, W. Y. (2002). An evaluation of two probiotic bacterial strains, *Enterococcus faecium* SF68 and *Bacillus toyoi*, for reducing edwardsiellosis in cultured European eel, *Anguilla anguilla* L. *Journal of Fish Diseases*, *25*, 311–315.
Chapman, J. S. (1977). *The atypical mycobacteria and human mycobacteriosis. Topics in infectious disease* (Vol. xvi). New York: Plenum Medical.
Cheaib, B., Seghouani, H., Ijaz, U. Z., & Derome, N. (2020). Community recovery dynamics in yellow perch microbiome after gradual and constant metallic perturbations. *Microbiome*, *8*(1), 14. Available from https://doi.org/10.1186/s40168-020-0789-0.
Cheng, H., & Leblond, C. P. (1974). Origin, differentiation and renewal of the four main epithelial cell types in the mouse small intestine. I. Columnar cell. *The American Journal of Anatomy*, *141*, 461–474.

Clandinin, M. T., Cheema, S., Field, C. J., Garg, M. L., Venkatraman, J., & Clandinin, T. R. (1991). Dietary fat: Exogenous determination of membrane structure and cell function. *The FASEB Journal*, *5*, 2761–2769.

Croucher, S. C., Houston, A. P., Bayliss, C. E., & Turner, R. J. (1983). Bacterial populations associated with different regions of the human colon wall. *Applied and Environmental Microbiology*, *45*, 1025–1033.

Crump, K. L., & Trudeau, V. L. (2009). Mercury-induced reproductive impairment in fish. *Environmental Toxicology and Chemistry: An International Journal*, *28*(5), 895–907.

Cummins, C. P. (1994). Acid solutions. In P. Calow (Ed.), *Handbook of ecotoxicology* (Vol. 2, pp. 21–44). Oxford: Blackwell Scientific Publications.

Dallinger, R., Prosi, F., Segner, H., & Back, H. (1987). Contaminated food and uptake of heavy metals by fish: A review and a proposal for further research. *Oceologia (Berlin)*, *73*, 91–98.

Daoust, P. Y., & Ferguson, H. W. (1983). Gill diseases of cultured salmonids in Ontario. *Canadian Journal of Comparative Medicine (Gardenvale, Quebec)*, *47*(3), 358–362.

Dar, G. H., Bhat, R. A., Kamili, A. N., Chishti, M. Z., Qadri, H., Dar, R., & Mehmood, M. A. (2020). *Correlation between pollution trends of freshwater bodies and bacterial disease of fish fauna. Fresh water pollution dynamics and remediation* (pp. 51–67). Singapore: Springer.

Dar, G. H., Dar, S. A., Kamili, A. N., Chishti, M. Z., & Ahmad, F. (2016). Detection and characterization of potentially pathogenic *Aeromonas sobria* isolated from fish *Hypophthalmichthys molitrix* (Cypriniformes: Cyprinidae). *Microbial Pathogenesis*, *91*, 136–140.

Dar, G. H., Kamili, A. N., Chishti, M. Z., Dar, S. A., Tantry, T. A., & Ahmad, F. (2016). Characterization of *Aeromonas sobria* isolated from fish rohu (*Labeo rohita*) collected from polluted pond. *Journal of Bacteriology & Parasitology*, *7*(3), 1–5. Available from https://doi.org/10.4172/2155-9597.1000273.

Daughton, C. G., & Ternes, T. A. (1999). Special report. Pharmaceuticals and personal care products in the environment: Agents of subtle change. *Environmental Health Perspectives*, *107*, 907–938.

Diep, C. Q., Ma, D., Deo, R. C., Holm, T. M., Naylor, R. W., Arora, N., Wingert, R. A., Bollig, F., Djordjevic, G., Lichman, B., Zhu, H., Ikenaga, T., Ono, F., Englert, C., Cowan, C. A., Hukriede, N. A., Handin, R. I., & Davidson, A. J. (2011). Identification of adult nephron progenitors capable of kidney regeneration in zebrafish. *Nature*, *470*(7332), 95–100.

Doherty, P. J. (1982). Some effects of density on the juveniles of two species of tropical, territorial damselfishes. *Journal of Experimental Marine Biology and Ecology*, *65*, 249–261.

Doka, S. E., McNicol, D. K., Mallory, M. L., Wong, I., Minns, C. K., & Yan, N. D. (2003). Assessing potential for recovery of biotic richness and indicator species due to changes in acidic deposition and lake pH in five areas of southeastern Canada. *Environmental Monitoring and Assessment*, *88*, 53–101.

Domingo, J. L., de la Torre, A., Bellés, M., Mayayo, E., Llobet, J. M., & Corbella, J. (1997). Comparative effects of the chelators sodium 4,5-dihydroxybenzene-1,3-disulfonate (Tiron) and diethylenetriaminepentaacetic acid (DTPA) on acute uranium nephrotoxicity in rats. *Toxicology*, *118*, 49–59.

Durakoviae, A. (1999). Medical effects of internal contamination with uranium. *Croatian Medical Journal*, *40*, 1–17.

Dziewulska, K., & Domagala, J. (2003). Histology of salmonid testes during maturation. *Reproductive Biology and Endocrinology*, *3*, 47–61.

Eckert, G. L. (1985). Settlement of coral reef fishes to different substrata and at different depths. *Proceedings of the 5th International Coral Reef Congress*, *5*, 385–390.

Ellis, L. J. (2009). *Fishing murky waters: China's aquaculture challenges—upstream and downstream* (pp. 1–2). Business forum China.

Fabry, V. J., Seibel, B. A., Feely, R. A., & Orr, J. C. (2008). Impacts of ocean acidification on marine fauna and ecosystem processes. *Ices Journal of Marine Science*, *65*, 414–432.

Faktorovich, K. A. (1969). *Histological changes in the liver, kidneys, skin and brain of fish sick with red rot. Infectious diseases of fish and their control* (pp. 83–101). Washington, DC: Division of Fisheries Research, Bureau of Sport Fisheries and Wildlife, Translated from the Russian by R. M. Howland.

FAO. (2003). *Overview of fish production, utilization, consumption and trade*. Rome, Italy: FAQ.

FAO. (2006). *Fisheries statistics: Aquaculture production, 88/2* (p. 12) Rome, Italy: FAO.

FAO. (2016). *FAO global aquaculture production database updated to 2013—summary information*. Available from: http://www.fao.org/fishery/statistics/en. Accessed 02.01.15.

Farkas, J. (1985). Filamentous *Flavobacterium* sp. isolated from fish with gill diseases in cold water. *Aquaculture (Amsterdam, Netherlands)*, *44*(1), 1–10.

Feller, S. E., & Gawrisch, K. (2005). Properties of docosahexaenoic-acid-containing lipids and their influence on the function of rhodopsin. *Current Opinion in Structural Biology*, *15*, 416–422.

Ferguson, H. W., Ostland, V. E., Byrne, P., & Lumsden, J. S. (1991). Experimental production of bacterial gill disease in trout by horizontal transmission and by bath challenge. *Journal of Aquatic Animal Health*, *3*, 118–123.

Fitzgerald, W. F., Engstrom, D. R., Mason, R. P., & Nater, E. A. (1998). The case for atmospheric mercury contamination in remote areas. *Environmental Science & Technology*, *32*, 1–7.

Fjellheim, A., & Raddum, G. G. (1990). Acid precipitation: Biological monitoring of streams and lakes. *The Science of the Total Environment*, *96*, 57–66.

Francis, A. J., Quinby, H. L., & Hendrey, G. R. (1984). Effect of lake pH on microbial decomposition of allochthonous litter. In G. R. Hendrey (Ed.), *Early biotic responses to advancing lake acidification* (pp. 1–21). Boston: Butterworth Publishers.

Fuentes-Gandara, F., Pinedo-Hernández, J., Marrugo-Negrete, J., & Diez, S. (2018). Human health impacts of exposure to metals through extreme consumption of fish from the Colombian Caribbean Sea. *Environmental Geochemistry and Health*, *40*(1), 229–242.

Gad, N. S. (2009). Determination of glutathione-related enzyme and cholinesterase activities in *Oreochromis niloticus* and *Clariasgariepinus* as bioindicator for pollution in lake Manzala. *Global Veterinaria*, *3*, 37–44.

GISSI-Prevenzione Investigators. (1999). Dietary supplementation with n-3 poly unsaturated fatty acids and vitamin E in 11, 324 patients with myocardial infarction. Results of the GISSI-Prevenzione trial. *Lancet*, *354*, 447–455.

Gomez, J. M., Weil, C., Ollitrault, M., Le Bail, P. Y., Breton, B., & Le Gac, F. (1999). Growth hormone (GH) and gonadotropin subunit gene expression and pituitary and plasma changes during spermatogenesis and oogenesis in rainbow trout (*Oncorhynchus mykiss*). *General and Comparative Endocrinology*, *113*, 413–428.

Griffin, B. R. (1992). A simple procedure for identification of *Cytophagacolumnaris*. *Journal of Aquatic Animal Health*, *4*, 63–66.

Groff, J. M., & LaPatra, S. E. (2001). In C. E. Lim, & C. D. Webster (Eds.), *Nutrition and fish health* (pp. 11–78). New York: Haworth Press.

Gulland, F. M. D. (1995). The impact of infectious diseases on wild animal populations—A review. In B. T. Grenfell, & A. P. Dobson (Eds.), *Ecology of infectious diseases in natural populations* (pp. 20–51). Cambridge, UK: Cambridge University Press.

Harvell, C. D., Kim, K., & Burkholder, J. M. (1999). Emerging marine diseases—Climate links and anthropogenic factors. *Science (New York, N.Y.)*, *285*(5433), 1505–1510.

Havas, M. (1985). Aluminum bioaccumulation and toxicity to Daphnia magna in soft water at low pH. *Canadian Journal of Fisheries and Aquatic Sciences*, *42*, 1741–1748.

Havas, M. (1986). Effects of acidic deposition on aquatic ecosystems. In A. Stem (Ed.), *Air pollution* (Vol. VI, pp. 351–389). Academic Press.

Hawke, J. P., & Thune, R. L. (1992). Systemic isolation and antimicrobial susceptibility of Cytophagacolumnaris from commercially reared channel catfish. *Journal of Aquatic Animal Health*, *4*, 109–113.

Hendrey, G. R. (1982). Effects of acidification on aquatic primary producers and decomposers. In R. E. Johnson (Ed.), *Acid rain/fisheries. Proceedings of the International Symposium on Acidic Precipitation and Fish: impacts in Northeastern North America* (pp. 125–134). Bethesda, MD: American Fisheries Society.

Herikstad, H., Hayes, P. S., Hogan, J., Floyd, P., Snyder, L., & Ángulo, F. J. (1997). Ceftriaxone-resistant salmonella in the United States. *The Pediatric Infectious Disease Journal*, *16*, 904–905.

Herrmann, J. (1993). Strategies for Lake Ecosystems beyond 2000. In: Proceedings/Gussani, G. and C. Callieri (Eds.), *5th International Conference on the Conservation and Management of Lakes*, Stresa, Italy, 17–21 May 1993, pp. 418–421.

He, K., Song, Y., Daviglus, M. L., Liu, K., Van Horn, L., Dyer, A. R., Goldbourt, U., & Greenland, P. (2004). Fish consumption and incidence of stroke: A meta-analysis of cohort studies. *Stroke; A Journal of Cerebral Circulation*, *35*, 1538–1542.

Heys, C., Cheaib, B., Busetti, A., Kazlauskaite, R., Maier, L., Sloan, W. T., & Llewellyn, M. S. (2020). Neutral processes dominate microbial community assembly in Atlantic salmon, Salmo salar. *Applied and Environmental Microbiology*, *86*(8). Available from https://doi.org/10.1128/AEM.02283-19, e02283-19.

Hoppe, B., Pietsch, S., Franke, M., Engel, S., Groth, M., Platzer, M., & Englert, C. (2015). MiR-21 is required for efficient kidney regeneration in fish. *BMC Developmental Biology*, *15*(1), 1–10.

Hoseini, S. M., & Yousefi, M. (2019). Beneficial effects of thyme (*Thymus vulgaris*) extract on oxytetracyclineinduced stress response, immunosuppression, oxidative stress and enzymatic changes in rainbow trout (*Oncorhynchus mykiss*). *Aquaculture Nutrition*, *25*, 298–309.

Hoshina, T., Sano, T., & Morimoto, Y. (1958). A Streptococ – Cus pathogenic to fish. *Journal of Tokyo University of Fisheries*, *44*, 57–68.

Huckabee, J. W., Mattice, J. S., Pitelka, L. F., Porcella, D. B., & Goldstein, R. A. (1989). An assessment of the ecological effects of acidic deposition. *Archives of Environmental Contamination and Toxicology*, *18*, 3–27.

Humphreys, B. D., Valerius, M. T., Kobayashi, A., Mugford, J. W., Soeung, S., Duffield, J. S., McMahon, A. P., & Bonventre, J. V. (2008). Intrinsic epithelial cells repair the kidney after injury. *Cell Stem Cell*, *2* (3), 284–291.

Jacobs, J. M., Lazur, A., & Baya, A. (2004). *Prevention and disinfection of mycobacterium sp. in aquaculture*. Maryland Sea Grant Extension Publication, Publication number UM-SG-SGEP, Finfish Worksheet 9.

Jacobson, L. D., De Oliveira, J. A. A., Barange, M., Cisneros-Mata, M. A., Felix-Uraga, R., Hunter, J. R., Kim, J. Y., Matsuura, Y., Niquen, M., Porteiro, C., Rothschild, B., Sanchez, R. P., Serra, R., Uriarte, A., & Wada, T. (2001). Surplus production, variability, and climate change in the great sardine and anchovy fisheries. *Canadian Journal of Fisheries and Aquatic Sciences*, *58*, 1891–1903.

Janda, J. M., & Abbott, S. L. (1998). Evolution concepts regarding the genus Aeromonas: An expanding panorama of species, disease presentation, and unanswered questions. *Clinical Infectious Diseases*, *27*, 332–344.

Japan Environmental Agency. (1997). Progress report of the 3rd acid rain monitoring. In Japan Environmental Agency (Ed.), *Whereabouts of global environment-acid rain* (pp. 203–252). Tokyo: Chuohoki Publications, [In Japanese].

Jones, G. P. (1987). Some interactions between residents and recruits in two coral reef fishes. *Journal of Experimental Marine Biology and Ecology*, *114*, 169–182.

Karunasagar, I., Pai, R., Malahti, G. R., & Karunasagar, I. (1994). Mass mortality of *Penaeus monodon* larvae due to antibiotic-resistant *Vibrio harveyi* infection. *Aquaculture (Amsterdam, Netherlands)*, *128*, 203–209.

Kelly, C. A., Rudd, J. W. M., Cook, R. B., & Schindler, D. W. (1982). The potential importance of bacterial processes in regulating rate of lake acidification. *Limnology and Oceanography*, *27*, 868–882.

Kime, D. E. (1998). *Endocrine disruption in fish*. Boston, MA: Kluwer Academic.

Kimura, N., Wakabayashi, H., & Kudo, S. (1992). Studies on bacterial gill disease in salmonids. 1. Selection of bacterium transmitting gill disease. *Fish Pathology*, *12*, 233–242.

Knutton, S. (1995). Electron microscopical methods in adhesion. *Methods of Microbiology*, *253*, 145–158.

Krause-Dellin, D., & Steinberg, C. (1986). Cladoceran remains as indicators of lake acidification. *Hydrobiologia*, *143*, 129–134.

Kusuda, R., Kawai, K., Salati, F., Banner, C. R., & Fryer, J. L. (1991). *Enterococcus seriolicida* sp. nov., a fish pathogen. *International Journal of Systematic Bacteriology*, *41*, 406–409.

Laake, M. (1976). Effekteravlav pH pa produksjon, nedbrytingog stoffkretsl0p ilittoralsonen. Internal report IR 29/76. Sur NedborsVirkningPla Skog og Fisk, Oslo, Norway.

Lee, F. D., Kraszewski, A., Gordon, J., Howie, J. G. R., McSeveney, D., & Harland, W. A. (1971). Intestinal spirochaetosis. *Gut*, *12*, 126–133.

Lee, L. A., Puhr, N. D., Maloney, K., Bean, N. H., & Tauxe, R. V. (1994). Increase in antimicrobial-resistant Salmonella infections in the United States, 1989-1990. *The Journal of Infectious Diseases*, *170*, 128–134.

Lehmann, J., Mock, D., Stürenberg, F.-J., & Bernardet, J.-F. (1991). First isolation of *Cytophaga psychrophila* from a systemic disease in eel and cyprinids. *Diseases of Aquatic Organisms*, *10*, 217–220.

Lin, P. Y., & Su, K. P. (2007). A meta-analytic review of double-blind, placebo-controlled trials of antidepressant efficacy of omega-3 fatty acids. *The Journal of Clinical Psychiatry*, *68*, 1056–1061.

Linde-Arias, A. R., Inácio, A. F., Novo, L. A., de Alburquerque, C., & Moreira, J. C. (2008). Multibiomarker approach in fish to assess the impact of pollution in a large Brazilian river, Paraiba do Sul. *Environmental Pollution*, *156*, 974–979.

Linell, F., & Norden, A. (1954). Mycobacterium balnei. A new acidfast Bacillus occurring in swimming pools and capable of producing skin lesions in humans. *Acta Tuberculosea Scandinavica*, *33*, 1–84.

Linell, F. N. M. (1954). A new acid-fast bacilli occurring in swimming pools and capable of producing skin lesions in humans, balnei: *Acta Tuberculosea Scandinavica*, *33*, 1–84.

Ling, S. H. M., Wang, X. H., Xie, L., Lim, T. M., & Leung, K. Y. (2000). Use of green fluorescent protein (GFP) to study the invasion pathways of *Edwardsiella tarda* in in vivo and in vitro fish models. *Microbiology (Reading, England)*, *146*(1), 7–19.

Lulijwa, R., Rupia, E. J., & Alfaro, A. C. (2020). Antibiotic use in aquaculture, policies and regulation, health and environmental risks: A review of the top 15 major producers. *Reviews in Aquaculture*, *12*, 640–663.

Macfarlane, S., & Dillon, J. F. (2007). Microbial biofilms in the human gastrointestinal tract. *Journal of Applied Microbiology*, *102*(5), 1187–1196.

Macfarlane, S., Furrie, E., Cummings, J. H., & Macfarlane, G. T. (2004). Chemotaxonomic analysis of bacterial populations colonizing the rectal mucosa in patients with ulcerative colitis. *Clinical Infectious Diseases: An Official Publication of the Infectious Diseases Society of America*, *38*, 1690–1699.

McCormick, M. I. (1993). Development and changes at settlement in the barbel structure of the reef fish *Upeneus tragula* (Mullidae). *Environmental Biology of Fishes*, *37*, 269–282.

McDonald, K. L., Cohen, M. L., Hargrett-Bean, N., Wells, J. G., Puhr, N. D., Collin, S. F., & Blake, P. A. (1987). Changes in antimicrobial resistance of Salmonella isolated from humans in the United States. *The Journal of the American Medical Association*, *258*, 1496–1499.

McFarlane, G. A., Smith, P. E., Baumgartner, T. R., & Hunter, J. R. (2002). Climate variability and Pacific sardine populations and fisheries. *American Fisheries Society Symposium*, *32*, 195–214.

McLaughlin, R. L., Smyth, E. R., Castro-Santos, T., Jones, M. L., Koops, M. A., Pratt, T. C., & Vélez-Espino, L. A. (2013). Unintended consequences and trade-offs of fish passage. *Fish and Fisheries*, *14*(4), 580–604.

Meriläinen, J. J., & Hynynen, J. (1990). Benthic invertebrates in relation to acidity in Finnish forest lakes. In P. Kauppi, P. Anttila, & K. Kenttämies (Eds.), *Acidification in Finland* (pp. 1029–1049). Berlin: Springer-Verlag.

Meyer, F. P. (1991). Aquaculture disease and health management. *Journal of Animal Science, 69*(10), 4201–4208.

Mayon, N., Bertrand, A., Leroy, D., Malbrouck, C., Mandiki, S. N. M., Silvestre, F., Goffart, A., Thome, J. P., & Kestemont, P. (2006). Multiscale approach of fish responses to different types of environmental contaminations: A case study. *The Science of the Total Environment, 367*, 715–731.

Miyazaki, T., & Jo, Y. (1985). A histopathological study on motile aeromonad disease in ayu. *Fish Pathology, 20*, 55–59.

Morris, R., Taylor, E. W., Brown, D. J. A., & Brown, J. A. (Eds.), (1989). *Acid toxicity and aquatic animals*. Cambridge: Cambridge University Press.

Morris, M. C., Evans, D. A., Tangney, C. C., Bienias, J. L., & Wilson, R. S. (2005). Fish consumption and cognitive decline with age in a large community study. *Archives of Neurology, 62*, 1849–1853.

Mozaffarian, D. (2009). Fish, mercury, selenium and cardiovascular risk: Current evidence and unanswered questions. *International Journal of Environmental Research and Public Health, 6*(6), 1894–1916.

Mozaffarian, D., Psaty, B. M., Rimm, E. B., Lemaitre, R. N., Burke, G. L., Lyles, M. F., Lefkowitz, D., & Siscovick, D. S. (2004). Fish intake and risk of incident atrial fibrillation. *Circulation, 110*, 368–373.

Mozaffarian, D., & Rimm, E. B. (2006). Fish intake, contaminants, and human health: Evaluating the risks and the benefits. *JAMA: the Journal of the American Medical Association, 296*, 1885–1899.

Muniz, I. P. & Leivestad, H. (1980). Ecological impact of acid precipitation. In: Drablos, D. and A. Tollan (Eds.), *Proceedings of an International Conference*, Sandefjord, Norway, March 11–14, 1980, SNSF Project, pp. 84–92.

Muzquiz, J. L., Royo, F. M., Ortega, C., De Blas, I., Ruiz, I., & Alonso, J. L. (1999). Pathogenicity of streptococcosis in rainbow trout (*Oncorhynchus mykiss*): dependence on age of diseased fish. *Bulletin – European Association of Fish Pathologists, 19*, 114–119.

Norden, A., & Linell, F. (1951). A new type of pathogenic mycobacterium. *Nature, 168*, 826.

Novotny, L., Dvorska, L., Lorencova, A., Beran, V., & Pavlik, I. (2004). Fish: A potential source of bacterial pathogens for human beings. *Veterinary Medicine, 49*, 343–358.

Nussey, G., Van Vuren, J. H. J., & du Preez, H. H. (1995). Effect of copper on the haematology and osmoregulation of the *Mozambique tilapia, Oreochromis mossambicus* (Cichlidae). *Comparative Biochemistry and Physiology C, 111*(3), 369–380.

Ohwada, K., Tabor, P. S., & Colwell, R. R. (1980). Species composition and barotolerance of gut microflora of deep-sea benthic macrofauna collected at various depths in the Atlantic Ocean. *Applied and Environmental Microbiology, 40*, 746–755.

Ormerod, S. J., Weatherley, N. S., & Merrett, W. J. (1990). Restoring acidified streams in upland Wales: a modelling comparison of the chemical and biological effects of liming and reduced sulphate deposition. *Environmental Pollution (Barking, Essex: 1987), 64*, 67–85.

Orr, J. C., Fabry, V. J., Aumont, O., Bopp, L., Doney, S. C., Feely, R. A., Gnanadesikan, A., Gruber, N., Ishida, A., Joos, F., Key, R. M., Lindsay, K., Maier-Reimer, E., Matear, R., Monfray, P., Mouchet, A., Najjar, R. G., Plattner, G. K., Rodgers, K. B., & Yool, A. (2005). Anthropogenic ocean acidification over the twenty-first century and its impact on calcifying organisms. *Nature, 437*, 681–686.

Otte, E. (1963). Die heutigenAnsichtentiber die Atiologie der InfektiosenBauchwassersucht der Karpfen. *Wiener Tierärztliche Monatsschrift, 50*(11), 996–1005.

Plant, K. P., & LaPatra, S. E. (2011). Advances in fish vaccine delivery. *Developmental and Comparative Immunology, 35*, 1253–1259.

Pond, M. J., Stone, D. M., & Alderman, D. J. (2006). Comparison of conventional and molecular techniques to investigate the intestinal microflora of rainbow trout (*Oncorhynchus mykiss*). *Aquaculture (Amsterdam, Netherlands), 261*, 194–203.

Poureetezadi, S. J., & Wingert, R. A. (2016). Little fish, big catch: Zebrafish as a model for kidney disease. *Kidney International*, *89*(6), 1204–1210.

Ranzani-Paiva, M. J. T., Ishikawa, C. M., Acd, E., & Silveira, V. R. D. (2004). Effects of an experimental challenge with *Mycobacterium marinum* on the blood parameters of Nile tilapia, *Oreochromis niloticus* (Linnaeus, 1757). *Brazilian Archives of Biology and Technology*, *47*(6), 945–953.

Reichley, S. R., Ware, C., Steadman, J., Gaunt, P. S., & others. (2017). Comparative phenotypic and genotypic analysis of Edwardsiella spp. isolates from different hosts and geographic origins, with an emphasis on isolates formerly classified as *E. tarda* and an evaluation of diagnostic methods. *Journal of Clinical Microbiology*, *55*, 3466–3491.

Rhodes, M. W., Kator, H., Kaattari, I., Gauthier, D., Vogelbein, W., & Ottinger, C. A. (2004). Isolation and characterization of mycobacteria from striped bass Morone saxatilis from the Chesapeake Bay. *Diseases of Aquatic Organisms.*, *61*(1-2), 41–51. Available from https://doi.org/10.3354/dao061041.

Riley, L. W., Cohen, M. L., Seals, J. E., Blaser, M. J., Birkness, K. A., Hargrett, N. T., Martin, S. M., & Feldman, R. A. (1984). Importance of host factors in human salmonellosis caused by multiresistant strains of Salmonella. *The Journal of Infectious Diseases*, *149*, 878–883.

Ringø, E., & Gatesoupe, F. J. (1998). Lactic acid bacteria in fish: A review. *Aquaculture (Amsterdam, Netherlands)*, *160*, 177–203.

Robertson, D. R., Green, D. G., & Victor, B. C. (1988). Temporal coupling of reproduction and recruitment of larvae of a Caribbean reef fish. *Ecology*, *69*, 370–381.

Rosseland, B. O., & Staurnes, M. (1994). Physiological mechanisms for toxic effects and resistance to acidic water: An ecophysiological and ecotoxicological approach. In C. E. W. Steinberg, & R. W. Wright (Eds.), *Acidification of freshwater ecosystems: Implications for the future* (pp. 227–246)). John Wiley & Sons, Ltd.

Rudd, J. W. M., Kelly, C. A., Schindler, D. W., & Turner, M. A. (1988). Disruption of the nitrogen cycle in acidified lakes. *Science (New York, N.Y.)*, *240*, 1515–1517.

Rudd, J. W. M., Kelly, C. A., Schindler, D. W., & Turner, M. A. (1990). A comparison of the acidification efficiencies of nitric and sulfuric acids by two whole-lake addition experiments. *Limnology and Oceanography*, *35*, 663–679.

Sankhla, M. S., Kumari, M., Sharma, K., Kushwah, R. S., & Kumar, R. (2018). Water contamination through pesticide & their toxic effect on human health. *International Journal for Research in Applied Science & Engineering Technology (IJRASET)*, *6*(1), 967–970.

Santos, L., & Ramos, F. (2016). Analytical strategies for the detection and quantification of antibiotic residues in aquaculture fishes: A review. *Trends in Food Science & Technology*, *52*, 16e30.

Sapkota, A., Sapkota, A. R., Kucharski, M., Burke, J., McKenzie, S., Walker, P., & Lawrence, R. (2008). Aquaculture practices and potential human health risks: Current knowledge and future priorities. *Environment International*, *34*, 1215–1226.

Schachte, J. H. (1983). Bacterial gill disease. In F. P. Meyer, J. W. Warren, & T. G. Carey (Eds.), *A Guide to integrated fish health management in the great lakes basin (Special publication)* (pp. 181–184). Ann Arbor, MI: Great Lakes Fishery Commission.

Schindler, D. W., Kasian, S. E. M., & Hesslein, R. H. (1989). Losses of biota from America aquatic communities due to acid rain. *Environmental Monitoring and Assessment*, *12*, 269–285.

Schindler, D. W., Wagemann, R., Cook, R. B., Ruszczynski, T., & Prokopowich, J. (1980). Experimental acidification of Lake 223, Experimental Lakes Area: Background data and the first three years of acidification. *Canadian Journal of Fisheries and Aquatic Sciences*, *37*, 342–354.

Scott, G. R., & Sloman, K. A. (2004). The effects of environmental pollutants on complex fish behaviour: integrating behavioural and physiological indicators of toxicity. *Aquatic Toxicology*, *68*(4), 369–392.

Shulman, M. J. (1984). Resource limitation and recruitment patterns in coral reef fish assemblages. *Journal of Experimental Marine Biology and Ecology.*, *74*, 85, 109.

Singh, R. B., Niaz, M. A., Sharma, J. P., Kumar, R., Rastoqi, V., & Moshiri, M. (1997). Randomized, double-blind, placebo controlled trial of fish oil and mustard oil in patients with suspected acute myocardial infarction. The Indian experiment of infarct survival-4. *Cardiovascular Drugs and Therapy / Sponsored by the International Society of Cardiovascular Pharmacotherapy*, *11*, 485–491.

Snieszko, S. F. (1973). Recent advances in scientific knowledge and developments pertaining to diseases of fishes. *Advances in Veterinary Science and Comparative Medicine*, *17*, 291–314.

Sonone, S. S., Jadhav, S., Sankhla, M. S., & Kumar, R. (2020). Water contamination by heavy metals and their toxic effect on aquaculture and human health through food chain. *Letters in Applied NanoBioScience*, *10*, 2148–2166.

Spanggaard, B., Huber, I., Nielsen, J., Nielsen, T., Appel, K. F., & Gram, L. (2000). The microflora of rainbow trout intestine: A comparison of traditional and molecular identification. *Aquaculture (Amsterdam, Netherlands)*, *182*, 1–15.

Srinivasa Rao, P. S., Yamada, Y., & Leung, K. Y. (2003). A major catalase (KatB) that is required for resistance to H_2O_2 and phagocytemediated killing in *Edwardsiella tarda*. *Microbiology (Reading, England)*, *149*, 2635–2644.

Stankov, B., Fraschinin, F., & Reiter, R. J. (1993). The melatonin receptor: Distribution, biochemistry, and pharmacology. In H.-S. You, & R. J. Reiter (Eds.), *Melatonin: Biosynthesis, physiological effects, and clinical applications* (pp. 155–186). Boca Raton: CRC Press.

Subasinghe, R. P., Bondad-Reantaso, M. G., & McGladdery, S. E. (2001). Aquaculture development, health and wealth. In R. P. Subasinghe, P. Bueno, M. J. Phillips, C. Hough, S. E. McGladdery, & J. R. Arthur (Eds.), *Aquaculture in the Third Millennium. Technical Proceedings of the Conference on Aquaculture in the Third Millennium* (pp. 167–191). NACA and FAO.

Sugumar, G., Karunasagar, I., & Karunasagar, I. (2002). Mass mortality of fingerlings of *Labeo rohita* (Hamilton) in a nursery pond. *Indian Journal of Fisheries*, *49*, 305–309.

Sun, Q., Ma, J., Campos, H., Rexrode, K. M., Albert, C. M., Mozaffarian, D., & Hu, F. B. (2008). Blood concentrations of individual long-chain n-3 fatty acids and risk of nonfatal myocardial infarction. *The American Journal of Clinical Nutrition*, *88*, 216–223.

Swain, E. B., Engstrom, D. R., Brigham, M. E., Henning, T. A., & Brezonik, P. L. (1992). Increasing rates of atmospheric mercury deposition in midcontinental North America. *Science (New York, N.Y.)*, *257*, 784–787.

Tolimieri, N. (1995). Effects of microhabitat characteristics on the settlement and recruitment of a coral reef fish at two spatial scales. *Oecologia*, *102*, 52–63.

Tonjum, T., Welty, D. B., Jantzen, E., & Small, P. L. (1998). Diꞁerentiation of *Mycobacterium ulcerans*, *M. marinum*, and *M. haemophilum*: Mapping of their relationships to *M. tuberculosis* by fatty acid pro¢le analysis, DNA-DNA hybridization, and 16S rRNA gene sequence analysis. *Journal of Clinical Microbiology*, *36*, 918–925.

Toranzo, A. E., Magariños, B., & Romalde, J. L. (2005). A review of the main bacterial fish diseases in mariculture systems. *Aquaculture (Amsterdam, Netherlands)*, *246*(1-4), 37–61.

Traaen, T. S. (1980). Effects of acidity on decomposition of organic matter in aquatic environments. In: D. Drabløs and A. Tollan (Eds.), *Proceedings of the International Conference on the Ecological Impact of Acid Precipitation*. Sur NedborsVirkning Pa Skog og Fisk, Oslo, Norway (pp. 340–341).

Tytler, P., & Blaxter, J. H. S. (1988).). Drinking in yolk-sac stage larvae of the halibut, *Hippoglossus hippoglossus* (L.). *Journal of Fish Biology*, *32*, 493–494.

Van der Oost, R., Beyer, J., & Vermeulen, N. P. (2003). Fish bioaccumulation and biomarkers in environmental risk assessment: A review. *Environmental Toxicology and Pharmacology, of of*, *13*, 57–149.

Vanden Heuvel, J. P. (2004). Diet, fatty acids, and regulation of genes important for heart disease. *Current Atherosclerosis Reports*, *6*, 432–440.

Verschuere, L., Rombaut, G., Sorgeloos, P., & Verstraete, W. (2000). Probiotic bacteria as biocontrol agents in aquaculture. *Microbiology and Molecular Biology Reviews: MMBR*, *64*, 655–671.

Victor, B. C. (1983). Settlement and larval metamorphosis produce distinct marks on the otoliths of the slippery dick, *Haltehoeres bivittatus*. In: M. L. Reakaj (Ed.), *The Ecology of Deep and Shallozv Cora! Reefs, Symp. Ser. Undersea Res.* 1, 47–51. National Oceanic and Atmospheric Administration, Rockville, MD.

Virgona, J. L. (1992). Environmental factors influencing the prevalence of a cutaneous ulcerative disease (red spot) in the sea mullet, *Mugil cephalus* L., in the Clarence River, New South Wales, Australia. *Journal of Fish Diseases*, *15*, 363–378.

Von Betegh, L. (1910). More reviews on experimental tuberclosis of marine fish. Zentrablatt fur Bakteriologie, Parasiten- € kunde, Infektionskrankheiten und Hygiene. *Abteilung*, *153*, 54.

Von Graevenitz, A. (1990). Revised nomenclature of *Campylobacter laridis*, *Enterobacter intermedium* and "*Flavobacterium branchiophila*". *International Journal of Systematic Bacteriology*, *40*(2), 211.

Wakabayashi, H., & Egusa, S. (1973). Edwardsiella tarda (*Paracolobactrum anguillimortiferum*) associated with pondcultured eel diseases. *Bulletin of the Japanese Society for the Science of Fish*, *39*, 931–936.

Wakabayashi, H., Huh, G. J., & Kimura, N. (1989). *Flavobacterium branchiophila* sp. nov., a causative agent of bacterial gill disease of freshwater fishes. *International Journal of Systematic Bacteriology*, *39*(3), 213–216.

Wakabayashi, H., & Iwado, T. (1985). Effects of a bacterial gill disease on the respiratory functions ofjuvenile rainbow trout. In A. E. Ellis (Ed.), *Fish and shellfish pathology* (pp. 153–160). London: Academic Press, Inc. (London), Ltd.

Wellington, G. M. (1992). Habitat selection and juvenile persistence control the distribution of two closely related Caribbean damselfishes. *Oecologia*, *90*, 500–508.

Wiernasz, N., Leroi, F., Chevalier, F., Cornet, J., Cardinal, M., Rohloff, J., & Pilet, M.-F. (2020). Salmon gravlax biopreservation with lactic acid bacteria: A polyphasic approach to assessing the impact on organoleptic properties, microbial ecosystem and volatilome composition. *Frontiers in Microbiology*, *10*, 3103. Available from https://doi.org/10.3389/fmicb.2019.03103.

Wiklund, T., Nystedt, S., & Kroon, K. (1997). Flexibacterpsychrophilus: överlevnadivattenoch sediment (*Flexibacter psychrophilus*: survival in water and sediment). *Laxforskningsinstitutet, Meddelande*, *2*, 13–17.

Wong, C. K., Wong, P. P. K., & Chu, L. M. (2001). Heavy metal concentration in marine fish collected from culture sites in Hong Kong. *Archives of Environmental Contamination and Toxicology*, *40*, 60–69.

World Health Organization (WHO). (1999). Joint FAO/NACA/WHO Study Group on food safety issues associated with products from aquaculture. WHO Technical Report Series no. 883. Geneva: WHO.

Yoshimizu, M., Kimura, T., & Sakai, M. (1976). Studies on the intestinal microflora of salmonids: V. The intestinal microflora of the anadromous salmon. *Bulletin of the Japanese Society for the Science of Fish*, *42*, 1291–1298.

CHAPTER

Understanding the pathogenesis of important bacterial diseases of fish

9

Fernanda Maria Policarpo Tonelli[1,2], Moline Severino Lemos[1], Flávia Cristina Policarpo Tonelli[3], Núbia Alexandre de Melo Nunes[4] and Breno Luiz Sales Lemos[1]

[1]Institute of Biological Science, Federal University of Minas Gerais, Belo Horizonte, Brazil [2]Pitágoras College, Divinópolis, Brazil [3]Estácio de Sá University, Ribeirão Preto, Brazil [4]Institute of Biological Science, Federal University of Minas Gerais, Belo Horizonte, Brazil

9.1 Introduction

9.1.1 The importance of fisheries

As the world population increases, a global concern for developing strategies (preferentially sustainable ones) to provide human beings with healthy sources of nutrition is also increasing. The potential of fisheries to address this issue cannot be overstated. Fishes possess a high nutritional value and are an important source of human dietary protein (Sampantamit et al., 2020).

According to the Food and Agriculture Organization (FAO) of the United Nations, there has been an increase over the years of aquaculture in both marine and inland waters and also an increase in capture fisheries in both waters (FAO, 2020).

However, two main concerns have arisen regarding sanitary aspects related to fish cultivation and consumption: fish health, especially when raised in captivity (Kousar et al., 2020), and consequently human health of those who consume this protein source (Novoslavskij et al., 2016).

When it comes to fishes that are raised to be commercialized, special attention should be dedicated to water quality and to the quantity of fish confined in each tank. Stress, deficient diet, variations in temperature, dissolved oxygen, pH, and presence of contaminants are factors that can affect fishes' health, making them more vulnerable to diseases (Bhatnagar & Garg, 2000; Su et al., 2020). Among the contaminants are biological ones.

Biological contaminants involve bacteria, fungi, and viruses that can present themselves as pathogens if they are able to cause diseases. Some species are native to natural freshwater habitats and some species are associated with water pollution. In fish, contamination of food and water can lead to digestive tract colonization by pathogens, and due to water exposure skin and gills are the main sites of colonization. Muscle contamination is less common, occurring mainly when the animal's immunological status is already unstable (Guzman et al., 2004; Novoslavskij et al., 2016).

Some of these pathogens can only infect and do damage to fish, but some species can also infect humans (being responsible for zoonoses), causing additional concern (Haenen et al., 2013; Sudheesh et al., 2012). There are also bacteria species, such as the ones from the *Salmonella* genus, that can infect humans but are not biological contaminants originally reported in fish; they were

Bacterial Fish Diseases. DOI: https://doi.org/10.1016/B978-0-323-85624-9.00007-5

introduced through anthropic action resulting from improper handling or contaminated water (Sant'ana, 2012). However, these last two categories will not be the focus of this chapter.

This chapter will address the pathogenesis of bacterial fish diseases that possess the potential to negatively affect the aquaculture field.

9.1.2 Main bacteria species capable of causing fish diseases

Among fish pathogens, bacteria are an important group that threatens their health. Fish species possess defense mechanisms against these organisms that involve physical and chemical barriers. The mucus that covers fish skin contains antimicrobial peptides, enzymes, and lectins, for example, that help to avoid pathogen attachment. Blood also possesses factors that act against bacteria and iron-binding proteins [iron is important to various pathogens' mechanism of action (Magnadottir, 2006)].

Adult fish are in general more resistant than larvae, in which the adaptive immune system is not fully developed yet (but there are strategies aiming to "train" their innate immune system, for example with ß-glucans (Zhang et al., 2019)). Adult fish can use their adaptive immune responses against pathogens, as the complement system can enhance phagocytic activity and/or cause bacterial cell lysis (Holland & Lambris, 2002).

Microbiologists classify bacteria into two groups (gram-positive and gram-negative types) regarding characteristics related to their cell surface accessed through the Gram staining method (Beveridge & Davies, 1983). Bacteria are fixed on microscope slides and exposed to the colorant crystal violet. All bacteria will be stained bluish purple. The mordant substance (iodine) is added to complex with the colorant and destaining is later performed using water and ethanol. Counter staining is performed with a colorant with different color (pinkish) and a wash step finalizes the process. Gram-positive bacteria end up presenting the color from the first colorant, and gram-negative bacteria from the second one (Fig. 9.1). The thick structure of peptidoglycan outside the cell membrane contributes to the retention of the purple crystal violet-iodine complex from the ethanol wash step in gram-positive bacteria (O'Toole, 2016).

In fish diseases caused by bacteria, gram-negative organisms are the main concern. A large number of gram-negative bacteria are related to diseases that cause large losses in fisheries/aquaculture (Table 9.1). Vibriosis is the most prevalent group of fish diseases and is caused by gram-negative bacteria belonging to the genus *Vibrio*. The affected fish may present various symptoms such as lethargy with necrosis of skin and appendages, body malformation, slow growth, internal organ liquefaction, blindness, muscle opacity, and mortality (Ina-Salwany et al., 2019). It is not a local problem; vibriosis affects fish species worldwide and causes great economic loss to the aquaculture industry. The most commonly affected species are saltwater ones, but occasionally it can occur in freshwater fish (Austin & Austin, 1999; Austin & Zhang, 2006; Bergh et al., 2001).

A large number of gram-positive aerobic and anaerobic bacteria are also involved in fish diseases, such as *Lactococcus garvieae*, *Streptococcus agalactiae*, and Streptococcus *iniae* (Evans et al., 2002; Figueiredo et al., 2012; Karami et al., 2019) (Table 9.1).

A pathogenic bacterium possesses a virulence factor, important in causing infection. Gram-negative bacteria, for example, possess lipopolysaccharide (LPS), an important virulence factor, and they also commonly contain quorum sensing systems (signaling cell-to-cell communication systems) regulating virulence gene expression and influencing stress adaptation (e.g., secretion systems), development of biofilm, and production of secondary metabolites (Bruhn et al., 2005; Deep

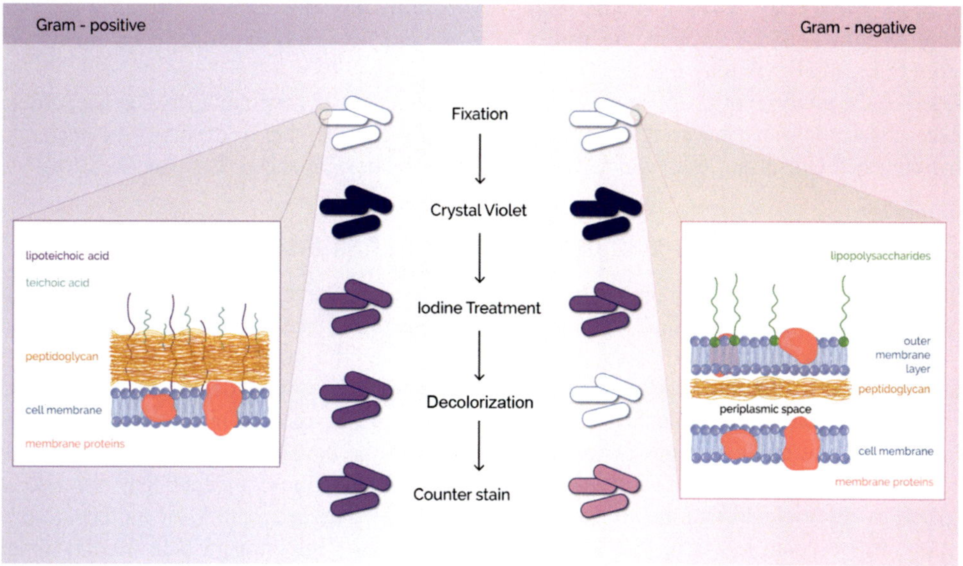

FIGURE 9.1

Gram staining method.

Table 9.1 Classification of main bacterium species that cause fish disease.

Bacterium species	Gram staining	Reference
Vibrio ordalii	Negative	Ruiz, Poblete-Morales et al. (2016)
Vibrio anguillarum	Negative	Gao et al. (2018)
Aliivibrio salmonicida	Negative	Toranzo et al. (2017)
Vibrio vulnificus	Negative	Linkous and Oliver (1999)
Aliivibrio wodanis	Negative	Hjerde et al. (2015)
Moritella viscosa	Negative	Hjerde et al. (2015)
Vibrio alginolyticus	Negative	Fu et al. (2016)
Yersinia enterocolitica	Negative	Fredriksson et al. (2006)
Yersinia ruckeri	Negative	Kumar et al. (2015)
Aeromonas hydrophila	Negative	Xia et al. (2017)
Photobacterium damselae subsp. *damselae*	Negative	Hassanzadeh et al. (2015)
Clostridium botulinum	Positive	Barash and Arnon (2014)
Lactococccus piscium	Positive	Williams et al. (1990)
Lactococcus garvieae	Positive	Karami et al. (2019)
Streptococcus agalactiae	Positive	Evans et al. (2002)
Streptococcus iniae	Positive	Figueiredo et al. (2012)

et al., 2011; Pena et al., 2019). Another common feature from this bacteria group is the type VI secretion system (T6SS); through the use of this protein secretion apparatus, the toxins can be delivered to host cells (Salomon et al., 2015).

However, the presence of a virulence factor coding sequence in a bacterium's DNA is not sufficient to guarantee disease development capacity. The organism must also be able to express these factors under the environmental circumstances to which it is subjected in the host organism (Lages et al., 2019).

9.1.3 Important diagnosis methods for bacterial fish diseases

Different methods are used to diagnose bacterial fish diseases. Some diseases can lead to the development of characteristic signs that facilitate the identification, but some possess signs common to other diseases, making the diagnosis a challenging task. For the latter ones, clinical signs are not sufficient to lead to diagnosis. Histological analysis and analysis that depends upon cell culture (microbiological analysis and biochemical identification) can be used to make the diagnosis possible. However, these methods require a great deal of time to be developed and usually time is a crucial factor when animal is suffering. Thus methods that are culture independent are interesting tools, such as serological (e.g., enzyme-linked immunosorbent assay, ELISA) or molecular analysis (e.g., polymerase chain reaction, PCR) (Austin, 2019). *S. iniae* infection in Nile tilapia, for example, can be detected through PCR (aiming to detect the pathogen's genes) and ELISA (aiming to detect tilapia's contact with the pathogen through specific immunoglobulins produced by the host as a response) (Kayansamruaj et al., 2017).

PCR is a relevant diagnosis method and, although it is not optimal for field use, there are target genes capable of allowing the efficient differentiation even of genetically very similar fish pathogens. *Vibrio anguillarum* and *V. ordalii*, for example, share approximately 99% sequence identity of 16 S rDNA, a common target for species identification. However, in this particular situation there are other target genes that allow secure pathogen identification, such as *rpoN*, *rpoS* (Osorio & Toranzo, 2002), *empA* (Xiao et al., 2009), *amiB* (Hong et al., 2007), and *groEL* (Kim et al., 2010).

New experimental and field environmental analysis methods are continuously being proposed; loop-mediated isothermal amplification is a rapid strategy suitable for both. For example, *V. alginolyticus* can be detected by three primer pairs targeting the *toxR* gene in a real-time PCR device and also in portable fluorescence readers such as ESE's tube scanners (Fu et al., 2016).

9.1.4 Vaccines to prevent fish bacterial diseases

Since fishes are an important part of the diet for a large number of people worldwide, special attention should be paid when choosing strategies to treat fish disease, as humans may consume treated fish organisms as food. Besides problems that can arise from the transfer of antibiotic resistance genes among bacterial species, generating more resistant organisms (Park et al., 2009), the indiscriminate use of antibiotics can select for the antibiotic-resistant bacteria that are enormous threats to human health (Yildirim-Aksoy & Beck, 2017). There are also species for which chemotherapeutic agents have limited efficiency in field conditions (Meyburgh et al., 2017). So, prevention strategies are preferable to treatment strategies.

Since fish infection largely depends on pathogen capacity to adhere to fish skin to later penetrate the organism, there are several surface proteins of fish pathogens that are fundamental to disease development (Dar et al., 2020; Dar, Dar et al., 2016; Dar, Kamili et al., 2016). These are interesting targets for vaccine development against these bacteria, for example, *S. galactiae* (Carey-Ann et al., 2003; Lindahl et al., 2005).

Therefore there are interesting alternatives such as fish vaccination and use of biological substances with antimicrobial activity to treat/prevent fish diseases. Chitosan, obtained from the exoskeleton of crustaceans and insects and also from some microorganisms, can be used as an antimicrobial agent to inhibit warm-water fish bacterial pathogens (Samayanpaulraj et al., 2020; Zheng & Zhu, 2003).

Vaccination can prevent damage to animals as well as economic losses, and decreases the use of antibiotics and other antimicrobials (Kayansamruaj et al., 2020; Moraes et al., 2016). It is considered an effective and long-term prevention measure (Jeong et al., 2020). It can be performed by immersion (immersing the animals in the diluted vaccine solution), oral administration (with the food), injection, and through other routes (Somamoto & Nakanishi, 2020). It is common against lactococcosis (Meyburgh et al., 2017), *Aeromonas hydrophila* (Ford & Thune, 1992; Lamers et al., 1985; Nayak, 2020), streptococcosis (Vos et al., 2015), and vibriosis, as these diseases may cause enormous losses in aquaculture. This strategy allows a diminished incidence and frequency of disease outbreaks, but can also allow the increase of infections caused by opportunistic bacterial species (Austin, 1999; Zorrilla et al., 2003).

Vaccines can be, for example, modified versions of the pathogen: attenuated vaccines usually involve deletions and specific mutations in genes associated with metabolism, virulence, and resistance, thus reducing the pathogenicity of the bacteria (Vos et al., 2015). However, this is not the only option; there are, for example, inactivated and genetically engineered live vaccines (Wang et al., 2020); the potential of DNA vaccines is also of great interest to the aquaculture field (Chang, 2020).

The intraperitoneal vaccine is considered the most effective and safest method and it has fewer side effects due to the frequent use of inactivated vaccines (considered safe because replication does not occur); it causes no damage due to the release of toxin in fish, decreasing the cases of fish mortality (Gomes, Afonso, & Gartner, 2006). Tilapia vaccinated against *S. agalactiae* with intraperitoneal application, inoculated with two doses of vaccine, showed survival rates of more than 95%, and fish inoculated with one dose presented survival rates of around 83%, highlighting the importance of preparing the fish's immune system for possible outbreaks of disease capable of causing high mortality and morbidity (Pretto-Giordano et al., 2010).

Other alternatives for prophylactic immunoprotection have been developed, such as passive immunization. It offers immediate protection (differently from vaccines that require a period of approximately 1 month for the immune response to be established): immunoglobulins are inoculated directly into fish (Fagundes et al., 2016).

It is important to highlight that it is always necessary in order to avoid diseases to provide proper maintenance of environmental conditions for the fish, good nutrition, correct disposal of dead organisms, separation of symptomatic animals, and periodic performance of laboratory tests, to make sure that the production is free of bacterial pathogens (Figueiredo et al., 2007).

9.2 Main pathogenesis of bacterial diseases in fish

Due to its high prevalence, vibriosis is studied by different research groups. The *Vibrio* genus possesses diverse species related to the fish diseases known as vibriosis (Novoslavskij et al., 2016;

Vezzulli et al., 2013). *V. ordalii*, for example, has the potential to result in significant economic losses by causing an atypical vibriosis that results in hemorrhagic septicemia in salmonids. *V. ordalii* virulence mechanisms still are not fully known, but interesting features have already been revealed (Naka et al., 2011). Different strains can resist the killing activity of Atlantic salmon mucus and serum, and it seems that virulence relates at some level with the capacity of a strain to cause hemagglutination, but not with hydrophobicity (Ruiz et al., 2015). Autochthonous microbiota seem to interfere with *V. ordalii* survival: in sterile seawater the bacteria remained infective for 60 days, but 2 days of natural seawater exposure were sufficient to make them nonculturable (Ruiz, Poblete-Morales et al., 2016). *V. ordalii* can use heme and hemoglobin as iron sources and can also produce siderophores (piscibactin can act as a siderophore) to capture iron ions as same as diverse other pathogenic gram-negative bacteria can produce (Ruiz, Balado et al., 2016). This behavior can also influence virulence, since iron is not only important to cell metabolism but also is a regulator of other virulence gene expression (Hood & Skaar, 2012).

V. anguillarum is a species very similar to *V. ordalii* from the genetic aspects (Steinum et al., 2016) and also causes septicemia with hemorrhage, exophthalmia, and corneal opacity in diverse marine fish and freshwater species (Frans et al., 2011), causing economically significant losses (Toranzo et al., 2017). It can survive in the intestinal environment (skin and oral contamination routes are feasible) (Wang, Lauritz et al., 2003) and starvation reduces this pathogen's infectious capacity. However, it also increases chemotactic response, especially to serine presence (Larsen et al., 2004). It presents a higher virulence at 15°C (expressing more toxin T6SS2 and enzymes related to exopolysaccharide synthesis to favor biofilm formation on fish colonizations) and at low iron levels expresses more virulence-related factors, such as lipopolysaccharide, hemolysins, lysozyme, and siderophores (Lages et al., 2019). It can cross intestinal epithelium, achieving the lamina propria, entering the bloodstream to reach diverse organs (Grisez et al., 1996).

V. vulnificus can cause a warm-water vibriosis in fish species and can also cause vibrioses in human beings, as it is an opportunistic pathogen responsible for a large number of seafood-related deaths (Jones & Oliver, 2009). Most data available regarding its pathogenesis is related to human organisms. These bacteria can resist acid environments, breaking amino acids to yield amines and CO_2 (Rhee et al., 2002). The capsule makes it possible for the pathogen to resist opsonization and phagocytosis (Kashimoto et al., 2003). The lipopolysaccharide from the external membrane is an important virulence factor directly related to lethality rate (McPherson et al., 1991). This species also suffers the influence of iron atoms on their capacity to do damage (Alice et al., 2008; Stelma et al., 1992; Wright et al., 1981). Higher iron concentrations favor the survival of more virulent strains (Bogard & Oliver, 2007) and this species can secrete hemolytic factors to help to release iron from hemoglobin (Wright & Morris, 1991). Proteases can also be secreted to contribute to the development of tissue lesions and favor bacteria penetration into tissues (Miyoshi & Shinoda, 1988). Toxins are also produced; RtxA1, for example, is highly similar to the RtxA from *V. cholerae* and very powerful in promoting cell death (Kim et al., 2008).

The *Aliivibrio* genus contains some species previously classified into the *Vibrio* genus and then reclassified due to the development of genome analysis methodologies. *Aliivibrio salmonicida*, formerly known as *V. salmonicida* (Urbanczyk et al., 2007), can cause cold-water vibriosis in fishes. Exophthalmos, swollen vent, pinpoint hemorrhaging along the abdomen and at the bases of fins, and dark color on fish skin are signs (Kent & Poppe, 2002). Research on the virulence factors of this species indicates that the bacterium may initially use them to fight the fish's innate immune system until it collapses and infection occurs (Bjelland et al., 2013). Flagellar proteins that favor the bacteria to access fish cells are produced in larger amount in the presence of the mucus that covers fish skin (Raeder et al., 2007). After entering the animals' blood, bacteremia increases rapidly. The establishment in the intestine is slower, leading scientists to believe that the host intestine would be a reservoir, favoring survival and transmission (Bjelland et al., 2012). This species contains genes of secretable extracellular toxins and enzymes (Bjelland et al., 2013) and a siderophore-based iron sequestration mechanism (Colquhoun & Sørum, 2001).

Aliivibrio wodanis, previously known as *V. wodanis* (Urbanczyk et al., 2007), is often isolated from ulcers and internal organs from fish suffering from winter-ulcer disease (Hjerde et al., 2015), together with *Moritella viscosa* [previously known as *V. viscosus* (Benediktsdóttir et al., 2000)], an etiological agent. Studies involving both species at the same time revealed that *A. wodanis* could adhere to fish surfaces, but no invasion was detected, although it seems that it could secrete toxins (Karlsen et al., 2014). *M. viscosa* extracellular cytotoxic products (ECPs) are cytotoxic and hemolytic to Atlantic salmon tissues and they help to understand the pathology caused by this bacterium (MacKinnon et al., 2019). The presence of *A. wodanis* affects *M. viscosa*'s trascriptome in an inhibitory, not contact-dependent, manner (Hjerde et al., 2015).

V. alginolyticus (previously classified as *V. parahaemolyticus*) (Buchanan & Gibbons, 1974) is an opportunist fish pathogen that can also infect human beings, causing seafood poisoning or fatal extraintestinal infections (Fu et al., 2016). It can release enzymes to the extracellular environment such as casein and also presents hemolytic activity (Sedano et al., 1996).

Another genus of bacteria that causes concern to the aquaculture field due to the fact that it can cause diseases in fish is *Yersinia*. There are various species of *Yersinia* that cause infections in humans and animals, including *Yersinia pestis*, *Y. pseudotuberculosis*, *Y. enterocolitica*, and *Y. ruckeri*. The main ones that affect fish are *Y. enterocolitica* and *Y. ruckeri* (Francis & Auerbuch, 2019). The presence of *Yersinia* in fish reflects not only the condition and safety of aquatic environments but also increases public health concerns regarding the safety of fish used for human consumption (Novoslavskij et al., 2016).

Y. enterocolitica is a gram-negative, rod-shaped, nonsporulating, facultatively anaerobic c-proteobacteria. These bacteria can be classified using biochemical characteristics along with serotyping (Fredriksson et al., 2006; Kot et al., 2010) and it is a classic enteric pathogen able to cause infections in humans and animals. Among the signs of disease are diarrhea, terminal ileitis, intestinal intussusception, mesenteric lymphadenitis, arthritis, and septicemia (Imoto et al., 2012). Its main effects are due to colonization of the intestine, and factors such as temperature and calcium concentration regulate the infection, virulence, and survival of this pathogen (Asadishad et al., 2013; Fàbrega & Vila, 2012). Crucian carp (*Carassius carassius*) is one of the most important freshwater fishes for aquaculture in China once it has several advantageous characteristics, such as good survival rate and disease resistance. However, these fishes can be infected by *Y. enterocolitica*, compromising their production (Xiao et al., 2011). *C. carassius* when infected by *Y. enterocolitica* strain G6029 presents clinical signs such as slowness, extensive dermatorrhagia (mainly around jaw and abdomen), abdominal swelling, and red swelling of the anus. Fishes present a high mortality rate days after infection, due to the effects of septicemia, and the most efficient antibiotics against this pathogen are florfenicol, vibramycin, cephaloridine, ciprofloxacin, streptomycin, and ampicillin (Wang, 2016).

Y. ruckeri, an optional intracellular bacterium capable of surviving within macrophages, can infect several species of fish, such as carp, catfish, sturgeon, perch and burbot, and mainly salmonids (Kumar et al., 2015; Ryckaert et al., 2010; Tobback et al., 2007). This pathogen is classified into four different strains, with the O1 serotype causing the most damage in pisciculture (Romalde et al., 1993, 2003). The most commonly used enteric red mouth (ERM) disease biotype 1 vaccine, developed in the 1980s, is very effective in preventing this disease (Amend et al., 1983). Nevertheless, in the last few years several outbreaks worldwide have been caused by other negative lipase strains belonging to biotype II, able to escape the action of the vaccine. This has motivated the development and subsequent commercialization of a new vaccine, which simultaneously provides protection against both biotypes (Arias et al., 2007; Austin et al., 2003; Calvez et al., 2014; Fouz et al., 2006; Tinsley et al., 2011). The acute phase of the disease is highlighted by high lethality due to the onset of septicemia. Mortality rates are proportionally higher the smaller the age (size) of the fish (Rodgers, 1991). In adult fish, the condition manifests itself as a chronic infection, and the clinical signs are capillary and venous congestion of the brain and ocular vessels. Hemorrhagic lesions are seen in internal organs, muscle, intestine, and on the entire body surface. The reddish hue in the fish's mouth, from which the name of this pathology derives, is a clinical sign observed with low frequency, since it seems to manifest late, in the last stages of the chronic phase of the disease. Mortality due to ERM, in most cases, starts between the fifth and tenth days of contact with a bacterium. It is known that the health status of animals can also have an influence, since stress can accelerate or delay the outbreak of the disease (Avci & Bürüncüoúlu, 2005; Fuhrmann et al., 1983). Several studies have already been carried out in order to discover *Y. ruckeri*'s route of entry into fish and it seems that the gills are the main entrance portal for the bacteria, being the first organ to be infected (Khimmakthong et al., 2013; Ohtani et al., 2014, 2015; Tobback et al., 2009). This event is followed by an intestinal spread, which is the main feature of the pathology (Méndez & Guijarro, 2013; Ohtani et al., 2014). However, there are studies that report the possibility of infection through skin surface on the lateral line, from which the pathogen penetrates the inner cell layers (Khimmakthong et al., 2013). The entry through the digestive system is not discarded; the bacteria may enter through three different ways and spread inside the host, infecting other organs and causing the death of the fish (Guijarro et al., 2018). Through genomic and proteomic analyses, virulence factors of *Y. ruckeri* have been investigated. These analyses showed that iron is an essential nutrient for the bacteria and is difficult to obtain during the infection process; therefore, a limited iron environment induces the expression of a set of genes, some of which are related to pathogenesis and iron obtaining. These studies also demonstrated that in virulent strains, a total of 16 proteins were shown to be upregulated, including expression of hemolysis HtrA regulators, DegQ proteases, an antisigma regulatory factor, different transcriptional regulators belonging to LuxR, AsnC, and PhoP families, Hfq RNA binding protein, invasion protein Inv, and Cu-Zn superoxide dismutase, among others (Kumar et al., 2016, 2017). Genomic analysis of *Y. ruckeri* strains argues that, although they share approximately 75% of their genes, genetic differences have been detected depending on the serotype or virulence of these strains (Cascales et al., 2017). For example, the main component of LPS was exclusively present in strains of serotype O1, a component involved in virulence and also present in different aquatic pathogenic bacteria, such as *V. vulnificus*, *Aeromonas salmonicida*, and *V. fischeri* (Cascales et al., 2017). The virulence of *Y. ruckeri* is certainly multifactorial and depends not only on the presence of certain genes in each strain, but also on environmental factors, particularly temperature. A subideal temperature seems to be important to the expression of some of the genes that encode virulence factors (Guijarro et al., 2018). It is an area of study that deserves attention for a better understanding of infection mechanisms and factors involved.

Lactococcccus piscium is a fish pathogen isolated in 1990 from a diseased salmonid fish (Sakala et al., 2002; Williams et al., 1990). The disease caused by this gram-positive bacteria is not able to cause significant loss in aquaculture and, due to that, it does not receive too much attention. The recent research on this species is related to its potential to inhibit a seafood poisoning human pathogen *Listeria monocytogenes* (Rodrigues et al., 2017), capable of causing listeriosis [that can end up killing the host (Saraoui et al., 2016)]. *Lactococcus garvieae*, however, can infect fish (especially rainbow trout culture) and human beings (Meyburgh et al., 2017). Lactococcosis in fish is characterized by signs such as widespread hemorrhaging, melanosis, anorexia, bilateral exophthalmia, erratic swimming, and anal prolapses (Austin & Austin, 2012; Bekker et al., 2011; Eldar & Ghittino, 1999). This species is sensitive to fluorophenicol (preferentially), lincospectin, gentamicin, and tylosin (Karami et al., 2019).

Clostridium botulinum is another example of bacteria that can infect human beings and fish species. This gram-positive, saprophytic, anaerobic, spore-forming bacillus is commonly associated with foodborne botulism. It is widely distributed in nature and can be found in soil and aquatic environments. Botulism occurs due to the production of botulinum neurotoxin, of which there are eight currently known types (A, B, C, D, E, F, G, and H) (Barash & Arnon, 2014; Smith & Sugiyama, 1988). *Clostridium botulinum* can generate spores highly resistant to heat, light, and drying and they possess the capacity to germinate under anaerobic conditions at temperatures of 15°C–45°C. Toxins may be released from vegetative cells (by cell lysis) or by diffusion through the cell wall within several days of germination. After the release, the single-chain toxins are inactive until cleaved by bacterial or tissue proteases to the active form dichain neurotoxin. All serotypes of botulinum toxin are composed of a heavy chain with a higher molecular weight and a light chain connected by a single disulfide bond (Aston & Beeching, 2013). Toxin serotypes are differently distributed around the globe; for example, serotype E is dominant in sediments from arctic and subarctic regions, while serotype B is more prevalent in soil (Hielm et al., 1998; Huss, 1980; Johannsen, 1963; Leclair et al., 2013; Miller, 1975). In Japan, the presence of botulinum toxin serotypes B, C, and E has been documented (Umeda et al., 2013; Yamakawa & Nakamura, 1992; Yamakawa et al., 1988) and serotype E is also more common in the Great Lakes areas and the Pacific Northwest of North America (Shapiro et al., 1998). In spite of this, in general, outbreaks of environmental botulism are associated with serotypes C, mosaic C/D and E. Food studies have indicated that a high percentage of fish are contaminated with *C. botulinum* type E, consistent with the Type E coverage in the aquatic environment (Hauschild, 1989). Given the outbreaks of botulism reported in the literature in North America, Europe, and Asia, and for the first time in Egypt in 1991, the importance of effective health surveillance is shown (Weber et al., 1993). In human beings, botulism should be suspected if the patient has eaten traditional Alaskan food and if any of the clinical signs such as nausea or vomiting, dysphagia (difficulty swallowing), diplopia (double vision), dilated pupils, and dry throat or mouth are present (Horowitz, 2010). As the disease affects humans, the pathogenesis is well studied regarding the human organism. Botulinum toxins after ingestion are actively transported through the lumen of the intestinal tract via endocytosis and transcytosis. Then, they gain systemic circulation and reach sites of neurotransmission mediated by acetylcholine, causing toxicity (Caya et al., 2004; Simpson, 2004); once inside the neuron, the botulinum toxin inhibits the release of acetylcholine, as it ages in the SNARE proteins (fusion sensitive to soluble *N*-ethylmaleimide protein fixation receptor). Without acetylcholine, muscle contraction does not occur, causing paralysis (Arnon et al., 2001; Shapiro et al., 1998; Simpson, 2004). Type E botulism occurs mainly in arctic regions, since this bacterium has a high resistance to low temperatures. The fishes from these areas may eat or have their skin contaminated with *Clostridium* spores. The heptavalent antitoxin (H-BAT) introduced in 2010 covers all botulism serotypes, including type E, and is being used in Alaska and Canada; there are also trivalent antitoxins and tetravalent ones available in other countries (Dolman & Iida, 1963; Dolman, 1960; Horowitz, 2010).

A. hydrophila is also responsible for significant economic loss in commercial aquaculture and can act as an opportunistic pathogen to humans (Figueras, 2005; Samayanpaulraj et al., 2020). It causes motile aeromonad septicemia in freshwater fish (Chu & Lu, 2005). After defeating the defense provided by the fish skin and mucus, the pathogen triggers the fish innate immune system defense mechanism. Neutrophils increase the lysozyme production and release, and the levels of alkaline phosphatase are also increased. The pathogen induces a reduction of superoxide dismutase and catalase activity, making the host more prone to the damage caused by reactive oxygen species (Xia et al., 2017). *A. hydrophila* produces two types of hemolysin (Wang, Clark et al., 2003), cytotoxic enterotoxins (Sha et al., 2002), metalloproteases, and an iron-capturing system, and the presence of lipase and aerolysin genes makes the strain more virulent (Aoki & Holland, 1985; Nawaz et al., 2010; Nieto & Ellis, 1986). The nuclease product of the *ahn* gene is important for this pathogen's virulence, making it able to evade fish's innate immune clearance in vivo (Ji et al., 2015).

Photobacterium damselae subsp. *damselae* is a marine fish gram-negative emerging pathogen from which highly hemolytic strains possess a characteristic plasmid known as pPHDD1: that encodes two hemolysins (Hassanzadeh et al., 2015; Terceti et al., 2016). It can cause ulcers and septicemia in wild and cultured fish (Rivas et al., 2011), inducing as main clinical signs, besides ulcerative lesions in skin, hemorrhages in mouth, eyes, and muscles (Han et al., 2009). The extracellular products released by this pathogen can increase the host's respiratory frequency and mucus production, cause internal hemorrhage and ascitic liquid accumulation, and also lethargy (Labella et al., 2010). *P. damselae* possesses genes related to important virulence factors, that allow the use of hemin and hemoglobin as iron sources (Río et al., 2005).

Streptococcosis is a group of important bacterial infections that affect fish and have the potential to cause great economic loss. The main species responsible for infecting fishes are *S. agalactiae* and *S. iniae* (Evans et al., 2002; Figueiredo et al., 2012) and the most common signs of disease are meningitis and septicemia (Eldar et al., 1996). *Streptococcus* spp. are catalase negative, gram-positive cocci-shaped bacteria that can be either α-, β-, or nonhemolytic. Several factors can be favorable to the occurrence of outbreaks of streptococcosis, for example, high stocking densities, poor water quality (including high ammonia or nitrite concentrations and low levels of dissolved oxygen), and even stress conditions (Evans et al., 2002; Yanong & Francis-Floyd, 2013). Another factor that can contribute to bacterial proliferation is high temperatures (Amal et al., 2012). Since streptococcosis starts with pathogens being able to maintain contact with fish skin, controlling the external environment is a good option to prevent the spread of the disease (Mishra et al., 2018). Among all species of the *Streptococcus* genus, *S. agalactiae* is the main one found in warm climates (Dar, Dar et al., 2016; Dar, Kamili et al., 2016; Pretto-Giordano et al., 2010).

In the bacterial species, *S. agalactiae,* the main virulence factors already described in the literature are adhesins, such as fibrinogen binding proteins (FbsA and FbsB), fibronectin binding protein pavA, laminin-binding protein, pill proteins, β-hemolysin/cytolysin, CAMP factor, C-α protein and rib surface protein, proteins with immune system evasion functions such as c-β protein, capsule genes, C5a peptidase, and cspA serine protease (Lin et al., 2011). Some studies have isolated the *S. agalactiae* strain from tilapia (*Oreochromis niloticus*) and injected them into the intraperitoneal cavity for further analysis of the pathogenesis caused; the fish presented septicemia and most of the lesions were epicarditis and meningitis (Filho et al., 2009).

Similar to the result of the infection caused in fish by *S. agalactiae*, fish affected by *S. iniae* have meningoencephalitis and septicemia and a high mortality rate is observed. There are reports in the literature of outbreaks of infections caused by *S. iniae* in different regions of the world, including North America, the Middle East, and Asia-Pacific. There are also reports on infections in Brazil (Agnew & Barnes, 2007; Figueiredo et al., 2012; Mian et al., 2009). *S. iniae* is a pathogen that, in addition to infectious potential regarding fish, in many regions of the world is also a zoonotic bacterium (that can infect humans). The first infections in humans were reported in the 1990s, with the majority of cases being reported in people of Asian origin. Common signs are endocarditis, meningitis, arthritis, sepsis, pneumonia, osteomyelitis, and toxic shock. Once infected, these patients can be treated with antimicrobial drugs such as penicillin, ampicillin, amoxicillin, cloxacillin, cefazolin and/or gentamicin, doxycycline, and sulfamethoxazole (Koh et al., 2009; Sun et al., 2007; Weinstein et al., 1997).

9.3 Conclusions

Studies of bacterial diseases that affect fish species demonstrate how diseases are a challenge for producers and researchers from the aquaculture field. This theme raises important aspects that require attention, such as the need to carry out routine monitoring of possible infections, the implementation of protective behaviors, and continuous research, in order to understand the pathologies and propose efficient treatments/prevention protocols, and to better understand the economic losses that can be caused by sick fish, as well as the impacts on human health. Gram-negative and gram-positive bacteria pathogens are threats that should receive attention not only due to their potential to do damage, but also because there are still many gaps in our knowledge of pathogenesis and consequently prophylaxis and treatment.

9.4 Future perspectives

Bacterial fish diseases can be caused by different species of bacteria and they have the potential to cause major economic losses in aquaculture. The bacterial species for which the greatest amount of information on disease-causing processes when infection occurs is available are those capable of becoming zoonoses (thus threatening human health). Fish bacterial pathogens can be gram-positive or gram-negative. For the latter, for example, it is known that the iron transport system consists of a relevant virulence factor commonly present. Therefore it is necessary and important that more studies be performed on the pathogenesis of these diseases caused by bacteria, aiming at greater understanding of the infection process until the onset of clinical signs. These studies are relevant not only for proposing treatments, but also for the development of prevention strategies, such as vaccination.

References

Agnew, W., & Barnes, A. C. (2007). *Streptococcus iniae*: An aquatic pathogen of global veterinary significance and challenging candidate for reliable vaccination. *Veterinary Microbiology*, *122*, 1–15.

Alice, A. F., Naka, H., & Crosa, J. H. (2008). Global gene expression as a function of the iron status of the bacterial cell: Influence of differentially expressed genes in the virulence of the human pathogen Vibrio vulnificus. *Infection and Immunity*, *76*, 4019–4037.

Amal, M. N. A., Zamri-Saad, M., Iftikhar, A. R., SitiZahrah, A., Aziel, S., & Fahmi, S. (2012). An outbreak of *Streptococcus agalactiae* infection in cage-cultured golden pompano, Trachinotus blochii (Lacápède), in Malaysia. *Journal of Fish Diseases*, *35*(11), 849–852.

Amend, D. F., Johnson, K. A., Croy, T. R., & McCarthy, D. H. (1983). Some factors affecting the potency of *Yersinia ruckeri* bacterins. *Journal of Fish Diseases*, *6*, 337–344.

Aoki, T., & Holland, B. I. (1985). The outer membrane proteins of the fish pathogens *Aeromonas hydrophila, Aeromonas salmonicida* and *Edwardsiella tarda*. *FEMS Microbiology Letters*, *27*, 299–305.

Arias, C. R., Olivares-Fuster, O., Hayden, K., Shoemaker, C. A., Grizzle, J. M., & Klesius, P. H. (2007). First report of *Yersinia ruckeri* biotype 2 in the USA. *Journal of Aquatic Animal Health*, *19*, 35–40.

Arnon, S. S., Schechter, R., Inglesby, T. V., Henderson, D. A., Bartlett, J. G., Ascher, M. S., Eitzen, E., Fine, A. D., Hauer, J., Layton, M., Lillibridge, S., Osterholm, M. T., O'Toole, T., Parker, G., Perl, T. M., Russell, P. K., Swerdlow, D. L., & Tonat, K. (2001). Botulism toxin as a biological weapon: Medical and public health management. *JAMA: The Journal of the American Medical Association*, *285*, 1059–1070.

Asadishad, B., Ghoshal, S., & Tufenkji, N. (2013). Role of cold climate and freeze-thaw on the survival, transport, and virulence of *Yersinia enterocolitica*. *Environmental Science and Technology*, *47*(24), 14169–14177.

Aston, S. J., & Beeching, N. J. (2013). *Hunter's tropical medicine and emerging infectious disease* (9th ed.). Saunders.

Austin, B. (1999). Emerging bacterial fish pathogens. *Bulletin – European Association of Fish Pathologists*, *19*, 231–234.

Austin, B. (2019). Methods for the diagnosis of bacterial fish diseases. *Marine Life Science & Technology*, *1*, 41–49.

Austin, B., & Austin, D. A. (1999). *Bacterial fish pathogens: Disease in farmed and wild fish* (3rd ed.). New York, NY: Springer.

Austin, B., & Austin, D. A. (2012). *Bacterial fish pathogens: Diseases of farmed and wild fish* (5th ed.). Chichester: Springer/Praxis Publishing.

Austin, B., & Zhang, X. H. (2006). *Vibrio harveyi*: A significant pathogen of marine vertebrates and invertebrates. *Letters in Applied Microbiology*, *43*(2), 119–124.

Austin, D. A., Robertson, P. A., & Austin, B. (2003). Recovery of a new biogroup of *Yersinia ruckeri* from diseases rainbow trout (*Oncorhynchus mykiss*, Walbaum). *Systematic and Applied Microbiology*, *26*, 127–131.

Avci, H., & Bürüncüoúlu, S. (2005). Pathological findings in rainbow trout (*Oncorhynchus mykiss* Walbaum, 1792) experimentally infected with *Yersinia ruckeri*. *Turkish Journal of Veterinary and Animal Sciences*, *29*, 1321–1328.

Barash, J. R., & Arnon, S. S. (2014). A novel strain of *Clostridium botulinum* that produces type B and type H botulinum toxins. *The Journal of Infectious Diseases*, *209*, 183–191.

Bekker, A., Hugo, C., Albertyn, J., Boucher, C. E., & Bragg, R. R. (2011). Pathogenic Gram-positive cocci in South African rainbow trout, *Oncorhynchus mykiss* (Walbaum). *Journal of Fish Diseases*, *34*, 483–487.

Benediktsdóttir, E., Verdonck, L., Spröer, C., Helgason, S., & Swings, J. (2000). Characterization of *Vibrio viscosus* and *Vibrio wodanis* isolated at different geographical locations: A proposal for reclassification of *Vibrio viscosus* as Moritella viscosa comb. nov. *International Journal of Systematic and Evolutionary Microbiology*, *50*(2), 479–488.

Bergh, O., Nilsen, F., & Samuelsen, O. B. (2001). Diseases, prophylaxis and treatment of the Atlantic halibut *Hippoglossus hippoglossus*: A review. *Diseases of Aquatic Organisms*, *48*(1), 57–74.

Beveridge, T. J., & Davies, J. A. (1983). Cellular responses of *Bacillus subtilis* and *Escherichia coli* to the Gram stain. *Journal of Bacteriology*, *156*, 846–858.

Bhatnagar, A., & Garg, S. K. (2000). Causative factors of fish mortality in still water fish ponds under sub-tropical conditions. *Aquaculture (Amsterdam, Netherlands)*, *1*, 91–96.

Bjelland, A. M., Fauske, A. K., Nguyen, A., Orlien, I. E., Østgaard, I. M., & Sørum, H. (2013). Expression of *Vibrio salmonicida* virulence genes and immune response parameters in experimentally challenged Atlantic salmon (*Salmo salar* L.). *Frontiers in Microbiology*, *4*, 401.

Bjelland, A. M., Johansen, R., Brudal, E., Hansen, H., Winther-Larsen, H. C., & Sørum, H. (2012). *Vibrio salmonicida* pathogenesis analyzed by experimental challenge of Atlantic salmon (*Salmo salar*). *Microbial Pathogenesis*, *52*(1), 77–84.

Bogard, R. W., & Oliver, J. D. (2007). Role of iron in human serum resistance of the clinical and environmental *Vibrio vulnificus* genotypes. *Applied and Environmental Microbiology*, *73*, 7501–7505.

Bruhn, J. B., Dalsgaard, I., Nielsen, K. F., Buchholtz, C., Larsen, J. L., et al. (2005). Quorum sensing signal molecules (acylated homoserine lactones) in Gram negative fish pathogenic bacteria. *Diseases of Aquatic Organisms*, *65*, 43–52.

Buchanan, R. E., & Gibbons, N. E. (1974). *Bergey's manual of determinative bacteriology* (8th ed.). Baltimore: Williams & Wilkins Company.

Calvez, S., Gantelet, H., Blanc, G., Douet, D. G., & Daniel, P. (2014). *Yersinia ruckeri* biotypes 1 and 2 in France: Presence and antibiotic susceptibility. *Diseases of Aquatic Organisms*, *109*, 117–126.

Carey-Ann, D., Burnhama., & Gregory, J. T. (2003). Virulence factors of group B streptococci. *Reviews in Medical Microbiology*, *14*, 109–118.

Cascales, D., Guijarro, J. A., García-Torrico, A. I., & Méndez, J. (2017). Comparative genome analysis reveals important genetic differences among serotype O1 and serotype O2 strains of *Yersina ruckeri* and provides insights into host adaptation and virulence. *Microbiologyopen*, *6*, e00460.

Caya, J. G., Agni, R., & Miller, J. E. (2004). Clostridium botulinum and the clinical laboratorian: A detailed review, including biologic warfare ramifications of botulinum toxin. *Archives of Pathology & Laboratory Medicine*, *128*, 653–662.

Chang, C. J. (2020). Immune sensing of DNA and strategies for fish DNA vaccine development. *Fish and Shellfish Immunology*, *101*(2020), 252–260.

Chu, W. H., & Lu, C. P. (2005). Multiplex PCR assay for the detection of pathogenic *Aeromonas hydrophila*. *Journal of Fish Diseases*, *28*(7), 437–441.

Colquhoun, D. J., & Sørum, H. (2001). Temperature dependent siderophore production in *Vibrio salmonicida*. *Microbial Pathogenesis*, *31*, 213–219.

Dar, G. H., Bhat, R. A., Kamili, A. N., Chishti, M. Z., Qadri, H., Dar, R., & Mehmood, M. A. (2020). *Correlation between pollution trends of freshwater bodies and bacterial disease of fish fauna. Fresh water pollution dynamics and remediation* (pp. 51–67). Singapore: Springer.

Dar, G. H., Dar, S. A., Kamili, A. N., Chishti, M. Z., & Ahmad, F. (2016). Detection and characterization of potentially pathogenic *Aeromonas sobria* isolated from fish *Hypophthalmichthys molitrix* (Cypriniformes: Cyprinidae). *Microbial Pathogenesis*, *91*, 136–140.

Dar, G. H., Kamili, A. N., Chishti, M. Z., Dar, S. A., Tantry, T. A., & Ahmad, F. (2016). Characterization of *Aeromonas sobria* isolated from fish rohu (*Labeo rohita*) collected from polluted pond. *Journal of Bacteriology & Parasitology*, *7*(3), 1–5. Available from https://doi.org/10.4172/2155-9597.1000273.

Deep, A., Chaudhary, U., & Gupta, V. (2011). Quorum sensing and bacterial pathogenicity: From molecules to disease. *Journal of Laboratory Physicians*, *3*(1), 4–11.

Dolman, C. E. (1960). Type E botulism: A hazard of the north. *Arctic*, *13*, 230–256.

Dolman, C. E., & Iida, H. (1963). Type E botulism: Its epidemiology, prevention and specific treatment. *Canadian Journal of Public Health*, *54*, 293–308.

Eldar, A., & Ghittino, C. (1999). *Lactococcus garvieae* and *Strepto coccus iniae* infections in rainbow trout *Oncorhynchus mykiss*:similar, but different diseases. *Diseases of Aquatic Organisms*, *36*, 227–231.

Eldar, A., Ghittino, C., Asanta, L., Bozzetta, E., Goria, M., Prearo, M., & Bercovier, H. (1996). *Enterococcus seriolicida* is a junior synonym of *Lactococcus garvieae*, a causative agent of septicemia and meningoencephalitis in fish. *Current Microbiology*, *32*, 85–88.

Evans, J. J., Klesius, P. H., Gilbert, P. M., Shoemaker, C. A., Al Sarawi, M. A., Landsberg, J., & Al Zenki, S. (2002). Characterization of β-haemolytic group B *Streptococcus agalactiae* in cultured sea bream, Sparusauratus L., and wild mullet, *Liza klunzingeri* (Day), in Kuwait. *Journal of Fish Diseases*, *25*(9), 505–513.

Fàbrega, A., & Vila, J. (2012). *Yersinia enterocolitica*: Pathogenesis, virulence and antimicrobial resistance. *Enfermedades Infecciosas y Microbiologia Clinica*, *30*(1), 24–32.

Fagundes, L. C., Eto, S. F., Marcusso, P. F., Fernandes, D. C., Marinho Neto, F. A., Claudiano, G. S., & Salvador, R. (2016). Passive transfer of hyperimmune serum anti *Streptococcus agalactiae* and its prophylactic effect on Nile tilapia experimentally infected. Arquivo Brasileiro de Medicina Veterinária e Zootecnia. *68*, *2*, 379–386.

FAO. (2020). *The State of World Fisheries and Aquaculture (SOFIA) 2020. Sustainability in action*. Rome. https://doi.org/10.4060/ca9229en.

Figueiredo, H. C. P., et al. (2007). Estreptococose em tilápia do Nilo – parte 2. *Panorama da Aqüicultura*, *17* (104), 42–45.

Figueiredo, H. C. P., Netto, L. N., Leal, C. A. G., Pereira, U. P., & Mian, G. F. (2012). *Streptococcus iniae* outbreaks in Brazilian Nile Tilapia (*Oreochromis niloticus* L.:) farms. *Brazilian Journal of Microbiology*, *43*(2), 576–580.

Figueras, M. J. (2005). Clinical relevance of *Aeromonas* sM503. *Reviews in Medical Microbiology*, *16*, 145–153.

Filho, I. C., Ernst, E. M., Pretto-Giordano, G. L., & Bracarense, F. R. L. A. (2009). Histological findings of experimental *Streptococcus agalactiae* infection in nile tilapias (*Oreochromis niloticus*). *Brazilian Journal of Veterinary Pathology*, *2*(1), 12–15.

Ford, L. A., & Thune, R. L. (1992). Immunization of channel catfish with a crude, acid-extracted preparation of motile aeromonad S-layer protein. *Biomedical Letters*, *47*(188), 355–362.

Fouz, B., Zarza, C., & Amaro, C. (2006). First description of nonmotile *Yersinia ruckeri* serovar I strains causing disease in rainbow trout, *Oncorhynchus mykiis* (Walbaum), cultured in Spain. *Journal of Fish Diseases*, *29*, 339–346.

Francis, M. S., & Auerbuch, V. (2019). Editorial: The pathogenic yersiniae—Advances in the understanding of physiology and virulence, second edition. *Frontiers in Cellular and Infection Microbiology*, *9*, 119.

Frans, I., Michiels, C. W., Bossier, P., Willems, K. A., Lievens, B., & Rediers, H. (2011). *Vibrio anguillarum* as a fish pathogen: Virulence factors, diagnosis and prevention. *Journal of Fish Diseases*, *34*, 643–661.

Fredriksson, A. M., Stolle, A., & Korkeala, H. (2006). Molecular epidemiology of *Yersinia enterocolitica* infections. *FEMS Immunology and Medical Microbiology*, *47*(3), 315–329.

Fu, K., Li, J., Wang, Y., Liu, J., Yan, H., Shi, L., & Zhou, L. (2016). An innovative method for rapid identification and detection of *Vibrio alginolyticus* in different infection models. *Frontiers in Microbiology*, *7*, 651.

Fuhrmann, H., Bohm, K. H., & Schlotfeldt, H. J. (1983). An outbreak of enteric redmouth disease in West Germany. *Journal of Fish Diseases*, *6*, 309–311.

Gao, X., Pi, D., Chen, N., Li, X., Liu, X., Yang, H., Wei, W., & Zhang, X. (2018). Survival, virulent characteristics, and transcriptomic analyses of the pathogenic *Vibrio anguillarum* under starvation stress. *Frontiers in Cellular and Infection Microbiology*, *8*, 389.

Gomes, S., Afonso, A., & Gartner, F. (2006). Vacinação em peixes contra infecções por espécies de Streptococcus e ocaso particular da Lactococcose. *Revista Portuguesa Ciências Veterinária*, *101*, 25–35.

Grisez, L., Chair, M., Sorgeloos, P., & Ollevier, F. (1996). Mode of infection and spread of *Vibrio anguillarum* in turbot *Scophthalmus maximus* larvae after oral challenge through live feed. *Diseases of Aquatic Organisms*, *26*, 181–187.

Guijarro, J. A., García-Torrico, A. I., Cascales, D., & Méndez, J. (2018). The infection process of *Yersinia ruckeri*: Reviewing the pieces of the jigsaw puzzle. *Frontiers in Cellular and Infection Microbiology*, *8*, 218.

Guzman, M. C., Bistoni, M. A., Tamagninii, L. M., & Gonzales, R. D. (2004). Recovery of *Escherichia coli* in fresh water fish, jenynsia multidentata and *Bryconamericus iheringi*. *Water Research*, *38*, 2368–2374.

Haenen, O. L. M., Evans, J. J., & Berthe, F. (2013). Bacterial infections from aquatic species: Potential for and prevention of contact zoonoses. *Revue Scientifique et Technique (International Office of Epizootics)*, *32*(2), 497–507.

Han, J. E., Gomez, D. K., Kim, J. H., Choresca, C. H., Jr, Shin, S. P., Baeck, G. W., & Park, S. C. (2009). Isolation of *Photobacterium damselae* subsp. damselae from Zebra shark *Stegostoma fasciatum*. *Korean Journal of Veterinary Research*, *49*, 35–38.

Hassanzadeh, Y., Bahador, N., & Baseri-Salehi, M. (2015). First time isolation of *Photobacterium damselae* subsp. *damselae* from *Caranx sexfasciatus* in Persian Gulf, Iran. *Iranian Journal of Microbiology*, *7*(3), 178–184.

Hauschild, A. H. W. (1989). Clostridium botulinum. In M. P. Doyle (Ed.), *Foodborne bacterial pathogens*. New York, NY: Marcel Dekker, Inc., p. Ill.

Hielm, S., Hyytia, E., Andersin, A. B., & Korkeala, H. (1998). A high prevalence of *Clostridium botulinum* type E in finnish freshwater and Baltic Sea sediment samples. *Journal of Applied Microbiology*, *84*, 133–137.

Hjerde, E., Karlsen, C., Sørum, H., Parkhill, J., Willassen, N. P., & Thomson, N. R. (2015). Co-cultivation and transcriptome sequencing of two co-existing fish pathogens Moritella viscosa and *Aliivibrio wodanis*. *BMC Genomics*, *16*(1), 447.

Holland, M. C. H., & Lambris, J. D. (2002). The complement system in teleosts. *Fish & Shellfish Immunology*, *12*, 399–420.

Hong, G. E., Kim, D. G., Bae, J. Y., Ahn, S. H., Bai, S. C., & Kong, I. S. (2007). Species-specific PCR detection of the fish pathogen, *Vibrio anguillarum*, using the amiB gene, which encodes *N*-acetylmuramoyl-l-alanine amidase. *FEMS Microbiology Letters*, *269*, 201–206.

Hood, M. I., & Skaar, E. P. (2012). Nutritional immunity: Transition metals at the pathogen-host interface. *Nature Reviews. Microbiology*, *10*, 525–537.

Horowitz, B. Z. (2010). Type E botulism – Review. *Clinical Toxicology*, *48*(2010), 880–895.

Huss, H. H. (1980). Distribution of *Clostridium botulinum*. *Applied and Environmental Microbiology*, *39*, 764–769.

Imoto, A., Murano, M., Hara, A., Narabayashi, K., Ogura, T., Ishida, K., Abe, Y., Masuda, D., Inoue, T., Tokioka, S., Takii, M., Umegaki, E., Egashira, Y., & Higuchi, K. (2012). Adult intussusception caused by *Yersinia enterocolitica* enterocolitis. *Internal Medicine*, *51*(18), 2545–2549.

Ina-Salwany, M. Y., Al-Saari, N., Mohamad, A., Mursidi, F. A., Mohd-Aris, A., Amal, M. N. A., Kasai, H., Mino, S., Sawabe, T., & Zamri-Saad, M. (2019). Vibriosis in fish: A review on disease development and prevention. *Journal of Aquatic Animal Health*, *31*(1), 3–22.

Jeong, K. H., Kim, H. J., & Kim, H. J. (2020). Current status and future directions of fish vaccines employing virus-like particles. *Fish and Shellfish Immunology*, *100*, 49–57.

Ji, Y., Li, J., Qin, Z., Li, A., Gu, Z., Liu, X., Lin, L., & Zhou, Y. (2015). Contribution of nuclease to the pathogenesis of *Aeromonas hydrophila*. *Virulence*, *6*(5), 515–522.

Johannsen, A. (1963). *Clostridium botulinum* in Sweden and adjacent waters. *The Journal of Applied Bacteriology*, *26*, 43–47.

Jones, M. K., & Oliver, J. D. (2009). *Vibrio vulnificus*: Disease and pathogenesis. *Infection and Immunity*, *77* (5), 1723–1733.

Karami, E., Alishahi, M., Molayemraftar, T., Ghorbanpour, M., Tabandeh, M. R., & Mohammadian, T. (2019). Study of pathogenicity and severity of *Lactococcus garvieae* isolated from rainbow trout (*Oncorhynchus mykiss*) farms in Kohkilooieh and Boyerahmad province. *Fisheries and Aquatic Sciences*, *22*, 21.

Karlsen, C., Vanberg, C., Mikkelsen, H., & Sørum, H. (2014). Co-infection of Atlantic salmon (*Salmo salar*), by *Moritella viscosa* and *Aliivibrio wodanis*, development of disease and host colonization. *Veterinary Microbiology*, *171*(1–2), 112–121.

Kashimoto, T., Ueno, S., Hanajima, M., Hayashi, H., Akeda, Y., Miyoshi, S., Hongo, T., Honda, T., & Susa, N. (2003). *Vibrio vulnificus* induces macrophage apoptosis in vitro and in vivo. *Infection and Immunity*, *71*, 533–535.

Kayansamruaj, P., Areechon, N., & Unajak, S. (2020). Development of fish vaccine in Southeast Asia: A challenge for the sustainability of SE Asia aquaculture. *Fish and Shellfish Immunology*, *103*(2020), 73–87.

Kayansamruaj, P., Dong, H. T., Pirarat, N., Nilubol, D., & Rodkhum, C. (2017). Efficacy of α-enolase-based DNA vaccine against pathogenic *Streptococcus iniae* in Nile tilapia (*Oreochromis niloticus*). *Aquaculture (Amsterdam, Netherlands)*, *468*, 102–106.

Kent, M. L., & Poppe, T. T. (2002). Infectious diseases coldwater fish in marine and brackish water. In P. T. K. Woo, D. W. Bruno, & H. S. Lim (Eds.), *Diseases and disorders of finfish in cage culture* (1st ed., pp. 61–105). Wallingford, UK: CAB International.

Khimmakthong, U., Deshmukh, S., Chettri, J. K., Bojesen, A. M., Kania, P. W., Dalsgaard, I., & Buchmann, K. (2013). Tissue specific uptake of inactivated and live *Yersina ruckeri* in rainbow trout (*Oncorhynchus mykiss*): Visualization by immunohistochemistry and in situ hybridization. *Microbial Pathogenesis*, *59–60*, 33–41.

Kim, D. G., Kim, Y. R., Kim, E. Y., Cho, H. M., Ahn, S. H., & Kong, I. S. (2010). Isolation of the groESL cluster from *Vibrio anguillarum* and PCR detection targeting groEL gene. *Fisheries Science*, *76*, 803–810.

Kim, Y. R., Lee, S. E., Kook, H., Yeom, J. A., Na, H. S., Kim, S. Y., Chung, S. S., Choy, H. E., & Rhee, J. H. (2008). *Vibrio vulnificus* RTX toxin kills host cells only after contact of the bacteria with host cells. *Cellular Microbiology*, *10*, 848–862.

Koh, T. H., Sng, L. H., Yuen, S. M., Thomas, C. K., Tan, P. L., Tan, S. H., & Wong, N. S. (2009). Streptococcal cellulitis following preparation of fresh raw seafood. *Zoonoses and Public Health*, *56*, 206–810.

Kot, B., Piechota, M., & Jakubczak, A. (2010). Analysis of occurrence of virulence genes among *Yersinia enterocolitica* isolates belonging to different biotypes and serotypes. *Polish Journal of Veterinary Sciences*, *13*(1), 13–19.

Kousar, R., Shafi, N., Andleeb, S., Ali, N. M., Akhtara, T., & Khalida, S. (2020). Assessment and incidence of fish associated bacterial pathogens at hatcheries of Azad Kashmir, Pakistan. *Brazilian Journal of Biology = Revista Brasleira de Biologia*, *80*(3), 607–614.

Kumar, G., Hummel, K., Ahrens, M., Mennateau-Ledoble, S., Welch, T. J., Eisenacher, M., Razzazi-Fazeli, E., & El-Matbouli, M. (2016). Shotgun proteomic analysis of *Yersinia ruckeri* strains under normal and iron-limited conditions. *Veterinary Research*, *47*, 100–113.

Kumar, G., Hummel, K., Welch, T. J., Razzazi-Fazeli, W., & El-Matbouli, M. (2017). Global proteomic profiling of *Yersinia ruckeri* strains. *Veterinary Research*, *48*, 55–66.

Kumar, G., Menanteau-Ledouble, S., Saleh, M., & El-Matbouli, M. (2015). *Yersinia ruckeri*, the causative agent of enteric redmouth disease in fish. *Veterinary Microbiology*, *46*, 103.

Labella, A., Sanchez-Montes, N., Berbel, C., Aparicio, M., Castro, D., Manchado, M., & Borrego, J. J. (2010). Toxicity of *Photobacterium damselae* subsp. *damselae* strains isolated from new cultured marine fish. *Diseases of Aquatic Organisms*, *92*(1), 31–40.

Lages, M. A., Balado, M., & Lemos, M. L. (2019). The expression of virulence factors in *Vibrio anguillarum* is dually regulated by iron levels and temperature. *Frontiers in Microbiology*, *10*, 2335.

Lamers, C. H. J., Haas, M. J. H., & Muiswinkel, W. B. (1985). The reaction of the immune system of fish to vaccination: Development of immunological memory in carp, *Cyprinus carpio* L., following direct immersion in *Aeromonas hydrophila* bacterin. *Journal of Fish Diseases*, *8*(3), 253–262.

Larsen, M. H., Blackburn, N., Larsen, J. L., & Olsen, J. E. (2004). Influences of temperature, salinity and starvation on the motility and chemotactic response of *Vibrio anguillarum*. *Microbiology (Reading, England)*, *150*, 1283–1290.

Leclair, D., Farber, J. M., Doidge, B., Blanchfield, B., Suppa, S., Pagotto, F., & Austin, J. W. (2013). Distribution of *Clostridium botulinum* type E strains in Nunavik, Northern Quebec, Canada. *Applied and Environmental Microbiology*, *79*, 646–654.

Lin, F. P. Y., Lan, R., Sintchenko, V., Gilbert, G. L., Kong, F., & Coiera, E. (2011). Computational bacterial genome-wide analysis of phylogenetic profiles reveals potential virulence genes of *Streptococcus agalactiae*. *PLoS One*, *6*(4), e17964.

Lindahl, G., Stålhammar-Carlemalm, M., & Areschoug, T. (2005). Surface proteins of *Streptococcus agalactiae* and related proteins in other bacterial pathogens. *Clinical Microbiology Reviews*, *18*, 102–127.

Linkous, D. A., & Oliver, J. D. (1999). Pathogenesis of *Vibrio vulnificus*. *FEMS Microbiology Letters*, *174*(2), 207–214.

MacKinnon, B., Groman, D., Fast, M. D., Manning, A. J., Jones, P., & St-Hilaire, S. (2019). Atlantic salmon challenged with extracellular products from *Moritella viscosa*. *Diseases of Aquatic Organisms*, *133*, 119–126.

Magnadottir, B. (2006). Innate immunity of fish. *Fish & Shellfish Immunology*, *20*, 137–151.

McPherson, V. L., Watts, J. A., Simpson, L. M., & Oliver, J. D. (1991). Physiological effects of the lipopolysaccharide on *Vibrio vulnificus* on mice and rats. *Microbios*, *67*, 141–149.

Méndez, J., & Guijarro, J. A. (2013). In vivo monitoring of *Yersinia ruckeri* in fish tissues: Progression and virulence gene expression. *Environmental Microbiology Reports*, *5*, 179–185.

Meyburgh, C. M., Bragg, R. R., & Boucher, C. E. (2017). *Lactococcus garvieae*: An emerging bacterial pathogen of fish. *Diseases of Aquatic Organisms*, *123*, 67–79.

Mian, G. F., Godoy, D. T., Leal, C. A., Costa, G. M., & Figueiredo, H. C. (2009). Aspects of the natural history and virulence of *S. agalactiae* infection in Nile tilapia. *Veterinary Microbiology*, *136*, 180–183.

Miller, L. G. (1975). Observations on distribution and ecology of *Clostridium botulinum* type E in Alaska. *Canadian Journal of Microbiology*, *21*, 920–926.

Mishra, A. (2018). Current Challenges of Streptococcus Infection and Effective Molecular, Cellular, and Environmental Control Methods in Aquaculture. *Molecules and Cells*, *41*(6), 495–505.

Miyoshi, S., & Shinoda, S. (1988). Role of the protease in the permeability enhancement by *Vibrio vulnificus*. *Microbiology and Immunology*, *32*, 1025–1032.

Moraes, F. R., Salvador, R., & Marcusso, P. F. (2016). Vacina para peixes e uso da mesma.

Naka, H., Dias, G. M., Thompson, C. C., Dubay, C., & Thompson, F. L. (2011). Complete genome sequence of the marine fish pathogen *Vibrio anguillarum* harboring the pJM1 virulence plasmid and genomic comparison with other virulent strains of *V. anguillarum* and *V. ordalii*. *Infection and Immunity*, *79*, 2889–2900.

Nawaz, M., Khan, S. A., Khan, A. A., Sung, K., Tran, Q., Kerdahi, K., & Steele, R. (2010). Detection and characterization of virulence genes and integrons in *Aeromonas veronii* isolated from catfish. *Food Microbiology*, *27*(3), 327–331.

Nayak, S. K. (2020). Current prospects and challenges in fish vaccine development in India with special reference to *Aeromonas hydrophila* vaccine. *Fish and Shellfish Immunology*, *100*(2020), 283–299.

Nieto, T. P., & Ellis, A. E. (1986). Characterization of extracellular metallo- and serineproteases of *Aeromonas hydrophila* (strain B51). *Journal of General Microbiology*, *132*, 1975–1979.

Novoslavskij, A., Terentjeva, M., Eizenberga, I., Valciņa, O., Bartkevičs, V., & Bērziņš, A. (2016). Major foodborne pathogens in fish and fish products: A review. *Annals of Microbiology*, *66*, 1–15.

Ohtani, M., Villumsen, R., Koppang, E. O., & Raida, M. K. (2015). Global 3D imaging of *Yersinia ruckeri* bacterin uptake in rainbow trout fry. *PLoS One*, *10*, e0117263.

Ohtani, M., Villumsen, R., Kragelund, H., & Raida, M. K. (2014). 3D visualization of the initial *Yersinia ruckeri* infection route in rainbow trout (*Oncorhynchus mykiss*) by optical projection tomography. *PLoS One*, *9*, e89672.

Osorio, C., & Toranzo, A. E. (2002). DNA-based diagnostics in sea farming. In M. Fingerman, & R. Nagabhushanam (Eds.), *Recent advances in marine biotechnology, Volume 7: Seafood Safety and Human Health* (pp. 253–310). Plymouth, UK: Science Publishers.

O'Toole, G. A. (2016). Classic spotlight: How the Gram stain works. *Journal of Bacteriology*, *198*, 3128.

Park, Y. K., Nho, S. W., Shin, G. W., Park, S. B., Jang, H. B., Cha, I. S., Ha, M. A., Kim, Y. R., Dalvi, R. S., & Kang, B. J. (2009). Antibiotic susceptibility and resistance of *Streptococcus iniae* and *Streptococcus parauberis* isolated from olive flounder (*Paralichthys olivaceus*). *Veterinary Microbiology*, *136*, 76–81.

Pena, R. T., Blasco, L., Ambroa, A., González-Pedrajo, B., Fernández-García, L., López, M., Bleriot, I., Bou, G., García-Contreras, R., Wood, T. K., & Tomás, M. (2019). Relationship between quorum sensing and secretion systems. *Frontiers in Microbiology*, *10*, 1100.

Pretto-Giordano, L. G., Müller, E. E., Freitas, J. C., & Silva, V. G. (2010). Evaluation on the pathogenesis of *Streptococcus agalactiae* in Nile Tilapia (*Oreochromis niloticus*). *Brazilian Archives of Biology and Technology*, *53*, 87–92.

Raeder, I. L., Paulsen, S. M., Smalås, A. O., & Willassen, N. P. (2007). Effect of skin mucus on the soluble proteome of *Vibrio salmonicida* analysed by 2-D gel electrophoresis and tandem mass spectrometry. *Microbial Pathogenesis*, *42*, 36–45.

Rhee, J. E., Rhee, J. H., Ryu, P. Y., & Choi, S. H. (2002). Identification of the cadBA operon from *Vibrio vulnificus* and its influence on survival to acid stress. *FEMS Microbiology Letters*, *208*, 245–251.

Río, S. J., Osorio, C. R., & Lemos, M. L. (2005). Heme uptake genes in human and fish isolates of Photobacterium damselae: Existence of hutA pseudogenes. *Archives of Microbiology*, *183*(5), 347–358.

Rivas, A. J., Balado, M., Lemos, M. L., & Osorio, C. R. (2011). The *Photobacterium damselae* subsp. *damselae* hemolysins damselysin and HlyA are encoded within a new virulence plasmid. *Infection and Immunity*, *79*(11), 4617–4627.

Rodgers, C. J. (1991). The control of enteric redmouth disease in fish. Maff, fish diseases laboratory, Weymouth, Dorset. *Trouts News*, *12*, 27–30.

Rodrigues, C. S., de Sá, C. V. G. C., & de Melo, C. B. (2017). An overview of *Listeria monocytogenes* contamination in ready to eat meat, dairy and fishery foods. *Ciencia Rural*, *47*(2), 1–11.

Romalde, J. L., Magariños, B., Barja, J. L., & Toranzo, A. E. (1993). Antigenic and molecular characterization of *Yersinia ruckeri* proposal for a new intraspecies classification. *Systematic and Applied Microbiology*, *16*, 411–419.

Romalde, J. L., Planas, E., Sotelo, J. M., & Toranzo, A. E. (2003). First description of *Yersinia ruckeri* serotype O2 in Spain. *Bulletin of the European Association of Fish Pathologists*, *23*, 135–138.

Ruiz, P., Balado, M., Toranzo, A. E., Poblete-Morales, M., Lemos, M. L., & Avendaño-Herrera, R. (2016). Iron assimilation and siderophore production by *Vibrio ordalii* strains isolated from diseased Atlantic salmon *Salmo salar* in Chile. *Diseases of Aquatic Organisms*, *118*, 217–226.

Ruiz, P., Poblete, M., Yáñez, A. J., Irgang, R., Toranzo, A. E., & Avendaño-Herrera, R. (2015). Cell-surface properties of *Vibrio ordalii* strains isolated from Atlantic salmon *Salmo salar* in Chilean farms. *Diseases of Aquatic Organisms*, *113*(1), 9–23.

Ruiz, P., Poblete-Morales, M., Irgang, R., Toranzo, A. E., & Avendaño-Herrera, R. (2016). Survival behaviour and virulence of the fish pathogen *Vibrio ordalii* in seawater microcosms. *Diseases of Aquatic Organisms*, *120*, 27–38.

Ryckaert, J., Bossier, P., D'Herde, K., Diez-Fraile, A., Sorgeloos, P., Hasebrouck, F., & Pasmans, F. (2010). Persistence of *Yersinia ruckeri* in trout macrophages. *Fish & Shellfish Immunology*, *29*, 648–655.

Sakala, R. M., Hayashidani, H., Kato, Y., Kaneuchi, C., & Ogawa, M. (2002). Isolation and characterization of *Lactococcus piscium* strains from vacuum-packaged refrigerated beef. *Journal of Applied Microbiology*, *92* (1), 173–179.

Salomon, D., Klimko, J. A., Trudgian, D. C., Kinch, L. N., Grishin, N. V., Mirzaei, H., & Orth, K. (2015). Type VI secretion system toxins horizontally shared between marine bacteria. *PLoS Pathogens*, *11*(8), e1005128.

Samayanpaulraj, V., Sivaramapillai, M., Palani, S. N., Govindaraj, K., Velu, V., & Ramesh, U. (2020). Identification and characterization of virulent *Aeromonas hydrophila* Ah17 from infected *Channa striata* in river Cauvery and *in vitro* evaluation of shrimp chitosan. *Food Science & Nutrition*, *8*(2), 1272–1283.

Sampantamit, T., Ho, L., Lachat, C., Sutummawong, N., Sorgeloos, P., & Goethals, P. (2020). Aquaculture production and its environmental sustainability in Thailand: Challenges and potential solutions. *Sustainability*, *12*, 1–17.

Sant'ana, A. (2012). Introduction to the special issue: Salmonella in foods: Evolution, strategies and challenges. *Food Research International*, *45*, 451–454.

Saraoui, T., Fall, P. A., Leroi, F., Antignac, J. P., Chéreau, S., & Pilet, M. F. (2016). Inhibition mechanism of *Listeria monocytogenes* by a bioprotective bacteria *Lactococcus piscium* CNCM I-4031 Author links open overlay panel. *Food Microbiology*, *53*, 70–78.

Sedano, J., Zorrilla, I., Moriñigo, M., Balebona, M., Vidaurreta, A., Bordas, M., & Borrego, J. (1996). Microbial origin of the abdominal swelling affecting farmed larvae of gilt-head seabream, *Sparus aurata* L. *Aquaculture Research*, *27*(5), 323–333.

Sha, J., Kozlova, E. V., & Chopra, A. K. (2002). Role of various enterotoxins in *Aeromonas hydrophila*-induced gastroenteritis: Generation of enterotoxin gene-deficient mutants and evaluation of their enterotoxic activity. *Infection and Immunity*, *70*(4), 1924–1935.

Shapiro, R. L., Hatheway, C., & Swerdlow, D. L. (1998). Botulism in the United States: A clinical and epidemiologic review. *Annals of Internal Medicine*, *129*, 221–228.

Simpson, L. L. (2004). Identification of the major steps in botulinum toxin action. *Annual Review of Pharmacology and Toxicology*, *44*, 167–193.

Smith, L. D. S., & Sugiyama, H. (1988). *Botulism: The organism, its toxins, the disease* (2nd ed., p. 139) Springfield, IL: Thomas.

Somamoto, T., & Nakanishi, T. (2020). Mucosal delivery of fish vaccines: Local and systemic immunity following mucosal immunisations. *Fish and Shellfish Immunology*, *99*(2020), 199–207.

Steinum, T. M., Karatas, S., Martinussen, N. T., Meirelles, P. M., Thompson, F. L., & Colquhoun, D. J. (2016). Multilocus sequence analysis of close relatives *Vibrio anguillarum* and *Vibrio ordalii*. *Applied and Environmental Microbiology*, *82*, 5496–5504.

Stelma, G. N., Reyes, A. L., Peter, J. T., Johnson, C. H., & Spaulding, P. L. (1992). Virulence characteristics of clinical and environmental isolates of *Vibrio vulnificus*. *Applied and Environmental Microbiology*, *58*, 2776–2782.

Su, X., Sutarlie, L., & Loh, X. J. (2020). Sensors, biosensors, and analytical technologies for aquaculture water quality. *Research: A Journal of Science and its Applications*, *2020*, 8272705.

Sudheesh, P. S., Al-Ghabshi, A., Al-Mazrooei, N., & Al-Habsi, S. (2012). Comparative pathogenomics of bacteria causing infectious diseases in fish. *International Journal of Evolutionary Biology*, *2012*, 1–16.

Sun, J. R., Yan, J. C., Yeh, C. Y., Lee, S. Y., & Lu, J. J. (2007). Invasive infection with *Streptococcus iniae* in Taiwan. *Journal of Medical Microbiology*, *56*(Pt 9), 1246–1249.

Terceti, M. S., Ogut, H., & Osorio, C. R. (2016). *Photobacterium damselae* subsp. *damselae*, an emerging fish pathogen in the Black Sea: Evidence of a multiclonal origin. *Applied and Environmental Microbiology*, *82* (13), AEM.00781–16.

Tinsley, J. W., Lyndon, A. R., & Austin, B. (2011). Antigenic and crossprotection studies of biotype 1 and biotype 2 isolates of *Yersinia ruckeri* in rainbow trout, *Oncorhynchus mykiss* (Walbaum). *Journal of Applied Microbiology*, *111*, 8–16.

Tobback, E., Decostere, A., Hermans, K., Haesebrouck, F., & Chiers, K. (2007). *Yersinia ruckeri* infections in salmonid fish. *Journal of Fish Diseases*, *30*, 257–268.

Tobback, E., Decostere, A., Hermans, K., Ryckaert, J., Duchateau, L., Haesebrouck, F., & Chiers, K. (2009). Route of entry and tissue distribution of *Yersinia ruckeri* in experimentally infected rainbow trout *Oncorhynchus mykiss*. *Diseases of Aquatic Organisms*, *84*, 219–228.

Toranzo, A. E., Magariños, B., & Avendaño-Herrera, R. (2017). Vibriosis: *Vibrio anguillarum*, *V. ordalii* and *Aliivibrio salmonicida*. In P. T. K. Woo, & R. C. Cipriano (Eds.), *Fish viruses and bacteria: Pathobiology and protection* (pp. 314–333). Wallingford: CABI.

Umeda, K., Wada, T., Kohda, T., & Kozaki, S. (2013). Multi-locus variable number tandem repeat analysis for *Clostridium botulinum* type B isolates in Japan: Comparison with other isolates and genotyping methods. *Infection, Genetics and Evolution*, *16*, 298–304.

Urbanczyk, H., Ast, J. C., Higgins, M. J., Carson, J., & Dunlap, P. V. (2007). Reclassification of *Vibrio fischeri*, *Vibrio logei*, *Vibrio salmonicida* and *Vibrio wodanis* as *Aliivibrio fischeri* gen. nov., comb. nov., *Aliivibrio salmonicida* comb nov. and *Aliivibrio wodanis* comb. nov. *International Journal of Systematic and Evolutionary Microbiology*, *57*, 2823–2829.

Vezzulli, L., Colwell, R. R., & Pruzzo, C. (2013). Ocean warming and spread of pathogenic vibrios in the aquatic environment. *Microbial Ecology*, *65*, 817–825.

Vos, A. F., Dessing, M. C., Lammers, A. J., Porto, A. P., Florquin, S., Boer, Q. J., Beer, R., Terpstra, S., Bootsma, H. J., Hermans, P. W., Veer, C., & Poll, T. (2015). The polysaccharide capsule of *Streptococcus pneumonia* partially impedes MyD88-mediated immunity during pneumonia in mice. *PLoS One*, *550*(2), e0118181, 10.

Wang, G., Clark, C. G., Liu, C., Pucknell, C., Munro, C. K., Kruk, T. M. A. C., Caldeira, R., Woodward, D. L., & Rodgers, F. G. (2003). Detection and characterization of the hemolysin genes in *Aeromonas hydrophila* and *Aeromonas sobria* by multiplex PCR. *Journal of Clinical Microbiology*, *41*(3), 1048–1054.

Wang, L. (2016). *Yersinia enterocolitica* as a cause of septicemia in crucian carp (*Carassius carassius*). *Iranian Journal of Fisheries Sciences*, *15*(1), 402–414.

Wang, Q., Ji, W., & Xu, Z. (2020). Current use and development of fish vaccines in China. *Fish and Shellfish Immunology*, *96*, 223–234.

Wang, S. Y., Lauritz, J., Jass, J., & Milton, D. L. (2003). Role for the major outer-membrane protein from *Vibrio anguillarum* in bile resistance and biofilm formation. *Microbiology (Reading, England)*, *149*, 1061–1071.

Weber, J. T., Hibbs, R. G., Jr, Darwish, A., Mishu, B., Corwin, A. L., Rakha, M., Hatheway, C. L., Sharkawy, S. e., el-Rahim, S. A., & al-Hamd, M. F. (1993). A massive outbreak of type E botulism associated with traditional salted fish in Cairo. *The Journal of Infectious Diseases*, *167*(2), 451–454.

Weinstein, M. R., Litt, M., Kertesz, D. A., Wyper, P., Rose, D., Coulter, M., McGeer, A., Facklam, R., Ostach, C., Willey, B. M., Borczyk, A., & Low, D. E. (1997). Invasive infections due to a fish pathogen, *Streptococcus iniae*. *S. iniae* study group. *The New England Journal of Medicine*, *337*, 589–594.

Williams, A. M., Fryer, J. L., & Collins, M. D. (1990). Lactococcus piscium sp. nov. a new Lactococcus species from salmonid fish. *FEMS Microbiology Letters*, *68*, 109–114.

Wright, A. C., & Morris, J. G., Jr (1991). The extracellular cytolysin of *Vibrio vulnificus*: Inactivation and relationship to virulence in mice. *Infection and Immunity*, *59*, 192–197.

Wright, A. C., Simpson, L. M., & Oliver, J. D. (1981). Role of iron in the pathogenesis of *Vibrio vulnificus* infections. *Infection and Immunity*, *34*, 503–507.

Xia, H., Tang, Y., Lu, F., Luo, Y., Yang, P., Wang, W., Jiang, J., Li, N., Han, Q., Liu, F., & Liu, L. (2017). The effect of *Aeromonas hydrophila* infection on the non-specific immunity of blunt snout bream (*Megalobrama amblycephala*). *Central European Journal of Immunology*, *42*(3), 239–243.

Xiao, J., Zou, T., Chen, Y., Chen, L., Liu, S., Tao, M., Zhang, C., Zhao, R., Zhou, Y., Long, Y., You, C., Yan, J., & Liu, Y. (2011). Coexistence of diploid, triploid and tetraploid crucian carp (*Carassius auratus*) in natural waters. *BMC Genetics*, *12*(1), 20.

Xiao, P., Mo, Z. L., Mao, Y. X., Wang, C. L., Zou, Y. X., & Li, J. (2009). Detection of *Vibrio anguillarum* by PCR amplification of the empA gene. *Journal of Fish Diseases*, *32*, 293–296.

Yamakawa, K., Kamiya, S., Nishida, S., Yoshimura, K., Yu, H., Lu, D. Y., & Nakamura, S. (1988). Distribution of *Clostridium botulinum* in Japan and in Shinkiang district of China. *Microbiology and Immunology*, *32*, 579–587.

Yamakawa, K., & Nakamura, S. (1992). Prevalence of *Clostridium botulinum* type E and coexistence of *C. botulinum* nonproteolytic type B in the river soil of Japan. *Microbiology and Immunology*, *36*, 583–591.

Yanong, R. P. E., & Francis-Floyd, R. (2013). *Streptococcal Infections of Fish*. This document is Circular 57, one of a series of the Fisheries and Aquatic Sciences Department, Florida Cooperative Extension Service, Institute of Food and Agricultural Sciences, University of Florida. Original publication date April 2002. Revised August 2006. Reviewed June 2013. Visit the EDIS website at http://edis.ifas.ufl.edu.

Yildirim-Aksoy, M., & Beck, B. H. (2017). Antimicrobial activity of chitosan and a chitosan oligomer against bacterial pathogens of warmwater fish. *Journal of Applied Microbiology*, *122*(6), 1570–1578. Available from https://doi.org/10.1111/jam.13460.

Zhang, Z., Chi, H., & Dalmo, R. A. (2019). Trained innate immunity of fish is a viable approach in larval aquaculture. *Frontiers in Immunology*, *10*, 42.

Zheng, L. Y., & Zhu, J. F. (2003). Study on antimicrobial activity of chitosan with different molecular weights. *Carbohydrate Polymers*, *54*(4), 527–530.

Zorrilla, I., Moriñigo, M. A., Castro, D., Balebona, M. C., & Borrego, J. J. (2003). Intraspecific characterization of *Vibrio alginolyticus* isolates recovered from cultured fish in Spain. *Journal of Applied Microbiology*, *95*(5), 1106–1116.

CHAPTER

10 Evaluation of the Fish Invasiveness Scoring Kit (FISK v2) for pleco fish or devil fish

Nahum Andrés Medellin-Castillo[1,2], Hilda Guadalupe Cisneros-Ontiveros[3], Candy Carranza-Álvarez[4], Cesar Arturo Ilizaliturri-Hernandez[5], Leticia Guadalupe Yánez-Estrada[6] and Andrea Guadalupe Rodríguez-López[7]

[1]Professor of Graduate Studies and Research Center, Faculty of Engineering, Autonomous University of San Luis Potosi, San Luis Potosi, Mexico [2]Multidisciplinary Graduate Program in Environmental Sciences, Autonomous University of San Luis Potosi, San Luis Potosi, Mexico [3]Student of Graduate Studies and Research Center, Faculty of Engineering, Autonomous University of San Luis Potosi, San Luis Potosi, Mexico [4]Professor of Faculty of Professional Studies Huasteca Zone, Autonomous University of San Luis Potosi, San Luis Potosi, Mexico [5]Professor of Multidisciplinary Graduate Program in Environmental Sciences, Autonomous University of San Luis Potosi, San Luis Potosi, Mexico [6]Professor of Faculty of Medicine, Autonomous University of San Luis Potosi, San Luis Potosi, Mexico [7]Faculty of Medicine, Autonomous University of San Luis Potosi, San Luis Potosi, Mexico

10.1 Introduction

Biodiversity in Mexico and other regions of the world is under constant threat due to anthropogenic activities, such as loss of habitat caused by land-use changes; air, water and soil pollution; overexploitation of natural resources; climate change; and, finally, the introduction of exotic species. The introduction of nonnative species in an ecosystem generates an environmental problem, since they become invasive (plagues), causing a significant loss of native species (Comisión Nacional para el Conocimiento y Uso de la Biodiversidad (CONABIO), 2020). In addition, the manner of accidental or deliberate introduction of exotic species is irresponsible and these species are generating problems, such as predation of native species, more rigorous competition between species, and transmission of diseases or parasites. The magnitude of the damage that the presence of these species can cause to ecosystems, their environmental services, and human, animal, and plant health has not been sufficiently studied, as well as the environmental and social impacts. Likewise, its scale is not appreciated, nor the economic losses caused by the presence of these species (Barba, 2010).

This chapter focuses on the identification of the invasion risk of pleco or devil fish (family Loricariidae) in the Huasteca region of the state of San Luis Potosi (SLP) in Mexico, an invasive species in the United States and Mexico. The common name in aquariums of this species is "glass cleaner" or "fish tank cleaner," but it is also known as pleco fish in the place of origin. This exotic species is native to the Amazon basin in South America. They are also very common in the North American aquarium trade, and it is believed that the introduction of this species was through accidental or intentional escape or release from upstream ornamental fish hatcheries

Bacterial Fish Diseases. DOI: https://doi.org/10.1016/B978-0-323-85624-9.00021-X

(Mendoza-Alfaro et al., 2009). Due to the characteristics of the devil fish, its presence in Mexico has become an environmental problem because its morphology and physiology make its establishment rapid and very successful, especially in ecosystems similar to its origin.

Since their first sighting in Mexico in 1995 in the Mezcala River, in the Balsas River basin, devil fish have been a significant threat to traditional and commercial fisheries. Possible adverse economic effects include losses of indirect income derived from fishing and losses in the related fish processing and commercialization industries. Also, there is an impact on the economy of families and communities that depend on fishermen's income. Currently, in the Huasteca, the ecological and economic impacts generated by the presence of this species have been identified as unfavorable. Fishers have suffered economic losses due to the displacement of tilapia, catfish, and silverfish, which are fish of commercial value in the area's communities.

Protecting biodiversity, economy, and health from the negative impacts of invasive alien species requires setting priorities and taking preventive actions to reduce intentional introduction of these species. Invasion risk analysis is a tool for identifying risks of nonnative species, which allows the estimation of the probability of being an invasive species and, therefore, susceptible to detailed risk analysis ((Baptiste et al., 2010; Dar et al., 2020; Dar, Dar et al., 2016; Dar, Kamili et al., 2016).

In this study, the use of the Fish Invasiveness Scoring Kit (FISK) was proposed as an invasion detection tool for freshwater fish. This tool is easy to obtain and evaluates the species efficiently through biogeographic, ecological, and climatic factors over the area to be studied. Similarly, the need for the use of risk analysis tools for exotic aquatic species was established to determine the risk of invasion and potential impacts caused by the invasion in the Huasteca.

10.2 Invasive risk analysis

A risk analysis assessment is a tool that allows decisions about actions related to the issue of species introduction and evaluates the potential for establishment, impact, and control that an introduced species may have, related to biodiversity, economy, culture, or human health. This type of tool mainly identifies the probability that the species will become invasive and determines the potential impact that they can have if they become invasive and the viability of control and eradication (Baptiste et al., 2010; Dar, Dar et al., 2016; Dar, Kamili et al., 2016).

González-Martínez et al. (2017) describe the fact that risk analyses represent the first point of control or filter for the prevention of the introduction of species that can be potentially harmful; they constitute the mechanism for the classification and categorization of the risk of species entry, to prioritize actions for the prevention, management, and control of exotic species in a specific country or region.

10.2.1 Fish invasiveness scoring kit

FISK is an adapted version of the original Australian model for Weed Risk Assessment for freshwater fish by Pheloung et al. (1999). FISK was developed by Copp et al. (2005a) and is based on programming code in Excel's Visual Basic for Applications (VBA). FISK has been used to detect the potential invasiveness of nonnative freshwater fish and has become a popular model for risk identification, with applications worldwide (Copp et al., 2005b).

The assessment method is semiquantitative and provides a scoring framework for a species' biogeographic, historical, biological, and ecological information. Higher scores indicate a high risk, and threshold values are set to classify species as low, medium, or high risk. FISK was explicitly designed to comply with international standards, such as the World Trade Organization Agreement on the Application of Sanitary and Phytosanitary Measures and the Convention on Biological Diversity, and has been incorporated as a detection mechanism under the European Nonnative Species in Aquaculture Risk Assessment Scheme (Copp et al., 2008). Its first application and the calibration of FISK_v1 (version 1) occurred in England and Wales by Copp et al. (2009).

FISK is an additive spreadsheet model available on the internet at the UK Centre for Environment Fisheries and Aquaculture Science (CEFAS) website at https://www.cefas.co.uk/services/research-advice-and-consultancy/non-native-species/decision-support-tools-for-the-identification-and-management-of-invasive-non-native-aquatic-species/ and is described in Copp et al. (2005b).

10.3 Zone of study

San Luis Potosi is divided into four geographical zones: Altiplano, Center, Media, and Huasteca. The zone of study of this work corresponds to the Huasteca, which is located to the east of the state. Fig. 10.1 shows the zones of the state, where the green shaded area represents our study zone. This zone includes 19 of the 58 municipalities of SLP, with a territorial area of approximately 12,900 km^2, which corresponds to 17.3% of the total area of the state (Instituto Nacional de Estadística y Geografía (INEGI), 2017).

The Huasteca is located in the hydrological region Panuco No. 26, which is divided into two portions: Upper Panuco and Lower Panuco. This region is located in the lower portion, where a significant number of runoffs are generated, affluents that in some way are of great importance for the Panuco River. This river has its origin in the State of Mexico; it enters in the Sierra Madre Oriental on an irregular topography, which is more notable as the current descends, until the confluence of the Tempoal and Tampaon rivers. From there, it receives the name of the Panuco River, and continues with that designation until its mouth in the Gulf of Mexico, downstream from the city of Tampico, Tamaulipas (INEGI & Cadena, 1985). Its relief is slightly undulated, with altitudes that vary approximately between 50 and 3000 m above sea level. Four basins subdivide the hydrological region No. 26: Panuco River, Tamuin R., Tamesi R., and Moctezuma R. In Table 10.1, the subbasins that are located over the study zone are shown. The set of subbasins comprises an area of 28,162 of 88,811.5 km^2 of the Pánuco hydrological region (INEGI, 2017). Fig. 10.2 shows a map of the subbasins, which includes the study area and their surface currents.

On the other hand, the type of climate in the Huasteca is a tropical wet and dry (Aw) and tropical monsoon climate (Am) (Comisión Nacional para el Conocimiento y Uso de la Biodiversidad (CONABIO) & García, 1998); its annual average temperature is 24°C. The annual precipitation interval is between 290 and 3196 mm (CICESE, 2013).

The predominant geomorphological units are the karst system and the folding mountain (Monroy & Calvillo, 1997).

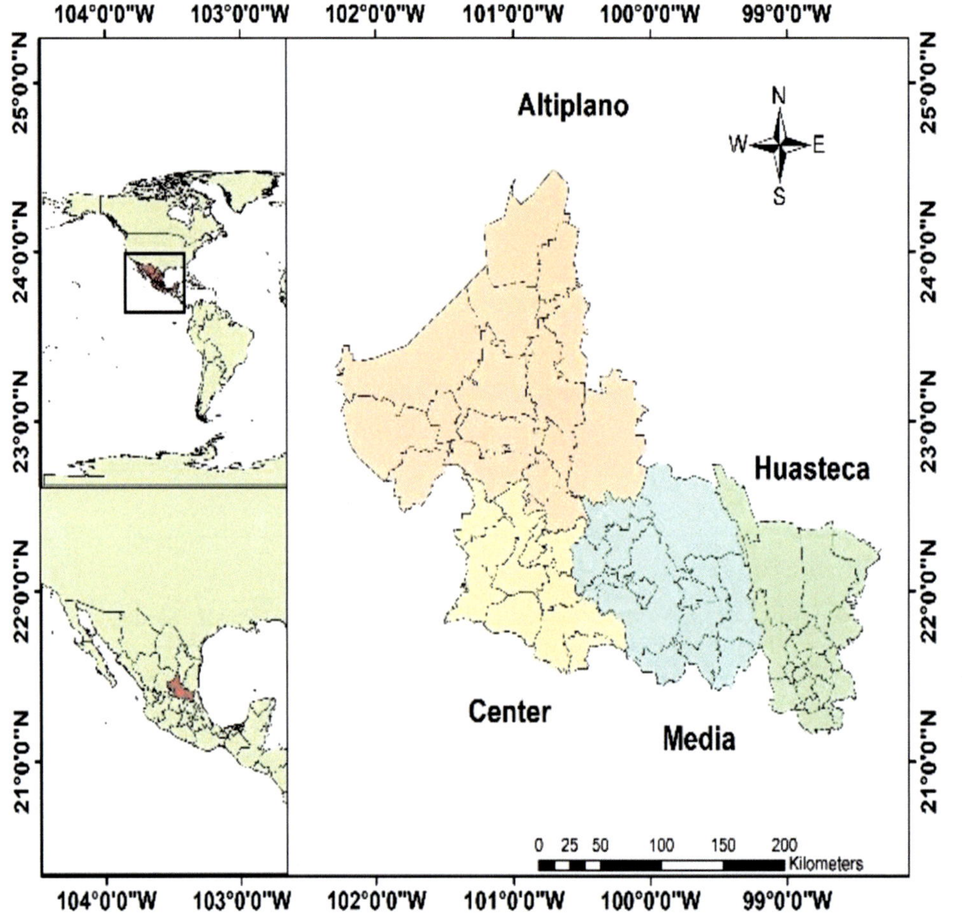

FIGURE 10.1

Classification of the areas of the State of San Luis Potosi, Mexico.

The primary use of soil and vegetation is agriculture, forests, and vegetation (CONABIO & INEGI, 2016).

One of the main economic activities of the Huasteca is the development of livestock and extensive agriculture. The main crops are sugar cane, coffee, and fruits such as citrus and mango (Cabrera & Betancourt, 2002).

The vegetation included in the Huasteca is the sub-evergreen medium forest, deciduous forest, tropical oak forest, cold temperate forest, sub-montane scrub, and secondary vegetation (Puig, 1991).

In this area, many animal inhabitants native to the location, particularly mammals, are in danger of extinction. There are also turtles, iguanas, lizards, snakes, and poisonous snakes (Monroy & Calvillo, 1997).

Table 10.1 Subbasins in the Huasteca, San Luis Potosi, Mexico.

Hydrological region	Basin	Subbasin	Identification
Panuco (RH26)	Panuco River (A)	Panuco R.	RH26Aa
	Tamesi R. (B)	Tamesi R.	RH26Ba
	Tamuin R. (C)	Tamuin or Tampaon R.	RH26Ca
		Valles R.	RH26Cb
		Puerco R.	RH26Cc
		Mesillas R.	RH26Cd
		Naranjos R.	RH26Ce
		Gallinas R.	RH26Cg
		Subterranean Drainage	RH26Ck
	Moctezuma R. (D)	Moctezuma R.	RH26Da
		Axtla R.	RH26Db
		Amajac R	RH26Ds
		San Pedro R.	RH26Dz

The Huasteca is a markedly rural region that concentrates 94% of the state's indigenous population, highlighting two linguistic-cultural groups: Nahuas and Teenek (García-Marmolejo, 2013). This area contains a great wealth of natural resources; however, it has the highest average marginal poverty rate at the state level. The population is mainly dedicated to productive activities in the primary sector (García-Marmolejo, 2013; INEGI, 2007, 2017).

10.4 Pleco fish or devil fish (Loricariidae)

10.4.1 Taxonomic category

According to the Fish Base database (Fish, 2020), 957 species of the family Loricariidae have been identified, of which 148 are species of the genus Hypostomus and 16 are species of Pterygoplichthys. Table 10.2 shows the taxonomy of the family Loricariidae, to which the devil fish belongs. The scientific name of the species and the author who discovered the taxonomic branch are listed.

The identification of the species is complicated, and its taxonomic identification is confusing. Currently, in Mexico, the species *Hypostomus spp.* and *Pterygoplichthys spp.* have been identified. However, there is the possibility that there are still unidentified species; in addition, there is a high probability that these individuals can hybridize (Mendoza-Alfaro et al., 2007).

10.4.2 Native and current distribution

The devil fish is native to South America, in the Amazon River basin of Brazil and Peru (Amador-Del Ángel and Wakida-Kusunoki, 2014; Amador-Del Ángel et al., 2016; Mendoza-Alfaro et al., 2009). Fig. 10.3 shows a map with the native locations in South America (Global Biodiversity Information Facility (GBIF) Loricariidae, 2020).

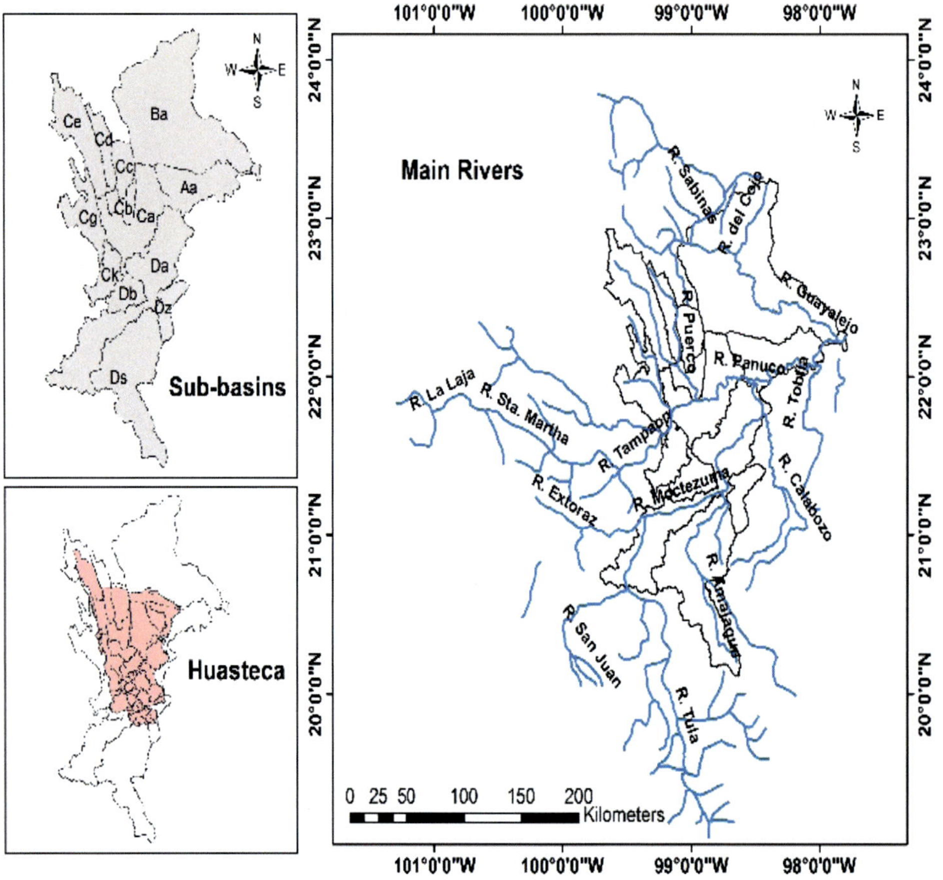

FIGURE 10.2

Subbasins in Huasteca, San Luis Potosi, Mexico.

Current world distribution of this species of the family Loricariidae is shown in Fig. 10.4, where it is observed that these species are found worldwide, but predominantly in America.

Pleco fishes were expanding around the world in the mid-20th century, the first invasion being in the natural waters of North and Central America and later in the islands of the Pacific Ocean (Hoover et al., 2004). Species of the genus Pterygoplichthys have become established in Hawaii, Mexico, and the United States, as well as in Malaysia, Indonesia, Taiwan, and Singapore (Ayala-Pérez et al., 2014; Nico & Martin, 2001).

10.4.3 Description of the species

Fish of the family Loricariidae in their adult life stage have a pigmentation with spots similar to those of a leopard, as shown in Fig. 10.5, which includes various photographs of species in the

Table 10.2 Devil fish taxonomy.

Taxonomic branch	Scientific name	Species authors
Kingdom	Animalia,	Linnaeus (1758)
Phylum	Chordata,	Bateson (1885)
Class	Osterchthyes,	Huxley (1880)
Order	Siluriformes,	Cuvier (1815)
Family	Loricariidae,	Rafinesque (1815)
Species	*Pterygoplichthys pardalis; Hypostomus plecostomus; Pterygoplichthys disjunctivus; Pterygoplichthys anisitsi; Pterygoplichthys multiradiatus*	Castlenau (1855), Linnaeus (1758), Weber (1991), Eigenmann and Kennedy (1903), Hancock (1828)

family Loricariidae, showing the similarities in the pigmentation of their skin. The flexible bony plates and ventral sucking mouth distinguish loricariids from native catfish, which have a terminal mouth and are missing bony plates and a spine (Hoover et al., 2004). The size depends on the species; it has been reported by Hoover et al. (2004) that frequently its size ranges from 14 to 50 cm and it can grow more than 35 cm in the first 2 years (Nico & Martin, 2001). Fish of the family Loricariidae are found in the demersal zone of freshwater bodies, in a pH range of 5.5–8.5 in a tropical climate with temperatures of 20°C–30°C (Amador-Del Ángel et al., 2014; Mendoza-Alfaro et al., 2009).

It has been reported that these fish prefer rocky habitats. Some prefer sandy, shallow lagoons where woody debris abounds, shallow streams in the jungle, or deep regions of larger rivers. Smaller fish are usually collected only in tributary streams, while larger fish are generally found in the mainstream (Cano-Salgado, 2011; Liang et al., 2005; Nico & Martin, 2001). This family of fish can withstand extreme concentrations in the habitat, as they are very tolerant of polluted water and can adapt without difficulty to varying water quality conditions, although they may be intolerant of low water temperatures (Nico & Martin, 2001). They can survive up to 30 hours out of the water and exhibit a high tolerance to cardiac hypoxia (Amador-Del Ángel et al., 2014; Mendoza-Alfaro et al., 2009; Nico & Martin, 2001).

They have moderately high fertility, as females produce 500–3000 eggs (Mendoza-Alfaro et al., 2009). Sex differentiation, at first sight, has been very complicated. One of the most accurate ways to identify the sex of the fish is to extract the eggs from pregnant females at spawning times; measuring plasma vitellogenin may also be useful if a laboratory is available (Liang et al., 2005; Mendoza-Alfaro et al., 2009).

These fish feed mainly on organic matter from the bottom of water bodies; they can also consume worms, benthic insect larvae, and fish eggs; most of their diet is composed of detritus, algae, and vegetal matter (Amador-Del Ángel et al., 2014; Ayala-Pérez et al., 2014; Hoover et al., 2004; Mendoza-Alfaro et al., 2007, 2009, 2011; Nico & Martin, 2001).

The behavior of this species is territorial since they can be aggressive in defense of their territory and competitive when obtaining food. It has been observed that when resources are less limited at high population densities, such behavior is reduced (Mendoza-Alfaro et al., 2009).

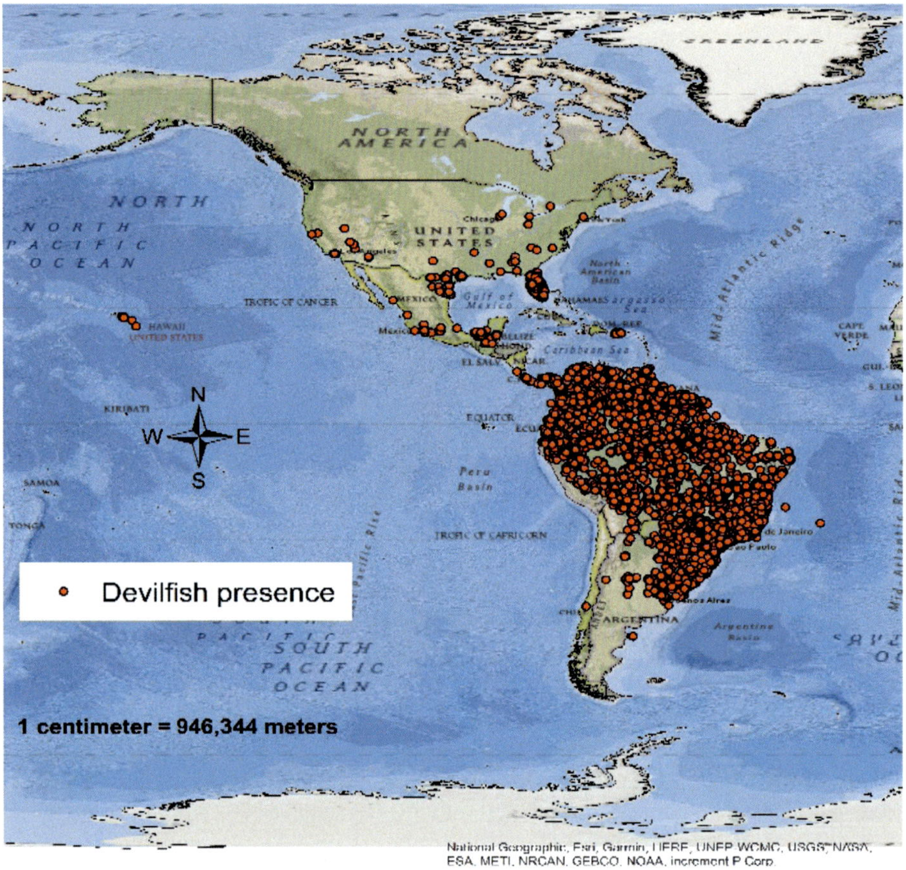

FIGURE 10.3

Devil fish native distribution (GBIF Loricariidae, 2020).

Natural predators include the crocodile *Crocodylus moreletti*, the cormorant *Phalacrocorax brasilianus*, the osprey *Pandion haliaetus*, the otter *Lutra longicaudis annectens*, larger carnivorous fish such as the sea bass *Centropomus undecimalis*, and the shad *Megalops atlacticus* (Wakida Kusunoki & Toro-Ramírez, 2016).

10.4.4 Devil fish in Mexico and the Huasteca potosina

In Mexico, large aquaculture producers are considered responsible for the introduction of this species since, due to carelessness or intentionality, the fish could have been thrown into bodies of water. These fish arrived at the Balsas River where, without natural enemies, they became a dense population that caused the displacement of native species with significant commercial value (Mendoza-Alfaro et al., 2009). The first report of sightings of this fish in Mexico occurred in 1995

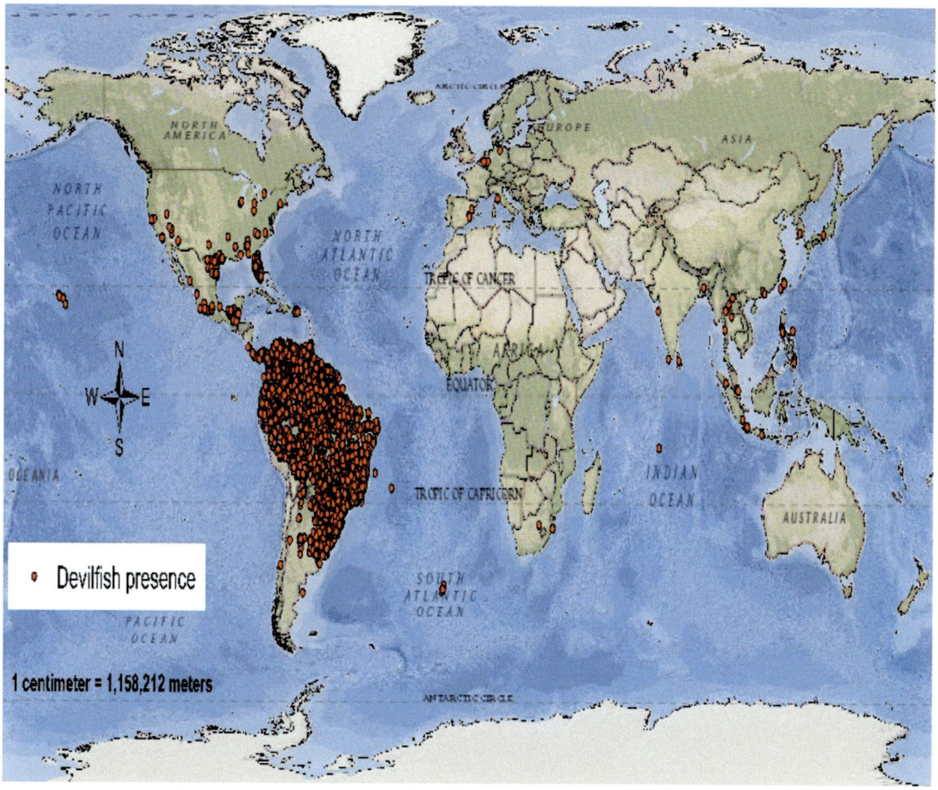

FIGURE 10.4

Present distribution of fishes of the family Loricariidae in the world (GBIF Loricariidae, 2020).

in the Mezcala River, in the Balsas River basin (Guzmán & Barragán, 1997). Subsequently, its presence was recorded at the Infiernillo Dam , Michoacan, Chiapas, and Tabasco. Fig. 10.6 shows the map of the devil fish's current distribution in Mexico, which was created in this study based on the collection of reports from fishers in the study area, literature, and journalistic notes.

Also, Fig. 10.7 shows the distribution of devil fish in the Huasteca. The points of presence were confirmed through sampling conducted in the study area, where interviews were conducted with fishers, presidents of fishing cooperatives, and families affected by the devil fish's presence. In addition, a series of captures were made in water bodies with the most abundance of devil fish.

Fig. 10.8 shows different images of devil fish from different locations in the Huasteca, where there are similarities in the characteristics and patterns in their physical appearance to those shown in Fig. 10.5.

10.4.5 Environmental and socioeconomic effects

Some of the documented effects caused by the introduction of localized fish are problems with siltation, shoreline instability, and erosion in reservoirs and channels caused by nests built by males

FIGURE 10.5

Images of fishes of the family Loricariidae (Fish Base, 2020).

to care for their young (Hoover et al., 2004; Nico & Martin, 2001), as well as the alteration of food chain dynamics, the displacement of native species, the destruction of fishing gear, and consequently a decrease in commercial fish catches (Amador-Del Ángel et al., 2016; Mendoza-Alfaro et al., 2009, 2011; Wakida Kusunoki et al., 2007). The displacement of native species by devil fish is due to competition with other species that feed on algae (Nico & Martin, 2001). This species' morphology is a consumption risk for other natural predators such as pelicans, so it is not easy to control naturally in its habitat.

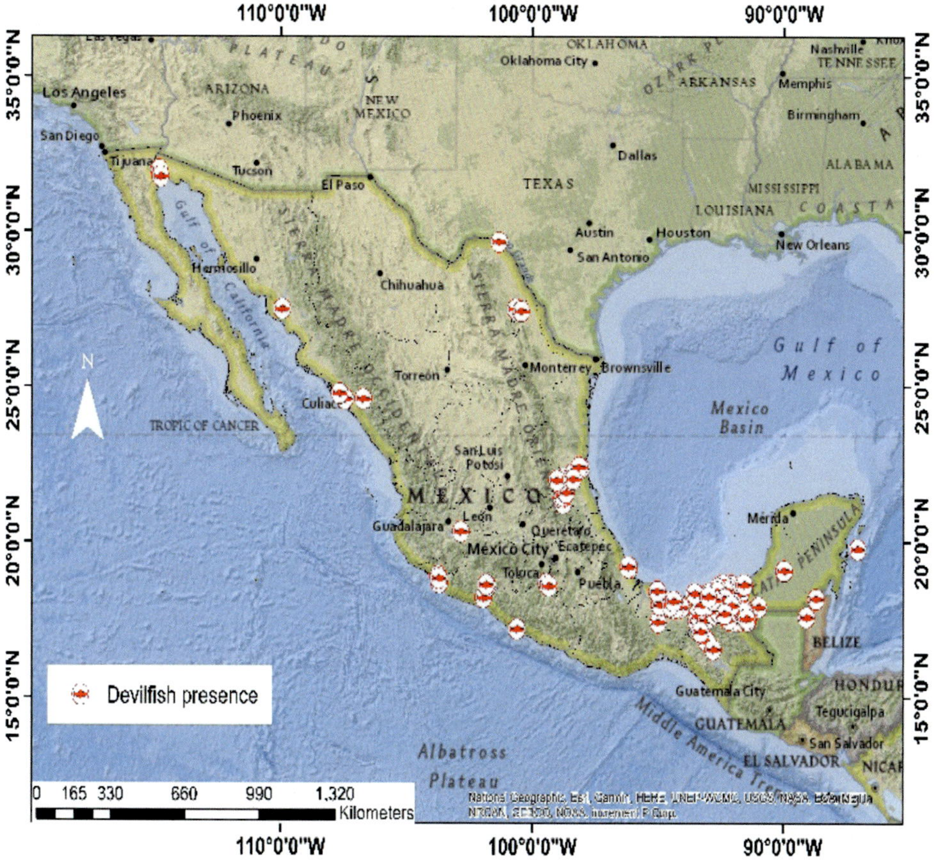

FIGURE 10.6

Devil fish distribution in Mexico.

At the Infiernillo Dam, Michoacan, where the devil fish was first reported as an invasive species that generated an environmental problem, an economic study was conducted of the losses caused by the devil fish since its establishment, where it was reported that, in 2007, 70% and 80% of the tilapia catch were replaced by at least three devil fish species and some probable hybrids. This loss reached an approximate amount of 36 million Mexican pesos (1.65 million USD) per year, along with a high social cost, by leaving 3600 fishers unemployed, which, together with the processors and their families, added up to a total of 46,000 people (Mendoza-Alfaro et al., 2007).

Amador-Del Ángel et al. (2016), Ayala-Pérez et al. (2014), González-Martínez et al. (2017), and Velázquez-Velázquez et al. (2013) reported negative environmental, economic, and social impacts such as monetary losses due to the effects on the fishing market, a decrease in native

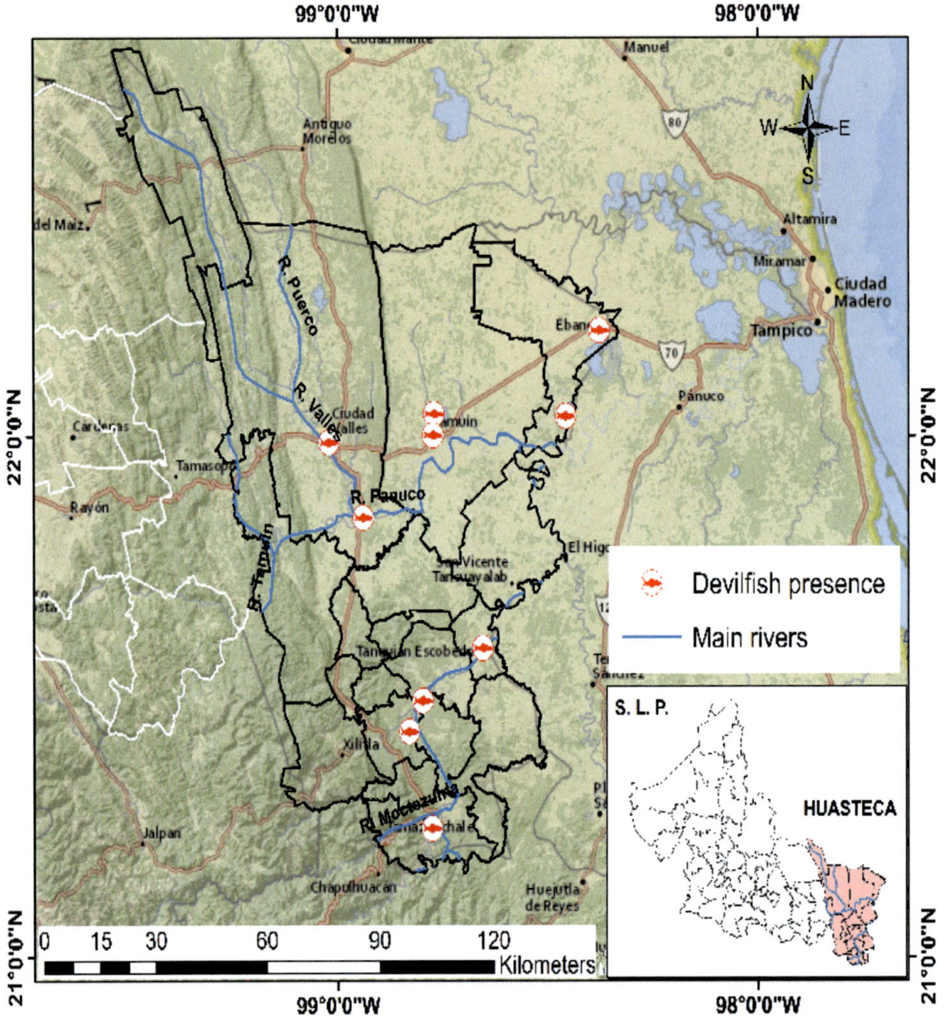

FIGURE 10.7

Devil fish distribution in the Huasteca.

species, loss of fishing equipment such as fishing nets, and an increase in the number of hours worked in fishing in Mexico.

The health impact of devil fish is due to their being discarded and abandoned on the banks of bodies of water by fishers, in addition to causing hand injuries to fishers when they try to remove the fish from the nets. They consider it a waste of time to bury and burn the fish, so their abandonment becomes a focus of infection adjacent to the bodies of water and communities in the area (Amador-Del Ángel et al., 2014; Domínguez-Lemus, 2018; Mendoza-Alfaro et al., 2007).

FIGURE 10.8

Images of devil fish from the Huasteca.

10.5 Evaluation of the fish Invasiveness Screening Kit (FISK v2)

10.5.1 Methodology

From October 2018 to August 2019, visits were made to the main rivers and lagoons of the Huasteca, where the objective was to identify devil fish presence through their capture, testimonies of fishers, or to observe their bone remains or nests. Table 10.3 shows the location of the sites sampled, with the geographic coordinates and municipalities of water bodies given.

Fig. 10.9A shows the main window of the FISK test. This test consists of an Excel sheet with programming in VBA. In this program, a new entry ("New species") was generated in the species

Table 10.3 Geographic location of the sites sampled in the Huasteca.

Municipality	Water body	Geographical coordinates Latitude	Longitude
Tamazunchale	Amajac River	21°14′46″	98°46′33″
Matlapa	Tancuilín River	21°21′25.3″	98°51′51.2″
Axtla de Terrazas	Axtla River	21°26′02″	98°52′39″
Axtla de Terrazas	Moctezuma River	21°25′55.89″	98°49′46.80″
Tampamolon Corona	Moctezuma and Claro Rivers	21°29′28.67″	98°47′51.86″
Tanquian de Escobedo	Moctezuma River	21°35′32.94″	98°39′22.09″
Tamuin	Patitos Lagoon	22°2′30.7″	98°46′17.6″
Tamuin	Tampaon River	22°00′04.17″	98°46′25.27″
Ebano	Plan de Iguala Lagoon	22°02′16.76″	98°27′28.3″
Veracruz	Panuco River	22°02′12.24″	98°24′0.0″
Cd. Valles	Valles River	21°59′12.81″	99°01′15.64″
Tamasopo	Puente de Dios River	21°55′48.48″	99°24′59.54″
Ebano	Plan de Iguala Lagoon	22°02′17.9988″	98°27′29.0016″
Ebano	Marland Lagoon	22°12′12.9996″	98°22′39″
Cd. Valles	Tampaon-Valles, Pujal Coy River	21°50′33.7992″	98°56′23.0028″
Veracruz	Tamiahua Lagoon	22°06′23.9976″	97°47′00.0024″
Tamaulipas	Playa virgen, Tampico Alto	22°06′07.992″	97°54′30.0240″
Cd. Valles	Las Lajillas Damn	22°41′51″	99°2′37.597″
El Naranjo	El Salto River	22°21′45.454″	99°15′58.327″

list for the family Loricariidae (see Fig. 10.9B), and then 49 sequential questions were answered (see Table 10.4), each of which requires a confidence level of certainty and a justification. As shown in Fig. 10.9C, the questions portal has three requirements (question, certainty, and justification) needed in each answer of the FISK questionnaire. Each question requires a "yes/no" answer accordingly, or "do not know" when the information is not available, and a level of certainty. For each FISK answer, a certainty score is assigned as (1 = very uncertain; 2 = mostly uncertain; 3 = mostly certain; 4 = very certain), and from these values, a certainty factor (CF) is defined as each question requires a response "yes/no" accordingly or "don't know" when information is not available and a level of certainty. As each response of FISK for a given species is allocated a certainty score (1 = very uncertain; 2 = mostly uncertain; 3 = mostly certain; 4 = very certain), the certainty factor (CF) was computed as:

$$CF = \sum \frac{CQ_i}{4 \times 49} (i = 1, \ldots, 49)$$

where CQ_i is the certainty for question i, 4 is the maximum achievable value for certainty (i.e., "very safe"), and 49 is the total number of questions that comprise the FISK tool. Therefore, the CF value can vary from a minimum of 0.25 (when all 49 questions have a certainty equal to 1) to a maximum of 1 (that is, when all 49 questions have a certainty equal to 4) (Almeida et al., 2013; Lawson et al., 2015; Simonović et al., 2013).

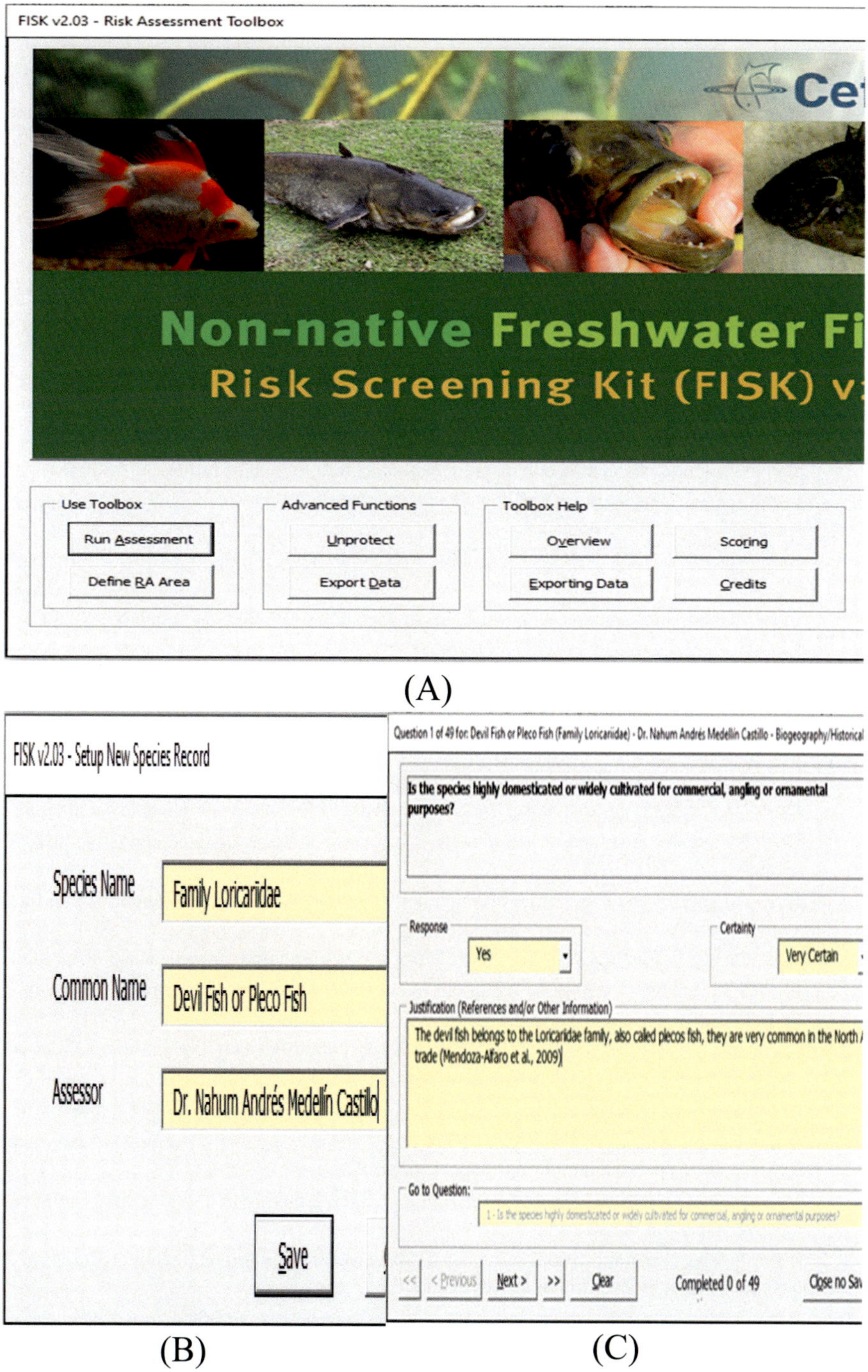

FIGURE 10.9

Fish Invasiveness Screening Kit (Fisk v2). (A) Initial program window; (B) Set up new species; (C) Sample question.

Table 10.4 FISK questionnaire for invasion risk assessment.

#	Subset	Question
1	1.01	Is the species highly domesticated or widely cultivated for commercial, angling or ornamental purposes?
2	1.02	Has the species established self-sustaining populations where introduced?
3	1.03	Does the species have invasive races/varieties/sub-species?
4	2.01	What is the level of matching between the species' reproductive tolerances and the climate of the RA area?
5	2.02	What is the quality of the climate match data?
6	2.03	Does the species have self-sustaining populations in three or more (Köppen-Geiger) climate zones?
7	2.04	Is the species native to, or has established self-sustaining populations in, regions with similar climates to the RA area?
8	2.05	Does the species have a history of being introduced outside its natural range?
9	3.01	Has the species established one or more self-sustaining populations beyond its native range?
10	3.02	In the species' introduced range, are there impacts to wild stocks of angling or commercial species?
11	3.03	In the species' introduced range, are there impacts to aquacultural, aquarium or ornamental species?
12	3.04	In the species' introduced range, are there impacts to rivers, lakes or amenity values?
13	3.05	Does the species have invasive congeners?
14	4.01	Is the species poisonous/venomous, or poses other risks to human health?
15	4.02	Does the species out-compete with native species?
16	4.03	Is the species parasitic of other species?
17	4.04	Is the species unpalatable to, or lacking, natural predators?
18	4.05	Does the species prey on a native species previously subjected to low (or no) predation?
19	4.06	Does the species host, and/or is it a vector, for one or more recognized nonnative infectious agents?
20	4.07	Does the species achieve a large ultimate body size (i.e. >15 cm total length) (more likely to be abandoned)?
21	4.08	Does the species have a wide salinity tolerance or is euryhaline at some stage of its life cycle?
22	4.09	Is the species able to withstand being out of water for extended periods (e.g., minimum of one or more hours)?
23	4.10	Is the species tolerant of a range of water velocity conditions (e.g., versatile in habitat use)
24	4.11	Does feeding or other behaviors of the species reduce habitat quality for native species?
25	4.12	Does the species require minimum population size to maintain a viable population?
26	5.01	If the species is mainly herbivorous or piscivorous/carnivorous (e.g., amphibia), then is its foraging likely to have an adverse impact in the RA area?
27	5.02	If the species is an omnivore (or a generalist predator), then is its foraging likely to have an adverse impact in the RA area
28	5.03	If the species is mainly planktivorous or detritivorous or algivorous, then is its foraging likely to have an adverse impact in the RA area?
29	5.04	If the species is mainly benthivorous, then is its foraging likely to have an adverse impact in the RA area?

Table 10.4 *(Continued)*

#	Subset	Question
30	6.01	Does the species exhibit parental care and/or is it known to reduce age-at-maturity in response to environment?
31	6.02	Does the species produce viable gametes?
32	6.03	Is the species likely to hybridize with native species (or use males of native species to activate eggs) in the RA area?
33	6.04	Is the species hermaphroditic?
34	6.05	Is the species dependent on the presence of another species (or specific habitat features) to complete its life cycle?
35	6.06	Is the species highly fecund (> 10,000 eggs/kg), iteropatric or has an extended spawning season relative to native species?
36	6.07	What is the species' known minimum generation time (in years)?
37	7.01	Are life stages likely to be dispersed unintentionally?
38	7.02	Are life stages likely to be dispersed intentionally by humans (and suitable habitats abundant nearby)?
39	7.03	Are life stages likely to be dispersed as a contaminant of commodities?
40	7.04	Does natural dispersal occur as a function of egg dispersal?
41	7.05	Does natural dispersal occur as a function of dispersal of larvae (along linear and/or "stepping stone" habitats)?
42	7.06	Are juveniles or adults of the species known to migrate (spawning, smolting, feeding)?
43	7.07	Are eggs of the species known to be dispersed by other animals (externally)?
44	7.08	Is dispersal of the species density dependent?
45	8.01	Are any life stages likely to survive out of water transport?
46	8.02	Does the species tolerate a wide range of water quality conditions, especially oxygen depletion and temperature extremes?
47	8.03	Is the species readily susceptible to piscicides at the doses legally permitted for use in the risk assessment area?
48	8.04	Does the species tolerate or benefit from environmental disturbance?
49	8.05	Are there effective natural enemies of the species present in the risk assessment area?

For this study, all responses were answered according to the information available from the literature and that generated in the interviews with fishers in the region. The sources of information were available scientific articles and books, online databases, and specialized fisheries websites.

To answer the questions related to climate similarity, a series of ecological niche modeling was carried out using the maximum entropy method (MaxEnt), which is not shown in this study. This modeling revealed the ideal areas of Mexico and the Huasteca that meet the requirements of the ecological niche of the devil fish, that is, the potential distribution of the species.

Table 10.4 shows the questions evaluated by the FISK tool, which comprises a set of 49 questions divided into eight subsets: domestication/cultivation (1), climate and distribution (2), invasive elsewhere (3), undesirable traits (4), feeding guild (5), reproduction (6), dispersal mechanisms (7), and persistence attributes (8).

10.5.2 Results and discussion

As part of the sampling results conducted in the Huasteca, 18 interviews were conducted with fishers, presidents of fishing societies and cooperatives, and land shareholders. According to the responses obtained, the fishers stated that since their first sightings in 2016, they had gradually noticed a decrease in native species such as tilapia, silver carp, snook, and catfish, among others. Regarding the socioeconomic impacts, the fishers' population has decreased; some are farmers or traders, and others have migrated to the United States. The fishers also stated that the severe drought of the last 3 years has further accelerated the decline of fish with a commercial interest and that the devil fish are the only species that have managed to survive and have rarely been consumed.

The FISK test score for the devil fish (family Loricariidae) in the Huasteca was 37, which means that it is a species with a very high risk of invasion. This value corresponds to biogeography/historical 15, undesirable traits of the species 9, and characteristics of biology/ecology 13, resulting in a recommendation to "reject" the invasive species in the Huasteca. The calculated value of the certainty factor (CF) was 0.8878 (>0.5), which is a reliable value for the certainty of the result.

Table 10.5 shows the FISK score and the certainty factor (CF) value of the devil fish (family Loricariidae) from this study and includes the FISK results reported by other authors.

The FISK score of this study was higher than that reported in other papers, and the CF values are in the range of 0.83–0.92. The FISK score could have been different because the set of family species was evaluated in general, resulting in a higher risk of establishment. A species that is evaluated as being at high risk is considered problematic for the establishment site. That means that devil fish are a high-risk species that can have several environmental and social impacts in the Huasteca.

In the study carried out by Mendoza-Alfaro et al. (2015), the risk of invasion was determined through the analysis of ornamental fish species risk assessment in Mexico, where 368 species of

Table 10.5 FISK and CF values for species of the family Loricariidae.

Species	FISK score	Certainty factor (CF)	Location	Reference
Devil fish (family Loricariidae)	37	0.88	Huasteca, SLP, Mexico	This study
Hypostomus spp.	23	0.83	Florida, USA	Lawson (2014)
Hypostomus plecostomus	23	0.91	Mexico	Mendoza-Alfaro et al. (2015)
Pterygoplichthys disjuntivus	34	0.90	Mexico	Mendoza-Alfaro et al. (2015)
	24	0.86	Florida, USA	Lawson (2014)
Pterygoplichthys anisitsi	20	0.85	Florida, USA	Lawson (2014)
Pterygoplichthys multiradiatus	17	0.92	Florida, USA	Lawson (2014)
Pterygoplichthys pardalis	17	0.91	Mexico	Mendoza-Alfaro et al. (2015)

Table 10.6 Risk classification according to FISK by various authors.

Risk		FISK score				
		Copp et al. (2009)	Almeida et al. (2013)	Simonović et al. (2013)	Piria et al. (2016)	Uyan et al. (2020)
	n =	67	89	43	40	57
	AUC ROC =	0.807	0.8807	0.67	0.6752	0.7397
Low risk	Low	>1	−15 to 1	−15 to 1	−15 to 1	−20 to 1
Medium risk	Medium	1 to 19	1 to 20.25	1 to 9.5	1 to 11.75	1 to 5.5
High risk sensu lato	Moderately high	≥ 19	20.25 to 25	9.5 to 25	11.75 to 25	5.5 to 68
	High		25 to 30	25 to 30	25 to 30	
	Very high		30 to 57	30 to 57	30 to 57	

freshwater aquarium fish commonly commercialized in Mexico were considered, carrying out a calibration of the FISK test with 30 species with greater invasive importance in the country. Within this list are species belonging to the family Loricariidae: *Hypostomus plecostomus*, *Pterygoplichthys disjuntivus*, and *Pterygoplichthys pardalis*. Their AUC value was 0.829, so their calibration was validated. Table 10.5 shows the comparison of FISK and CF score values of species of the family Loricariidae with other authors.

In order to locate in which range of invasiveness the devil fish of the present study are located, it was necessary to obtain from bibliographic reports the risk classification from the FISK score. Table 10.6 shows the ranges of invasiveness determined by different authors from evaluations of n native and nonnative species where the ranges of low, medium, and high invasion were determined. These authors used receiver operating characteristic (ROC) curves and demonstrated that FISK accurately distinguishes between potentially invasive and nonnative fish species by obtaining a statistically appropriate threshold score for species scores (> 0.5). Also, the values of Area Under the ROC curve (AUC ROC) and the number of species reported in each study are shown.

10.6 Conclusions

The use of FISK allows simple detection of the range of invasiveness of exotic aquatic species. The certainty responses and justifications determined in this study for the devil fish provided information to understand the potential risk of invasion that this species represents in the Huasteca.

Devil fish (family Loricariidae) were evaluated as a species with a high risk of invasion in the Huasteca with a score of 37 and a CF of 0.8878. It was also established that the species must be rejected from any place where it is considered exotic. Due to the lack of control and limits that the site offers, it is necessary to restrict it from natural water bodies. In other words, the recommendation offered is to keep it as an ornamental species and not to release it into natural waters and channels of ecosystems.

The most critical impacts in the Huasteca due to devil fish are the decreases of commercial value fish such as tilapia and carp, in addition to a decrease in fishing activities and an important desertion of fishers.

Regarding the devil fish's current presence, it is considered an invasive species in the Huasteca, and it is necessary to carry out an environmental and health risk analysis due to its presence. Similarly, it is necessary to give a detailed report on the environmental and health problems to state and national institutions, which breaks down the relationship between heavy metals and pesticides in organisms and using them for consumption or application in other uses.

Acknowledgments

This work was supported by Consejo Nacional de Ciencia y Tecnología, CONACyT, Mexico, under grants Nos. CB-2016–286990 and PN- 2016–3947.

References

Almeida, D., Ribeiro, F., Leunda, P. M., Vilizzi, L., & Copp, G. (2013). Effectiveness of FISK, an invasiveness screening tool for non-native freshwater fishes, to perform risk identification assessments in the Iberian Peninsula. *Risk Analysis*, *33*(8), 1404–1413. Available from https://doi.org/10.1111/risa.12050.

Amador-Del Ángel, L. E., Guevara-Carrió, E., Del, C., Brito-Pérez, R., & Endañú-Huerta, E. (2014). *Aspectos biológicos e impacto socio-económico de los plecos del género Pterygoplichthys y dos cíclidos no nativos en el sistema fluvio lagunar deltaico Río Palizada, en el Área Natural Protegida Laguna de Términos, Campeche. Centro de Investigación de Ciencias Ambientales. Facultad de Ciencias Naturales* (p. 21) Informe Final SNIB-CONABIO.

Amador-Del Ángel, L. E. & Wakida-Kusunoki, A. (2014). Especies acuáticas exóticas e invasoras del estado de Tabasco, México. In: A. Low Pfeng, P. Quijon and E. Peters Recagno (Eds.), *Especies invasoras acuáticas: Casos de estudio en ecosistemas de México* (1st edn.) Mexico (pp. 178–198).

Amador-Del Ángel, L. E., Wakida-Kusunoki, A. T., Sánchez-Martínez, M. A., & Hernández-Nava, J. (2016). Consideraciones económicas para el manejo del pez diablo en el area protegida Laguna de Términos. El Pez diablo en México. In: *Protocolo de prevención, respuesta rápida y control* (pp. 144–156).

Ayala-Pérez, L. A., Pineda-Peralta, A. D., Álvarez-Guillen, H., & Amador-Del Ángel, L. E. (2014). El pez diablo (*Pterygoplichthys* spp.) en las cabeceras estuarinas de la Laguna de Términos, Campeche. In: *Especies invasoras acuáticas: Casos de estudio en ecosistemas de México* (pp. 313–336).

Baptiste, M. P., Castaño, N., Lasso, C. A., Cárdenas, D., Gutiérrez, F. D. P., & Gil, D. L. (2010). *Análisis de riesgo y propuesta de categorización de especies introducidas para Colombia*. Bogotá, DC, Colombia: Instituto de Investigación de Recursos Biológicos Alexander von Humboldt.

Barba, E. (2010). Situación actual de los recursos acuáticos en Tabasco: Impacto económico y social de los plecos (Loricáridos). In: Comité Asesor Nacional sobre Especies Invasoras (Ed.), *Estrategia nacional sobre especies invasoras en México, prevención, control y erradicación., Comisión Nacional para el Conocimiento y Uso de la Biodiversidad* (1st ed.). Mexico, p. 58.

Cabrera, A. J. & Betancourt, I. (2002). *La Huasteca Potosina: Ligeros apuntes sobre este país. Centro de Investigaciones y Estudios Superiores en Antropología Social*. El Colegio de San Luís.

Cano-Salgado, M. (2011). *El Plecos (*Pterygoplichthys *spp.): Su invasión y el abordaje de las cooperativas balancanenses* (p. 164) Tesis de Doctorado. El Colegio de la Frontera Sur.

CICESE, Centro de Investigación Científica y de Educación Superior de Ensenada, Baja California. (2013). *Datos climáticos diarios del CLICOM del SMN*. Available at: http://clicom-mex.cicese.mx/ Accessed 25 .03.19.

Comisión Nacional para el Conocimiento y Uso de la Biodiversidad (CONABIO). (2020). Available at: https://www.biodiversidad.gob.mx/biodiversidad/porque Accessed 9.06.20.

Comisión Nacional para el Conocimiento y Uso de la Biodiversidad (CONABIO), & García, E. (1998). *Geoportal del Sistema Nacional de Información sobre Biodiversidad [12,674 mapas] – CONABIO*. Available at: http://www.conabio.gob.mx/informacion/gis/ Accessed 9.06.20.

Comisión Nacional para el Conocimiento y Uso de la Biodiversidad (CONABIO), & Instituto Nacional de Estadística y Geografía (INEGI). (2016). *Geoportal del Sistema Nacional de Información sobre Biodiversidad [12,674 mapas] – CONABIO*. Available at: http://www.conabio.gob.mx/informacion/gis/ Accessed 9.06.20.

Copp, G., Britton, R. J., Jeney, G., Joly, J.-P., Gherardi, F., Gollasch, S., Gozlan, R. E., Jones, G., MacLeod, A., Midtlyng, P. J., Miossec, L., Nunn, A. D., Occhipinti Ambrogi, A., Oidtmann, B., Olenin, S., Peeler, E., Russell, I. C., Savini, D., Tricarico, E., & Thrush, M. (2008). Risk assessment protocols and decision making tools for use of alien species in aquaculture and stock enhancement. *IMPASSE: Environmental Impacts of Alien Species in Aquaculture, 044142*, 85.

Copp, G., Garthwaite, R., & Gozlan, R. (2005a). Risk identification and assessment of non-native freshwater fishes: A summary of concepts and perspectives on protocols for the UK. *Journal of Applied Ichthyology, 21*(4), 371–373. Available from https://doi.org/10.1111/j.1439-0426.2005.00692.x.

Copp, G., Garthwaite, R., & Gozlan, R. (2005b). Risk identification and assessment of non-native freshwater fishes: Concepts and perspectives on protocols for the UK. Science Series Technical Report. *Cefas Lowestoft, 129*, 1–32.

Copp, G., Vilizzi, L., Mumford, J., Fenwick, G., Godard, M., & Gozlan, R. (2009). Calibration of FISK, an invasiveness screening tool for nonnative freshwater fishes. *Risk Analysis, 29*(3), 457–467. Available from https://doi.org/10.1111/j.1539-6924.2008.01159.x.

Dar, G. H., Bhat, R. A., Kamili, A. N., Chishti, M. Z., Qadri, H., Dar, R., & Mehmood, M. A. (2020). *Correlation between pollution trends of freshwater bodies and bacterial disease of fish fauna. In Fresh water pollution dynamics and remediation* (pp. 51–67). Singapore: Springer.

Dar, G. H., Dar, S. A., Kamili, A. N., Chishti, M. Z., & Ahmad, F. (2016). Detection and characterization of potentially pathogenic *Aeromonas sobria* isolated from fish *Hypophthalmichthys molitrix* (Cypriniformes: Cyprinidae). *Microbial Pathogenesis, 91*, 136–140.

Dar, G. H., Kamili, A. N., Chishti, M. Z., Dar, S. A., Tantry, T. A., & Ahmad, F. (2016). Characterization of *Aeromonas sobria* isolated from fish rohu (*Labeo rohita*) collected from polluted pond. *Journal of Bacteriology & Parasitology, 7*(3), 1–5. Available from https://doi.org/10.4172/2155-9597.1000273.

Domínguez-Lemus, Y. C. (2018). *Impacto Socio-económico y Ambiental del Plecos (*Pterygoplichthys *spp.) en Humedales de Catazajá, Chiapas, México* (pp. 1–63). Universidad Autónoma de Chiapas.

Fish Base. (2020). *World Wide Web electronic publication*. Froese, R. and D. Pauly. Editors. Available at: http://www.fishbase.org Accessed 9.09.20.

García-Marmolejo, G., (2013). *Adecuación, uso y manejo del hábitat de artiodáctilos silvestres en bosque tropical caducifolio secundario*. Tesis de Doctorado. Instituto Potosino de Investigacion Cientifica y Tecnologica, A.C., p. 184.

Global Biodiversity Information Facility (GBIF) Loricariidae (2020) Available at: https://www.gbif.org/es/species/5158 Accessed 31.01.20.

González-Martínez, A., Barrios-Caballero, Y., & De Jesús, S. (2017). *El impacto de las especies exóticas invasoras en México*. Revista Legislativa de Estudios Sociales y de Opinión Pública (Vol. 103).

Guzmán, A. F., & Barragán, S. J. (1997). *Presencia de bagres sudamericanos (Osteichthyes: Loricariidae) en el Rio Mezcala* (pp. 1–4). Guerrero, México: Vertebrata Mexicana.

Hoover, J. J., Killgore, K. J., & Cofrancesco, A. F. (2004). Suckermouth catfishes: Threats to aquatic ecosystems of the United States? *Aquatic Nuisance Species Research Bulletin*, *4*, 1–9.

Instituto Nacional de Estadística y Geografía (INEGI). (2007). *VIII Censo Agrícola, Ganadero y Forestal.* INEGI.

Instituto Nacional de Estadística y Geografía (INEGI). (2017). *Anuario estadístico y geográfico de San Luis Potosí 2017.* INEGI.

Instituto Nacional de Estadística y Geografía (INEGI), & Cadena, A. (1985). *Hidrología. Síntesis Geográfica del Estado de San Luis Potosí* (p. 4). INEGI.

Lawson, L. L. (2014). *Evaluation of the fish invasiveness screening kit (Fisk V2) for identifying the invasiveness risk of non-native freshwat.* Thesis of Master Degree. University of Florida.

Lawson, L. L., Hill, J. E., Hardin, S., Vilizzi, L., & Copp, G. (2015). Evaluation of the fish invasiveness screening kit (FISK v2) for peninsular Florida. *Management of Biological Invasions*, *6*(4), 413–422. Available from https://doi.org/10.3391/mbi.2015.6.4.09.

Liang, S. H., Wu, H. P., & Shieh, B. S. (2005). Size structure, reproductive phenology, and sex ratio of an exotic armored catfish (*Liposarcus multiradiatus*) in the Kaoping River of Southern Taiwan. *Zoological Studies*, *44*(2), 252–259.

Mendoza-Alfaro, R., Contreras-Baldera, S., Ramírez, C., Koleff-Osorio, P., Álvarez, P., & Aguilar, V. (2007). Los peces diablo: Especies invasoras de alto impacto. *Biodiversitas*, *70*, 2–5.

Mendoza-Alfaro, R., Cudmore, B., Orr, R., & y Contreras, S. (2009). *Directrices trinacionales para la evaluación de riesgos de las especies acuáticas exóticas invasoras* (71, pp. 70–80). México: Comisión para la Cooperación Ambiental.

Mendoza-Alfaro, R., Koleff-Osorio, P., Ramírez-Martínez, C., Orbe-Mendoza, A., Álvarez-Torres, P., Arroyo-Damián, M., & Escalera-Gallardo, C. (2011). La evaluación de riesgos por especies acuáticas exóticas invasoras: Una visión compartida para Norteamérica. *Ciencia Pesquera*, *19*(2), 65–75.

Mendoza-Alfaro, R., Luna, S., & Aguilera, C. (2015). Risk assessment of the ornamental fish trade in Mexico: Analysis of freshwater species and effectiveness of the FISK (Fish invasiveness screening kit). *Biological Invasions*, *17*(12), 3491–3502. Available from https://doi.org/10.1007/s10530-015-0973-5.

Monroy, M. I., & Calvillo, T. (1997). *Breve historia de San Luis Potosí (Serie Breves Historias de los Estados de la República Mexicana).* San Luis Potosí, México: Fondo de Cultura Económica.

Nico, L. G., & Martin, T. (2001). The South American Suckermouth Armored Catfish, Pterygoplichthys anisitsi (Pisces: Loricaridae), in Texas, with Comments on Foreign Fish Introductions in the American Southwest. *The Southwestern Naturalist*, *46*(1), 98–104.

Pheloung, P. C., Williams, P. A., & Halloy, S. R. (1999). A weed risk assessment model for use as a biosecurity tool evaluating plant introductions. *Journal of Environmental Management*, *57*(4), 239–251. Available from https://doi.org/10.1006/jema.1999.0297.

Piria, M., Povž, M., Vilizzi, L., Zanella, D., Simonović, P., & Copp, G. (2016). Risk screening of non-native freshwater fishes in Croatia and Slovenia using the Fish Invasiveness Screening Kit. *Fisheries Management and Ecology*, *23*(1), 21–31. Available from https://doi.org/10.1111/fme.12147.

Puig, H. (1991). *Vegetacion de la Huasteca, México. Estudio fitogeográfico y ecológico.* México: Instituto de Ecología, INECOL.

Simonović, P., Tošić, A., Vassilev, M., Apostolou, A., Mrdak, D., Ristovska, M., Kostov, V., Nikolić, V., Škraba, D., Vilizzi, L., & Copp, G. (2013). Risk assessment of non-native fishes in the Balkans Region using FISK, the invasiveness screening tool for non-native freshwater fishes. *Mediterranean Marine Science*, *14*(2), 369–376. Available from https://doi.org/10.12681/mms.337.

Uyan, U., Oh, C. W., Tarkan, A. S., Top, N., Copp, G. H., & Vilizzi, L. (2020). Risk screening of the potential invasiveness of non-native marine fishes for South Korean coastal waters. *Marine Pollution Bulletin*, *153*, 111018. Available from https://doi.org/10.1016/j.marpolbul.2020.111018.

Velázquez-Velázquez, E., López-Vila, J. M., & Romero-Berny, E. I. (2013). El pez diablo: Especie invasora en Chiapas. *Lacandonia*, *7*(1), 99–104.

Wakida Kusunoki, A. T., Ruiz-Carus, R., & Amador-Del-Ángel, L. E. (2007). Amazon sailfin catfish, *Pterygoplichthys Pardalis* (Castelnau, 1855) (Loricariidae), another exotic species established in Southeastern Mexico. *The Southwestern Naturalist*, *52*(1), 141–144.

Wakida Kusunoki, A. T., & Toro-Ramírez, A. (2016). El robalo prieto (Centropomus poeyi), nuevo depredador del pez diablo (Pterygoplichthys pardalis). *Hidrobiológica*, *26*(1), 147–149.

CHAPTER

Profiling of common bacterial pathogens in fish

11

Tariq Oluwakunmi Agbabiaka[1,2], Ismail Abiola Adebayo[3], Kamoldeen Abiodun Ajijolakewu[1] and Toyin Olayemi Agbabiaka[1]

[1]Department of Microbiology, Faculty of Life Sciences, University of Ilorin, Ilorin, Nigeria [2]Microbiology Unit, Department of Science Laboratory Technology, School of Science and Technology, Federal Polytechnic, Damaturu, Nigeria [3]Department of Microbiology and Immunology, Faculty of Biomedical Sciences, Kampala International University, Kampala, Uganda

11.1 Introduction

Fish represent an important piece of the ecosystem, not only in terms of bioconservation (Fernández-Alacid et al., 2018; Lenders, 2017; Maureaud et al., 2019) and environmental preservation (Alam et al., 2020; Giakoumi et al., 2019; Vermeij et al., 2019), but also in their provision of a cheap and essential source of animal protein (Béné et al., 2015; Comerford & Pasin, 2016; Kobayashi et al., 2015; Mustapha et al., 2014; Tacon & Metian, 2013). Global fish production was estimated globally at 140 million tons (Halim et al., 2016), which is an important nutrition source (Tacon & Metian, 2013)—representing almost 16.6% consumption of animal protein (Kobayashi et al., 2015) and 40% in adults living in rural areas (Grema et al., 2011)—mainly due to its low saturated fat and high nutrient content (Thurstan & Roberts, 2014). This represents a valuable scheme against hunger and malnutrition (Béné et al., 2015; Committee on Considerations for the Future of Animal Science et al., 2015). This probably explains the immense demand for fish, projected to reach 186 million tons by 2030 (Kobayashi et al., 2015), from aquaculture (Béné et al., 2015), which will contribute 62% (Kobayashi et al., 2015) since the fish catch count is not likely to fulfill demands (Vermeij et al., 2019) due to several reasons (Adewumi, 2015; Basra et al., 2018; Hunt et al., 2011; Miller & Atanda, 2011; Solomon et al., 2018; Vermeij et al., 2019). This is true globally as well as in Nigeria (Miller & Atanda, 2011; Mustapha et al., 2014; Solomon et al., 2018), a major fish-producing nation (Alawode & Oluwatayo, 2019), being the largest in sub-Saharan Africa (Allen et al., 2017) and second in Africa (Dauda et al., 2018), as the estimation of the 2020 fish requirement was expected to be 600,000 tonnes higher than the mid-2010s catch (Allen et al., 2017), estimated at 120,000 Mt in 2011 for catfish alone (Miller & Atanda, 2011). In the same vein, the country is the largest consumer of fish in Africa, as some consider it more nutritious than meat (Grema et al., 2011).

Bacterial Fish Diseases. DOI: https://doi.org/10.1016/B978-0-323-85624-9.00004-X

11.2 Fish production in Nigeria

Decreasing fish catches and an increasing demand mean that a deficit must be catered to. This has led to increased aquaculture practice in Nigeria. Aquaculture, the fastest growing agricultural sector (Peatman et al., 2013), unlike capture, relies on intensive farming practice, often as an economic endeavor, and has thus exponentially grown in recent times in Nigeria. This is reflected in the annual average between 2009 and 2014 increasing by 6.6%, in consonance with global growth (Adewumi, 2015), while capture decreased by a marginal 0.71% (Dauda et al., 2018), which reflects changing per capita consumption of animal protein (Allen et al., 2017).

Aquaculture practice predates independence (Adewumi, 2015; Dauda et al., 2018; Miller & Atanda, 2011), as numerous riverine communities exist in all regions (Adeyeye et al., 2016; Uduji & Okolo-Obasi, 2018) with informal practice that transcends capture into subsistent hatchery and harvest. A semblance of structure and organization began postindependence (Alawode & Oluwatayo, 2019; Allen et al., 2017) and in recent years, the government has instituted a number of frameworks (Adewumi, 2015; Alawode & Oluwatayo, 2019).

Freshwater fish represents the dominant fish product from aquaculture practice in Nigeria (Allen et al., 2017; Dauda et al., 2018). Significantly, while different fish including tilapia (Adeogun et al., 2020; Grema et al., 2011), reticulate knifefish, cyprinids, Nile perch (Allen et al., 2017), African bonytongue, obscure snakehead, common carp, tarpons (Adewuyi et al., 2010), and sole (Silva et al., 2011) are produced in Nigeria, the country appear to have niched to catfish and its variants (Grema et al., 2011; Ibor et al., 2020; Ikutegbe & Sikoki, 2014; Miller & Atanda, 2011), plausibly due to the success of the extensive artificial breeding schemes that were popular in the 1980s (Dauda et al., 2018), as well as their hardy nature (Miller & Atanda, 2011).

11.3 Impact of practice

Increase in aquaculture practice in the country has led to some impacts, which can be categorized into three major categories: health, economics, and environmental.

Fish are integral to the ecosystem (Siqueira et al., 2019), so aquaculture practices have a bidirectional impact on the environment. While a healthy fish population can aid environmental conservation, unhealthy practices can cause severe problems, particularly pollution. Effluents from farms cause deviations in the physicochemical parameters of receiving water bodies and soils (Amankwaah et al., 2014; Famoofo & Adeniyi, 2020), causing eutrophication (Adewumi, 2015) and structural and biotic disturbances (Omofunmi et al., 2018; Yoboué et al., 2020). These effects are often due to the chemical nature of effluents (Amankwaah et al., 2014; Ansah et al., 2013), which makes them toxic (Carballeira et al., 2018). Personal communication with residents around fish farms has revealed that air quality may intermittently be plagued with foul odor. This has been corroborated by Bukola et al. (2015). An associated impact is rapid changes in land use occasioned by rapid growth of aquaculture practices (Kobayashi et al., 2015). This implies that there exist avenues and need to develop more sustainable and environmentally friendly practice in the industry.

Aquaculture practice is often motivated by economic interests because of the perceived profitability (Thompson & Mafimisebi, 2014). This appears to be true, as prices have risen by more than

25% in some areas, with farmers earning up to USD$1 per kilogram, which can translate into more than 30% investment return (Miller & Atanda, 2011). The profitability of the endeavor, however, strongly depends on proper management (Gomna et al., 2020; Kobayashi et al., 2015), which is influenced by education and age (Gomna et al., 2020). Lower costs and enhanced technology is also noted to enhance growth by up to 50% (Kobayashi et al., 2015). Aquaculture, in 2005, contributed 162.61 billion Nigerian naira to the GDP (Grema et al., 2011).

Finally, the health impact of fish can be considered from the perspective of the fish itself and from that of its consumers. Fish are susceptible to bacteria (Iaria et al., 2019; Ina-Salwany et al., 2019; Leung et al., 2019; Pulkkinen et al., 2010; Soliman et al., 2019), viruses (Iaria et al., 2019; Lin et al., 2019), parasites (Iaria et al., 2019; Wuensch et al., 2018), and fungi (Bizjak-Mali et al., 2018; Coleman et al., 2018; Sharma & Sihag, 2013). Some follow-on effects of diseases in fish are deterioration and mortality, which often translate into economic loss (Grema et al., 2011). Since fish of aquacultural origin are primarily intended for consumption, inadequate cooking can predispose consumers to foodborne infections and toxicities—commonly referred to as food poisoning—from fish sources, and empirical reports exist to that effect (Awan et al., 2018; Barkham et al., 2019; Clemence & Guerrant, 2015; Rodríguez-Morales, 2015; Villazanakretzer et al., 2016). Further, farmed fished appear to be more predisposed to infection due to the intensive nature and stress-packed conditions of farming (Dias et al., 2016). An allied health impact is that of the presence, ubiquity, interaction, and ultimate fate of antibiotics. Antibiotics are heavily used as prophylaxis (Hoseinifar et al., 2018) and therapeutics (Bojarski et al., 2020), and this has led to increased incidence of antibiotic resistance (Yuan et al., 2019) resulting from the pressure assertion (Santos & Ramos, 2018).

Because the major fish pathogen is bacteria (Ben Hamed et al., 2014; Ben Hamed et al., 2018; El-Bahar et al., 2019), this chapter discusses three of the prominent bacterial pathogens.

11.4 Selected common pathogens of fish in Nigeria

11.4.1 Aeromonas hydrophila

A. hydrophila, a rod-shaped (Jin et al., 2020), facultatively anaerobic, nonspore forming, chemoorganoheterotroph (Xie et al., 2018), gram-negative aquatic bacterium, previously regarded as generally free living but opportunistically pathogenic (Harikrishnan & Balasundaram, 2005), is now regarded as an emergent pathogen of humans, reptiles, and aquatic animals (Awan et al., 2018; Li et al., 2019) capable of singly or coinfecting its hosts (Fernández-Bravo et al., 2019) at different sites (Li et al., 2019) manifesting as different diseases (Singh et al., 2008), generally called "motile aeromonad septicemia" (Zhou et al., 2019) or "hemorrhagic septicemia" (Yun et al., 2020). Other symptoms include ulceration, edema, red sores, dropsy, and necrosis (Poobalane et al., 2010). Its aquatic origin, with attendant drastic fluctuations in physicochemical conditions that translate into sublethal stressors (Pajares et al., 2013; Patin et al., 2018; Righetto et al., 2012), may be a significant variable that has contributed to its high adaptability (Awan et al., 2018; Li et al., 2019), as it is also known to form biofilm communities (Li et al., 2019) in the presence of a suitable matrix (Awan et al., 2018), which is implicated in virulence (Zhou et al., 2019). These make *A. hydrophila* a hardy organism with extended occurrence across habitats (Dias et al., 2016; Rasmussen-Ivey, Figueras et al., 2016; Rasmussen-Ivey, Hossain et al., 2016).

A. hydrophila has grown to become a successful aquaculture pathogen. The array of virulence factors, the expression of which is often influenced by environmental conditions (Awan et al., 2018; Mateos et al., 1993; Pianetti et al., 2012), makes an overview of virulence factors imperative. These factors include serine protease; adhesins (Rasmussen-Ivey, Figueras et al., 2016; Rasmussen-Ivey, Hossain et al., 2016) such as S-layer protein and fimbriae; aerolysin (Dias et al., 2016; Zhou et al., 2019); enterotoxin; cytotoxin; hemolysin (Khajanchi et al., 2012; Rasmussen-Ivey, Figueras et al., 2016; Rasmussen-Ivey, Hossain et al., 2016; Zhou et al., 2019); signaling proteins (Khajanchi et al., 2012; Rasmussen-Ivey, Figueras et al., 2016; Rasmussen-Ivey, Hossain et al., 2016; Xie et al., 2018; Zhang, Qin et al., 2019; Zhang, Zhuang et al., 2019; Zhou et al., 2019); extracellular proteins like proteases, lipases (Rasmussen-Ivey, Figueras et al., 2016; Rasmussen-Ivey, Hossain et al., 2016), chitinase, gelatinase, and amylase (Dias et al., 2016); type II (Wang, Hu et al., 2019; Wang, Yan et al., 2019) and type III secretion systems (T3SS); and ADP-ribosylating toxin (Dias et al., 2016). Hypervirulent strains have been reported to possess a truncated type VI secretion system (T6SS) (Rasmussen-Ivey, Figueras et al., 2016; Rasmussen-Ivey, Hossain et al., 2016). Association of virulence with *A. hydrophila* has, however, required reevaluation since the previous association is predicated on the characterization of *A. hydrophila* SSU and AH-3 strains has been reclassified as *A. dhakensis* SSU and *A. piscicola* AH-3, respectively, by current molecular characterization (Rasmussen-Ivey, Figueras et al., 2016; Rasmussen-Ivey, Hossain et al., 2016). A number of genes have been associated with the expression of these virulence factors. Li et al. (2019) link the cytochrome c4 gene (*cyt-c4*) to biofilm competence and silencing this gene resulted in decreased competence in biofilm formation, drug resistance, adhesion, and pathogenicity in *A. hydrophila* subsp. *hydrophila* (ATCC 7966). El-Bahar et al. (2019) identified the presence of *aer*, *act*, and *hylA* genes regulating the expression of aerolysin, cytotoxic enterotoxin, and hemolysin production in *A. hydrophila* isolated from infected *Oreochromis niloticus* (Nile tilapia), just as it has been reported to be present in *Mugil cephalus* (mullet) (Ramadan et al., 2018). Significantly, all three genes correlated though *hly* and *ast* genes were more strongly correlated than *hly* and *aer* (Ramadan et al., 2018). Similarly, virulence factors present in strains vary with genotypes (El-Bahar et al., 2019), implying that the array of virulence factors present in a strain is a function of presence, expression, and regulation of the virulence genes present. In reports of multiple antibiotic resistance (Elbehiry et al., 2019; Mao et al., 2020; Sekizuka et al., 2019; Zhang et al., 2018), its genes (Ramadan et al., 2018) and/or plasmids potentiate the virulence of *A. hydrophila* (Awan et al., 2018). However, results from Ramadan et al. (2018) hint at a weak relationship between the presence of β-lactam resistance genes and virulence. On the other hand, Mao et al. (2020) suggests that the transcriptional factors $luxR_{05}$ and $luxR_{164}$ are influential in the susceptibility of *A. hydrophila* to different families of drugs.

In an experimental intradermal inoculation of *A. hydrophila* into *Arapaima gigas*, the lethal dose LD_{50-96h} was determined to be 1.8×10^8 CFU/mL with 91.6% after 23 h suggesting the first 24 h of infection may be crucial (Dias et al., 2016), an observation that had earlier been made by Rey et al. (2009). Samayanpaulraj et al. (2019) estimated the LD_{50-96h} of strain Ah17 at 4.1×10^8 CFU/mL in *Channa striata* (snake-head fish) while a value of 1.6×10^6 CFU/mL was determined for mandarin fish infection by strain G3 (Chen et al., 2018). The results from Dias et al. (2016) imply that *A. gigas* generally had a high level of tolerance to *A. hydrophila* relative to a few other fish, suggesting that pathogenicity varies with factors such as susceptibility of host, infectious dose, strain, fish type, and other variables that may influence the triad of interactions between pathogen, host, and their environment. Symptoms of infection include altered behavior, including anorexia, loss of balance, reduced respiratory movements, lesions of varying sizes and dimensions, depigmentation of different sites, abrasions, and necrotic hemorrhage of vital organs (Dias et al., 2016), including the muscular layers and nervous plexus of the intestinal tracts (Rey et al., 2009). Marinho-Neto et al. (2019) reports that sepsis in *Piaractus mesopotamicus* resulting from intraperitoneal inoculation of *A. hydrophila* is characterized by congestion, hemorrhage, vascular damage, and necrosis between 6 and 9 h after inoculation. Results also show severe damage to vital organs such as the kidney and liver, which Rey et al. (2009) attributes to systemic dissemination of the pathogen in a manner similar to that observed in mammalian septicemia. Table 11.1 shows some reported *A. hydrophila* outbreaks.

The progression of pathogenesis appears to be multifocal. Physical manifestations of infection, in the acute stage, include loss of scales and depigmentation (Algammal et al., 2020). Upon inoculation, *A. hydrophila* causes hemolysis, hemorrhage (due to vascular permeability), degeneration of the mucosa, and necrosis in the viscera before it is dispersed, via blood circulation, to the liver between 4–6 h postinfection, leading to aggregation of erythrocytes, leukocytes, and thrombocytes (AlYahya et al., 2018; Rey et al., 2009). This stimulates the recruitment of macrophages, which in turn causes systemic inflammation aggravated by proliferation of the pathogen in intestinal muscles, pancreas, mesenteries, spleen, liver, kidney, and white blood cells. The significant damage

Table 11.1 ***A. hydrophila*** **outbreaks.**

Host	Strain/serotype	Location	References
Channel catfish	ZC1, ML-09–119, ML10–51K, S04–690, S14–296, and S14–452	United States of America	Rasmussen-Ivey, Figueras et al. (2016), Rasmussen-Ivey, Hossain et al. (2016)
Channel catfish	ML09–121	United States of America	Griffin et al. (2013)
Crucian carp, large-mouth buffalo, rice eels and Wuchang bream	ATCC 7966, O9, O13, O64 and O97	China	Nielsen et al. (2001)
Nile tilapia, African catfish	Diverse strains	Uganda	Wamala et al. (2018)
Asian/grass carp and catfish	Virulent *A. hydrophila* (VAh) isolates: ZC1 and S04–690	China and United States of America	Hossain et al. (2014)
Cyprinid fish	*A. hydrophila* subsp. *Hydrophila*	China	Zhang et al. (2014)
Channel catfish	L09–71, AL09–72, and AL09–73	United States of America	Pridgeon and Klesius, (2011a, 2011b)
Tilapia	ND	Egypt	Aboyadak et al. (2015)
Nile tilapia and red hybrid tilapia	AH62	Thailand	Nicholson et al. (2020)
Yellow perch and lake whitefish	ND	Canada	Scott and Bollinger (2014)

Key: ND, Not determined.

caused to the intestinal muscle, resulting in intussusception, is caused by intense local nervous activity (Rey et al., 2009), suggesting that the neurons along with immune cells are involved in the pathogenic response (Dar et al., 2020; Dar, Dar et al., 2016; Dar, Kamili et al., 2016). The implication of the nervous system seems substantiated by observations made by Bandeira Junior et al. (2019) when *Rhamdia quelen* (silver catfish) showed increased locomotor activity and overexpressions of the *crh* (which produces an adrenocorticotropic hormone) and *hspa12a* (which encodes the heat shock protein70, member 12) in the fish brain.

The reports of severe effect on muscles have been corroborated by Samayanpaulraj et al. (2019), though the liver appear to harbor more of the pathogen (Algammal et al., 2020), which has been supported by AlYahya et al. (2018) while others (Chen et al., 2018) have suggested gill, spleen, and/or kidney, though Meng et al. (2019) suggest the liver as a major immune response site on the basis of high C3 titer. Some of the pathogenic progressions appear to be due to the individual and/or combined effects of bacterial extracellular proteins such as AST, ALT, and ACT enterotoxins, aerolysin, α- and β-hemolysins, RNases and proteases (AlYahya et al., 2018; Rey et al., 2009). The (degree of) pathogenicity appears to be linked to $luxR_{05}$—responsible for the regulation of biofilm formation, movement, and adhesion—and $luxR_{164}$, which appears to be responsible for growth, extracellular protein activity, and adhesion (Mao et al., 2020). The course and severity of pathogenicity may differ, reflecting the arsenal of virulence genes harbored by the specific infecting strain (El-Bahar et al., 2019), as the *icmF* has been shown to influence motility, biofilm formation, adhesion, and acid resistance competence (Wang, Hu et al., 2019; Wang, Yan et al., 2019).

The onset of pathogenesis elicits immune responses even from the point of entry. The first barrier, the skin, a major innate immunity component, may have a significant role in eventual mucosal immune response. Analysis of the transcriptome in *A. hydrophila*-infected *Carassius auratus* (crucian carp) showed differential expression with variation in time postinfection. Significantly increased expression was observed in the acute phase of infection for genes encoding the complement components crl, crmp, c1, c1ql2, ce, cg, cfh, c7; the complement receptor cr1; some interferon-inducible genes; interferon regulatory factor-1 (irf-1); interleukins (IL); apolipoprotein, transferrin, ceruloplasmin, vitellogenin, alpha-2-macro-globulin; and others were upregulated as opposed to downregulation of genes putatively linked to the MHC antigen processing and presenting functions. Other differentially expressed components include mitogen-activated protein kinases (mapkg1, map2k2, map3k13, and mapk5); heat shock proteins (hsp) 70 and 90; caspases responsible for proteolysis and transport of exogenous and endogenous proteins during antigen presentation; and the genes (LOC100332375, LOC100038770, LOC100698102, LOC393377, LOC550603, LOC767810, zgc:171497 and zgc:77326) of unknown function(s) (Wang, Hu et al., 2019; Wang, Yan et al., 2019). Similar reports have been made elsewhere (Chen et al., 2018; Mohammadian et al., 2018). Several proteins, peptides, carbohydrates, sugars, glycoproteins, heat shock proteins, extracellular RNA, lipids, histones, complement factors, and several metabolites present in the mucus of the fish skin are presumed to be active in the immune response (Brinchmann, 2016).

Another first defense line and innate immune component, the alternative pathway of the complement system, is presumed to be involved in the immune response. Upon entry, C3 cleaves into C3a and C3b to generate C5, which in turn triggers the development of the membrane-attack complex (MAC) from the C5b, C6, C7, C8, and C9 assemblage. The MAC can then lysis the *A. hydrophila* cell membrane. The pathogen, however, evades this fate using proteases (especially metalloprotease elastase) by denaturing the C3 proteins at a dose-dependent rate over time. The degradation of C3 leads to the inhibition of complement cascades of the alternative complement system (Chen et al., 2019). This report has been corroborated by others (Meng et al., 2019), though variability of C6 may be instrumental to resistance of *A. hydrophila* (Shen et al., 2013).

Upon hemolysis and aggregation of erythrocytes, macrophages are recruited (Rey et al., 2009) to liberate intermediate reactive oxygen species (ROS) with antimicrobial effect. Significantly, *A. hydrophila* can circumvent phagocytosis by producing enzymes such as catalase, peroxynitritase, and peroxidase that can mend the oxidative destruction of affected cellular organelles and/or reverse the toxic oxidative processes, thus making the phagocytes a repository for evading the fish immune system (Zhang et al., 2018). A number of mechanisms have been suggested for the evasion process. Zhang et al. (2018) suggests that the *KatG* gene in *A. hydrophila* B11, with a rated 93% homologous relationship with other *KatG* of other *Aeromonas*, is highly conserved; encodes catalase, peroxidase, and peroxynitritase activity; and highly increased the survival rate of *A. hydrophila* B11 by resisting the peroxide oxidative strain before it is disseminated to cause septicemia. Other physiological fitness mechanisms include the expression of *icmF*, a core T6SS gene, that serves as a T6SS exciter while also encoding a protein that forms integral transmembrane channels that transport secretory proteins through the cell membrane. Silencing of the *icmF* with an *icmF*-RNAi resulted in a 71.8% survival rate and 12.5% escape rate reduction relative to the wild-type strain (Wang, Hu et al., 2019; Wang, Yan et al., 2019). Similarly, the superoxide dismutase genes (*sodA* and *sodB*) that encode metalloenzymes involved in degrading ROS have been shown to also be involved in the survival and escape of *A. hydrophila* from macrophages (Zhang, Qin et al., 2019; Zhang, Zhuang et al., 2019). *A. hydrophila* also possesses two-component regulatory systems (TCSs) that are instrumental in the response to environmental changes that can be physical stressors and cause differential expression (Awan et al., 2018).

Apart from the action of macrophages, humoral adaptive immunity plays a functional role in response to *A. hydrophila* infection. Wang et al. (2017) reports that the complete length of cDNA for the polymeric immunoglobulin receptor (pIgR), which facilitates the trafficking of IgA and IgM, was upregulated in healthy *C. auratus*, peaking in the liver [which reinforces previous reports (Meng et al., 2019) of the liver as a bastion of immune response to *A. hydrophila*] and mediated rapid response to infection. The sequence alignment of CapIgR showed high similarity to those of other fish, suggesting a similar mechanism of humoral response in other fish, though minimal homology (28.5%) was achieved with mammals, reinforcing previous reports (Kong et al., 2018).

The control of *A. hydrophila* infection is varied. Vaccinations, however, are gaining attention due to their effectiveness and potential to make aquaculture a sustainable practice (Wang et al., 2020). Formalin-destroyed whole-cell vaccine for crucian carp against *A. hydrophila* resulted in an 87% relative percentage survival (RPS) (Wang et al., 2020). Zhang et al. (2020) reports that a novel vaccine produced by sonication-mediated inactivation of the whole cell and further conjugated with the bacterial lysate using single-walled carbon nanotubes (SWCNTs) carrier inoculated (10 mg/L for bath immersion and 10 μg/fish for intraperitoneal injection) into grass carp resulted in elevated antibody titers; upregulation of tumor necrosis factor-α (TNF-α), type I interferons (IFN-1), immunoglobulin M (IgM), and C-reactive protein (CRP) genes in the injection group; and ultimately RPS 28 days postinjection, thus suggesting effective immune response induction (Zhang et al., 2020). In addition to the aforementioned genes, Liu et al. (2015) reported that DNA vaccine with SWCNTs carriers against *A. hydrophila* in juvenile grass carp also upregulated CD8-α, MHC I, and IL-8 and possessed impressive prophylactic immune effect (Liu et al., 2016). Other candidates including live-attenuated ones (Pridgeon, & Klesius, 2011a, 2011b); killed cells (Shoemaker, LaFrentz et al., 2018; Shoemaker, Mohammed et al., 2018; Yun et al., 2017); futuristic recombinant (Liu et al., 2015), fimbrial (Abdelhamed et al., 2016), membrane (Abdelhamed et al., 2017), β-glucan binding (Anjugam et al., 2018) proteins; also, bivalent (Bastardo et al., 2012) vaccines have been used. Other unconventional approaches to achieve control include the use of herbal formulations (Harikrishnan et al., 2010) and medicinal plants (Acar et al., 2018; Baba et al., 2018; Kesbiç et al., 2020; Nya & Austin, 2011) in the diet have similarly been pursued. Other nontraditional antimicrobial agents like thymol (Dong et al., 2020), morin (Dong et al., 2018), magnolol (Dong et al., 2017), and curcumin (Tanhay Mangoudehi et al., 2020) have been shown to possess inhibitory effects.

While there are studies from Nigeria, they are largely limited to incidence and prevalence (Ashiru et al., 2011; Dashe et al., 2014). This represents a dearth that needs to be filled to enhance locally applicable and feasible solutions, since the country is a major producer of fish in Africa.

11.4.2 Flavobacterium

Flavobacterium, a category of rod-shaped, gram-negative, aerobic (Vaikuntapu et al., 2018), or facultatively anaerobic (Zhang, Thongda et al., 2017; Zhang, Li et al., 2017; Zhang, Zhao et al., 2017) bacteria, that may be motile or otherwise (Enisoglu-Atalay et al., 2018), contains the species *F. columnare* (Guo et al., 2018) and *F. psychrophilum* (Duchaud et al., 2007; Rochat et al., 2019), both of which are major pathogens of freshwater fish (Cai et al., 2019; Declercq et al., 2013) even though other species in the genus are free-living (Duchaud et al., 2007).

A number of hydrolytic enzymes (Yu et al., 2006), including elastinase (Rochat et al., 2019), protease (Osorio et al., 2019; Secades et al., 2003), cellulases (Herrera et al., 2019), glycosidases (Garbe & Collin, 2012; Vaikuntapu et al., 2018), and peptidase (Kitazono et al., 1996), of diverse metabolic capability (Gangwar et al., 2011) have been reported from the genus, which may be instrumental to its pathogenesis in fish. Apart from possessing plasmids (Duchaud et al., 2018; Ngo et al., 2017; Zhang, Thongda et al., 2017; Zhang, Li et al., 2017; Zhang, Zhao et al., 2017), Duchaud et al. (2007) reports that the JIP02/86 (ATCC 49511) strain of *F. psychrophilum*, with virulent properties, has a circular chromosome bearing 2,861,988 base pair (bp), which putatively encodes 2432 coding sequences differentially expressed (Paneru et al., 2016) according to physicochemical properties of the environment (LaFrentz et al., 2009); this relatively small genome size is presumed to be responsible for its limited distribution, which restricts it to cold habitats (Jarau et al., 2018) with reports of differential expression of the genes in *F. columnare* (Peatman et al., 2013) relative to the physicochemical nature of habitats also available (Cai et al., 2019; Lange et al., 2018). *F. psychrophilum* infection is referred to as bacterial cold-water disease (BCWD) (Lange et al., 2018; Sundell et al., 2019), tail-rot, peduncle disease (Jarau et al., 2018), or rainbow trout fry syndrome (RTFS) (Duchaud et al., 2018; Jarau et al., 2018), with columnaris disease being the nomenclature for *F. columnare* infection (Cai et al., 2019) describing the characteristic column-like lesions first described in the Mississippi River fish population (Declercq et al., 2013) of different types (LaFrentz et al., 2018). The genome of *F. columnare* possesses 3,171,081 bp-bearing circular DNA with a forecast of 2784 protein-encoding genes (Zhang, Thongda et al., 2017; Zhang, Li et al., 2017; Zhang, Zhao et al., 2017).

The arsenal of virulence factors includes biofilm formation (Cai et al., 2019; Lange et al., 2018), enzymes (Herrera et al., 2019; Rochat et al., 2019), efflux pump, gliding motility, secretory protein (Zhang, Thongda et al., 2017; Zhang, Li et al., 2017; Zhang, Zhao et al., 2017), antibiotic resistance (Jarau et al., 2019; Mata et al., 2018), outer membrane proteins (Omp) (Kayansamruaj et al., 2017), and type IX secretion system (T9SS) (Barbier et al., 2020; Kayansamruaj et al., 2017), among other factors (Li et al., 2015; Zhang, Thongda et al., 2017; Zhang, Li et al., 2017; Zhang, Zhao et al., 2017). The genetic diversity of *F. psychrophilum* (Duchaud et al., 2018; Marancik & Wiens, 2013; Ngo et al., 2017; Rochat et al., 2019; Sundell et al., 2019) and *F. columnare* (Barony et al., 2015; Dong et al., 2015; LaFrentz et al., 2018; LaFrentz et al., 2019) means that the possibility of discovering other factors is highly plausible.

Pathogenesis begins with contact and colonization of the gills and/or scales (Shoemaker & LaFrentz, 2015), which are in direct contact with the water body that may harbor the pathogens. Consequently, the mucus on the scales plays an important immunomodulatory role (Papadopoulou et al., 2017; Shoemaker, LaFrentz et al., 2018; Shoemaker, Mohammed et al., 2018) in the initial phase of contact. Significantly, however, *F. columnare* (Shoemaker, LaFrentz et al., 2018; Shoemaker, Mohammed et al., 2018) and *F. psychrophilum* (Madsen & Dalsgaard, 2008; Papadopoulou et al., 2017) are able to circumvent this innate defense to survive and grow within the mucus matrix with the chemical nature of the mucus and the proteolytic competence (Beck et al., 2012; Shoemaker, LaFrentz et al., 2018; Shoemaker, Mohammed et al., 2018) of the pathogens instrumental to its successful colonization and transmission. *F. columnare* cells aggregate around and inside the mucus pores, with this interaction mediated by a lectin with rhamnose-binding ability called "rhamnose-binding lectin (RBL)" (Declercq et al., 2013; Peatman et al., 2013; Shoemaker, LaFrentz et al., 2018; Shoemaker, Mohammed et al., 2018). Reported elastinase production in JIP 02/86 strain (Rochat et al., 2019), cell-surface adhesins, peptidases, and other enzymes (Barbier et al., 2020) may be influential in subsequent tissue permeation and dispersal to trigger systemic hemorrhage (Dar et al., 2016). This interaction stimulates chemotactic, cellular, and molecular responses (Declercq et al., 2013; Peatman et al., 2013; Shoemaker, LaFrentz et al., 2018; Shoemaker, Mohammed et al., 2018).

While disease manifestation differs depending on the virulence (Declercq et al., 2013; Rochat et al., 2019) and host resistance and/or susceptibility index (Declercq et al., 2016; Karvonen et al., 2019; Peatman et al., 2013), the major clinical signs are open lesions on body surfaces, pyknosis, necrosis in vital organs, loss of appetite, discolored and eroded fin, pale gills, exophthalmia, abdominal distension, vacuolar degeneration, dermal infiltration by lymphocyte (Starliper, 2011), and cardiac alterations, with the skin and gills appearing to be disproportionately affected (Declercq et al., 2013; Rochat et al., 2019). Table 11.2 outlines selected reports of *Flavobacterium* outbreaks.

Susceptibility to flavobacterial infection is influenced by some physiological factors. Mohammed and Arias (2015) report that the usage of the surface-active disinfectant $KMnO_4$, popularly called potassium permanganate (PP), resulted in dysbiosis, which consequently resulted in greater susceptibility of channel catfish to *F. columnare*. High organic matter content has led to similar results (Srisapoome & Areechon, 2017). Stress is also influential, as Declercq et al. (2016) reports that higher cortisol level led to higher phenotypic virulence. This preliminary observation necessitates *in vivo* assays for confirmation. Cortisol has been shown to enhance biofilm formation and dispersal ability, while *sprT*—a T9SS component that encodes mobile surface adhesion

Table 11.2 Selected *Flavobacterium* outbreaks.

Host	Species/strain	Location	References
Farmed salmonid fish	*Flavobacterium psychrophilum*	Sweden	Söderlund, Hakhverdyan, Aspán, and Jansson (2018)
Rainbow trout	*F. psychrophilum*	Canada	Jarau et al. (2018)
North American salmonids	*F. psychrophilum*	United States and Canada	Knupp et al. (2019)
Rainbow trout, coho salmon, Chinook salmon	*F. psychrophilum*	United States	Van Vliet, Wiens, Loch, Nicolas, and Faisal (2016)
Rainbow trout	*F. psychrophilum*	Denmark, Finland, Norway, and Sweden	Nilsen et al. (2014)
Rainbow trout, brown trout, lacustrine trout	*F. psychrophilum*	Switzerland	Strepparava, Nicolas, Wahli, Segner, and Petrini (2013)
Rainbow trout	*F. psychrophilum*	France	Fraslin et al. (2019)
Striped catfish	*Flavobacterium columnare*	Vietnam	Tien, Dung, Tuan, and Crumlish (2012)
Amazon catfish and pacamã	*F. columnare*	Brazil	Barony et al. (2015)
Catfish and sport fish species	*F. columnare*	United States	Mohammed and Arias (2014)
Catla	*F. columnare*	India	Verma and Rathore (2013)
Salmonids, centrarchids, cyprinids, percids, and esocids	*F. columnare*	United States	Faisal et al. (2017)
Yellow perch and lake whitefish	*F. columnare*	Canada	Scott and Bollinger (2014)
Perch	*F. columnare*	United Kingdom	Morley and Lewis (2010)

competence—was markedly upregulated (Declercq et al., 2019). These reports suggest that stress and predisposing factors have a major impact on virulence and susceptibility.

There also appears to be a difference in immune response between resistant and susceptible hosts. Peatman et al. (2013) reports the differential expression of essential innate immune constituents. Resistant *Ictalurus punctatus* (channel catfish) had the inducible nitric oxide synthase 2b (iNOS2b), lysozyme C, IL-8, and TNF-alpha significantly highly expressed in resistant fish relative to susceptible fish, in which a glyco-protein4(MFAP4)-like molecule associated to microfibril, NK-lysin, CD103, CD8, IL-17RA, IL-17A/F2 expression was upregulated. MHC class I and class II genes were differentially expressed in both fish cohorts (Peatman et al., 2013). Starliper (2011) suggests that, in rainbow trout, innate immunity is associated with the spleen size, which has been corroborated elsewhere (Zwollo et al., 2017). Thus acute serum amyloid A (A-SAA) may be essential in immune response; however, the size may not confer any protective advantage (Moore et al., 2019), although Zwollo et al. (2017) implied that this may not be conclusive yet.

The immune cells of myeloid origin are similarly involved in immune response to *F. psychrophilum* infection in rainbow trout. Macrophage colony stimulating factor-receptor 1 (M-CSFR) was ominously overexpressed in anterior and posterior kidneys. The cells include neutrophils, early B-cell factor-expressing (EBF), macrophages, and monocytes, and the proteins Pu1, myeloperoxidase (MPO), IL-1β were identified in active bacterial challenge and differential expression was observed between resistant and susceptible fish (Moore et al., 2019). IgT/Z, a new teleost antibody isotype

(Du et al., 2016) functionally akin to IgA–IgD, IgM singly and/or coliberated by B cells, is also important in humoral response to infection (Zwollo et al., 2017).

A major control approach against flavobacterial infection involves diet. Florfenicol (FFC) (15 mg/kg)—an antibacterial agent that inhibits protein synthesis by binding the 50S ribosomal RNA—and copper sulfate (2.1 mg/L) infused into a commercially approved fish feed resulted in significantly higher survival rates (90% and 88%, respectively) in sunshine bass against experimentally infected *F. columnare* (Darwish et al., 2012). The use of antibiotic-supplemented feeds comes with the danger of hypothetically promoting antibiotic resistance; further discouraging the extensive prophylactic usage of antibiotics in agriculture represents a cornerstone of antibiotic stewardship. Consequently, the use of pre- and probiotic agents represents a viable alternative. Results from Zhao et al. (2015) has validated the benefits of commercial prebiotic dietary components as good immune priming agents that enhance response to *F. columnare* infection in channel catfish. Laboratory and field trials have shown that probiotic controls using *Pseudomonas fluorescens* CP14 and CP23 (Seghouani et al., 2017), intraperitoneally administered C6-6 *Enterobacter* species from rainbow trout gut (Ghosh et al., 2016), *Luteimonas aestuarii*, *Leucobacter luti*, *Microbacterium oxydans*, *Rhodococcus cercidiphylli*, *R. qingshengii*, *Dietzia maris*, and *Sphingopyxis bauzanensis* isolated from brook charr skin (Boutin et al., 2012), 16 different bacterial isolates from rainbow trout gut (Burbank et al., 2012), *B. subtilis* AB01, AP102, AP193, AP219, AP254, AP301, and *B. methylotrophicus* (Ran et al., 2012) resulted in high survival rates.

At a dose-dependent rate until a threshold of 5%, *Nigella sativa* seed and oil infused into feeds led to lower mortality relative to controls, signifying a role for herbal agents (Mohammed & Arias, 2016). The *in vitro* and *in vivo* reports of strong effects of aqueous extract of Asiatic pennywort against *F. columnare* in Nile tilapia (Rattanachaikunsopon & Phumkhachorn, 2010) substantiates this.

Since vaccination is the best control mechanism against infectious diseases (Kitiyodom et al., 2019), vaccination remains a major means of controlling flavobacterial infection. While vaccine development against *F. psychrophilum* (Gómez et al., 2014) and *F. columnare* (Kayansamruaj et al., 2020) is a challenging undertaking, some success is being recorded, with *in silico* rational design also contributing to the endeavor (Mahendran et al., 2016). Lange et al. (2019) reports that both live attenuated and recombinant *F. columnare* DnaK protein vaccines are promising vaccine candidates that boost survival. Other promising candidates include polyvalent immersion (Hoare et al., 2017); polyvalent injectable (Hoare et al., 2019); intramuscularly administered divalent and polyvalent (Fredriksen et al., 2013); *F. psychrophilum* immersion collagenase (Nakayama et al., 2017); live attenuated (Ma et al., 2019) vaccines. A killed bacterial suspension vaccine complexed with chitosan improved the mucoadhesiveness for immersive administration and resulted in enhanced relative percentage survival (Kitiyodom et al., 2019). Other factors that improve vaccine efficacy include the selection of an appropriate parent strain (Mohammed et al., 2013), as the current commercially available vaccine in the United States does not provide sufficient protection against more virulent strains (Zhang, Thongda et al., 2017; Zhang, Li et al., 2017; Zhang, Zhao et al., 2017) as well as booster regimes and dosage (Sudheesh & Cain, 2016). Vaccination stimulates more than just antibody response; it has now been shown to mediate the sensitization and proliferation of resident secretory cells (Zhang, Thongda et al., 2017; Zhang, Li et al., 2017; Zhang, Zhao et al., 2017). Lytic phages are also used in effective control (Castillo et al., 2014; Christiansen et al., 2016; Madsen et al., 2013; Papadopoulou et al., 2019) because of their host specificity.

11.5 Conclusion

Declining fish capture and the population explosion, among several other factors, are driving the acceleration of aquaculture. However, bacterial pathogens represent a major threat to this social and economic development, the main ones being *A. hydrophila*, *F. psychrophilum*, and *F. columnare*, which are reviewed extensively in this chapter.

Consequently, a better understanding of predisposing abiotic and biotic factors, the course of pathogenesis, immune response and mediators, and various control methods is pertinent. Our review shows an extensive body of knowledge on epidemiology and diversity—originating from the global North—is supporting the development of vaccines and informing prospecting for organic and chemical antibiotics. Consequently, capacity building in this regard is essential for sub-Saharan African countries. Due to the observation of continuous evolution of these pathogens, active and real-time surveillance of the pathogens is critical to ensure an updated record at all times.

References

Abdelhamed, H., Ibrahim, I., Nho, S. W., Banes, M. M., Wills, R. W., Karsi, A., & Lawrence, M. L. (2017). Evaluation of three recombinant outer membrane proteins, OmpA1, Tdr, and TbpA, as potential vaccine antigens against virulent *Aeromonas hydrophila* infection in channel catfish (*Ictalurus punctatus*). *Fish & Shellfish Immunology*, *66*, 480–486. Available from https://doi.org/10.1016/j.fsi.2017.05.043.

Abdelhamed, H., Nho, S. W., Turaga, G., Banes, M. M., Karsi, A., & Lawrence, M. L. (2016). Protective efficacy of four recombinant fimbrial proteins of virulent *Aeromonas hydrophila* strain ML09–119 in channel catfish. *Veterinary Microbiology*, *197*, 8–14. Available from https://doi.org/10.1016/j.vetmic.2016.10.026.

Aboyadak, I., Ali, N., Goda, A., Aboelgalagel, W., & Salam, A. (2015). Molecular detection of *Aeromonas hydrophila* as the main cause of outbreak in tilapia farms in Egypt. *Journal of Aquaculture & Marine Biology*, *2*(5), 00045.

Acar, Ü., Parrino, V., Kesbiç, O. S., Lo Paro, G., Saoca, C., Abbate, F., & Fazio, F. (2018). Effects of different levels of pomegranate seed oil on some blood parameters and disease resistance against *Yersinia ruckeri* in rainbow trout. *Frontiers in Physiology*, *9*, 596. Available from https://doi.org/10.3389/fphys.2018.00596.

Adeogun, A. O., Ibor, O. R., Omiwole, R., Chukwuka, A. V., Adewale, A. H., Kumuyi, O., & Arukwe, A. (2020). Sex-differences in physiological and oxidative stress responses and heavy metals burden in the black jaw tilapia, *Sarotherodon melanotheron* from a tropical freshwater dam (Nigeria). *Comparative Biochemistry and Physiology. Toxicology & Pharmacology: CBP*, *229*, 108676. Available from https://doi.org/10.1016/j.cbpc.2019.108676.

Adewumi, A. (2015). Aquaculture in Nigeria: Sustainability issues and challenges. *Direct Resource Journal of Agriculture and Food Science*, *3*, 12.

Adewuyi, S., Phillip, B., Ayinde, I., & Akerele, D. (2010). Analysis of profitability of fish farming in Ogun State, Nigeria. *Journal of Human Ecology*, *31*(3), 179–184.

Adeyeye, S., Oyewole, O., Obadina, O., Adeniran, O., Oyedele, H., Olugbile, A., & Omemu, A. (2016). Effect of smoking methods on microbial safety, polycyclic aromatic hydrocarbon, and heavy metal concentrations of traditional smoked fish from Lagos State, Nigeria. *Journal of Culinary Science & Technology*, *14*(2), 91–106.

Alam, I., Khattak, M. N. K., Mulk, S., Dawar, F. U., Shahi, L., & Ihsanullah, I. (2020). Heavy metals assessment in water, sediments, algae and two fish species from River Swat, Pakistan. *Bulletin of Environmental Contamination and Toxicology*, *105*, 546–552.

Alawode, O., & Oluwatayo, I. (2019). Development outcomes of fadama III among fish farmers in Nigeria: Evidence from Lagos State. *Evaluation and Program Planning*, *75*, 10–19.

Algammal, A. M., Mohamed, M. F., Tawfiek, B. A., Hozzein, W. N., El Kazzaz, W. M., & Mabrok, M. (2020). Molecular typing, antibiogram and PCR-RFLP based detection of *Aeromonas hydrophila* complex isolated from *Oreochromis niloticus*. *Pathogens*, *9*(3), 238. Available from https://doi.org/10.3390/pathogens9030238.

Allen, K., Rachmi, A. F., & Cai, J. (2017). Nigeria: Faster aquaculture growth needed to bridge fish demand-supply gap. *FAO Aquaculture Newsletter*, *57*, 36–37.

AlYahya, S. A., Ameen, F., Al-Niaeem, K. S., Al-Sa'adi, B. A., Hadi, S., & Mostafa, A. A. (2018). Histopathological studies of experimental *Aeromonas hydrophila* infection in blue tilapia, *Oreochromis aureus*. *Saudi Journal of Biological Sciences*, *25*(1), 182–185. Available from https://doi.org/10.1016/j.sjbs.2017.10.019.

Amankwaah, D., Cobbina, S., Tiwaa, Y., Bakobie, N., & Millicent, E. (2014). Assessment of pond effluent effect on water quality of Asuofia Stream, Ghana. *African Journal of Environmental Science and Technology*, *8*(5), 306–311.

Anjugam, M., Vaseeharan, B., Iswarya, A., Gobi, N., Divya, M., Thangaraj, M. P., & Elumalai, P. (2018). Effect of β-1, 3 glucan binding protein based zinc oxide nanoparticles supplemented diet on immune response and disease resistance in *Oreochromis mossambicus* against *Aeromonas hydrophila*. *Fish & Shellfish Immunology*, *76*, 247–259. Available from https://doi.org/10.1016/j.fsi.2018.03.012.

Ansah, Y. B., Frimpong, E. A., & Amisah, S. (2013). Characterisation of potential aquaculture pond effluents, and physico-chemical and microbial assessment of effluent-receiving waters in central Ghana. *African Journal of Aquatic Science*, *38*(2), 185–192. Available from https://doi.org/10.2989/16085914.2013.767182.

Ashiru, A., Uaboi-Egbeni, P., Oguntowo, J., & Idika, C. (2011). Isolation and antibiotic profile of *Aeromonas* species from tilapia fish (*Tilapia nilotica*) and catfish (*Clarias betrachus*). *Pakistan Journal of Nutrition*, *10*(10), 982–986.

Awan, F., Dong, Y., Wang, N., Liu, J., Ma, K., & Liu, Y. (2018). The fight for invincibility: Environmental stress response mechanisms and *Aeromonas hydrophila*. *Microbial Pathogenesis*, *116*, 135–145. Available from https://doi.org/10.1016/j.micpath.2018.01.023.

Baba, E., Acar, Ü., Yılmaz, S., Zemheri, F., & Ergün, S. (2018). Dietary olive leaf (*Olea europea* L.) extract alters some immune gene expression levels and disease resistance to *Yersinia ruckeri* infection in rainbow trout *Oncorhynchus mykiss*. *Fish & Shellfish Immunology*, *79*, 28–33. Available from https://doi.org/10.1016/j.fsi.2018.04.063.

Bandeira Junior, G., de Freitas Souza, C., Descovi, S. N., Antoniazzi, A., Cargnelutti, J. F., & Baldisserotto, B. (2019). *Aeromonas hydrophila* infection in silver catfish causes hyperlocomotion related to stress. *Microbial Pathogenesis*, *132*, 261–265. Available from https://doi.org/10.1016/j.micpath.2019.05.017.

Barbier, P., Rochat, T., Mohammed, H. H., Wiens, G. D., Bernardet, J.-F., Halpern, D., & McBride, M. J. (2020). The type IX secretion system is required for virulence of the fish pathogen *Flavobacterium psychrophilum*. *Applied and Environmental Microbiology*, *18*, e00799-20.

Barkham, T., Zadoks, R. N., Azmai, M. N. A., Baker, S., Bich, V. T. N., Chalker, V., & Chen, S. L. (2019). One hypervirulent clone, sequence type 283, accounts for a large proportion of invasive *Streptococcus agalactiae* isolated from humans and diseased tilapia in Southeast Asia. *PLoS Neglected Tropical Diseases*, *13*(6), e0007421. Available from https://doi.org/10.1371/journal.pntd.0007421.

Barony, G. M., Tavares, G. C., Assis, G. B., Luz, R. K., Figueiredo, H. C., & Leal, C. A. (2015). New hosts and genetic diversity of *Flavobacterium columnare* isolated from Brazilian native species and Nile tilapia. *Diseases of Aquatic Organisms*, *117*(1), 1–11. Available from https://doi.org/10.3354/dao02931.

Basra, K., Fabian, M. P., & Scammell, M. K. (2018). Consumption of contaminated seafood in an environmental justice community: A qualitative and spatial analysis of fishing controls. *Environmental Justice*, *11*(1), 6–14. Available from https://doi.org/10.1089/env.2017.0010.

Bastardo, A., Ravelo, C., Castro, N., Calheiros, J., & Romalde, J. L. (2012). Effectiveness of bivalent vaccines against *Aeromonas hydrophila* and *Lactococcus garvieae* infections in rainbow trout *Oncorhynchus mykiss* (Walbaum). *Fish & Shellfish Immunology*, *32*(5), 756–761. Available from https://doi.org/10.1016/j.fsi.2012.01.028.

Beck, B. H., Farmer, B. D., Straus, D. L., Li, C., & Peatman, E. (2012). Putative roles for a rhamnose binding lectin in *Flavobacterium columnare* pathogenesis in channel catfish *Ictalurus punctatus*. *Fish & Shellfish Immunology*, *33*(4), 1008–1015. Available from https://doi.org/10.1016/j.fsi.2012.08.018.

Ben Hamed, S., Guardiola, F. A., Mars, M., & Esteban, M. (2014). Pathogen bacteria adhesion to skin mucus of fishes. *Veterinary Microbiology*, *171*(1–2), 1–12. Available from https://doi.org/10.1016/j.vetmic.2014.03.008.

Ben Hamed, S., Tavares Ranzani-Paiva, M. J., Tachibana, L., de Carla Dias, D., Ishikawa, C. M., & Esteban, M. A. (2018). Fish pathogen bacteria: Adhesion, parameters influencing virulence and interaction with host cells. *Fish & Shellfish Immunology*, *80*, 550–562. Available from https://doi.org/10.1016/j.fsi.2018.06.053.

Béné, C., Barange, M., Subasinghe, R., Pinstrup-Andersen, P., Merino, G., Hemre, G.-I., & Williams, M. (2015). Feeding 9 billion by 2050 – Putting fish back on the menu. *Food Security*, *7*(2), 261–274. Available from https://doi.org/10.1007/s12571-015-0427-z.

Bizjak-Mali, L., Zalar, P., Turk, M., Babicč, M. N., Kostanjšek, R., & Gunde-Cimerman, N. (2018). Opportunistic fungal pathogens isolated from a captive individual of the European blind cave salamander *Proteus anguinus*. *Diseases of Aquatic Organisms*, *129*(1), 15–30. Available from https://doi.org/10.3354/dao03229.

Bojarski, B., Kot, B., & Witeska, M. (2020). Antibacterials in aquatic environment and their toxicity to fish. *Pharmaceuticals (Basel)*, *13*(8). Available from https://doi.org/10.3390/ph13080189.

Boutin, S., Bernatchez, L., Audet, C., & Derôme, N. (2012). Antagonistic effect of indigenous skin bacteria of brook charr (*Salvelinus fontinalis*) against *Flavobacterium columnare* and *F. psychrophilum*. *Veterinary Microbiology*, *155*(2–4), 355–361. Available from https://doi.org/10.1016/j.vetmic.2011.09.002.

Brinchmann, M. F. (2016). Immune relevant molecules identified in the skin mucus of fish using-omics technologies. *Molecular Biosystems*, *12*(7), 2056–2063.

Bukola, D., Zaid, A., Olalekan, E. I., & Falilu, A. (2015). Consequences of anthropogenic activities on fish and the aquatic environment. *Poultry, Fisheries & Wildlife Sciences*, *3*, 1–12.

Burbank, D. R., Lapatra, S. E., Fornshell, G., & Cain, K. D. (2012). Isolation of bacterial probiotic candidates from the gastrointestinal tract of rainbow trout, *Oncorhynchus mykiss* (Walbaum), and screening for inhibitory activity against *Flavobacterium psychrophilum*. *Journal of Fish Diseases*, *35*(11), 809–816. Available from https://doi.org/10.1111/j.1365-2761.2012.01432.x.

Cai, W., De La Fuente, L., & Arias, C. R. (2019). Transcriptome analysis of the fish pathogen *Flavobacterium columnare* in biofilm suggests calcium role in pathogenesis. *BMC Microbiology*, *19*(1), 151. Available from https://doi.org/10.1186/s12866-019-1533-4.

Carballeira, C., Cebro, A., Villares, R., & Carballeira, A. (2018). Assessing changes in the toxicity of effluents from intensive marine fish farms over time by using a battery of bioassays. *Environmental Science and Pollution Research International*, *25*(13), 12739–12748. Available from https://doi.org/10.1007/s11356-018-1403-x.

Castillo, D., Christiansen, R. H., Espejo, R., & Middelboe, M. (2014). Diversity and geographical distribution of *Flavobacterium psychrophilum* isolates and their phages: Patterns of susceptibility to phage infection and phage host range. *Microbial Ecology*, *67*(4), 748–757.

Chen, D. D., Li, J. H., Yao, Y. Y., & Zhang, Y. A. (2019). *Aeromonas hydrophila* suppresses complement pathways via degradation of complement C3 in bony fish by metalloprotease. *Fish & Shellfish Immunology*, *94*, 739–745. Available from https://doi.org/10.1016/j.fsi.2019.09.057.

Chen, N., Jiang, J., Gao, X., Li, X., Zhang, Y., Liu, X., & Zhang, X. (2018). Histopathological analysis and the immune related gene expression profiles of mandarin fish (*Siniperca chuatsi*) infected with *Aeromonas hydrophila*. *Fish & Shellfish Immunology*, *83*, 410–415.

Christiansen, R. H., Madsen, L., Dalsgaard, I., Castillo, D., Kalatzis, P. G., & Middelboe, M. (2016). Effect of bacteriophages on the growth of *Flavobacterium psychrophilum* and development of phage-resistant strains. *Microbial Ecology*, *71*(4), 845–859.

Clemence, M. A., & Guerrant, R. L. (2015). Infections and intoxications from the ocean: Risks of the shore. *Microbiology Spectrum*, *3*(6). Available from https://doi.org/10.1128/microbiolspec.IOL5-0008-2015.

Coleman, D. J., Camus, A. C., Martínez-López, B., Yun, S., Stevens, B., & Soto, E. (2018). Effects of temperature on *Veronaea botryosa* infections in white sturgeon *Acipenser transmontanus* and fungal induced cytotoxicity of fish cell lines. *Veterinary Research*, *49*(1), 11. Available from https://doi.org/10.1186/s13567-018-0507-0.

Comerford, K. B., & Pasin, G. (2016). Emerging evidence for the importance of dietary protein source on glucoregulatory markers and type 2 diabetes: Different effects of dairy, meat, fish, egg, and plant protein foods. *Nutrients*, *8*(8), 446. Available from https://doi.org/10.3390/nu8080446.

Committee on Considerations for the Future of Animal Science., R., Science., Technology for Sustainability, P., Policy, Global, A., Board on, A., & National Research, C. (2015). *Critical role of animal science research in food security and sustainability. Critical role of animal science research in food security and sustainability*. Washington (DC): National Academies Press (US), Copyright 2015 by the National Academy of Sciences. All rights reserved.

Dar, G. H., Bhat, R. A., Kamili, A. N., Chishti, M. Z., Qadri, H., Dar, R., & Mehmood, M. A. (2020). *Correlation between pollution trends of freshwater bodies and bacterial disease of fish fauna. Fresh water pollution dynamics and remediation* (pp. 51–67). Singapore: Springer.

Dar, G. H., Dar, S. A., Kamili, A. N., Chishti, M. Z., & Ahmad, F. (2016). Detection and characterization of potentially pathogenic *Aeromonas sobria* isolated from fish *Hypophthalmichthys molitrix* (Cypriniformes: Cyprinidae). *Microbial Pathogenesis*, *91*, 136–140.

Dar, G. H., Kamili, A. N., Chishti, M. Z., Dar, S. A., Tantry, T. A., & Ahmad, F. (2016). Characterization of *Aeromonas sobria* isolated from fish rohu (*Labeo rohita*) collected from polluted pond. *Journal of Bacteriology & Parasitology*, *7*(3), 1–5. Available from https://doi.org/10.4172/2155-9597.1000273.

Darwish, A. M., Bebak, J. A., & Schrader, K. K. (2012). Assessment of Aquaflor(®), copper sulphate and potassium permanganate for control of *Aeromonas hydrophila* and *Flavobacterium columnare* infection in sunshine bass, *Morone chrysops* female × *Morone saxatilis* male. *Journal of Fish Diseases*, *35*(9), 637–647. Available from https://doi.org/10.1111/j.1365-2761.2012.01393.x.

Dashe, Y., Raji, M., Abdu, P., Oladele, B., & Olarinmoye, D. (2014). Isolation of *Aeromonas hydrophila* from commercial chickens in Jos Metropolis. *Nigeria. International Journal of Poultry Science*, *13*(1), 26–30.

Dauda, A. B., Natrah, I., Karim, M., Kamarudin, M. S., & Bichi, A. (2018). African catfish aquaculture in Malaysia and Nigeria: Status, trends and prospects. *Fisheries and Aquaculture Journal*, *9*(1), 1–5.

Declercq, A. M., Aerts, J., Ampe, B., Haesebrouck, F., De Saeger, S., & Decostere, A. (2016). Cortisol directly impacts *Flavobacterium columnare* in vitro growth characteristics. *Veterinary Research*, *47*(1), 84. Available from https://doi.org/10.1186/s13567-016-0370-9.

Declercq, A. M., Cai, W., Naranjo, E., Thongda, W., Eeckhaut, V., Bauwens, E., & Decostere, A. (2019). Evidence that the stress hormone cortisol regulates biofilm formation differently among *Flavobacterium columnare* isolates. *Veterinary Research*, *50*(1), 24. Available from https://doi.org/10.1186/s13567-019-0641-3.

Declercq, A. M., Haesebrouck, F., Van den Broeck, W., Bossier, P., & Decostere, A. (2013). Columnaris disease in fish: A review with emphasis on bacterium-host interactions. *Veterinary Research*, *44*(1), 27. Available from https://doi.org/10.1186/1297-9716-44-27.

Dias, M. K., Sampaio, L. S., Proietti-Junior, A. A., Yoshioka, E. T., Rodrigues, D. P., Rodriguez, A. F., & Tavares-Dias, M. (2016). Lethal dose and clinical signs of *Aeromonas hydrophila* in *Arapaima gigas* (Arapaimidae), the giant fish from Amazon. *Veterinary Microbiology*, *188*, 12–15. Available from https://doi.org/10.1016/j.vetmic.2016.04.001.

Dong, H. T., LaFrentz, B., Pirarat, N., & Rodkhum, C. (2015). Phenotypic characterization and genetic diversity of *Flavobacterium columnare* isolated from red tilapia, *Oreochromis* sp., in Thailand. *Journal of Fish Diseases*, *38*(10), 901–913. Available from https://doi.org/10.1111/jfd.12304.

Dong, J., Ding, H., Liu, Y., Yang, Q., Xu, N., Yang, Y., & Ai, X. (2017). Magnolol protects channel catfish from *Aeromonas hydrophila* infection via inhibiting the expression of aerolysin. *Veterinary Microbiology*, *211*, 119–123. Available from https://doi.org/10.1016/j.vetmic.2017.10.005.

Dong, J., Liu, Y., Xu, N., Yang, Q., & Ai, X. (2018). Morin protects channel catfish from *Aeromonas hydrophila* infection by blocking aerolysin activity. *Frontiers in Microbiology*, *9*, 2828. Available from https://doi.org/10.3389/fmicb.2018.02828.

Dong, J., Zhang, L., Liu, Y., Xu, N., Zhou, S., Yang, Q., & Ai, X. (2020). Thymol protects channel catfish from *Aeromonas hydrophila* infection by inhibiting aerolysin expression and biofilm formation. *Microorganisms*, *8*(5), 636. Available from https://doi.org/10.3390/microorganisms8050636.

Du, Y., Tang, X., Zhan, W., Xing, J., & Sheng, X. (2016). Immunoglobulin Tau heavy chain (IgT) in flounder, *Paralichthys olivaceus*: Molecular cloning, characterization, and expression analyses. *International Journal of Molecular Sciences*, *17*(9), 1571.

Duchaud, E., Boussaha, M., Loux, V., Bernardet, J.-F., Michel, C., Kerouault, B., & Benmansour, A. (2007). Complete genome sequence of the fish pathogen *Flavobacterium psychrophilum. Nature Biotechnology*, *25*(7), 763–769. Available from https://doi.org/10.1038/nbt1313.

Duchaud, E., Rochat, T., Habib, C., Barbier, P., Loux, V., Guérin, C., & Nicolas, P. (2018). Genomic diversity and evolution of the fish pathogen *Flavobacterium psychrophilum. Frontiers in Microbiology*, *9*, 138. Available from https://doi.org/10.3389/fmicb.2018.00138.

El-Bahar, H. M., Ali, N. G., Aboyadak, I. M., Khalil, S., & Ibrahim, M. S. (2019). Virulence genes contributing to *Aeromonas hydrophila* pathogenicity in *Oreochromis niloticus. International Microbiology: The Official Journal of the Spanish Society for Microbiology*, *22*(4), 479–490. Available from https://doi.org/10.1007/s10123-019-00075-3.

Elbehiry, A., Marzouk, E., Abdeen, E., Al-Dubaib, M., Alsayeqh, A., Ibrahem, M., & Hemeg, H. A. (2019). Proteomic characterization and discrimination of *Aeromonas* species recovered from meat and water samples with a spotlight on the antimicrobial resistance of *Aeromonas hydrophila. MicrobiologyOpen*, *8*(11), e782. Available from https://doi.org/10.1002/mbo3.782.

Enisoglu-Atalay, V., Atasever-Arslan, B., Yaman, B., Cebecioglu, R., Kul, A., Ozilhan, S., & Catal, T. (2018). Chemical and molecular characterization of metabolites from *Flavobacterium* sp. *PLoS One*, *13*(10), e0205817. Available from https://doi.org/10.1371/journal.pone.0205817.

Faisal, M., Diamanka, A., Loch, T. P., LaFrentz, B. R., Winters, A. D., García, J. C., & Toguebaye, B. S. (2017). Isolation and characterization of *Flavobacterium columnare* strains infecting fishes inhabiting the Laurentian Great Lakes basin. *Journal of Fish Diseases*, *40*(5), 637–648. Available from https://doi.org/10.1111/jfd.12548.

Famoofo, O. O., & Adeniyi, I. F. (2020). Impact of effluent discharge from a medium-scale fish farm on the water quality of Odo-Owa stream near Ijebu-Ode, Ogun State, Southwest Nigeria. *Applied Water Science*, *10*(2), 68. Available from https://doi.org/10.1007/s13201-020-1148-9.

Fernández-Alacid, L., Sanahuja, I., Ordóñez-Grande, B., Sánchez-Nuño, S., Viscor, G., Gisbert, E., & Ibarz, A. (2018). Skin mucus metabolites in response to physiological challenges: A valuable non-invasive method to study teleost marine species. *Science of The Total Environment*, *644*, 1323–1335.

Fernández-Bravo, A., Kilgore, P. B., Andersson, J. A., Blears, E., Figueras, M. J., Hasan, N. A., & Chopra, A. K. (2019). T6SS and ExoA of flesh-eating *Aeromonas hydrophila* in peritonitis and necrotizing fasciitis during mono- and polymicrobial infections. *Proceedings of the National Academy of Sciences of the United States of America*, *116*(48), 24084–24092. Available from https://doi.org/10.1073/pnas.1914395116.

Fraslin, C., Brard-Fudulea, S., D'Ambrosio, J., Bestin, A., Charles, M., Haffray, P., & Phocas, F. (2019). Rainbow trout resistance to bacterial cold water disease: Two new quantitative trait loci identified after a natural disease outbreak on a French farm. *Animal Genetics*, *50*(3), 293–297. Available from https://doi.org/10.1111/age.12777.

Fredriksen, B. N., Olsen, R. H., Furevik, A., Souhoka, R. A., Gauthier, D., & Brudeseth, B. (2013). Efficacy of a divalent and a multivalent water-in-oil formulated vaccine against a highly virulent strain of *Flavobacterium psychrophilum* after intramuscular challenge of rainbow trout (*Oncorhynchus mykiss*). *Vaccine*, *31*(15), 1994–1998. Available from https://doi.org/10.1016/j.vaccine.2013.01.016.

Gangwar, P., Alam, S. I., & Singh, L. (2011). Metabolic characterization of cold active *Pseudomonas, Arthrobacter, Bacillus*, and *Flavobacterium* spp. from Western Himalayas. *Indian Journal of Microbiology*, *51*(1), 70–75. Available from https://doi.org/10.1007/s12088-011-0092-7.

Garbe, J., & Collin, M. (2012). Bacterial hydrolysis of host glycoproteins—powerful protein modification and efficient nutrient acquisition. *Journal of Innate Immunity*, *4*(2), 121–131.

Ghosh, B., Cain, K., Nowak, B., & Bridle, A. (2016). Microencapsulation of a putative probiotic Enterobacter species, C6-6, to protect rainbow trout, *Oncorhynchus mykiss* (Walbaum), against bacterial coldwater disease. *Journal of Fish Diseases*, *39*(1), 1–11.

Giakoumi, S., Pey, A., Di Franco, A., Francour, P., Kizilkaya, Z., Arda, Y., & Guidetti, P. (2019). Exploring the relationships between marine protected areas and invasive fish in the world's most invaded sea. *Ecological Applications: A Publication of the Ecological Society of America*, *29*(1), e01809. Available from https://doi.org/10.1002/eap.1809.

Gómez, E., Méndez, J., Cascales, D., & Guijarro, J. A. (2014). *Flavobacterium psychrophilum* vaccine development: A difficult task. *Microbial Biotechnology*, *7*(5), 414–423. Available from https://doi.org/10.1111/1751-7915.12099.

Gomna, A., Pawa, D.-A., & Mamman, Z. (2020). *Profitability analysis of fish farming in Niger State, Nigeria*. Hill Publishing Group.

Grema, H., Geidam, Y., & Egwu, G. (2011). Fish production in Nigeria: An update. *Nigerian Veterinary Journal*, *32*(3), 226–229.

Griffin, M. J., Goodwin, A. E., Merry, G. E., Liles, M. R., Williams, M. A., Ware, C., & Waldbieser, G. C. (2013). Rapid quantitative detection of *Aeromonas hydrophila* strains associated with disease outbreaks in catfish aquaculture. *Journal of Veterinary Diagnostic Investigation*, *25*(4), 473–481. Available from https://doi.org/10.1177/1040638713494210.

Guo, Y. L., Wu, P., Jiang, W. D., Liu, Y., Kuang, S. Y., Jiang, J., & Feng, L. (2018). The impaired immune function and structural integrity by dietary iron deficiency or excess in gill of fish after infection with *Flavobacterium columnare*: Regulation of NF-κB, TOR, JNK, p38MAPK, Nrf2 and MLCK signalling. *Fish & Shellfish Immunology*, *74*, 593–608. Available from https://doi.org/10.1016/j.fsi.2018.01.027.

Halim, N. R. A., Yusof, H. M., & Sarbon, N. M. (2016). Functional and bioactive properties of fish protein hydrolysates and peptides: A comprehensive review. *Trends in Food Science & Technology*, *51*, 24–33. Available from https://doi.org/10.1016/j.tifs.2016.02.007.

Harikrishnan, R., & Balasundaram, C. (2005). Modern trends in *Aeromonas hydrophila* disease management with fish. *Reviews in Fisheries Science*, *13*(4), 281–320. Available from https://doi.org/10.1080/10641260500320845.

Harikrishnan, R., Balasundaram, C., & Heo, M. S. (2010). Herbal supplementation diets on hematology and innate immunity in goldfish against *Aeromonas hydrophila*. *Fish & Shellfish Immunology*, *28*(2), 354–361. Available from https://doi.org/10.1016/j.fsi.2009.11.013.

Herrera, L. M., Braña, V., Franco Fraguas, L., & Castro-Sowinski, S. (2019). Characterization of the cellulase-secretome produced by the Antarctic bacterium *Flavobacterium* sp. AUG42. *Microbiological Research, 223-225*, 13–21. Available from https://doi.org/10.1016/j.micres.2019.03.009.

Hoare, R., Jung, S. J., Ngo, T. P. H., Bartie, K. L., Thompson, K. D., & Adams, A. (2019). Efficacy of a polyvalent injectable vaccine against *Flavobacterium psychrophilum* administered to rainbow trout (*Oncorhynchus mykiss* L.). *Journal of Fish Diseases*, *42*(2), 229–236. Available from https://doi.org/10.1111/jfd.12919.

Hoare, R., Ngo, T. P. H., Bartie, K. L., & Adams, A. (2017). Efficacy of a polyvalent immersion vaccine against *Flavobacterium psychrophilum* and evaluation of immune response to vaccination in rainbow trout fry (*Onchorynchus mykiss* L.). *Veterinary Research*, *48*(1), 43. Available from https://doi.org/10.1186/s13567-017-0448-z.

Hoseinifar, S. H., Sun, Y. Z., Wang, A., & Zhou, Z. (2018). Probiotics as means of diseases control in aquaculture, a review of current knowledge and future perspectives. *Frontiers in Microbiology*, *9*, 2429. Available from https://doi.org/10.3389/fmicb.2018.02429.

Hossain, M. J., Sun, D., McGarey, D. J., Wrenn, S., Alexander, L. M., Martino, M. E., & Liles, M. R. (2014). An Asian origin of virulent *Aeromonas hydrophila* responsible for disease epidemics in United States-farmed catfish. *mBio*, *5*(3), e00848–00814. Available from https://doi.org/10.1128/mBio.00848-14.

Hunt, L. M., Arlinghaus, R., Lester, N., & Kushneriuk, R. (2011). The effects of regional angling effort, angler behavior, and harvesting efficiency on landscape patterns of overfishing. *Ecological Applications: A Publication of the Ecological Society of America*, *21*(7), 2555–2575. Available from https://doi.org/10.1890/10-1237.1.

Iaria, C., Saoca, C., Guerrera, M. C., Ciulli, S., Brundo, M. V., Piccione, G., & Lanteri, G. (2019). Occurrence of diseases in fish used for experimental research. *Laboratory Animals*, *53*(6), 619–629. Available from https://doi.org/10.1177/0023677219830441.

Ibor, O. R., Andem, A. B., Eni, G., Arong, G. A., Adeougn, A. O., & Arukwe, A. (2020). Contaminant levels and endocrine disruptive effects in *Clarias gariepinus* exposed to simulated leachate from a solid waste dumpsite in Calabar, Nigeria. *Aquatic Toxicology (Amsterdam, Netherlands)*, *219*, 105375. Available from https://doi.org/10.1016/j.aquatox.2019.105375.

Ikutegbe, V., & Sikoki, F. (2014). Microbiological and biochemical spoilage of smoke-dried fishes sold in West African open markets. *Food Chemistry*, *161*, 332–336. Available from https://doi.org/10.1016/j.foodchem.2014.04.032.

Ina-Salwany, M. Y., Al-Saari, N., Mohamad, A., Mursidi, F. A., Mohd-Aris, A., Amal, M. N. A., & Zamri-Saad, M. (2019). Vibriosis in fish: A review on disease development and prevention. *Journal of Aquatic Animal Health*, *31*(1), 3–22. Available from https://doi.org/10.1002/aah.10045.

Jarau, M., Di Natale, A., Huber, P. E., MacInnes, J. I., & Lumsden, J. S. (2018). Virulence of *Flavobacterium psychrophilum* isolates in rainbow trout *Oncorhynchus mykiss* (Walbaum). *Journal of Fish Diseases*, *41*(10), 1505–1514. Available from https://doi.org/10.1111/jfd.12861.

Jarau, M., MacInnes, J. I., & Lumsden, J. S. (2019). Erythromycin and florfenicol treatment of rainbow trout *Oncorhynchus mykiss* (Walbaum) experimentally infected with *Flavobacterium psychrophilum*. *Journal of Fish Diseases*, *42*(3), 325–334. Available from https://doi.org/10.1111/jfd.12944.

Jin, L., Chen, Y., Yang, W., Qiao, Z., & Zhang, X. (2020). Complete genome sequence of fish-pathogenic *Aeromonas hydrophila* HX-3 and a comparative analysis: Insights into virulence factors and quorum sensing. *Scientific Reports*, *10*(1), 15479. Available from https://doi.org/10.1038/s41598-020-72484-8.

Karvonen, A., Fenton, A., & Sundberg, L. R. (2019). Sequential infection can decrease virulence in a fish-bacterium-fluke interaction: Implications for aquaculture disease management. *Evolutionary Applications*, *12*(10), 1900–1911. Available from https://doi.org/10.1111/eva.12850.

Kayansamruaj, P., Areechon, N., & Unajak, S. (2020). Development of fish vaccine in Southeast Asia: A challenge for the sustainability of SE Asia aquaculture. *Fish & Shellfish Immunology*, *103*, 73–87. Available from https://doi.org/10.1016/j.fsi.2020.04.031.

Kayansamruaj, P., Dong, H. T., Hirono, I., Kondo, H., Senapin, S., & Rodkhum, C. (2017). Comparative genome analysis of fish pathogen *Flavobacterium columnare* reveals extensive sequence diversity within the species. *Infection, Genetics and Evolution*, *54*, 7–17. Available from https://doi.org/10.1016/j.meegid.2017.06.012.

Kesbiç, O. S., Acar, Ü., Yilmaz, S., & Aydin, Ö. D. (2020). Effects of bergamot (*Citrus bergamia*) peel oil-supplemented diets on growth performance, haematology and serum biochemical parameters of Nile tilapia (*Oreochromis niloticus*). *Fish Physiology and Biochemistry*, *46*(1), 103–110. Available from https://doi.org/10.1007/s10695-019-00700-y.

Khajanchi, B. K., Kozlova, E. V., Sha, J., Popov, V. L., & Chopra, A. K. (2012). The two-component QseBC signalling system regulates *in vitro* and *in vivo* virulence of *Aeromonas hydrophila*. *Microbiology (Reading, England)*, *158*(1), 259–271. Available from https://doi.org/10.1099/mic.0.051805-0.

Kitazono, A., Kabashima, T., Huang, H. S., Ito, K., & Yoshimoto, T. (1996). Prolyl aminopeptidase gene from *Flavobacterium meningosepticum*: Cloning, purification of the expressed enzyme, and analysis of its sequence. *Archives of Biochemistry and Biophysics*, *336*(1), 35–41. Available from https://doi.org/10.1006/abbi.1996.0529.

Kitiyodom, S., Kaewmalun, S., Nittayasut, N., Suktham, K., Surassmo, S., Namdee, K., & Yata, T. (2019). The potential of mucoadhesive polymer in enhancing efficacy of direct immersion vaccination against *Flavobacterium columnare* infection in tilapia. *Fish & Shellfish Immunology*, *86*, 635–640. Available from https://doi.org/10.1016/j.fsi.2018.12.005.

Knupp, C., Wiens, G. D., Faisal, M., Call, D. R., Cain, K. D., Nicolas, P., & Loch, T. P. (2019). Large-scale analysis of *Flavobacterium psychrophilum* multilocus sequence typing genotypes recovered from North American salmonids indicates that both newly identified and recurrent clonal complexes are associated with disease. *Applied and Environmental Microbiology*, *85*, e02305-18. Available from https://doi.org/10.1128/aem.02305-18.

Kobayashi, M., Msangi, S., Batka, M., Vannuccini, S., Dey, M. M., & Anderson, J. L. (2015). Fish to 2030: The role and opportunity for aquaculture. *Aquaculture Economics & Management*, *19*(3), 282–300.

Kong, X., Wang, L., Pei, C., Zhang, J., Zhao, X., & Li, L. (2018). Comparison of polymeric immunoglobulin receptor between fish and mammals. *Veterinary Immunology and Immunopathology*, *202*, 63–69. Available from https://doi.org/10.1016/j.vetimm.2018.06.002.

LaFrentz, B. R., García, J. C., & Shelley, J. P. (2019). Multiplex PCR for genotyping *Flavobacterium columnare*. *Journal of Fish Diseases*, *42*(11), 1531–1542. Available from https://doi.org/10.1111/jfd.13068.

LaFrentz, B. R., García, J. C., Waldbieser, G. C., Evenhuis, J. P., Loch, T. P., Liles, M. R., & Chang, S. F. (2018). Identification of four distinct phylogenetic groups in *Flavobacterium columnare* with fish host associations. *Frontiers in Microbiology*, *9*(452). Available from https://doi.org/10.3389/fmicb.2018.00452.

LaFrentz, B. R., LaPatra, S. E., Call, D. R., Wiens, G. D., & Cain, K. D. (2009). Proteomic analysis of *Flavobacterium psychrophilum* cultured in vivo and in iron-limited media. *Diseases of Aquatic Organisms*, *87*(3), 171–182. Available from https://doi.org/10.3354/dao02122.

Lange, M. D., Abernathy, J., & Farmer, B. D. (2019). Evaluation of a recombinant *Flavobacterium columnare* DnaK protein vaccine as a means of protection against columnaris disease in channel catfish (*Ictalurus punctatus*). *Frontiers in Immunology*, *10*, 1175. Available from https://doi.org/10.3389/fimmu.2019.01175.

Lange, M. D., Farmer, B. D., & Abernathy, J. (2018). Catfish mucus alters the *Flavobacterium columnare* transcriptome. *FEMS Microbiology Letters*, *365*(22). Available from https://doi.org/10.1093/femsle/fny244.

Lenders, H. J. R. (2017). Fish and fisheries in the lower rhine 1550-1950: A historical-ecological perspective. *Journal of Environmental Management*, *202*(Pt 2), 403–411. Available from https://doi.org/10.1016/j.jenvman.2016.09.011.

Leung, K. Y., Wang, Q., Yang, Z., & Siame, B. A. (2019). *Edwardsiella piscicida*: A versatile emerging pathogen of fish. *Virulence*, *10*(1), 555–567. Available from https://doi.org/10.1080/21505594.2019.1621648.

Li, H., Qin, Y., Mao, X., Zheng, W., Luo, G., Xu, X., & Zheng, J. (2019). Silencing of cyt-c4 led to decrease of biofilm formation in *Aeromonas hydrophila. Bioscience, Biotechnology, and Biochemistry*, *83*(2), 221–232. Available from https://doi.org/10.1080/09168451.2018.1528543.

Li, N., Qin, T., Zhang, X. L., Huang, B., Liu, Z. X., Xie, H. X., & Nie, P. (2015). Development and use of a gene deletion strategy to examine the two chondroitin lyases in virulence of *Flavobacterium columnare. Applied and Environmental Microbiology*, *81*, 7394–7402.

Lin, Y. F., He, J., Zeng, R. Y., Li, Z. M., Luo, Z. Y., Pan, W. Q., & He, J. G. (2019). Deletion of the Infectious spleen and kidney necrosis virus ORF069L reduces virulence to mandarin fish *Siniperca chuatsi. Fish & Shellfish Immunology*, *95*, 328–335. Available from https://doi.org/10.1016/j.fsi.2019.10.039.

Liu, L., Gong, Y. X., Liu, G. L., Zhu, B., & Wang, G. X. (2016). Protective immunity of grass carp immunized with DNA vaccine against *Aeromonas hydrophila* by using carbon nanotubes as a carrier molecule. *Fish & Shellfish Immunology*, *55*, 516–522. Available from https://doi.org/10.1016/j.fsi.2016.06.026.

Liu, L., Gong, Y. X., Zhu, B., Liu, G. L., Wang, G. X., & Ling, F. (2015). Effect of a new recombinant *Aeromonas hydrophila* vaccine on the grass carp intestinal microbiota and correlations with immunological responses. *Fish & Shellfish Immunology*, *45*(1), 175–183. Available from https://doi.org/10.1016/j.fsi.2015.03.043.

Ma, J., Bruce, T. J., Sudheesh, P. S., Knupp, C., Loch, T. P., Faisal, M., & Cain, K. D. (2019). Assessment of cross-protection to heterologous strains of *Flavobacterium psychrophilum* following vaccination with a live-attenuated coldwater disease immersion vaccine. *Journal of Fish Diseases*, *42*(1), 75–84. Available from https://doi.org/10.1111/jfd.12902.

Madsen, L., & Dalsgaard, I. (2008). Water recirculation and good management: Potential methods to avoid disease outbreaks with *Flavobacterium psychrophilum. Journal of Fish Diseases*, *31*(11), 799–810. Available from https://doi.org/10.1111/j.1365-2761.2008.00971.x.

Madsen, L., Bertelsen, S. K., Dalsgaard, I., & Middelboe, M. (2013). Dispersal and survival of *Flavobacterium psychrophilum* phages *in vivo* in rainbow trout and *in vitro* under laboratory conditions: Implications for their use in phage therapy. *Applied and Environmental Microbiology*, *79*(16), 4853–4861.

Mahendran, R., Jeyabaskar, S., Sitharaman, G., Michael, R. D., & Paul, A. V. (2016). Computer-aided vaccine designing approach against fish pathogens *Edwardsiella tarda* and *Flavobacterium columnare* using bioinformatics softwares. *Drug Design, Development and Therapy*, *10*, 1703–1714. Available from https://doi.org/10.2147/dddt.S95691.

Mao, L., Qin, Y., Kang, J., Wu, B., Huang, L., Wang, S., & Yan, Q. (2020). Role of LuxR-type regulators in fish pathogenic *Aeromonas hydrophila. Journal of Fish Diseases*, *43*(2), 215–225. Available from https://doi.org/10.1111/jfd.13114.

Marancik, D. P., & Wiens, G. D. (2013). A real-time polymerase chain reaction assay for identification and quantification of *Flavobacterium psychrophilum* and application to disease resistance studies in selectively bred rainbow trout *Oncorhynchus mykiss. FEMS Microbiology Letters*, *339*(2), 122–129. Available from https://doi.org/10.1111/1574-6968.12061.

Marinho-Neto, F. A., Claudiano, G. S., Yunis-Aguinaga, J., Cueva-Quiroz, V. A., Kobashigawa, K. K., Cruz, N. R. N., & Moraes, J. R. E. (2019). Morphological, microbiological and ultrastructural aspects of sepsis by *Aeromonas hydrophila* in *Piaractus mesopotamicus. PLoS One*, *14*(9), e0222626. Available from https://doi.org/10.1371/journal.pone.0222626.

Mata, W., Putita, C., Dong, H. T., Kayansamruaj, P., Senapin, S., & Rodkhum, C. (2018). Quinolone-resistant phenotype of *Flavobacterium columnare* isolates harbouring point mutations both in gyrA and parC but not in gyrB or parE. *Journal of Global Antimicrobial Resistance*, *15*, 55–60. Available from https://doi.org/10.1016/j.jgar.2018.05.014.

Mateos, D., Anguita, J., Naharro, G., & Paniagua, C. (1993). Influence of growth temperature on the production of extracellular virulence factors and pathogenicity of environmental and human strains of *Aeromonas hydrophila. The Journal of Applied Bacteriology*, *74*(2), 111–118. Available from https://doi.org/10.1111/j.1365-2672.1993.tb03003.x.

Maureaud, A., Hodapp, D., Denderen, P. D. v, Hillebrand, H., Gislason, H., Dencker, T. S., & Lindegren, M. (2019). Biodiversity–ecosystem functioning relationships in fish communities: Biomass is related to evenness and the environment, not to species richness. *Proceedings of the Royal Society B: Biological Sciences*, *286*(1906), 20191189. Available from https://doi.org/10.1098/rspb.2019.1189.

Meng, X., Shen, Y., Wang, S., Xu, X., Dang, Y., Zhang, M., & Li, J. (2019). Complement component 3 (C3): An important role in grass carp (*Ctenopharyngodon idella*) experimentally exposed to *Aeromonas hydrophila*. *Fish & Shellfish Immunology*, *88*, 189–197. Available from https://doi.org/10.1016/j.fsi.2019.02.061.

Miller, J. W., & Atanda, T. (2011). The rise of peri-urban aquaculture in Nigeria. *International Journal of Agricultural Sustainability*, *9*(1), 274–281.

Mohammadian, T., Alishahi, M., Tabandeh, M. R., Ghorbanpoor, M., & Gharibi, D. (2018). Changes in immunity, expression of some immune-related genes of shabot fish, *Tor grypus*, following experimental infection with *Aeromonas hydrophila*: Effects of autochthonous probiotics. *Probiotics and Antimicrobial Proteins*, *10*(4), 616–628. Available from https://doi.org/10.1007/s12602-017-9373-8.

Mohammed, H. H., & Arias, C. R. (2014). Epidemiology of columnaris disease affecting fishes within the same watershed. *Diseases of Aquatic Organisms*, *109*(3), 201–211. Available from https://doi.org/10.3354/dao02739.

Mohammed, H. H., & Arias, C. R. (2015). Potassium permanganate elicits a shift of the external fish microbiome and increases host susceptibility to columnaris disease. *Veterinary Research*, *46*(1), 82. Available from https://doi.org/10.1186/s13567-015-0215-y.

Mohammed, H. H., & Arias, C. R. (2016). Protective efficacy of *Nigella sativa* seeds and oil against columnaris disease in fishes. *Journal of Fish Diseases*, *39*(6), 693–703. Available from https://doi.org/10.1111/jfd.12402.

Mohammed, H., Olivares-Fuster, O., LaFrentz, S., & Arias, C. R. (2013). New attenuated vaccine against columnaris disease in fish: Choosing the right parental strain is critical for vaccine efficacy. *Vaccine*, *31*(45), 5276–5280. Available from https://doi.org/10.1016/j.vaccine.2013.08.052.

Moore, C., Hennessey, E., Smith, M., Epp, L., & Zwollo, P. (2019). Innate immune cell signatures in a BCWD-Resistant line of rainbow trout before and after in vivo challenge with *Flavobacterium psychrophilum*. *Developmental and Comparative Immunology*, *90*, 47–54. Available from https://doi.org/10.1016/j.dci.2018.08.018.

Morley, N. J., & Lewis, J. W. (2010). Consequences of an outbreak of columnaris disease (*Flavobacterium columnare*) to the helminth fauna of perch (*Perca fluviatilis*) in the Queen Mary reservoir, south-east England. *Journal of Helminthology*, *84*(2), 186–192. Available from https://doi.org/10.1017/s0022149x09990459.

Mustapha, M. K., Ajibola, T. B., Salako, A. F., & Ademola, S. K. (2014). Solar drying and organoleptic characteristics of two tropical African fish species using improved low-cost solar driers. *Food Science & Nutrition*, *2*(3), 244–250. Available from https://doi.org/10.1002/fsn3.101.

Nakayama, H., Mori, M., Takita, T., Yasukawa, K., Tanaka, K., Hattori, S., & Amano, K. (2017). Development of immersion vaccine for bacterial cold-water disease in ayu *Plecoglossus altivelis*. *Bioscience, Biotechnology, and Biochemistry*, *81*(3), 608–613. Available from https://doi.org/10.1080/09168451.2016.1268041.

Ngo, T. P. H., Bartie, K. L., Thompson, K. D., Verner-Jeffreys, D. W., Hoare, R., & Adams, A. (2017). Genetic and serological diversity of *Flavobacterium psychrophilum* isolates from salmonids in United Kingdom. *Veterinary Microbiology*, *201*, 216–224. Available from https://doi.org/10.1016/j.vetmic.2017.01.032.

Nicholson, P., Mon-on, N., Jaemwimol, P., Tattiyapong, P., & Surachetpong, W. (2020). Coinfection of tilapia lake virus and *Aeromonas hydrophila* synergistically increased mortality and worsened the disease severity in tilapia (*Oreochromis* spp.). *Aquaculture (Amsterdam, Netherlands)*, *520*, 734746. Available from https://doi.org/10.1016/j.aquaculture.2019.734746.

Nielsen, M. E., Høi, L., Schmidt, A. S., Qian, D., Shimada, T., Shen, J. Y., & Larsen, J. L. (2001). Is *Aeromonas hydrophila* the dominant motile *Aeromonas* species that causes disease outbreaks in aquaculture production in the Zhejiang Province of China? *Diseases of Aquatic Organisms*, *46*(1), 23–29. Available from https://doi.org/10.3354/dao046023.

Nilsen, H., Sundell, K., Duchaud, E., Nicolas, P., Dalsgaard, I., Madsen, L., & Wiklund, T. (2014). Multilocus sequence typing identifies epidemic clones of *Flavobacterium psychrophilum* in Nordic countries. *Applied and Environmental Microbiology*, *80*(9), 2728–2736. Available from https://doi.org/10.1128/aem.04233-13.

Nya, E. J., & Austin, B. (2011). Development of immunity in rainbow trout (*Oncorhynchus mykiss*, Walbaum) to *Aeromonas hydrophila* after the dietary application of garlic. *Fish & Shellfish Immunology*, *30*(3), 845–850. Available from https://doi.org/10.1016/j.fsi.2011.01.008.

Omofunmi, O., Ilesanmi, O., & Alli, A. (2018). Assessing catfish pond effluent on soil physicochemical properties and its suitability for crop production. *Arid Zone Journal of Engineering, Technology and Environment*, *14*(3), 355–366.

Osorio, C. E., Wen, N., Mejias, J. H., Liu, B., Reinbothe, S., von Wettstein, D., & Rustgi, S. (2019). Development of wheat genotypes expressing a glutamine-specific endoprotease from barley and a prolyl endopeptidase from *Flavobacterium meningosepticum* or *Pyrococcus furiosus* as a potential remedy to celiac disease. *Functional & Integrative Genomics*, *19*(1), 123–136. Available from https://doi.org/10.1007/s10142-018-0632-x.

Pajares, S., Eguiarte, L. E., Bonilla-Rosso, G., & Souza, V. (2013). Drastic changes in aquatic bacterial populations from the Cuatro Cienegas Basin (Mexico) in response to long-term environmental stress. *Antonie Van Leeuwenhoek*, *104*(6), 1159–1175. Available from https://doi.org/10.1007/s10482-013-0038-7.

Paneru, B., Al-Tobasei, R., Palti, Y., Wiens, G. D., & Salem, M. (2016). Differential expression of long non-coding RNAs in three genetic lines of rainbow trout in response to infection with *Flavobacterium psychrophilum*. *Scientific Reports*, *6*, 36032. Available from https://doi.org/10.1038/srep36032.

Papadopoulou, A., Dalsgaard, I., Lindén, A., & Wiklund, T. (2017). In vivo adherence of *Flavobacterium psychrophilum* to mucosal external surfaces of rainbow trout (*Oncorhynchus mykiss*) fry. *Journal of Fish Diseases*, *40*(10), 1309–1320. Available from https://doi.org/10.1111/jfd.12603.

Papadopoulou, A., Dalsgaard, I., & Wiklund, T. (2019). Inhibition activity of compounds and bacteriophages against *Flavobacterium psychrophilum* biofilms in vitro. *Journal of Aquatic Animal Health*, *31*(3), 225–238. Available from https://doi.org/10.1002/aah.10069.

Patin, N. V., Pratte, Z. A., Regensburger, M., Hall, E., Gilde, K., Dove, A. D. M., & Stewart, F. J. (2018). Microbiome dynamics in a large artificial seawater aquarium. *Applied and Environmental Microbiology*, *84*(10), e00179. Available from https://doi.org/10.1128/aem.00179-18, 00118.

Peatman, E., Li, C., Peterson, B. C., Straus, D. L., Farmer, B. D., & Beck, B. H. (2013). Basal polarization of the mucosal compartment in *Flavobacterium columnare* susceptible and resistant channel catfish (*Ictalurus punctatus*). *Molecular Immunology*, *56*(4), 317–327. Available from https://doi.org/10.1016/j.molimm.2013.04.014.

Pianetti, A., Battistelli, M., Barbieri, F., Bruscolini, F., Falcieri, E., Manti, A., & Citterio, B. (2012). Changes in adhesion ability of *Aeromonas hydrophila* during long exposure to salt stress conditions. *Journal of Applied Microbiology*, *113*(4), 974–982. Available from https://doi.org/10.1111/j.1365-2672.2012.05399.x.

Poobalane, S., Thompson, K. D., Ardó, L., Verjan, N., Han, H.-J., Jeney, G., & Adams, A. (2010). Production and efficacy of an *Aeromonas hydrophila* recombinant S-layer protein vaccine for fish. *Vaccine*, *28*(20), 3540–3547. Available from https://doi.org/10.1016/j.vaccine.2010.03.011.

Pridgeon, J. W., & Klesius, P. H. (2011a). Development and efficacy of novobiocin and rifampicin-resistant *Aeromonas hydrophila* as novel vaccines in channel catfish and Nile tilapia. *Vaccine*, *29*(45), 7896–7904. Available from https://doi.org/10.1016/j.vaccine.2011.08.082.

Pridgeon, J. W., & Klesius, P. H. (2011b). Molecular identification and virulence of three *Aeromonas hydrophila* isolates cultured from infected channel catfish during a disease outbreak in west Alabama (USA) in 2009. *Diseases of Aquatic Organisms*, *94*(3), 249–253.

Pulkkinen, K., Suomalainen, L. R., Read, A. F., Ebert, D., Rintamäki, P., & Valtonen, E. T. (2010). Intensive fish farming and the evolution of pathogen virulence: The case of columnaris disease in Finland. *Proceedings. Biological Sciences / the Royal Society*, *277*(1681), 593–600. Available from https://doi.org/10.1098/rspb.2009.1659.

Ramadan, H., Ibrahim, N., Samir, M., Abd El-Moaty, A., & Gad, T. (2018). *Aeromonas hydrophila* from marketed mullet (Mugil cephalus) in Egypt: PCR characterization of β-lactam resistance and virulence genes. *Journal of Applied Microbiology*, *124*(6), 1629–1637. Available from https://doi.org/10.1111/jam.13734.

Ran, C., Carrias, A., Williams, M. A., Capps, N., Dan, B. C., Newton, J. C., & Liles, M. R. (2012). Identification of *Bacillus* strains for biological control of catfish pathogens. *PLoS One*, *7*(9), e45793. Available from https://doi.org/10.1371/journal.pone.0045793.

Rasmussen-Ivey, C. R., Figueras, M. J., McGarey, D., & Liles, M. R. (2016). Virulence factors of *Aeromonas hydrophila*: In the wake of reclassification. *Frontiers in Microbiology*, *7*(1337). Available from https://doi.org/10.3389/fmicb.2016.01337.

Rasmussen-Ivey, C. R., Hossain, M. J., Odom, S. E., Terhune, J. S., Hemstreet, W. G., Shoemaker, C. A., & Liles, M. R. (2016). Classification of a hypervirulent *Aeromonas hydrophila* pathotype responsible for epidemic outbreaks in warm-water fishes. *Frontiers in Microbiology*, *7*(1615). Available from https://doi.org/10.3389/fmicb.2016.01615.

Rattanachaikunsopon, P., & Phumkhachorn, P. (2010). Use of asiatic pennywort *Centella asiatica* aqueous extract as a bath treatment to control columnaris in Nile tilapia. *Journal of Aquatic Animal Health*, *22*(1), 14–20. Available from https://doi.org/10.1577/H09-021.1.

Rey, A., Verján, N., Ferguson, H. W., & Iregui, C. (2009). Pathogenesis of *Aeromonas hydrophila* strain KJ99 infection and its extracellular products in two species of fish. *The Veterinary Record*, *164*(16), 493–499. Available from https://doi.org/10.1136/vr.164.16.493.

Righetto, L., Casagrandi, R., Bertuzzo, E., Mari, L., Gatto, M., Rodriguez-Iturbe, I., & Rinaldo, A. (2012). The role of aquatic reservoir fluctuations in long-term cholera patterns. *Epidemics*, *4*(1), 33–42. Available from https://doi.org/10.1016/j.epidem.2011.11.002.

Rochat, T., Pérez-Pascual, D., Nilsen, H., Carpentier, M., Bridel, S., Bernardet, J. F., & Duchaud, E. (2019). Identification of a novel elastin-degrading enzyme from the fish pathogen *Flavobacterium psychrophilum*. *Applied and Environmental Microbiology*, *85*(6). Available from https://doi.org/10.1128/aem.02535-18.

Rodríguez-Morales, A. J. (2015). Challenges for fish foodborne parasitic zoonotic diseases. *Recent Patents on Anti-Infective Drug Discovery*, *10*(1), 3–5. Available from https://doi.org/10.2174/1574891x10666150504120415.

Samayanpaulraj, V., Velu, V., & Uthandakalaipandiyan, R. (2019). Determination of lethal dose of *Aeromonas hydrophila* Ah17 strain in snake head fish *Channa striata*. *Microbial Pathogenesis*, *127*, 7–11. Available from https://doi.org/10.1016/j.micpath.2018.11.035.

Santos, L., & Ramos, F. (2018). Antimicrobial resistance in aquaculture: Current knowledge and alternatives to tackle the problem. *International Journal of Antimicrobial Agents*, *52*(2), 135–143. Available from https://doi.org/10.1016/j.ijantimicag.2018.03.010.

Scott, S. J., & Bollinger, T. K. (2014). Flavobacterium columnare: An important contributing factor to fish die-offs in southern lakes of Saskatchewan, Canada. *Journal of Veterinary Diagnostic Investigation*, *26*(6), 832–836. Available from https://doi.org/10.1177/1040638714553591.

Secades, P., Alvarez, B., & Guijarro, J. A. (2003). Purification and properties of a new psychrophilic metalloprotease (Fpp2) in the fish pathogen *Flavobacterium psychrophilum*. *FEMS Microbiology Letters*, *226*(2), 273–279. Available from https://doi.org/10.1016/s0378-1097(03)00599-8.

Seghouani, H., Garcia-Rangel, C. E., Füller, J., Gauthier, J., & Derome, N. (2017). Walleye autochthonous bacteria as promising probiotic candidates against *Flavobacterium columnare*. *Frontiers in Microbiology*, *8*, 1349. Available from https://doi.org/10.3389/fmicb.2017.01349.

Sekizuka, T., Inamine, Y., Segawa, T., Hashino, M., Yatsu, K., & Kuroda, M. (2019). Potential KPC-2 carbapenemase reservoir of environmental *Aeromonas hydrophila* and *Aeromonas caviae* isolates from the effluent of an urban wastewater treatment plant in Japan. *Environmental Microbiology Reports*, *11*(4), 589–597. Available from https://doi.org/10.1111/1758-2229.12772.

Sharma, P., & Sihag, R. C. (2013). Pathogenicity test of bacterial and fungal fish pathogens in *Cirrihinus mrigala* infected with EUS disease. *Pakistan Journal of Biological Sciences*, *16*(20), 1204–1207. Available from https://doi.org/10.3923/pjbs.2013.1204.1207.

Shen, Y., Zhang, J., Xu, X., Fu, J., & Li, J. (2013). A new haplotype variability in complement C6 is marginally associated with resistance to *Aeromonas hydrophila* in grass carp. *Fish & Shellfish Immunology*, *34*(5), 1360–1365. Available from https://doi.org/10.1016/j.fsi.2013.02.011.

Shoemaker, C. A., & LaFrentz, B. R. (2015). Growth and survival of the fish pathogenic bacterium, *Flavobacterium columnare*, in tilapia mucus and porcine gastric mucin. *FEMS Microbiology Letters*, *362*(4), 1–5. Available from https://doi.org/10.1093/femsle/fnu060.

Shoemaker, C. A., LaFrentz, B. R., Peatman, E., & Beck, B. H. (2018). Influence of native catfish mucus on *Flavobacterium columnare* growth and proteolytic activity. *Journal of Fish Diseases*, *41*(9), 1395–1402. Available from https://doi.org/10.1111/jfd.12833.

Shoemaker, C. A., Mohammed, H. H., Bader, T. J., Peatman, E., & Beck, B. H. (2018). Immersion vaccination with an inactivated virulent *Aeromonas hydrophila* bacterin protects hybrid catfish (*Ictalurus punctatus* X *Ictalurus furcatus*) from motile *Aeromonas* septicemia. *Fish & Shellfish Immunology*, *82*, 239–242. Available from https://doi.org/10.1016/j.fsi.2018.08.040.

Silva, B., Adetunde, O., Oluseyi, T., Olayinka, K., & Alo, B. (2011). Polycyclic aromatic hydrocarbons (PAHs) in some locally consumed fishes in Nigeria. *African Journal of Food Science*, *5*(7), 384–391.

Singh, V., Rathore, G., Kapoor, D., Mishra, B., & Lakra, W. (2008). Detection of aerolysin gene in *Aeromonas hydrophila* isolated from fish and pond water. *Indian Journal of Microbiology*, *48*(4), 453–458.

Siqueira, A. C., Bellwood, D. R., & Cowman, P. F. (2019). The evolution of traits and functions in herbivorous coral reef fishes through space and time. *Proceedings. Biological Sciences / The Royal Society*, *286*(1897), 20182672. Available from https://doi.org/10.1098/rspb.2018.2672.

Söderlund, R., Hakhverdyan, M., Aspán, A., & Jansson, E. (2018). Genome analysis provides insights into the epidemiology of infection with *Flavobacterium psychrophilum* among farmed salmonid fish in Sweden. *Microbial Genome*, *4*(12), e000241. Available from https://doi.org/10.1099/mgen.0.000241.

Soliman, W. S., Shaapan, R. M., Mohamed, L. A., & Gayed, S. S. R. (2019). Recent biocontrol measures for fish bacterial diseases, in particular to probiotics, bio-encapsulated vaccines, and phage therapy. *Open Veterinary Journal*, *9*(3), 190–195. Available from https://doi.org/10.4314/ovj.v9i3.2.

Solomon, S. G., Ayuba, V. O., Tahir, M. A., & Okomoda, V. T. (2018). Catch per unit effort and some water quality parameters of Lake Kalgwai Jigawa State, Nigeria. *Food Science & Nutrition*, *6*(2), 450–456. Available from https://doi.org/10.1002/fsn3.573.

Srisapoome, P., & Areechon, N. (2017). Efficacy of viable *Bacillus pumilus* isolated from farmed fish on immune responses and increased disease resistance in Nile tilapia (*Oreochromis niloticus*): Laboratory and on-farm trials. *Fish & Shellfish Immunology*, *67*, 199–210. Available from https://doi.org/10.1016/j.fsi.2017.06.018.

Starliper, C. E. (2011). Bacterial coldwater disease of fishes caused by *Flavobacterium psychrophilum*. *Journal of Advanced Research*, *2*(2), 97–108. Available from https://doi.org/10.1016/j.jare.2010.04.001.

Strepparava, N., Nicolas, P., Wahli, T., Segner, H., & Petrini, O. (2013). Molecular epidemiology of *Flavobacterium psychrophilum* from Swiss fish farms. *Diseases of Aquatic Organisms*, *105*(3), 203–210. Available from https://doi.org/10.3354/dao02609.

Sudheesh, P. S., & Cain, K. D. (2016). Optimization of efficacy of a live attenuated *Flavobacterium psychrophilum* immersion vaccine. *Fish & Shellfish Immunology*, *56*, 169–180. Available from https://doi.org/10.1016/j.fsi.2016.07.004.

Sundell, K., Landor, L., Nicolas, P., Jørgensen, J., Castillo, D., Middelboe, M., & Wiklund, T. (2019). Phenotypic and genetic predictors of pathogenicity and virulence in *Flavobacterium psychrophilum*. *Frontiers in Microbiology*, *10*, 1711. Available from https://doi.org/10.3389/fmicb.2019.01711.

Tacon, A. G. J., & Metian, M. (2013). Fish matters: Importance of aquatic foods in human nutrition and global food supply. *Reviews in Fisheries Science*, *21*(1), 22–38. Available from https://doi.org/10.1080/10641262.2012.753405.

Tanhay Mangoudehi, H., Zamani, H., Shahangian, S. S., & Mirzanejad, L. (2020). Effect of curcumin on the expression of ahyI/R quorum sensing genes and some associated phenotypes in pathogenic *Aeromonas hydrophila* fish isolates. *World Journal of Microbiology and Biotechnology*, *36*(5), 70. Available from https://doi.org/10.1007/s11274-020-02846-x.

Thompson, O., & Mafimisebi, T. (2014). Profitability of selected ventures in catfish aquaculture in Ondo State, Nigeria. *Fisheries and Aquaculture Journal*, *5*(2), 1.

Thurstan, R. H., & Roberts, C. M. (2014). The past and future of fish consumption: Can supplies meet healthy eating recommendations? *Marine Pollution Bulletin*, *89*(1-2), 5–11.

Tien, N. T., Dung, T. T., Tuan, N. A., & Crumlish, M. (2012). First identification of *Flavobacterium columnare* infection in farmed freshwater striped catfish *Pangasianodon hypophthalmus*. *Diseases of Aquatic Organisms*, *100*(1), 83–88. Available from https://doi.org/10.3354/dao02478.

Uduji, J. I., & Okolo-Obasi, E. N. (2018). Does corporate social responsibility (CSR) impact on development of women in small-scale fisheries of sub-Saharan Africa? Evidence from coastal communities of Niger Delta in Nigeria. *Marine policy*, *118*.

Vaikuntapu, P. R., Mallakuntla, M. K., Das, S. N., Bhuvanachandra, B., Ramakrishna, B., Nadendla, S. R., & Podile, A. R. (2018). Applicability of endochitinase of *Flavobacterium johnsoniae* with transglycosylation activity in generating long-chain chitooligosaccharides. *International Journal of Biological Macromolecules*, *117*, 62–71. Available from https://doi.org/10.1016/j.ijbiomac.2018.05.129.

Van Vliet, D., Wiens, G. D., Loch, T. P., Nicolas, P., & Faisal, M. (2016). Genetic diversity of *Flavobacterium psychrophilum* isolates from three *Oncorhynchus* spp. in the United States, as revealed by multilocus sequence typing. *Applied and Environmental Microbiology*, *82*(11), 3246–3255. Available from https://doi.org/10.1128/aem.00411-16.

Verma, D. K., & Rathore, G. (2013). Molecular characterization of *Flavobacterium columnare* isolated from a natural outbreak of columnaris disease in farmed fish, *Catla catla* from India. *The Journal of General and Applied Microbiology*, *59*(6), 417–424. Available from https://doi.org/10.2323/jgam.59.417.

Vermeij, M. J. A., Latijnhouwers, K. R. W., Dilrosun, F., Chamberland, V. F., Dubé, C. E., Van Buurt, G., & Debrot, A. O. (2019). Historical changes (1905-present) in catch size and composition reflect altering fisheries practices on a small Caribbean island. *PLoS One*, *14*(6), e0217589. Available from https://doi.org/10.1371/journal.pone.0217589.

Villazanakretzer, D. L., Napolitano, P. G., Cummings, K. F., & Magann, E. F. (2016). Fish parasites: A growing concern during pregnancy. *Obstetrical & Gynecological Survey*, *71*(4), 253–259. Available from https://doi.org/10.1097/ogx.0000000000000303.

Wamala, S. P., Mugimba, K. K., Dubey, S., Takele, A., Munang'andu, H. M., Evensen, Ø., & Sørum, H. (2018). Multilocus sequence analysis revealed a high genotypic diversity of *Aeromonas hydrophila* infecting fish in Uganda. *Journal of Fish Diseases*, *41*(10), 1589–1600. Available from https://doi.org/10.1111/jfd.12873.

Wang, L., Zhang, J., Kong, X., Pei, C., Zhao, X., & Li, L. (2017). Molecular characterization of polymeric immunoglobulin receptor and expression response to *Aeromonas hydrophila* challenge in *Carassius auratus*. *Fish & Shellfish Immunology*, *70*, 372–380. Available from https://doi.org/10.1016/j.fsi.2017.09.031.

Wang, Q., Ji, W., & Xu, Z. (2020). Current use and development of fish vaccines in China. *Fish & Shellfish Immunology*, *96*, 223–234. Available from https://doi.org/10.1016/j.fsi.2019.12.010.

Wang, R., Hu, X., Lü, A., Liu, R., Sun, J., Sung, Y. Y., & Song, Y. (2019). Transcriptome analysis in the skin of *Carassius auratus* challenged with *Aeromonas hydrophila*. *Fish & Shellfish Immunology*, *94*, 510–516. Available from https://doi.org/10.1016/j.fsi.2019.09.039.

Wang, S., Yan, Q., Zhang, M., Huang, L., Mao, L., Zhang, M., & Qin, Y. (2019). The role and mechanism of icmF in *Aeromonas hydrophila* survival in fish macrophages. *Journal of Fish Diseases*, *42*(6), 895–904. Available from https://doi.org/10.1111/jfd.12991.

Wuensch, A., Trusch, F., Iberahim, N. A., & van West, P. (2018). *Galleria melonella* as an experimental in vivo host model for the fish-pathogenic oomycete *Saprolegnia parasitica*. *Fungal Biology*, *122*(2-3), 182–189. Available from https://doi.org/10.1016/j.funbio.2017.12.011.

Xie, Q., Mei, W., Ye, X., Zhou, P., Islam, M. S., Elbassiony, K. R. A., & Zhou, Y. (2018). The two-component regulatory system CpxA/R is required for the pathogenesis of *Aeromonas hydrophila*. *FEMS Microbiology Letters*, *365*(22). Available from https://doi.org/10.1093/femsle/fny218.

Yoboué, K. P., Ouattara, N. I., Berté, S., Aboua, B. R. D., Coulibaly, J. K., & Kouamélan, E. P. (2020). Structure of benthic macroinvertebrates population in an area of Mopoyem Bay (Ebrie Lagoon, Côte d'Ivoire) exposed to the discharge of a fish farm effluents. *Environmental Monitoring and Assessment*, *192*(4), 203. Available from https://doi.org/10.1007/s10661-020-8167-8.

Yu, Y., Li, H. R., Chen, B., Zeng, Y. X., & He, J. F. (2006). Phylogenetic diversity and cold-adaptive hydrolytic enzymes of culturable psychrophilic bacteria associated with sea ice from high latitude ocean, Artic. *Wei Sheng Wu Xue Bao = Acta Microbiologica Sinica*, *46*(2), 184–190.

Yuan, J., Ni, M., Liu, M., Zheng, Y., & Gu, Z. (2019). Occurrence of antibiotics and antibiotic resistance genes in a typical estuary aquaculture region of Hangzhou Bay, China. *Marine Pollution Bulletin*, *138*, 376–384. Available from https://doi.org/10.1016/j.marpolbul.2018.11.037.

Yun, S., Jun, J. W., Giri, S. S., Kim, H. J., Chi, C., Kim, S. G., & Park, S. C. (2017). Efficacy of PLGA microparticle-encapsulated formalin-killed *Aeromonas hydrophila* cells as a single-shot vaccine against *A. hydrophila* infection. *Vaccine*, *35*(32), 3959–3965. Available from https://doi.org/10.1016/j.vaccine.2017.06.005.

Yun, S., Lee, S. J., Giri, S. S., Kim, H. J., Kim, S. G., Kim, S. W., & Chang Park, S. (2020). Vaccination of fish against *Aeromonas hydrophila* infections using the novel approach of transcutaneous immunization with dissolving microneedle patches in aquaculture. *Fish & Shellfish Immunology*, *97*, 34–40. Available from https://doi.org/10.1016/j.fsi.2019.12.026.

Zhang, B., Zhuang, X., Guo, L., McLean, R. J. C., & Chu, W. (2019). Recombinant N-acyl homoserine lactone-Lactonase AiiA(QSI-1) Attenuates *Aeromonas hydrophila* virulence factors, biofilm formation and reduces mortality in *Crucian carp*. *Marine Drugs*, *17*(9), 499. Available from https://doi.org/10.3390/md17090499.

Zhang, D., Thongda, W., Li, C., Zhao, H., Beck, B. H., Mohammed, H., & Peatman, E. (2017). More than just antibodies: Protective mechanisms of a mucosal vaccine against fish pathogen *Flavobacterium columnare*. *Fish & Shellfish Immunology*, *71*, 160–170. Available from https://doi.org/10.1016/j.fsi.2017.10.001.

Zhang, M., Qin, Y., Huang, L., Yan, Q., Mao, L., Xu, X., & Chen, L. (2019). The role of sodA and sodB in *Aeromonas hydrophila* resisting oxidative damage to survive in fish macrophages and escape for further infection. *Fish & Shellfish Immunology*, *88*, 489–495. Available from https://doi.org/10.1016/j.fsi.2019.03.021.

Zhang, M., Yan, Q., Mao, L., Wang, S., Huang, L., Xu, X., & Qin, Y. (2018). KatG plays an important role in *Aeromonas hydrophila* survival in fish macrophages and escape for further infection. *Gene*, *672*, 156–164. Available from https://doi.org/10.1016/j.gene.2018.06.029.

Zhang, X., Li, N., Qin, T., Huang, B., & Nie, P. (2017). Involvement of two glycoside hydrolase family 19 members in colony morphotype and virulence in *Flavobacterium columnare*. *Chinese Journal of Oceanology and Limnology*, *35*(6), 1511–1523.

Zhang, X., Yang, W., Wu, H., Gong, X., & Li, A. (2014). Multilocus sequence typing revealed a clonal lineage of *Aeromonas hydrophila* caused motile *Aeromonas* septicemia outbreaks in pond-cultured cyprinid fish in an epidemic area in central China. *Aquaculture (Amsterdam, Netherlands)*, *432*, 1–6. Available from https://doi.org/10.1016/j.aquaculture.2014.04.017.

Zhang, Y., Zhao, L., Chen, W., Huang, Y., Yang, L., Sarathbabu, V., & Lin, L. (2017). Complete genome sequence analysis of the fish pathogen *Flavobacterium columnare* provides insights into antibiotic resistance and pathogenicity related genes. *Microbial Pathogenesis*, *111*, 203–211. Available from https://doi.org/10.1016/j.micpath.2017.08.035.

Zhang, Z., Liu, G., Ma, R., Qi, X., Wang, G., Zhu, B., & Ling, F. (2020). The immunoprotective effect of whole-cell lysed inactivated vaccine with SWCNT as a carrier against *Aeromonas hydrophila* infection in grass carp. *Fish & Shellfish Immunology*, *97*, 336–343. Available from https://doi.org/10.1016/j.fsi.2019.12.069.

Zhao, H., Li, C., Beck, B. H., Zhang, R., Thongda, W., Davis, D. A., & Peatman, E. (2015). Impact of feed additives on surface mucosal health and columnaris susceptibility in channel catfish fingerlings, *Ictalurus punctatus*. *Fish & Shellfish Immunology*, *46*(2), 624–637. Available from https://doi.org/10.1016/j.fsi.2015.07.005.

Zhou, S., Yu, Z., & Chu, W. (2019). Effect of quorum-quenching bacterium Bacillus sp. QSI-1 on protein profiles and extracellular enzymatic activities of *Aeromonas hydrophila* YJ-1. *BMC Microbiology*, *19*(1), 135. Available from https://doi.org/10.1186/s12866-019-1515-6.

Zwollo, P., Hennessey, E., Moore, C., Marancik, D. P., Wiens, G. D., & Epp, L. (2017). A BCWD-resistant line of rainbow trout exhibits higher abundance of IgT(+) B cells and heavy chain tau transcripts compared to a susceptible line following challenge with *Flavobacterium psychrophilum*. *Developmental and Comparative Immunology*, *74*, 190–199. Available from https://doi.org/10.1016/j.dci.2017.04.019.

CHAPTER

Status of furunculosis in fish fauna 12

Abdul Baset

Department of Zoology, Bacha Khan University Charsadda, Charsadda, Pakistan

12.1 Introduction

Equilibrium regulation is an essential factor in an ecosystem, including balanced or limited populations of producers and consumers, prey and predators, hosts and parasites. Microparasites and macroparasites can also regulate the communities of the host they infect. The parasites (viruses, bacteria, protozoans, helminths, and arthropods, among others) can be considered microbial predators. The association between disease (infection of parasites) and its effects on population dynamics must be considered, including the dynamics of the infection process (Lafferty & Harvell, 2014). Pathogens affect their hosts and cause a series of diseases in aquatic fauna. A wide range of bacteria (both gram positive and gram negative) have been related to various conditions of freshwater and marine fish worldwide (Meyburgh, 2017). Bacteria is one of the pathogens that cause many infectious disease problems in fishes, such as furunculosis, columnaris, dropsy, vibriosis, tuberculosis, bacterial gill disease, and fin rot/tail rot (Austin & Austin, 2012).

Furunculosis is a major, severe, contagious, and economically important disease in fresh and seawater fishes worldwide. The disease was first recognized and described in hatchery fish in Germany by Emmerich and Weibel (1894; Cain & Polinski, 2014). They documented that the causative agent of the disease was a bacterium and named the bacterium *salmonicida*. After that, the disease was commonly recognized in trout hatcheries in Germany, where it was supposed to be mainly a hatchery disease. Flehn (1909, 1911) discovered the existence of furunculosis in wild trout and dispelled the belief that it was only a hatchery disease. In a short time, furunculosis was reported from other regions of Germany and then was found in Austria, Belgium, Ireland, France, Switzerland, and the United Kingdom (Kar, 2015). Currently, this septicemic disease is found worldwide. This debilitating and lethal disease is one of the oldest known bacterial infections, affecting various farmed and wild stock fish species, including salmonids and cyprinids, but mainly affecting Salmonidae (Noga, 2010).

Furunculosis is caused by a prime particular etiological agent, *Aeromonas salmonicida*. The name of the disease comes from the furuncles/ulcers, which are liquefactive muscle lesions or red patches on the skin, as well as darkening of the skin. *A. salmonicida* is a gram-negative, facultatively anaerobic, nonmotile rod. (There are three types of motile species: *Aeromonas caviae*, *A. hydrophila*, and *A. sobria*. They do not cause furunculosis in fish.) *A. salmonicida* is a nonsporulating psychrophilic, with size ranging from 0.8 to 1.3 mm (Austin & Austin, 2012), which

Bacterial Fish Diseases. DOI: https://doi.org/10.1016/B978-0-323-85624-9.00002-6

characteristically causes infection in freshwater fishes, that is, cyprinids, salmonids, anoplopomids, and serranids. *A. salmonicida* is the only bacterium that causes furunculosis in fish species, although furunculosis in other fauna and organisms is caused by other bacteria. There are two main types of diseases in fish, acute and chronic. Change/darkening in color, red patches, and furuncles/ulcers on the fins of the fish are the clinical signs of an acute critical case and the survival time is only a few days. In contrast, spleen enlargement and inflammation of the lower intestine are signs of chronically ill fish; furuncles can be found but sometimes no signs are seen (Austin & Austin, 2007; Dar, Dar et al., 2016; Dar, Kamili et al., 2016; Dar et al., 2020).

The pathogen that causes furunculosis is more active in a warm environment and is likely to be hidden in a colder climate. Generally, the furunculosis cases occur in summer (July and August), the season when the surrounding air and water temperatures increase, but can still occur when the weather is cooler. The optimum temperature for furunculosis survival is 12.8°C–21.1°C, but some studies revealed that the pathogens of the disease are always found in 0.5°C–1.6°C (Perry et al., 2004; Schachte, 2002). Furunculosis causes fish mortality in every season, but massive losses occur in spring and autumn in seawater farms. In salmonids, the disease can be transmitted to other individuals by horizontal transmission via the water column and the physical contact of fish to fish, contaminated water, and also by animal vectors such as birds and invertebrates. However, nonsalmonids can be affected by the ingestion of tissues of infected fish; the congested stocked forms are more susceptible and have an increased degree of infection (Dar, Dar et al., 2016; Dar, Kamili et al., 2016; Khoo, 2019). A significant aspect of furunculosis outbreaks and mortality is that they can be amplified by crowding, fright, high temperature, physical trauma, and low water quality, in other words, by stress factors. Pollution is considered to be a major factor in furunculosis. Various sources of pollutants are chemical fertilizer companies, oil refineries, petrochemical industry, paper industry, power stations, pesticides, sewage systems, and overfishing (Plumb & Hanson, 2010).

12.2 Signs of infection

Furunculosis occurs worldwide and primarily infects salmonids and some nonsalmonids, causing high mortality in fish and heavy losses in the aquaculture industry. Sudden death occurs in fish after symptoms appear. Fish can be seen with exophthalmos, and they also can exhibit sluggish and lethargic swimming. The diseased fish experience loss of appetite accompanied with respiratory problems, and try to jump out of the water. The general symptoms of furuncles include blisters on epithelial and muscular tissues, also making bowl-shaped lesions (Fig. 12.1). The symptoms can range from acute to chronic in mature salmonids. Other frequently observed symptoms include bleeding from skin and bases of fins, gradual destruction of pectoral fins and the oral cavity; erosion can also occur in internal organs and muscles. Paling of gills, alteration in body color, and bleeding from nasal openings and anus are also frequently observed symptoms of the disease. Increase in size of spleen, destruction of liver, and the filling of the stomach and intestine with epithelial cells accompanied with mucus and blood are also important symptoms of the disease. The microscopic symptoms of the disease include the blending of lamellae of gills, death of epithelial cells, and eosinophilic inflammation of gills. Formation of bacterial colonies in different tissues and the collapse of epithelial cells of renal tubes and intestine in the renal tube cavity and intestinal cavity are also important symptoms of the disease.

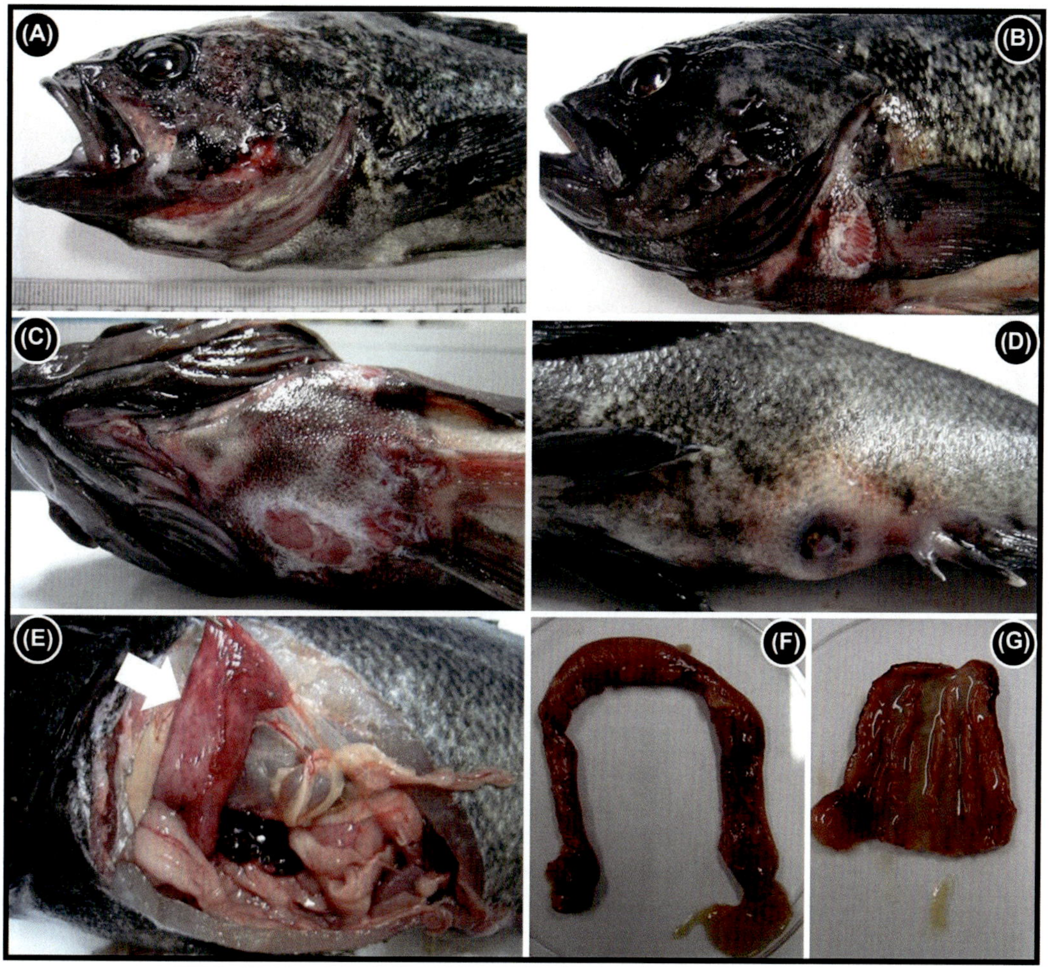

FIGURE 12.1

Clinical signs of the Korean rockfish infected with atypical *Aeromonas salmonicida*. (A–C) Hemorrhages at operculum, jaw area, and around the base of the pectoral and pelvic fins; (D) swelling of anus; (E–G) hemorrhages on the surface of the stomach and intestine and excessive secretion of mucus in the gastrointestinal tract (Kim et al., 2013).

12.3 Diagnosis

A. salmonicida is a nonmotile, fermentative, gram-negative bacteria that reduces nitrate and produces cytochrome oxidase. It can be isolated from blood, lesions, and other organs of an organism (Austin & Austin, 2012). The colonies are different in size, ranging from 0.5 to 3.0 mm in diameter within 72 h incubation, and the optimal temperature for growth is 18°C–22°C. Various diagnostic techniques have been used for furunculosis (Austin & Austin, 2012).

In clinical diagnosis, the bacterial agent can be investigated with simple cultured swabs (bacteriological culture). In this technique, a standard clinical examination can be performed. Investigations can be done not only from culture swabs of the lesions but also from the carrier sites of fish, for example, nostrils and perineum, etc. (Lagier et al., 2015). The vital clinical signs, gross pathology, and histopathological features have been described for the various forms of furunculosis, such as acute, chronic, peracute, and subacute. The gross pathological signs of peracute and acute furunculosis in young fish are often indistinguishable from other bacterial septicemias in an initial investigation. This may cause trouble for the inexpert diagnostician in differentiating the morphology and gram-staining behavior of *A. salmonicida* in fish tissue. Thus a firm diagnosis of clinical furunculosis needs to be based on isolation of infecting fish on agar media and identification by morphology, combined with either biochemical or serological tests, as *A. salmonicida* (Austin, 2019).

The diagnosis of covertly infected fish is an essential problem for diagnosticians and fish health workers. In covert furunculosis, the diagnosis of suspected fish in a population must be made on the basis of random sampling. The sampling must be done in such a way that it can be used as representative of the overall health of the population (Cipriano & Bullock, 2001). In the sampling procedure, the sampling tables obtained indicate the number of fish taken from a fish population of known number. The sample tables should be arranged in such a way as to be helpful in further plans for sampling and also for selection of appropriate methods for diagnosis and confirmation of diseases in fish populations. The detection of covert furunculosis was difficult to achieve by taking a statistical representative of the fish population (Ogut et al., 2005). A variety of methods for detecting covert infection in fish have been developed. Cohabitation studies that involve the release of *A. salmonicida* by diseased fish is an indirect approach to detection of the disease, while determining the presence of infectious organisms in the vicinity of a fish population is one of the direct approaches (Johansen et al., 2011). The transfer of furunculosis from clinically infected or covertly infected fish to a normal healthy one is called silent transmission (Cipriano & Bullock, 2001).

Rapid diagnostic techniques developed over the last few decades have played a significant role in the diagnosis of microbes. The rapid diagnostic techniques enable the direct detection of microbes in clinical samples and also in the habitat of the host without culturing the microbes. These techniques are frequently used for the detection of *A. salmonicida* (Austin, 2019; Austin & Austin, 2016). Besides genetic-based assays for *A. salmonicida*, bioassays detecting the immunology of the organisms have also been developed over the past few decades. However, a number of the techniques offered for clinical diagnosis of furunculosis that do not require the culture of microbes for the identification of *A. salmonicida* can be broadly classified as immune based-techniques and genome-based techniques in modern investigations (Austin, 2019).

12.4 Transmission

Furunculosis infection of *A. salmonicida* in fish can be transmitted to other individuals by the pathogen horizontally. Horizontal transmission can take place by either physical contact of fish to fish or shedding the bacterium into the water column (Bricknell, 2017; Cipriano & Bullock, 2001). The horizontal transmission of the pathogen represents a major threat to fish within marine and freshwater environments (Stride et al., 2014). Epizootics are predominantly troublesome at the start of marine culture if an *A. salmonicida* carrier is undetected and fish are transported into pens. Contagion contained by natural waterways can enhance disease in fish. The movement of infected fish from culture facilities or sea farms infect fish within the feral environment (Bricknell, 2017). The horizontal contact of infected fish among noninfected fish creates a way for *A. salmonicida* to enter a susceptible host. The reservoirs and source of furunculosis infection and transmission are probably similar for both types of strains, atypical *A. salmonicida* and typical strains (AQUAVETPLAN, 2009).

Another essential technique is vector transmission, such as through birds and invertebrates, which can cause the transmission of *A. salmonicida* infection. For example, with the use of wrasse as cleaner fish to clear sea lice from the culture pen, the wrasse may infect themselves with atypical *A. salmonicida* by orally ingesting infected sea lice or horizontally with other infected fish (Moksness et al., 2008). Therefore infected wrasse act as a different reservoir of contagion for salmonids or other fish. Another means of transmission of *A. salmonicida* is an aerosol transmission that is practically problematic in a culture system. The viable cells of *A. salmonicida* can be disseminated in aerosol droplets in the air and the water tanks exposed to aerosol spray downwind of the contaminant source (Nash et al., 2015).

12.5 Control of infection

Furunculosis infects salmonids primarily but can infect various hosts in a diversity of habitats. *A. salmonicida* is managed mainly as a geographically ubiquitous but obligate pathogen. Fish culturists, farmers, and resource managers should keep its pathogenic effects in mind. In fact, there is a strong relationship between the host, pathogen, and environment, and often it is in the conceptuality of the disease process. The prevention of furunculosis is critical in the fish population. Anadromous fish migrate into hatchery water supplies and infect fish culture stations. The fish managers have some control over such parameters, but they are also hard to prevent. The education of personnel is the best way to improve the significant sources of potential contamination and to prevent disease.

The managers should have a regular monitoring program to detect *A. salmonicida* in the water supply. The personnel have to keep their eyes open for such a condition. They should transfer only fertilized eggs or stocks of fish certified to be free of *A. salmonicida* infection to hatcheries and culture stations (Fig. 12.2). Preventive measures include disinfection of the culture stations before stocking, stress elimination, age class segregation, the use of water in the system that has not been exposed to wild or cultured fish, prior elimination of infected fish, and vaccination. Fish individuals with furunculosis should be removed from the facility before starting the treatment. An essential step in the prevention of severe communicable fish diseases is the adherence to a sound program of hatchery inspections.

12.6 Selection and breeding

Selective breeding plans are significant methods in salmonids, cyprinids, and other species to enhance experimental and field resistance to furunculosis. The fish can be selected based on growth

FIGURE 12.2

Separation of fish individuals with furunculosis, transferred into a vaccine dilution.

and the ability to survive natural epizootics of furunculosis. The subsequent progeny is continuously appraised for acquired resistance to the disease, so that the mortality can be reduced after a few successive generations. The selective resistant progeny grows faster than nonselective fish. The technique may require up to eight generations of selection but often *A. salmonicida* encounters through water bath, horizontal and nutritional transmission. The mortality can be reduced by a maximum percentage. Selective breeding can be planned based on size, coloration, and egg yield too. Acquired resistance to furunculosis is broadly acknowledged, and these types of fish are currently used throughout hatchery programs worldwide for their survival and performance in recreational fisheries where furunculosis is enzootic.

12.7 Immunization

As discussed in previous sections of this chapter, *Aeromonas salmonicida* causes a bacterial infection called furunculosis, a bacterial hemorrhagic septicemia, which affects both salmonids and nonsalmonids, such as carp and turbot, from both cultured and wild populations. The disease can be disturbing, harshly impacting the productivity of fish farming processes, and exhibiting clinical signs that include serosanguineous fluids, necrotic tissue, and the appearance of furuncles. The fish also exhibit other distinguishing features such as lethargy and darkening of the skin (Giraud et al., 2004).

Oil-adjuvanted injectable vaccines, which have been used for typical *A. salmonicida* for salmonids, are considered an historically verified positive and standard technique. However, immunization of farmed nonsalmonid fish types against *A. salmonicida* (typical and atypical) has been somewhat challenging; for instance, an experimental vaccine containing atypical strains protected Arctic charr (*Salvelinus alpinus*, L.) but not European grayling (*Thymallus thymallus*, L.) (Paredez et al., 2020). Vaccines of *A. salmonicida* combined in oil adjuvants and injected intraperitoneally are available for long-term defense, and their use is promoted in commercial aquaculture. Aqueous and oil-adjuvanted vaccines are both commercially made and may increase immunity at low water temperatures. It was noted that, even though an antibody response to *A. salmonicida* was late or highly inhibited at this temperature, protection was still acceptable at 18 weeks after vaccination. After 18 weeks, these researchers noted that protection was strongly reduced in those fish raised at 10°C and given the aqueous vaccine, but a similar reduction was not seen in those fish that were vaccinated with the oil-mixed vaccine. Though commercial furunculosis vaccines in mineral oil yield meaningfully higher protection than aqueous-based vaccines, the mineral oil adjuvant may produce adverse side effects, like adhesions between the internal organs and the abdominal wall near the site of injection. Adverse side effects may also include a reduction in the growth rate, the severity of which is unpredictable among fish farms and may or may not be significant when matched with the last growth weight of nonvaccinated fish. It must be remembered, though, that vaccination is not a total barrier treatment that guarantees no further expression of furunculosis or spread of furunculosis from covertly infected vaccinated carriers.

The first week after hatching, plasma cells were seen in the head kidney, revealing the existence of immunocompetent cells in this tissue. Eventually, plasma cells were seen in the spleen, and at the size of 80–90 mm (after 21 weeks after hatching) the extent of plasma cell was raised to 80–90 mm restricted in this stage in the lymphoid tissues of thymus and gut, but localized in the mucosal tissues of skin and gills (Grøntvedt & Espelid, 2004).

12.8 Treatment

Furunculosis of salmonids was the first disease of fishes to be treated with modern drugs containing both sulfonamides and nitrofurans, although other drugs successfully control this disease. Diseased fish often return to antibiotic treatment, but the use of antibiotics can be unwanted. This is mainly due to the tendency of microorganisms to become resistant to regularly recycled antibiotics and mostly involved frequently in treating of fish. Vaccines besides *A. salmonicida* have already been used for more than 30 years and have played an important role in cumulative aquaculture production. Intraperitoneal injection of vaccines is recognized as brilliant defensive possessions against *A. salmonicida* (Lim & Hong, 2020).

Vaccination (Table 12.1) is a robust treatment approach that decreases disease outbreaks and reduces the use of antibiotics. Presently, there are no active commercial vaccines established specifically for sablefish. Two promising vaccines are launching to be used in the sablefish business: one is ALPHA JECT micro 4 (Norway) commercialized for Atlantic salmon of 15 g or more in mass, containing *A. salmonicida*, *Vibrio salmonicida*, Listonella (Vibrio) *anguillarum serotypes* O1 and O2; the second is Forte Micro IV (Canada) developed for salmonids of 10 g or larger, which has deactivated infective *A. salmonicida*, salmon anemia virus, and *V. salmonicida* serotypes I and II, *V. anguillarum* serotypes I and II, *V. ordalii*, in the liquid mix with an oil-based adjuvant.

Table 12.1 Reported vaccinations of various fish species against atypical furunculosis with oil-adjuvant injection Bactrians.

Strain in bacterium	Fish species	Challenge strain	Protection	Source
Asa	Salmon	Asa	+	Gudmundsdóttir and Magnadóttir (1997)
	Salmon	Ass	−	Gudmundsdóttir and Gudmundsdottir (1997)
	Halibut	Asa	+	Gudmundsdóttir et al. (2003)
	Cod	Asa	−	Gudmundsdóttir et al. (2005a, 2005b)
aAs	Spotted wolffish	aAs	+	Lund et al. (2002)
	Halibut	aAs	+	Ingilæ et al. (2000)
	Cod	aAs	+	Mikkelsen et al. (2004)
	Arctic charr	aAs	+	Pylkkö (2004)
	European grayling	aAs	−	Pylkkö (2004)
Ass	Salmon	Asa	+	Gudmundsdóttir and Gudmundsdottir (1997)
	Halibut	Asa	+	Gudmundsdóttir et al. (2003)
	Cod	Asa	−	Gudmundsdóttir et al. (2005a, 2005b)
	Turbot	Asa	−	Björnsdóttir et al. (2004)
	Spotted wolffish	aAs	−	Lund et al. (2002)
	Halibut	aAs	−	Ingilæ et al. (2000)
	Arctic charr	aAs	−	Pylkkö (2004)
	European grayling	aAs	−	Pylkkö (2004)

Asa, A. salmonicida *subsp.* achromogenes*;* aAs, *atypical* A. salmonicida*;* Ass, A. salmonicida *subsp.* salmonicida*;* +, *significant protection was obtained;* −, *significant protection was not obtained.*

Recently, an autogenous polyvalent vaccine was successfully assessed in the treatment of *A. salmonicida* in sablefish in the United States, and it was described that an autogenous bacterin-researched vaccine injected intraperitoneally was more active than bath vaccination. Conversely, Canadian regulations do not permit the importation of autogenous vaccines, and current vaccines available in Canada against *A. salmonicida* have not been evaluated in sablefish (Vasquez et al., 2020).

12.9 Conclusion and future prospects

A bacterial infection called furunculosis caused by *A. salmonicida* affects both salmonids and non-salmonids of cultured and wild fish populations. The disease was first recognized in hatchery fish in Germany in 1894 and was then reported worldwide in a short time. The infection in fish can be either acute or chronic. The infected fish exhibit loss of appetite, respiratory distress, and/or blisters on skin or muscles causing cratered lesions, limited to the subacute or chronically in adult salmonids. The bacterial agent can be investigated with simple cultured swabs. The microorganisms can be detected directly in clinical samples or in the environment of the host, without the need for culture. Furunculosis fish infection of *A. salmonicida* can be transmitted to other individuals by the pathogen horizontally, either through physical contact of fish to fish or through shedding the bacteria into the water column. Furunculosis can be controlled by selective breeding, immunization, and treatment by modern vaccination. The disease can be disturbing and can harshly impact the productivity of fish farming processes. Fish culturists, farmers, managers, and stockholders should focus on the regular monitoring of fish to protect their stock from diseases such as furunculosis, because the prevention of furunculosis is critical in the fish population. The best way to protect their stock is by using modern tools and technology.

References

AQUAVETPLAN. (2009). Disease strategy furunculosis (*Aeromonas salmonicida* subsp. *salmonicida*). Australian Aquatic Veterinary Emergency Plan.

Austin, B. (2019). Methods for the diagnosis of bacterial fish diseases. *Marine Life Science & Technology*, *1*, 41–49.

Austin, B., & Austin, D. A. (2007). *Characteristics of the diseases. Bacterial fish pathogens: Diseases of farmed and wild fish* (pp. 15–46). Springer.

Austin, B., & Austin, D. A. (2012). *Aeromonadaceae representative (*Aeromonas salmonicida*). Bacterial fish pathogens* (pp. 147–228). Dordrecht: Springer.

Austin, B., & Austin, D. A. (2016). *Aeromonadaceae representatives (motile aeromonads). Bacterial fish pathogens* (pp. 161–214). Cham: Springer.

Björnsdóttir, B., Gudmundsdóttir, S., Bambir, S. H., Magnadóttir, B., & Gudmundsdóttir, B. K. (2004). Experimental infection of turbot, *Scophthalmus maximus* (L.), by *Moritella viscosa*, vaccination effort and vaccine-induced side-effects. *Journal of Fish Diseases*, *27*(11), 645–655.

Bricknell, I. (2017). *Types of pathogens in fish, waterborne diseases. Fish diseases* (pp. 53–80). Academic Press.

Cain, K. D., & Polinski, M. P. (2014). *Infectious diseases of coldwater fish in fresh water. Diseases and disorders of finfish in cage culture* (p. 60) CABI.

Cipriano, R. C., & Bullock, G. L. (2001). *Furunculosis and other diseases caused by* Aeromonas salmonicida. National Fish Health Research Laboratory.

Dar, G. H., Bhat, R. A., Kamili, A. N., Chishti, M. Z., Qadri, H., Dar, R., & Mehmood, M. A. (2020). *Correlation between pollution trends of freshwater bodies and bacterial disease of fish fauna. Fresh water pollution dynamics and remediation* (pp. 51–67). Singapore: Springer.

Dar, G. H., Dar, S. A., Kamili, A. N., Chishti, M. Z., & Ahmad, F. (2016). Detection and characterization of potentially pathogenic *Aeromonas sobria* isolated from fish *Hypophthalmichthys molitrix* (Cypriniformes: Cyprinidae). *Microbial Pathogenesis*, *91*, 136–140.

Dar, G. H., Kamili, A. N., Chishti, M. Z., Dar, S. A., Tantry, T. A., & Ahmad, F. (2016). Characterization of *Aeromonas sobria* isolated from fish rohu (*Labeo rohita*) collected from polluted pond. *Journal of Bacteriology & Parasitology*, *7*(3), 1–5. Available from https://doi.org/10.4172/2155-9597.1000273.

Emmerich, R., & Weibel, E. (1894). Ueber eine durch Bakterien erzengte Sauche unter den Fore1len. *Archiv fur Hygiene und Bakteriologie*, *21*, 1–21.
Flehn, M. (1909). Die Furuncu10se-epidemic der Salmoniden in, Su.ddeutschland. *Centralbl. f. Bakteriol. Abt. ' I. Orig.*, *LII*, 468. (After Fish, 1937).
Flehn, M. (1911). Die Furunculose del' salmoniden. *Centralbl. f. Bakteriol. Abt., I. Orig, V: Jl.*, *60*, 609–624, llPP.
Giraud, E., Blanc, G., Bouju-Albert, A., Weill, F. X., & Donnay-Moreno, C. (2004). Mechanisms of quinolone resistance and clonal relationship among *Aeromonas salmonicida* strains isolated from reared fish with furunculosis. *Journal of Medical Microbiology*, *53*(9), 895–901.
Grøntvedt, R. N., & Espelid, S. (2004). Vaccination and immune responses against atypical *Aeromonas salmonicida* in spotted wolffish (*Anarhichas minor* Olafsen) juveniles. *Fish & Shellfish Immunology*, *16*(3), 271–285.
Gudmundsdóttir, B. K., Björnsdóttir, B., Árnadóttir, H., Adalbjarnardóttiŕ, A., Magnadottir, B., & Gudmundsdóttir, S. (2005a). Experimental vaccination of cod against atypical furunculosis. In: *Proceedings of the European Association of Fish Pathologists 12th International Conference*, p. 104.
Gudmundsdóttir, B. K., Björnsdóttir, B., Árnadóttir, H., Adalbjarnardóttiŕ, A., Magnadottir, B., & Gudmundsdóttir, S., 2005b. Experimental vaccination of cod. In: *Abstract Book of Conference of Nordic Cod Farming Network in Co-operation with AVS Aquaculture Group on Cod Farming in Nordic Countries*, p. 22, Reykjavík.
Gudmundsdóttir, B. K., & Gudmundsdottir, S. (1997). Evaluation of cross protection by vaccines against atypical and typical furunculosis in Atlantic salmon, *Salmo salar* L. *Journal of Fish Diseases*, *20*(5), 343–350.
Gudmundsdóttir, B. K., & Magnadóttir, B. (1997). Protection of Atlantic salmon (*Salmo salar* L.) against an experimental infection of *Aeromonas salmonicida* ssp. achromogenes. *Fish & Shellfish Immunology*, *7*(1), 55–69.
Gudmundsdóttir, S., Lange, S., Magnadóttir, B., & Gudmundsdóttir, B. K. (2003). Protection against atypical furunculosis in Atlantic halibut, *Hippoglossus hippoglossus* (L.); comparison of a commercial furunculosis vaccine and an autogenous vaccine. *Journal of Fish Diseases*, *26*(6), 331–338.
Ingilæ, M., Arnesen, J. A., Lund, V., & Eggset, G. (2000). Vaccination of Atlantic halibut *Hippoglossus hippoglossus* L., and spotted wolffish *Anarhichas minor* L., against atypical *Aeromonas salmonicida*. *Aquaculture (Amsterdam, Netherlands)*, *183*(1–2), 31–44.
Johansen, L. H., Jensen, I., Mikkelsen, H., Bjørn, P. A., Jansen, P. A., & Bergh, Ø. (2011). Disease interaction and pathogens exchange between wild and farmed fish populations with special reference to Norway. *Aquaculture (Amsterdam, Netherlands)*, *315*(3–4), 167–186.
Kar, D. (2015). *Epizootic ulcerative fish disease syndrome*. Academic Press.
Khoo, L. (2019). *Renal diseases and disorders. Fish diseases and medicine* (pp. 211–229). CRC Press.
Kim, D. H., Choi, S. Y., Kim, C. S., Oh, M. J., & Jeong, H. D. (2013). Low-value fish used as feed in aquaculture were a source of furunculosis caused by atypical *Aeromonas salmonicida*. *Aquaculture (Amsterdam, Netherlands)*, *408*, 113–117.
Lafferty, K. D., & Harvell, C. D. (2014). *The role of infectious diseases in marine communities. Marine community ecology and conservation* (pp. 85–108). Sunderland, MA: Sinauer Associates, Inc.
Lagier, J. C., Edouard, S., Pagnier, I., Mediannikov, O., Drancourt, M., & Raoult, D. (2015). Current and past strategies for bacterial culture in clinical microbiology. *Clinical Microbiology Reviews*, *28*(1), 208–236.
Lim, J., & Hong, S. (2020). Characterization of *Aeromonas salmonicida* and *A. sobria* isolated from cultured salmonid fish in Korea and development of a vaccine against furunculosis. *Journal of Fish Diseases*, *43*(5), 609–620.
Lund, V., Arnesen, J. A., & Eggset, G. (2002). Vaccine development for atypical furunculosis in spotted wolffish *Anarhichas minor* O.: Comparison of efficacy of vaccines containing different strains of atypical *Aeromonas salmonicida*. *Aquaculture (Amsterdam, Netherlands)*, *204*(1–2), 33–44.

Meyburgh, C. M., Bragg, R. R., & Boucher, C. E. (2017). Lactococcus garvieae: An emerging bacterial pathogen of fish. *Diseases of Aquatic Organisms*, *123*(1), 67–79.

Mikkelsen, H., Schrøder, M. B., & Lund, V. (2004). Vibriosis and atypical furunculosis vaccines; efficacy, specificity and side effects in Atlantic cod, *Gadus morhua* L. *Aquaculture (Amsterdam, Netherlands)*, *242* (1–4), 81–91.

Moksness, E., Kjorsvik, E., & Olsen, Y. (Eds.), (2008). *Culture of cold-water marine fish*. John Wiley & Sons.

Nash, A. A., Dalziel, R. G., & Fitzgerald, J. R. (2015). *Mims' pathogenesis of infectious disease*. Academic Press.

Noga, E. J. (2010). *Fish disease: Diagnosis and treatment*. John Wiley & Sons.

Ogut, H. A. M. D. İ., LaPatra, S. E., & Reno, P. W. (2005). Effects of host density on furunculosis epidemics determined by the simple SIR model. *Preventive Veterinary Medicine*, *71*(1–2), 83–90.

Paredez, G. R., Verner-Jeffreys, D., Papadopoulou, A., Monaghan, S., Smith, L., Haydon, D., Wallis, T., Davie, A., Adams, A., & Migaud, H. (2020). *A commercial autogenous injection vaccine protects ballan wrasse (*Labrus bergylta, *Ascanius) against* Aeromonas salmonicida *vapA type V*. bioRxiv.

Perry, G. M., Tarte, P., Croisetiere, S., Belhumeur, P., & Bernatchez, L. (2004). Genetic variance and covariance for 0+ brook charr (*Salvelinus fontinalis*) weight and survival time of furunculosis (*Aeromonas salmonicida*) exposure. *Aquaculture (Amsterdam, Netherlands)*, *235*(1–4), 263–271.

Plumb, J. A., & Hanson, L. A. (2010). *Health maintenance and principal microbial diseases of cultured fishes*. John Wiley & Sons.

Pylkkö, P. (2004). *Atypical Aeromonas salmonicida-infection as a threat to farming of arctic charr (*Salvelinus alpinus *L.) and european grayling (*Thymallus thymallus *L.) and putative means to prevent the infection (No. 142)*. University of Jyväskylä.

Schachte, J. H. (2002). *Furunculosis (Chapter 25)*. Rome, NY: New York Department of Environmental Conservatory.

Stride, M. C., Polkinghorne, A., & Nowak, B. F. (2014). Chlamydial infections of fish: Diverse pathogens and emerging causes of disease in aquaculture species. *Veterinary Microbiology*, *170*(1–2), 19–27.

Vasquez, I., Cao, T., Hossain, A., Valderrama, K., Gnanagobal, H., Dang, M., Leeuwis, R. H., Ness, M., Campbell, B., Gendron, R., & Kao, K. (2020). *Aeromonas salmonicida* infection kinetics and protective immune response to vaccination in sablefish (*Anoplopoma fimbria*). *Fish & Shellfish Immunology*, *104*, 557–566.

CHAPTER

Bacterial gill disease and aquatic pollution: a serious concern for the aquaculture industry 13

Yahya Bakhtiyar, Tabasum Yousuf and Mohammad Yasir Arafat

Fish Biology and Limnology Research Laboratory, Department of Zoology, University of Kashmir, Hazratbal, Jammu and Kashmir, India

13.1 Introduction

Pollution is defined as the introduction and inclusion of contaminants in the natural environment, resulting in adverse effects. It is well-known that contaminants intrude into the aquatic environment at different times and cause changes in its constituents. Water is typically considered to be polluted when it is hampered by anthropogenic pollutants, for example, pesticides, heavy metals, and hydrocarbons, which are often released into the aquatic environment, rendering it unfit for survival of aquatic animals. Incorporation of large amounts of pollutants can lead to rapid effects, as demonstrated by the mortality of local organisms (Austin, 1999). There are a number of waterborne pollutants such as organochlorines, creosote, fecal debris, some heavy metals, inorganic compounds like nitrite, and others that are believed to be the causative agents of significant abnormalities in the aquatic biota, especially fish. Creosote is believed to cause epithelial lesions, fin erosion, or death in fish when they are exposed to high molecular weight creosote-bearing sediments, whereas the occurrence of head lesions is seen due to the impact of low molecular weight creosote (Sved et al., 1997). Organochlorines are associated with malformations of embryos of the common dab, *Platichthys flesus*, *Pleuronectes platessa*, and *Merlangius* sp. (Dethlefsen et al., 1996), and DDT has been associated with neoplasia of liver in *Pleuronectes americanus* (Moore et al., 1996). Poultry fecal matter used as manure in ponds caused large mortality in *Hypophthalmichthys molitrix* in Israel (Bejerano et al., 1979). Nitrites are believed to increase susceptibility of catfish to the bacterium *Aeromonas hydrophila* (Dar et al., 2020; Dar et al., 2016a,b; Hanson & Grizzle, 1985). The effect of a pollutant is proportional to the concentration of the pollutant experienced by aquatic animals, because high concentrations can cause mortalities and lower amounts of pollutants still result in severe damage (Austin, 1999), as shown diagrammatically in Fig. 13.1.

There are specific examples of fish diseases that directly show the impact of water pollution, like lesions on the surface of skin caused by *Serratia plymuthica* and red mouth disease from the activity of *Yersinia ruckeri* due to contamination of waterways by household effluents (Austin, 1999; Dar, Dar et al., 2016; Dar, Kamili et al., 2016). Gill disease is also attributed to *Flavobacterium* spp. due to organic load, decrease in oxygen content, and alteration of pH values

Bacterial Fish Diseases. DOI: https://doi.org/10.1016/B978-0-323-85624-9.00012-9

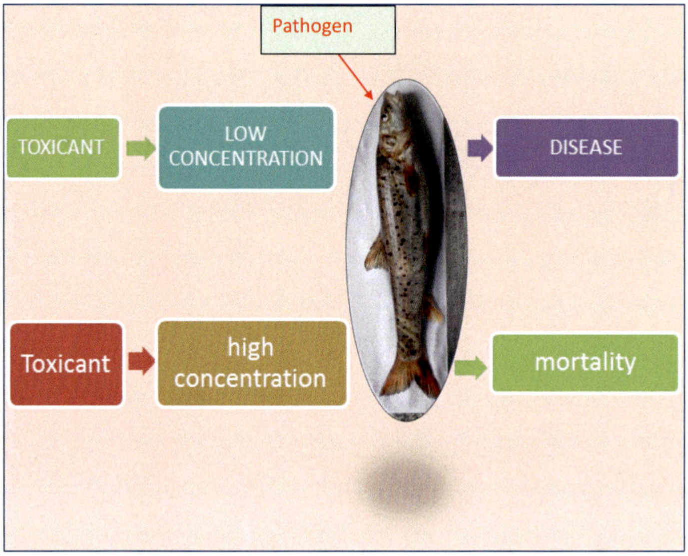

FIGURE 13.1

Diagrammatic illustration of effect of toxicants on fish.

(Austin, 1999). Therefore the role of pollution in causing disease in aquatic organisms, especially fishes, is quite significant, illustrated clearly in bacterial gill diseases (BGDs) of fishes, as shown in Fig. 13.2, which is in conformity with the equation Disease = Host + Pathogen + Environment.

13.2 Bacterial gill disease

The fish gill is a multifunctional organ that contributes to respiration, excretion, and nutrition/digestion by filtering of food particles. The outer covering of the gill is a chief site for various vital processes like gas exchange, acid base balance, ionic regulation, and metabolic waste excretion in fishes (Hoar & Randall, 1984). In addition, gills help the fish to acclimatize in diverse and fluctuating surroundings. The gills are designed in such a way that the physiological functions of the fish remain conserved due to adjustment of their gill morphology during adaptation in diverse environmental conditions (Laurent & Perry, 1990). Various aquatic contaminants have been found to have an impact on the fish gill, morphologically as well as physiologically. The gills are susceptible to parasites and bacterial and fungal diseases, so they may be used for studying the effect of toxic substances resulting in general epidermal pathologies (Evans, 1987). BGD is a peculiar form of pathology that is caused by not just one but several bacterial species, and that is significantly affected by ecological conditions (Snieszko, 1981), as its prevalence usually is related to the rise in concentration of lethal metabolic waste products as well as other significant factors, as illustrated in Fig. 13.3. The onset of BGD is usually linked to hostile environmental conditions in combination with overcrowding as well as ecological stress, which is believed to regulate the severity of disease.

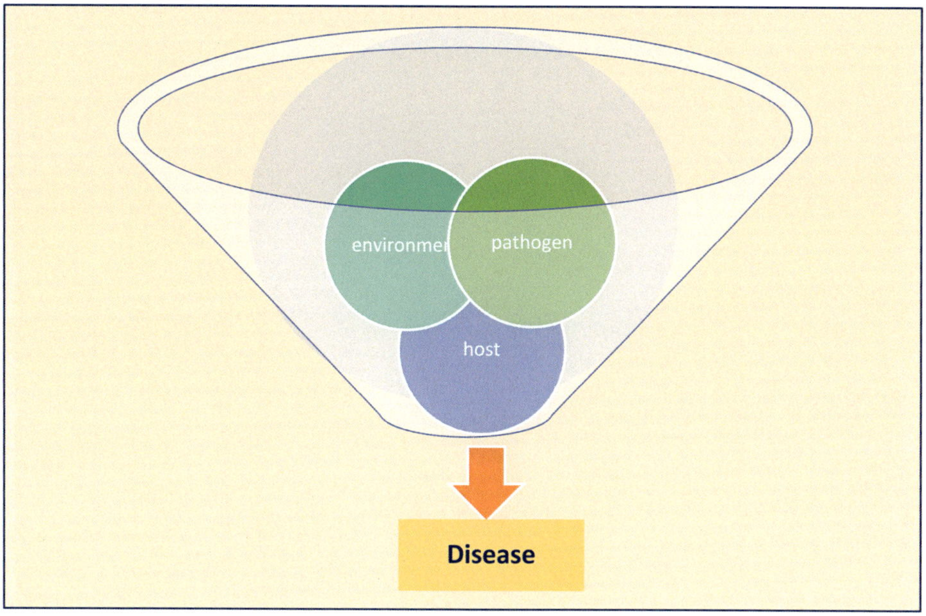

FIGURE 13.2

Interactions between host, pathogen, and environment causing outbreak of diseases (Bohl, 1989; Wood, 1974).

BGD is actually a condition of colonization of fishes by bacteria and its name reveals the pathological signs of infection by bacteria on the gill structure (Wood, 1974). It is considered to be a vital cause of morbidness and disambiguation in Salmonid fish reared in intensive conditions (Bullock, 1990; Daoust & Ferguson, 1983), as it decreases the gill capacity for providing oxygen to blood. The onset of disease depends on the size of the fish, as fish of size less than 90–100/lb are found to be the most prone to infection (Wood, 1974).

13.3 History and geographical range

Davis (1926) was the first to report BGD; however, Bullock (1972) proposed its pathogens as *Flavobacterium*, *Cytophaga*, and *Flexibactor*. Later, the etiological agent was described by Kimura et al. (1978) as *Flavobacterium branchiophilum*. Wakabayashi et al. (1989) isolated, characterized, and named it *F. branchiophila*. However, Von Graevenitz (1990) proposed the name *F. branchiophilum* as the proper nomenclature. Later, Ferguson et al. (1991) artificially induced *F. branchiophilum* in fingerlings that led to pathogenic disease. In the case of trout, MacPhee et al. (1995) reported spontaneous outbreaks of BGD while rearing them in controlled conditions. BGD is believed to prevail in low temperature-water dwelling fishes, posing a significant threat in juveniles of salmonids cultivated in hatcheries (Bullock, 1972, 1990; Daoust & Ferguson, 1983; Farkas, 1985; Schachte, 1983).

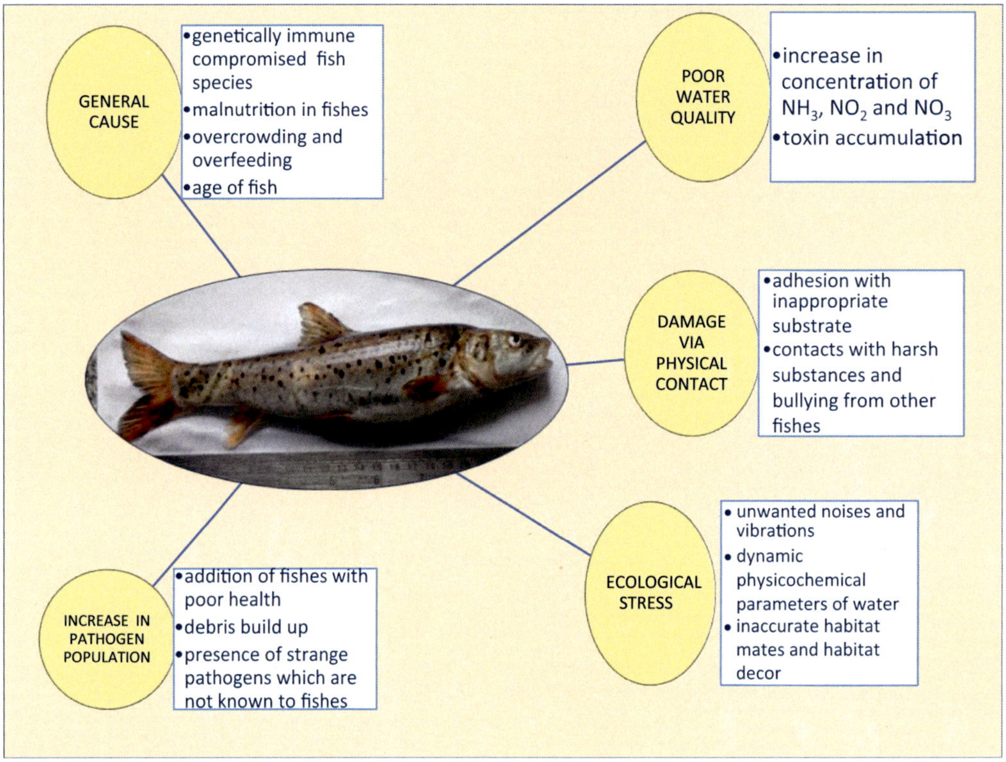

FIGURE 13.3

fffffCausative attributes of bacterial gill disease in fishes.

Generally, flavobacterial diseases cover a large distribution of hosts because of their ubiquitous nature and ability to acclimatize to a wide range of water temperatures (Starliper, 2012). However, the distribution of causative agents is determined by physicochemical parameters of water favorable for the growth of bacteria. BGD caused by *F. branchiophilum* is believed to have originated in North America (United States, Canada, Ontario), Asia (Japan, India, Korea), and Europe (Denmark, Hungary, Netherlands) (Dash et al., 2009; Farkas, 1985; Huh & Wakabayashi, 1989; Ko & Heo, 1997; Ostland et al., 1994; Swain et al., 2007; Wakabayashi et al., 1989).

13.4 Causative agents

BGD is most commonly caused by species of *Flavobacterium*, *Cytophaga*, and *Flexibacter*; however, the yellow-pigmented, filamentous, gram-negative filamentous bacteria belonging to genus *Flavobacterium* (most often *F. branchiophilum*) are considered the chief etiological agent (Farkas, 1985; Ferguson et al., 1991; Heo et al., 1990; Kimura et al., 1978; Ostland et al., 1994; Von Graevenitz, 1990; Wakabayashi et al., 1980, 1989) and are horizontally shared between fishes.

13.5 Host species of the disease

BGD is considered cosmopolitan, prevailing throughout the world, and almost all intensely reared inland fishes are susceptible to it (Snieszko, 1981). Coldwater fishes, mostly wild and reared Salmonids, are known to be affected by this disease, which includes rainbow trout (*Onchorhynchus mykiss*), Chinook salmon (*O. tshawytscha*), sockeye salmon (*O. nerka*), coho salmon (*O. kisutch*), brown trout (*Salmo trutta m. lacustris*), brook trout (*S. fontinalis*), lake trout (*S. namaycush*), Arctic charr (*S. alpinus L.*), sea trout (*Salmo trutta m. trutta*), and Atlantic salmon (*Salmo salar*) (Bernardet & Kerouault, 1989; Bullock, 1972; Bustos et al., 1995; Cipriano et al., 1996; Ostland et al., 1995; Rintamäki-Kinnunen et al., 1997; Schmidtke & Carson, 1995; Wiklund et al., 1994). The nonsalmonid hosts include ayu (*Plecoglossus altivelis*), eel (*Anguilla anguilla*), carp (*Cyprinus carpio*), silver carp (*Hypopthalmichthyes molitrix*), tench (*Tinca tinca*), and crucian carp (*Carassius carassius*) (Lehmann et al., 1991; Wakabayashi et al., 1994), along with tiger pike perciforms (walleye fingerlings) and tiger muskellunge, etc. (Schachte, 1983). Additionally, Ostland et al. (1995) observed that *F. branchiophilum* can cause death of *Luxilus cornutus* (common shiners).

13.6 Pathology and symptoms

Since gills are the primary organs of gaseous exchange, BGD causes disturbance in the normal gaseous exchange leading to fish mortalities. The incubation period ranges from 24 h to weeks (predominantly when water temperatures are below 5°C) and depends upon the age of the fish and virulence of the host (Snieszko, 1981). Transmission takes place through the water from fish to fish in a horizontal direction. Predisposing factors for epizootic outbreaks are suboptimal environmental conditions and suspended solids or abrasive feeds. Different strains of the bacteria have the ability to cause different symptoms in different fish species and the symptoms may vary from behavioral to microscopic, as illustrated in Table 13.1.

Table 13.1 Pathogenesis of bacterial gill diseases.

Strains	Host species	Symptoms	References
Flavobacterium branchiophillum	*Onchorhynchus mykiss, O. tshawytscha* *Salmo trutta* *Luxilus cornutus,* *Labeo rohita*	Lethargic fishes Loss of appetite Flaring of opercula Respiratory distress	Good et al. (2008, 2010), Ostland et al. (1995), Swain et al. (2007), Starliper (2012), Wakabayashi et al. (1980, 1989)
Flavobacterium columnare	Salmonids	Necrosis of gill filaments	Speare and Ferguson (1989)
Flavobacterium psychrophilum	*Onchorhynchus mykiss*	Renal tubular degeneration, pancreatitis, peritonitis, splenic edema. Fusion of secondary lamellae of gills	Starliper (2011)

13.7 Diagnosis

13.7.1 Diagnosis by direct observation

This type of direct diagnosis is based on presumption by experienced individuals who analyze prominent fish behavior, vulnerability of fish species, and preceding outbreaks of disease, as well as periodicity and yield cycle. It also includes expert microscopic observation of histopathological changes of gills and gill arches, which are positively correlated with association to pathogen organisms (Byrne et al., 1991; Ostland et al., 1990).

13.7.2 Other diagnostic procedures

This includes culture of pathogens, serological analysis, and PCR.

13.8 Control methods

The disease can be controlled by following proper prophylactic measures as well as some chemical treatments.

13.9 Prophylactic measures

- Pollution-free ponds with significant reduction of dead fishes and sufficient flow of clean water (Wood, 1974).
- Regulation of population density in aquaculture farms.
- Maintenance of good ecological condition and continuous monitoring of water quality in water bodies.
- Regulation of stress-causing factors, particularly environmental factors.

Table 13.2 Various compounds used for the treatment of bacterial gill disease.

Compound	Concentration	Reference
$KMnO_4$	1–2 ppm (1-h bath)	Wood (1974)
Hyamine-1622 (98.8%) Hyamine-3500 (50%) Roccal (on basis of active ingredient)	1–2 ppm (1-h bath)	Snieszko (1981)
Purina Four Power	3–4 ppm (1-h flush treatments)	Schachte (1983)
Chloramine –T	10 ppm (1-h flush treatments)	From (1980)
Diquat	8.4–16.8 ppm or 2–4 ppm of formulation	Snieszko (1981)
$CuSO_4.5H_2O$	1:2000 ppm (1-min bath) 1 ppm (1-h bath)	Davis (1926)

13.10 Chemical treatment

Various compounds have been found for treatment of BGDs, which are described in Table 13.2.

13.11 Conclusion

To meet the food security issues worldwide, aquaculture plays a significant role. However, the aquaculture industry is under the constant threat of various types of diseases that result from the deterioration of the aquatic ecosystem, mainly due to anthropogenic sources leading to disease outbreaks. The sole life of fish is water, and deterioration in the water quality due to pollution can be a conducive environment for various types of bacterial strains. Some cause gill disease in fishes, with significant effects due to disturbances in their normal gaseous exchange process and also due to secondary infections. To overcome this problem, proper prophylactic measures need to be followed when bacterial disease outbreaks occur, such as isolation of infected fish, decrease in overcrowding fish density, removal of stress factors, and proper chemical treatment. The pollution load in the water bodies must be mitigated on an individual level, so that extreme losses in the aquaculture industry are significantly reduced in order to continue to meet the never-ending food demands of the human population.

Acknowledgements

The authors are thankful to Head, Department of Zoology, University of Kashmir for providing necessary laboratory and library facilities and support in the form of financial assistance from DST-SERB, Govt. of India (File No: EMR/2017/003669).

References

Austin, B. (1999). The effects of pollution on fish health. *Journal of Applied Microbiology*, *85*, 234S–242S.

Bejerano, Y., Sarig, S., Horne, M. T., & Roberts, R. J. (1979). Mass mortalities in silver carp *Hypophthalmichthys molitrix* (Valenciennes) associated with bacterial infection following handling. *Journal of Fish Diseases*, *2*(1), 49–56.

Bernardet, J. F., & Kerouault, B. (1989). Phenotypic and genomic studies of *Cytophaga psychrophila* isolated from diseased rainbow trout (*Oncorhynchus mykiss*) in France. *Applied and Environmental Microbiology*, *55*(7), 1796–1800.

Bernardet, J. F., Segers, P., Vancanneyt, M., Berthe, F., Kersters, K., & Vandamme, P. (1996). Cutting a Gordian knot: Emended classification and description of the genus Flavobacterium, emended description of the family Flavobacteriaceae, and proposal of *Flavobacterium hydatis* nom. nov. (basonym, *Cytophaga aquatilis* Strohl and Tait 1978). *International Journal of Systematic and Evolutionary Microbiology*, *46*(1), 128–148.

Bohl, M. (1989). *Optimal water quality-basis of fish health and economical production. Current trends in fish therapy* (pp. 18–32). Giessen: Deutsche Veterinarmedizinische Gesellschaft ev.

Bullock, G. L. (1972). *Studies on selected myxobacteria pathogenic for fishes and on bacterial gill disease in hatchery-reared salmonids* (Vol. 60). US Department of the Interior, Fish and Wildlife Service, Bureau of Sport Fisheries and Wildlife.

Bullock, G. L. (1990). *Bacterial gill disease of freshwater fishes* (Vol. 84). US Department of the Interior, Fish and Wildlife Service, Division of Fisheries and Wetlands Research.

Bustos, P. A., Calbuyahue, J., Montana, J., Opazo, B., Entrala, P., & Solervicens, R. (1995). First isolation of *Flexibacter psychrophilus*, as causative agent of Rainbow Trout Fry Syndrome (RTFS), producing rainbow trout mortality in Chile. *Bulletin of the European Association of Fish Pathologists (United Kingdom), 15*, 162–164.

Byrne, P., Ferguson, H. W., Lumsden, J. S., & Ostland, V. E. (1991). Blood chemistry of bacterial gill disease in brook trout *Salvelinus fontinalis. Diseases of Aquatic Organisms, 10*, 1–6.

Cipriano, R. C., Schill, W. B., Teska, J. D., & Ford, L. A. (1996). Epizootiological study of bacterial cold-water disease in Pacific salmon and further characterization of the etiologic agent, *Flexibacter psychrophila. Journal of Aquatic Animal Health, 8*(1), 28–36.

Daoust, P. Y., & Ferguson, H. W. (1983). Gill diseases of cultured salmonids in Ontario. *Canadian Journal of Comparative Medicine, 47*(3), 358.

Dar, G. H., Bhat, R. A., Kamili, A. N., Chishti, M. Z., Qadri, H., Dar, R., & Mehmood, M. A. (2020). *Correlation between pollution trends of freshwater bodies and bacterial disease of fish fauna. Fresh water pollution dynamics and remediation* (pp. 51–67). Singapore: Springer.

Dar, G. H., Dar, S. A., Kamili, A. N., Chishti, M. Z., & Ahmad, F. (2016). Detection and characterization of potentially pathogenic *Aeromonas sobria* isolated from fish *Hypophthalmichthys molitrix* (Cypriniformes: Cyprinidae). *Microbial Pathogenesis, 91*, 136–140.

Dar, G. H., Kamili, A. N., Chishti, M. Z., Dar, S. A., Tantry, T. A., & Ahmad, F. (2016). Characterization of *Aeromonas sobria* isolated from fish rohu (*Labeo rohita*) collected from polluted pond. *Journal of Bacteriology & Parasitology, 7*(3), 1–5. Available from https://doi.org/10.4172/2155-9597.1000273.

Dash, S. S., Das, B. K., Pattnaik, P., Samal, S. K., Sahu, S., & Ghosh, S. (2009). Biochemical and serological characterization of *Flavobacterium columnare* from freshwater fishes of Eastern India. *Journal of the World Aquaculture Society, 40*(2), 236–247.

Davis, H. S. (1926). A new gill disease of trout. *Transactions of the American Fisheries Society, 56*(1), 156–160.

Dethlefsen, V., Von Westernhagen, H., & Cameron, P. (1996). Malformations in North Sea pelagic fish embryos during the period 1984–1995. *ICES Journal of Marine Science, 53*(6), 1024–1035.

Evans, D. H. (1987). The fish gill: Site of action and model for toxic effects of environmental pollutants. *Environmental Health Perspectives, 71*, 47–58.

Farkas, J. (1985). Filamentous *Flavobacterium* sp. isolated from fish with gill diseases in cold water. *Aquaculture (Amsterdam, Netherlands), 44*(1), 1–10.

Ferguson, H. W., Ostland, V. E., Byrne, P., & Lumsdsen, J. S. (1991). Experimental production of bacterial gill disease in trout by horizontal transmission and by bath challenge. *Journal of Aquatic Animal Health, 3* (2), 118–123.

From, J. (1980). Chloramine-T for control of bacterial gill disease. *The Progressive Fish- Culturist, 42*(2), 85–86.

Good, C. M., Thorburn, M. A., Ribble, C. S., & Stevenson, R. M. (2010). A prospective matched nested case–control study of bacterial gill disease outbreaks in Ontario, Canada government salmonid hatcheries. *Preventive Veterinary Medicine, 95*(1–2), 152–157.

Good, C. M., Thorburn, M. A., & Stevenson, R. M. (2008). Factors associated with the incidence of bacterial gill disease in salmonid lots reared in Ontario, Canada government hatcheries. *Preventive Veterinary Medicine, 83*(3–4), 297–307.

Hanson, L. A., & Grizzle, J. M. (1985). Nitrite induced predisposition of channel catfish to bacterial diseases. *The Progressive Fish-Culturist*, *47*(2), 98–101.

Heo, G. J., Kasai, K., & Wakabayashi, H. (1990). Occurrence of *Flavobacterium branchiophila* associated with bacterial gill disease at a trout hatchery. *Fish Pathology*, *25*(2), 99–105.

Hoar, W. S., & Randall, D. J. (Eds.), (1984). *Fish physiology* (Vols. XA, B). Orlando, FL: Academic Press.

Huh, G. J., & Wakabayashi, H. (1989). Serological characteristics of *Flavobacterium branchiophila* isolated from gill diseases of freshwater fishes in Japan, USA, and Hungary. *Journal of Aquatic Animal Health*, *1* (2), 142–147.

Kimura, N., Wakabayashi, H., & Kudo, S. (1978). Studies on bacterial gill disease in salmonids—I: Selection of bacterium transmitting gill disease. *Fish Pathology*, *12*(4), 233–242.

Ko, Y. M., & Heo, G. J. (1997). Characteristics of *Flavobacterium branchiophilum* isolated from rainbow trout in Korea. *Fish Pathology*, *32*(2), 97–102.

Laurent, P., & Perry, S. F. (1990). Effects of cortisol on gill chloride cell morphology and ionic uptake in the freshwater trout, *Salmo gairdneri*. *Cell and Tissue Research*, *259*(3), 429–442.

Lehmann, J. D. F. J., Mock, D., Stürenberg, F. J., & Bernardet, J. F. (1991). First isolation of *Cytophaga psychrophila* from a systemic disease in eel and cyprinids. *Diseases of Aquatic Organisms*, *10*(3), 217–220.

MacPhee, D. D., Ostland, V. E., Lumsden, J. S., Derksen, J., & Ferguson, H. W. (1995). Influence of feeding on the development of bacterial gill disease in rainbow trout *Oncorhynchus mykiss*. *Diseases of Aquatic Organisms*, *21*(3), 163–170.

Moore, M. J., Shea, D., Hillman, R. E., & Stegeman, J. J. (1996). Trends in hepatic tumours and hydropic vacuolation, fin erosion, organic chemicals and stable isotope ratios in winter flounder from Massachusetts, USA. *Marine Pollution Bulletin*, *32*(6), 458–470.

Ostland, V. E., Ferguson, H. W., Prescott, J. F., Stevenson, R. M. W., & Barker, I. K. (1990). Bacterial gill disease of salmonids; relationship between the severity of gill lesions and bacterial recovery. *Diseases of Aquatic Organisms*, *9*(1), 5–14.

Ostland, V. E., Lumsden, J. S., MacPhee, D. D., & Ferguson, H. W. (1994). Characteristics of *Flavobacterium branchiophilum*, the cause of salmonid bacterial gill disease in Ontario. *Journal of Aquatic Animal Health*, *6*(1), 13–26.

Ostland, V. E., MacPhee, D. D., Lumsden, J. S., & Ferguson, H. W. (1995). Virulence of *Flavobacterium branchiophilum* in experimentally infected salmonids. *Journal of Fish Diseases*, *18*(3), 249–262.

Rintamäki-Kinnunen, P., Bernardet, J. F., & Bloigu, A. (1997). Yellow pigmented filamentous bacteria connected with farmed salmonid fish mortality. *Aquaculture*, *149*(1-2), 1–14.

Schachte, J. H. (1983). *A guide to integrated fish health management in the Great Lakes Basin*. Ann Arbor, MI: Great Lakes Fishery Commission.

Schmidtke, L. M., & Carson, J. (1995). Characteristics of *Flexibacter psychrophilus* isolated from Atlantic salmon in Australia. *Diseases of Aquatic Organisms*, *21*(2), 157–161.

Snieszko, S. F. (1981). *Bacterial gill disease of freshwater fishes* (Vol. 62). US Department of the Interior, Fish and Wildlife Service, Division of Fishery Research.

Speare, D. J., & Ferguson, H. W. (1989). Clinical and pathological features of common gill diseases of cultured salmonids in Ontario. *The Canadian Veterinary Journal*, *30*(11), 882.

Starliper, C. E. (2011). Bacterial cold-water disease of fishes caused by *Flavobacterium psychrophilum*. *Journal of Advanced Research*, *2*(2), 97–108.

Starliper, C. E. (2012). *1.2. 1 Bacterial gill disease*. Available at: https://units.fisheries.org/fhs/wp-content/uploads/sites/30/2017/08/1.2.1-BGD_2014.pdf.

Sved, D. W., Roberts, M. H., Jr, & Van Veld, P. A. (1997). Toxicity of sediments contaminated with fractions of creosote. *Water Research*, *31*(2), 294–300.

Swain, P., Mishra, S., Dash, S., Nayak, S. K., Mishra, B. K., Pani, K. C., & Ramakrishna, R. (2007). Association of *Flavobacterium branchiophilum* in bacterial gill disease of Indian major carps. *The Indian Journal of Animal Sciences*, *77*(7), 646–649.

Von Graevenitz, A. (1990). Revised nomenclature of *Campylobacter laridis*, *Enterobacter intermedium*, and *Flavobacterium branchiophila*. *International Journal of Systematic and Evolutionary Microbiology*, *40*(2), 211, 211.

Wakabayashi, H. (1980). Bacterial gill disease of salmonid fish. *Fish Pathology*, *14*(4), 185–189.

Wakabayashi, H., Egusa, S., & Fryer, J. L. (1980). Characteristics of filamentous bacteria isolated from a gill disease of salmonids. *Canadian Journal of Fisheries and Aquatic Sciences*, *37*(10), 1499–1504.

Wakabayashi, H., Huh, G. J., & Kimura, N. (1989). *Flavobacterium branchiophila* sp. nov., a causative agent of bacterial gill disease of freshwater fishes. *International Journal of Systematic and Evolutionary Microbiology*, *39*(3), 213–216.

Wakabayashi, H., Toyama, T., & Iida, T. (1994). A study on serotyping of *Cytophaga psychrophila* isolated from fishes in Japan. *Fish Pathology*, *29*(2), 101–104.

Wiklund, T., Kaas, K., Lönnström, L., & Dalsgaard, I. (1994). Isolation of *Cytophaga psychrophila* (*Flexibacter psychrophilus*) from wild and farmed rainbow trout (*Oncorhynchus mykiss*) in Finland. *Bulletin of the European Association of Fish Pathologists*, *14*(2), 44–46.

Wood, J. W. (1974). *Diseases of Pacific salmon, their prevention and treatment*. State of Washington: Department of Fisheries, Hatchery Division.

CHAPTER

14

Common bacterial pathogens in fish: An overview

Podduturi Vanamala[1], Podduturi Sindhura[2], Uzma Sultana[1], Tanaji Vasavilatha[3] and Mir Zahoor Gul[4]

[1]Telangana Social Welfare Residential Degree College for Women, Kamareddy, India [2]Telangana Social Welfare Residential Degree College for Women, Bhupalapally, India [3]Telangana Social Welfare Residential Degree College for Women, Nizamabad, India [4]Department of Biochemistry, University College of Science, Osmania University, Hyderabad, India

14.1 Introduction

Fish has long been known as a safe and functional food and represents a valuable food source for humans, providing nearly 60% of the global supply of protein. Approximately 60% of the underdeveloped countries obtain 30% of their annual protein supply from fish (Abisoye et al., 2011). It is also the most significant source of high-quality protein in humans, supplying about 16% of the animal protein consumed by the global population (FAO, 1997). Fish is not only seen as nutritional but also as a viable source of revenue for the small-scale farming sector (Fig. 14.1). Fishes are, however, susceptible to a variety of pathogenic bacteria, which mostly can lead to infections and are considered saprophytic (Lipp & Rose, 1997). Most of the bacterial disease agents are part of the aquatic environment's common microflora and are usually regarded as a secondary or opportunistic pathogen. Nearly all the aquatic bacterial pathogens have the potential to live outside fish. There are very few obligatory pathogens, yet even these can remain in host tissues for a longer period, without inducing serious damage. Clinical infections and diseases typically occur after the development of any significant changes in the host's physiology or body. The number of bacterial species associated with fish diseases has been steadily increasing, with new pathogens consistently recognized in the published research (Austin & Austin, 2007). The microbial association with fish jeopardizes safety and human consumption quality, specifically when microorganisms are of an opportunistic and/or infective nature (Mhango et al., 2010). The microbial diversity of fresh fish muscles relies on the fishing grounds and surrounding environmental conditions (Cahill, 1990). Freshwater fishes may become infected from different sources, as bacteria are present in air, water, and in sediments too. It is also assumed that the temperature and various biotic and abiotic characteristics are the most significant factors favorable to the occurrence of infectious disease symptoms. It has been proposed that identifying a microorganism as associated with fish depends on the environment (Claucas & Ward, 1996). Most of the bacteria colonize the digestive tract, gills, and skin of the fish, living as parasites or commensals. The fish-associated bacterial pathogens are categorized as nonindigenous and indigenous (Kvenberg, 1991). The nonindigenous ones pollute the fish

Bacterial Fish Diseases. DOI: https://doi.org/10.1016/B978-0-323-85624-9.00010-5

FIGURE 14.1

Glimpses of a few freshwater fish.

or habitat one way or the other; examples include *Clostridium botulinum*, *Staphylococcus aureus*, *Shigella dysenteriae*, *Listeria monocytogens*, *Salmonella*, and *Escherichia coli*. The indigenous bacterial pathogens like *Aeromonas* and *Vibrio* are known to live naturally in the fish's habitat (Rodricks, 1991). Bacteria can become pathogenic only when fish are physiologically unstable, nutritionally deficient, or undergoing other causes of stress like water contamination or overstocking, ensuring opportunistic bacterial infections can thrive (Austin, 2011).

Most bacterial diseases display common symptoms, particularly in fish. Bacterial infection can occur in the muscles, in internal organs, on the skin, fins, exoskeleton, or crustacean appendages. The development of the disease is a complicated process, and various factors play a critical role in fish pathology. The changes in the freshwater ecosystem seem to be a significant factor responsible for emerging diseases (Johnson & Paull, 2011). Fish- and shellfish-related pathogenic and potentially pathogenic bacteria include *Aeromonas* spp., *Mycobacterium*, *Salmonella* spp., *Streptococcus* spp., *Vibrio* spp., and others (Lipp & Rose, 1997). A few bacterial species tend to be obligatory parasites responsible for mortality in both cultured and freshwater fishes. Most of the diseases in fish are caused by gram-negative bacteria such as *Acinetobacter* spp., *Aeromonas* spp., *Flavobacterium* spp., *Pseudomonas* spp., *Shewanella putrefacience*, *Sphingomonas paucimobilis*, and *Vibrios*. A few gram-positive bacterial species are also found in association with infections in fish, including *Lactococcus garvieae*, *Mycobacterium* spp., and *Streptococcus iniae*. Both freshwater and marine fishes are affected by different bacterial infections (Table 14.1). Rainbow trout and other salmonids from the Oncorhynchus, Cichlidae (tilapia), and Ictaluridae (catfish) family fishes are usually infected by different bacterial species (Austin & Austin, 2016).

Table 14.1 Common bacterial diseases in freshwater fishes.

S. no.	Pathogen	Disease
Gram-negative bacteria		
1	*Flavobacterium* species: *F. branchiophilum*; *F. columnare*; *F. psychrophilum.*	Bacterial gill disease—fin rot, Columnaris disease, Cold-water disease.
2	*Tenacibaculum maritimum*	Bacterial gill disease, Black patch necrosis, fin rot
3	*Edwardsiella tarda*; *E. ictaluri*	Edwardsiella septicemia, Enteric septicemia of catfish
4	*Yersinia ruckeri*	Enteric red mouth disease
5	*Vibrio anguillarum*; *V. ordalii*	Vibriosis
6	*Vibrio salmonicida*	Hitra disease: cold-water vibriosis
7	*Vibrio vulnificus*	Septicemia
8	*Aeromonadaceae members*	Furunculosis, Septicemia
9	*Moritella viscose*	Winter ulcer disease
10	*Photobacteriaceae members*	Pasteurellosis, Hemorrhagic septicemia
11	Pseudomonadaceae members: *Pseudomonas fluorescens*; *P. anguilliseptica*	Septicemia, Sekiten-byo, red spot
12	*Francisella* spp.	Granulomatosis.
13	*Rickettsia salmonis*	Strawberry disease
Gram positive bacteria		
1	*Renibacterium salmoninarum*	Bacterial kidney disease
2	*Carnobacterium piscicola*	Similar to BKD, pseudo- kidney disease
3	*Streptococcus iniae*; *S. faecalis;* *S. agalactiae*; *S. dysgalactiae*	Septicemia
4	*Lactococcus garvieae*	Septicemia
5	*Clostridium botulinum*	Type E botulism
6	*Mycobacterium marinum*; *M. fortium*; *M. chelonae*	Mycobacteriosis
7	*Nocardia asteroids*; *N. seriolae*	Nocardiosis

14.2 Gram-negative bacterial pathogens

The most commonly encountered bacterial infections in freshwater fishes are caused by *Aeromonas* (Verner-Jeffreys et al., 2009), *Edwardsiella* (Blazer et al., 1985; Park et al., 2012), *Pseudomonas* (Berthe et al., 1995; Trust & Barlett, 1974), and *Shewanella* (Korun et al., 2009). Bacterial diseases in both wild and cultured fish contribute to heavy mortality. The major fish pathogens belong to families Aeromonadaceae, Pseudomonadaceae, Enterobacteriaceae, and Vibrionaceae, which primarily consist of short gram-negative bacilli.

14.2.1 Aeromonadaceae

Infection that is caused by *Aeromonas* species is by far the most prevalent among bacterial infections of fishes and is primarily responsible for motile *Aeromonas* septicemia (MAS), furunculosis, and motile *Aeromonas* infections (Olesen & Vendramin, 2016). The genus *Aeromonas* comprises more than 30 different gram-negative bacterial species and is broadly distributed in diverse water ecosystems. The genus *Aeromonas* was at first included in Vibrionaceae but persuasive arguments have been made for the origin of a separate family (Colwell et al., 1986). Several species of *Aeromonas* are the major cause of septicemia in ornamental fish. Fourteen species of fish have been reported to be infected with *Aeromonas hydrophila* (*A. leuconosticus*, *Danio rerio*, *Gymnocorymbus ternetzi*, *Melanochromis auratus*, *Mikrogeophagus ramirezi*, *Paracheirodon innesi*, *Poecilia reticulata*, *Poecilia sphenops*, *Pseudotropheus saulosi*, *Pterygoplichthys gibbiceps*, *Puntigrus tetrazona*, *Trichogaster lalius*, *Xiphophorus hellerii*, and *X. maculatus*). *Aeromonas caviae* and *Aeromonas media* were primarily isolated from *Poecilia reticulata* and *P. saulosi* (Walczak et al., 2017).

14.2.1.1 Aeromonas hydrophila

14.2.1.1.1 Habitat

Aeromonas hydrophila is an opportunistic pathogen and is primarily found in clean water as well as organically polluted freshwater and is predominant in cosmopolitan aquatic ecosystems. *A. hydrophila* is also a part of a healthy fish intestine's physiological flora (Holmes et al., 1996; Newman, 1982).

14.2.1.1.2 Morphology

A. hydrophila is a freshwater gram-negative, straight rod, chemoorganotrophic bacterium with a monotrichous flagellum. They roughly measure 0.3–1.0 × 1.0–3.5 μm. The G + C% ratio of DNA is 57–63 mol (Huys et al., 2002; Larsen & Jensen, 1977; MacInnes et al., 1979).

14.2.1.1.3 Cultural and biochemical characteristics

A. hydrophila can be conveniently isolated from the infected fish's blood or kidneys and can be grown on traditional nutrient media. On NAM, colonies appear within 24 h at 22°C–28°C and are white to buff-colored. The Rimler Shotts Agar (R-S agar) is a selective media often used to isolate and presumptively distinguish bacteria from specimens prone to contamination with other bacteria. *A. hydrophila* can use a variety of substances such as D-mannitol, D-lactate, D-serine, D-sucrose, and putrescine, salicin as carbon and energy sources for their metabolism (Paterson, 1974). *A. hydrophila* shows positive for various biochemical tests and produces various enzymes like arginine hydrolase, catalase, cytochrome oxidase, phosphatases, ornithine decarboxylases, lysine decarboxylases, and phenylalanine or tryptophan deaminase. *A. hydrophila* is resistant to Vibriostatic Agent O/129.

14.2.1.1.4 Epizootiology

Epizootics in fish caused by *A. hydrophila* are mostly confined to the southeastern United States (Esch et al., 1976; Haley et al., 1967; Miller & Chapman, 1976). According to earlier studies, host stress could be a significant factor in the epizootiology of red pest disease or red sore disease. *A.*

hydrophila is associated with the development of Aeromonad septicemia or red pest in fishes (Esch & Hazen, 1978).

14.2.1.1.5 Pathogenicity and clinical signs

Infections with *A. hydrophila* in fishes result in the development of ulcers, tail rot, fin rot, and hemorrhagic septicemia. Virulence in *A. hydrophila* is multifactorial, with disease resulting from the production of diverse virulence factors that include adhesins, hemolysins, cytotoxin, enterotoxin, proteases, siderophores, acetylcholinesterase, hemagglutinin, and surface proteins (Cahill, 1990; Gosling, 1996; Howard et al., 1996). *O*-antigens are a class of structurally diverse lipopolysaccharides and help in bacterial colonization. Around eight distinct *O*-antigen gene clusters are present in *A. hydrophila*, and all epidemic strains isolated from channel catfish (*Ictalurus punctatus*) shared a homologous *O*-antigen gene cluster (Xu et al., 2019). The enzyme, protease protects the bacterium from the bactericidal effects of serum and enhance the ability of an organism to invade host tissues (Leung & Stevenson, 1988). Bacterial isolates with aerolysin (*aerA*), cytotoxic enterotoxin (*alt*), and serine protease genes (ahp) have been identified to be virulent with lower doses of LD_{50} compared to isolated with just one or two of these virulence genes in zebrafish (Li et al., 2011). In addition, acetylcholinesterase (15.5 kDa polypeptide) was established as a most important lethal agent with potential neurotoxic activity (Nieto et al., 1991; Rodriguez, Ellis et al., 1993; Rodriguez, Fernandez et al., 1993). The compound was given as a minimum lethal dose of 0.05 mg/g of fish. *A. hydrophila* is often present in conjunction with other pathogenic agents like *Aeromonas salmonicida*. Some of the clinical manifestations of *A. hydrophila* infections in freshwater ornamental fish in Sri Lanka were further described, which included the prevalence of eroded fins, sloughing scales, and hemorrhages on the skin, intestinal wall, and at the base of the caudal fin (Hettiarachchi & Cheong, 1994).

14.2.1.1.6 Treatment and control

A. hydrophila are resistant to a broad array of antimicrobial compounds, which includes ampicillin, erythromycin, novobiocin, nitrofurantoin, oxytetracycline, streptomycin, sulfonamides, etc. (Aoki, 1988; De Paola et al., 1988). Due to the widespread plasmid-mediated resistance in fish farms to oxytetracycline, the promising drug of choice is enrofloxacin and it is found to possess antimicrobial activity even at low dosages and the MIC of the compound is reported to be 0.002 μg/mL (Bragg & Todd, 1988).

14.2.2 Vibrionaceae

The family Vibrionaceae comprises gram-negative, curved, or straight rods that use polar flagella for motility. Vibriosis is among the most common disease and is a cause of significant mortality of fish, shellfish, and shrimp in Asia. The onset of the disease is affected by multiple factors, mainly the source of fish, organism virulence, and environmental conditions. Vibriosis is characterized by septicemia, dermal ulceration, hematopoietic necrosis, and hemorrhages at the base of fins. Vibriosis causes major economic damage in both fish farming and larviculture globally. The species *Vibrio anguillarum*, *V. alginolyticus*, *V. fluvialis*, *V. parahaemolyticus*, *V. salmonicida*, and *V. vulnificus* are known as the main pathogens associated with high mortality in many species of fish (Dar et al., 2020; Dar, Dar et al., 2016; Dar, Kamili et al., 2016; Toranzo et al., 2005). *Vibrio campbellii* is one of the most important species of Vibrionaceae contaminating farmed aquatic species (Dar, Dar et al., 2016; Dar, Kamili et al., 2016; Nor-Amalina et al., 2017). In the group of pomacentrid fish (damsels and clowns), Vibrio infections are well documented.

14.2.2.1 *Vibrio anguillarum*

Vibrio anguillarum, also called *Listonella anguillarum*, is the causative organism of classical vibriosis, a lethal hemorrhagic septicemia in warm- and cold-water species of fish. In 1893, Canestrini isolated this bacterium and named it *Bacillus anguillarum* and it was subsequently designated as *Vibrio anguillarum* by Bergman in 1909. High mortality and significant economic losses in aquaculture are induced by *V. anguillarum*. Around 23 serotypes (O1-O23, European serotype designation) have been identified, with the most virulent strains belonging to O1, O2, and to a lesser extent O3.

14.2.2.1.1 Habitat

V. anguillarum is typically present in marine and estuarine habitats and can grow over a variety of temperatures and salinity.

14.2.2.1.2 Morphology

V. anguillarum is a motile (through polar flagella), rod-shaped, gram-negative bacterium.

14.2.2.1.3 Cultural and biochemical characteristics

The bacterium grows well at 15°C–37°C and the growth is enhanced by 1%–2% NaCl. Thiosulfate-citrate-bile salts-sucrose agar (TCBS) is a selective medium employed for the isolation of infective vibrios (except *V. ordallii*). *V. anguillarum* forms yellow colonies on TCBS. By their specific sensitivity to O/129, *V. anguillarum* can be isolated from *Aeromonas* species and is inhibited by Novobiocin. On nonselective media, round, smooth, convex white-colored colonies are formed after an incubation period of 48 h. The organism can utilize arabinose, amygdalin, cellobiose, galactose, glycerol, maltose, mannose, sorbitol, and sucrose trehalose, resulting in the production of acids. The G + C content of DNA is 45.6–46.3 mol%.

14.2.2.1.4 Epizootiology

Vibriosis was first recognized in eels (*Anguilla anguilla*), where it was called a red bug, and it has now been documented in a wide range of telecasts. It is the most extreme epizootic condition in fish from salt or brackish waters. Outbreaks were also observed in freshwater ecosystems and were linked with offal feeding from marine fish (Hacking & Budd, 1971; Ross et al., 1968; Rucker, 1959). *V. anguillarum* is an alimentary microflora of healthy fish both wild and cultured and is recovered from tissues of rotifers as well as other invertebrates (Mizuki et al., 2006). The incubation period is greatly influenced by various factors like temperature virulence of the strain and the levels of stress under which fish live. Mortality rates in an outbreak are higher than 50% in cultured fish, especially in young fish. A sudden temperature change, especially in the midsummer period, salinity fluctuations, and inadequate diet for fish are the main factors that trigger an outbreak of vibriosis. In Japan, *V. anguillarum* is the main reason for extreme losses in freshwater cultured ayu and a major cause of severe casualties in freshwater grown ayu (Kitao et al., 1983). Moreover, cultured species such as ayu sweetish (*Plecoglossus altivezis*), Japanese eel (*Anguilla japonica*), and yellowtail (*Seriola quinqueradiata*) are highly susceptible to vibriosis.

14.2.2.1.5 Pathogenicity and clinical signs

The initial step in disease pathogenesis is by the attachment to the surface of the host cell and it is regarded as an essential stage in pathogenicity. Besides, the presence of multiflagella promotes the motility of bacteria required to reach to the host (Ormonde et al., 2000). The factors associated with virulence found so far in *V. anguillarum* are multifactorial (Hickey & Lee, 2018) and they include endotoxin (LPS), mortality and chemotaxis, extracellular product secretion with hemolytic and proteolytic activities (Li & Ma, 2017; Toranzo et al., 2017). Hemolytic activity is a major determinant of virulence and contributes to hemorrhagic septicemia characteristics of *V. anguillarum* infections (Hirono et al., 1996). Vibrio infections result in the formation of skin ulcers and if left untreated may lead to systemic infections (Sonia & Lipton, 2012). In chronic cases, the lytic toxin of the vibrio induces severe hematolytic anemia that results in a huge deposition of hemosiderin in the melanomacrophage centers of renal and splenic hemopoietic tissue.

Clinical manifestations of vibriosis were initially reported in eels in Norway and subsequently in the United Kingdom (McCarthy, 1976). These clinical signs involve anorexia, abdominal distension, abnormal swimming pattern, darkened skin, exophthalmia or pop-eye, gill necrosis, lethargy, ulcerative skin lesions, and death. Respiratory indications include piping, high rate of opercula, and respiratory distress (Ransangan & Mustafa, 2009).

14.2.2.1.6 Treatment and control

Antibiotics and certain antimicrobial agents can combat vibriosis, which could be used as feed additives in fish farms. Chemotherapeutic agents such as oxytetracycline, nitrofurazone, oxolinic acid, and sulphamerazine have shown great promise in the treatment of fish disease, including vibriosis (Austin & Austin, 1993). Genetic selection and immunization procedures were shown to increase salmonid tolerance to vibrio infections. Moreover, growing evidence suggests that the use of probiotics in aquaculture is effective against vibriosis by enhancing crop standards, improving nutrition, boosting host immunity, and impeding infectious agents (Iranto & Austin, 2002).

14.2.3 Pseudomonadaceae

The major fish pathogenic Pseudomonads include *P. anguilliseptica*, *P. bactica*, *P. chlororaphis*, *P. fluorescens*, *P. luteola*, *P. plecoglossicida*, *P. pseudoalcaligens*, and *P. putida* that are the causative agents of Sekiten byo (red spot). The *Pseudomonas* species are gram negative, slightly curved, or straight and are motile through polar flagellae. The Pseudomonads have been isolated from the gills of healthy juvenile rainbow trout (Nieto et al., 1984)

14.2.3.1 Pseudomonas fluorescens

Pseudomonas fluorescens is one of the main feature of freshwater ecosystems and is identified as one of the causative agents of bacterial hemorrhagic septicemia of fish. *P. fluorescens* has been suggested to cause diseases in fish species such as bighead (*Arisichthys nobilis*), silver carp (*Hypopthalmichthys molitrix*) (Csaba et al., 1981b; Markovic et al., 1996), black carp (*Mylopharyngodon piceus*) (Bauer et al., 1973), goldfish (*Carassius auratus*) (Bullock, 1965), grass carp (*Ctenopharyngodon idella*), rainbow trout (Li & Flemming, 1967; Li & Traxler, 1971; Sakai et al., 1989a), and tench (*Tinca tinca*) (Ahne & Popp Hoffmann, 1982).

14.2.3.1.1 Habitat

The bacterium *P. fluorescens* is discovered in soil and water and has been implicated as a cause of food spoilage, including fish.

14.2.3.1.2 Morphology

P. fluorescens is an aquatic rod-shaped pathogenic bacterium ($0.8 \times 2.0-3.0\ \mu m$) and is mobile by flagella.

14.2.3.1.3 Cultural and biochemical characteristics

On an ordinary nutrient medium, *P. fluorescens* grows well at 22°C–25°C. Cetrimide agar is used to separate *P. fluorescens* selectively from the infected samples. The bacterium is a strict aerobe and produces a fluorescent pigment (fluorescein). It can utilize arabinose, citrate, inositol, mannitol, maltose, sucrose, sorbitol, trehalose, and xylose and results in the production of acids. *P. fluorescens* is positive for oxidase and catalase, but negative for the Voges-Proskauer test and hydrolyzes gelatin.

14.2.3.1.4 Epizootiology

P. fluorescens are abundant in aquatic environments and are usually associated with hemorrhagic septicemia and have been reported as primary fish pathogens clinically indistinguishable from motile Aeromonad septicemias. It is considered to be an opportunistic pathogen causing disease in fish compromised by concomitant environment stress or other chronic viral diseases and following severe trauma (Roberts & Horne, 1978; Schaperclaus, 1926). Water temperatures of $10°C \pm 2°C$ or more favor these opportunistic pathogens.

14.2.3.1.5 Pathogenicity and clinical signs

Hemorrhagic septicemia is characterized by the presence of hemorrhages on the body surface, dark coloration of the skin, sloughing of scales, abdominal distension, and popping/protruding eyeballs (Abowei & Briyai, 2011; Austin & Austin, 2016; Eissa et al., 2010; Roberts, 2012). Hemorrhagic septicemia in fishes may be acute or chronic. Lethargy, large hemorrhagic lesions, and necrosis of the fins are the most observed signs. Abdominal distension with ascitic fluid and exophthalmia are also observed. Internally, the kidney and spleen may be enlarged.

14.2.3.1.6 Treatment and control

Poor environmental conditions potentially expose fish to disease outbreaks and therefore major improvements in the condition can be accomplished by improving water quality or reducing stock densities. Treatment with benzalkonium chloride (1 mg/h), malachite green (1–5 mg/L of water/h), and furanace (0.5–1 mg/L of water/5–10 min) can control early signs of the diseases. Antibiotics such as kanamycin, nalidixic acid, and tetracycline have been successful in the treatment of infections associated with *P. fluorescens* (Sakai et al., 1989a).

14.2.4 Flavobacteriaceae

Flavobacterial infections in fish are triggered by a large variety of bacterial species within the Flavobacteriaceae family and account for catastrophic damages in both cultured and wild fish populations worldwide. Three bacterial species from the family of Flavobacteriaceae were originally related to these diseases (Bernardet & Nakagawa, 2006). *Flavobacterium columnare* acts as the causative agent of columnaris disease that is also called "saddleback disease" (Hawke & Thune, 1992; Shotts & Starliper, 1999); *Flavobacterium branchiophilum* is the causative agent of bacterial gill disease, a significant contributor to young salmonid mortality (Shotts et al., 1972; Wakabayashi et al., 1989); and *Flavobacterium psychrophilum* induces bacterial cold-water disease and fry syndrome of rainbow trout (Nematollahi et al., 2003; Starliper, 2011). Another species of *Flavobacterium*, *F. hydatis*, is also suspected as a putative agent of fish disease (Bernardet et al., 1996; Strohl & Tait, 1978). For a variety of organisms (plants, invertebrates, amphibians, reptiles, birds, and mammals including humans), many flavobacteria are viewed as pathogenic. *Flavobacterium* species have been detected in intestines, gills, fins, reproductive fluids, mucus from eggs and other internal organs of cold- and warm-water fish (Austin & Austin, 2007; Bernardet & Bowman, 2006; Shotts & Starliper, 1999).

14.2.4.1 Flavobacterium columnare (Flexibacter/Cytophaga columnaris)

The bacterium was initially identified as *Bacillus columnaris*; however, it was later called *Chondrococcus columnaris* (Ordal & Rucker, 1944) and *Cytophaga columnaris* (Garnjobst, 1945). It was ultimately placed in the Flexibacter genus and named *Flavobacterium columnare* (Bernardet et al., 1996; Leadbetter, 1974).

14.2.4.1.1 Habitat

The bacterium, *F. columnare*, is widely spread in freshwater sources and is also part of the microbiota of freshwater fish, eggs, and rearing waters (Barker, et al., 1990).

14.2.4.1.2 Morphology

F. columnare is a gram-negative, slender flexible rod with the dimensions of $0.5 \times 4-12$ μm, and mobile via gliding. The bacterium appeared to be congregated in column-like masses in wet mount preparations of infected material, giving the bacillus its specific name.

14.2.4.1.3 Cultural and biochemical characteristics

Like all *Flavobacterium*, *F. columnare* needs relatively lower nutrient levels for growth. The organism does not grow on Tryticase soy agar (TSA); cytophaga agar and Bootsma and Clerx's medium are used for the isolation of the pathogen (Anacker & Ordal, 1959; Bootsma & Clerx, 1976). Growth is typically aerobic and occurs at 4°C–30°C. Colonies can vary widely from round and convex to rhizoid and flat and produce nondiffusible yellow to orange pigment. The DNA base ratio has been variously given as 29–32 mol% G + C (Bernardet & Grimont, 1989; Song et al., 1988). The organism is positive for catalase, oxidase, and nitrate reduction tests. The organism is negative for Voges-Proskauer tests and methyl red. Gelatin is not degraded and carbohydrates are not utilized.

14.2.4.1.4 Epizootiology

The causative agent of columnaris disease, *F. columnare*, is widely distributed worldwide in freshwater ecosystems and can affect both cultivated and freshwater species like carp, catfish, eel, goldfish, perch, salmon, and tilapia (Bernardet and Nakagawa, 2006; Figueiredo et al., 2005; Morley & Lewis, 2010; Řehulka & Minařík, 2007; Soto et al., 2008; Suomalainen et al., 2009). After *Edwardsiella ictalurid*, *F. columnare* is the most common bacteria to induce morbidity and death in the channel catfish (*Ictalurus punctatus*) industry in the United States, with an estimated loss of US$ 30 million per annum (Hawke & Thune, 1992; Shoemaker et al., 2011; Wagner et al., 2002). It was reported that rainbow trout surviving an *F. columnare* infection can release up to 5×10^3 CFU/mL/h of viable organisms into tank water (Fujihara & Nakatani, 1971). The most significant factors that decide the magnitude of the disease are water temperature and virulence. Strain virulence is variable and chondroitin lyase production, a temperature-dependent factor, plays a significant role in assessing infection rate (Suomalainen et al., 2006).

14.2.4.1.5 Pathogenicity and clinical signs

F. columnare causes acute to chronic diseases and usually affects gills, fins, and skin. The acute infection usually happens in young fish and mainly harms the gills. The disease can take an acute, subacute, or chronic path in the case of adults. In acute or subacute forms of the disease, yellowish areas of necrotic tissues are evident in the gills, eventually destroying the complete gills. Columnaris disease can develop acute ulcerative dermatitis that may spread into the muscle and hypodermis. Proteases also contribute to tissue damage or enhance microbial invasion (Dalsgaard, 1993). Due to the appearance of greyish discolored areas around the base of the dorsal fin, the disease is often referred to as saddleback disease (Pacha & Ordal, 1967). Fin rot is also often seen in the infected fish (Bernardet, 2006; Decostere & Haesebrouck, 1999). Histologically, epidermal spongiosis, tissue damage, and ulceration are observed with necrosis extension into the dermis and peripheral hyperemia and hemorrhage. Upon examination of rainbow trout gill tissue under a transmission electron microscope, the presence of long slender bacteria in close contact with gill tissue is revealed.

14.2.4.1.6 Treatment and control

Enhanced oxygenation, control of organic water addition, and water temperature reduction are the most important factors for supportive therapy. Antimicrobial agents such as chloramphenicol, nifurprazine, nifurpirinol, and oxolinic acid are commonly used for treatment for the diseases caused by *F. columnare* (Amend & Ross, 1970; Deufel, 1974; Endo et al., 1973; Ross, 1972; Shiraki et al., 1970; Snieszko, 1958; Soltani et al., 1995).

14.2.5 Photobacteria

14.2.5.1 *Photobacterium damsel*

Photobacterium damsel subsp. *piscicida* was found to be the most important pathogen of this family, associated with ulcerative lesions in blacksmith (*Chromis punctipinnis*). It is identified as a significant primary pathogen, surviving only for a limited period in seawater. It is also reported that the pathogen may survive in water for a long period but is found in the noncultivable form.

14.2.5.1.1 Habitat

Photobacterium is free-living or found in association with certain species of fishes. They have characteristic features and are investigated as a particular category of bacteria due to their unique capability to produce luminescence. They are also used as indicator organisms, referred to as biosensors in food and environmental monitoring of drowning victims.

14.2.5.1.2 Morphology

Photobacterium damsel bacteria are gram-negative, facultatively anaerobic, rods, measuring about $0.8-1.3 \times 1.4-4.0\,\mu m$. In older cultures, coccobacillary forms are isolated whereas filamentary cells may be found in fresh cultures. The bacteria are nonmotile or weakly motile and show bipolar staining with dyes like giemsa or methylene blue.

14.2.5.1.3 Cultural and biochemical characteristics

Pathogenic forms are easily isolated from spleen, kidney, and liver of the infected fish. The bacterium can be cultivated on minimal media, as it is non-fastidious and forms distinct dewdrop-like colonies after 48 h of incubation. It is positive for catalase and oxidase production, but does not produce β-galactosidase, H_2S, and indole. It is resistant to vibriostat 0/129 like Vibrionaceae.

14.2.5.1.4 Epizootiology

P. damsela was at first identified with ulcerative lesions among populations in late summer and fall. The disease, transmitted from fish to fish, causes summer epizootics of "pseudotuberculosis" with high mortality, especially in juvenile grown fishes. It is reported that a temperature of 20°C–25°C seems to be a crucial factor that leads to serious outbreaks (Toranzo et al., 1991).

14.2.5.1.5 Pathogenicity and clinical signs

The bacteria is isolated from infected sharks, turbot, and yellow-tail fish. The fishes affected generally appear dark in colour . Characteristic granulomata in the hemopoietic tissues of the spleen and the kidney are observed in histopathological studies. During the later stages, the bacteria release from granulomata to produce hemorrhagic septicemia (Kitao et al., 1979). A siderophore iron sequestration method is also identified as contributing to the organism's pathogenicity (Fouz et al., 1994, 1997).

14.2.5.1.6 Treatment and control

Various antibiotics are used in treating this disease in Japan, and transferable resistance conferred by R-plasmids has also been observed in strains recovered from infections (Aoki & Kitao, 1985). Several vaccines have been developed that provide a high degree of safety and are effective in controlling the disease. In general, avoidance of overcrowding in cultures and proper handling are essential to control outbreaks.

14.3 Gram-positive bacterial pathogens

A wide array of gram-positive aerobic and anaerobic bacteria have been identified as fish pathogens. The detailed investigations on these bacterial infections have shown that the anaerobes cause more widespread problems than aerobic forms. Gram-positive cocci belonging to genera Lactococcus and Streptococci and are important pathogens found in association with fish pathology. They are nonmotile, spherical, or ovoid cells occurring with a cell diameter of 0.6–0.9 μm. They are facultative anaerobes, oxidase- as well as catalase-negative in nature.

14.3.1 Streptococci

14.3.1.1 Streptococcus iniae

Streptococciosis was primarily defined in populations of Japanese rainbow trout (Hoshina et al., 1958). The disease, also known as pop-eye, is found in fishes from various parts of the world. Outbreaks are also observed in numerous fish species, including yellowtails (Kusuda et al., 1976). The most common strains that cause a potential threat to aquaculture include *Streptococcus agalactiae*, *S. diffilis*, *S. iniae*, *S. phocae* (Domeénech et al., 1996; Romalde et al., 2008), *Lactococcus garvieae*, *L. piscium* (Collins et al., 1983; Vendrill et al., 2006), *Carnobacterium piscicola*, and *Vagococcus salmoninarum* (Wallbanks et al., 1990), which causes infectious diseases of warm-water as well as cold-water fishes. The temperature of the water is a predisposing factor for the onset of the disease. *Streptococcus iniae* causes septicemia and meningoencephalitis in several cultured fishes. Fish streptococcosis is the most common disease in different geographical areas where the temperature in water bodies is adequate for an infection to start. *S. parauberis* is a causative agent of streptococcosis; common pathological symptoms of streptococcosis and lactococcosis are dark pigmentation, congestion, lethargy, hemorrhage, and exophthalmos with clouding of the cornea (Domeénech et al., 1996; Kusuda et al., 1991). *S. parauberis* is closely related to *S. uberis* and both are indistinguishable biochemically and serologically. Both the species were earlier considered as type I and type II of *S. uberis*; however, later they were demonstrated to be distinctive and the type I was renamed as *S. parauberis* (Williams & Collins, 1990). The *S. parauberis* genome encodes for a virulence factor that is like an M-like protein of *S. iniae* (Locke et al., 2008). It also possesses hasA and hasB genes that encode for a capsule that makes the bacteria resistant to phagocytosis. Recently, *S. iniae* and *S. garvieae* were recognized as potential pathogens of both fish and human beings (Austin & Austin, 1999; Elliott et al., 1991).

14.3.1.1.1 Habitat

S. iniae is a normal inhabitant of freshwater, isolated from rainbow trout, as well as from a few marine species associated with streptococcal infections.

14.3.1.1.2 Morphology

These are small, gram-positive facultative anaerobic forms in nature. Streptococci occur in straight long chains measuring about 0.3–0.5 μm.

14.3.1.1.3 Culture and biochemical characteristics

S. iniae grow on nutrient agar media and, as they are β-hemolytic, grow well on blood agar also. This bacterium cannot grow at the temperatures less than 10°C and greater than 45°C or pH 9.6, hence it can be easily distinguished from *Lactococcus garvieae*. It gives a negative result in Voges–Prausker tests. Furthermore, it does not yield sorbitol acid, but may ferment sucrose and hydrolyze starch.

14.3.1.1.4 Epizootiology

It is evident from the scientific literature that the pathogens survive in the water habitats in the proximity of fish all through the year (Kitao et al., 1979). The infective causative agents conceivably were released from infected fish and were passively retained in the water and underlying mud. Such environmental isolates are the markers of an unhygienic state and it implies the presence of serious pathogenic strains. It is reported, with several shreds of evidence, that streptococciosis spreads by contact with infected fish. In this regard, host specificity to gram-positive cocci in chains plays an important role (Boomker et al., 1979). It has been recognized that cell-free culture supernatants have significant pathogenicity (Kimura & Kusuda, 1982). Although oral administration may not be toxic, it produced visible damage following the percutaneous injection of yellowtails.

14.3.1.1.5 Pathogenicity and clinical signs

The infected fish exhibit bilateral exophthalmia and eye distension, rendering their heads into a strange shape. Internal and external hemorrhages are observed in most of the affected fishes. The skin is often darkened. They are found to be anorexic and swim listlessly.

14.3.1.1.6 Treatment and control

Antimicrobial compounds are used to control the infection in various concentrations, especially erythromycin; a 25 mg/kg dose was found effective in yellowtail (Kitao, 1982), and it was more effective than oxytetracycline or ampicillin (Shiomitsu et al., 1980). A novel therapeutic agent, sodium nifurstyrenate, at the dose of 50 mg/kg is more effective in controlling strepotococcicosis. With this drug, plasmid-mediated resistance has also been resolved. Effective control was also reported with ionophores, namely lasalocid, monensin in rainbow trout. A successful vaccination was applied to rainbow trout by injection with a formalized suspension of streptococcus (Sakai et al., 1987, 1989b).

14.3.2 Lactococci

14.3.2.1 Lactococcus garvieae

Lactococcus garvieae is a streptococcus-like bacteria that has been isolated from infected Japanese yellowtail culture over the years (Kusuda et al., 1991).

14.3.2.1.1 Habitat

L. garvieae is widely distributed all over the world and appear to be a normal inhabitant in freshwater fishes. It is isolated from occasional outbreaks of wild fishes. The bacteria has also been detected in extoparasites of unhealthy species (Madinabetia et al., 2009).

14.3.2.1.2 Morphology

L. garvieae is a gram-positive ovoid, nonmotile, facultative anaerobic cocci, usually appearing in chains, measuring $0.7 \times 1.4\ \mu m$.

14.3.2.1.3 Cultural and biochemical properties

Lactococcus garvieae grows on selective media supplemented with bile salts. Growth occurs at 10°C–45°C, but not at 50°C. Colonies are small, spherical, and white on the culture plate. As observed in most of the members of enterococci, it is capable of fermenting sugars, producing acid from various sugars. However, it does not digest casein or gelatin. It is α-hemolytic and does not produce catalase, oxidase, indole, or H_2S and Voges-Proskauer's reaction is positive.

14.3.2.1.4 Epizootiology

L. garvieae can be isolated from fish as well as benthos throughout the year. Most infections are related to a shift in feeding, storage, and transportation, which are predisposing factors.

14.3.2.1.5 Pathogenicity and clinical signs

Enterococcosis is a common hemorrhagic bacterial septicemia, followed by petechiae over the flanks or at the base of the fin and sometimes by hemorrhagic ascites, often found with ulceration over the abdominal region. Histopathological observation of the eye reveals retrobulbar edema, cellular infiltration, and hemorrhage of the orbit leads to the evulsion of the entire orbital contents. Encapsulated cultures are reported more virulent and less efficient in fixing complement (Barnes et al., 2002). *L. garvieae* is recognized in two forms: one is a capsulated form, whereas the other one shows a microcapsule associated with fimbriae type components on the cell surface (Ooyama et al., 2002).

14.3.2.1.6 Treatment and control

Long-term protection was ensured with intraperitoneal injection of formalin-inactivated cells of *L. garvieae*, which were tested in yellowtail. Antimicrobial compounds like erythromycin have a proven effect in controlling infection in fish, but in regions such as the Mediterranean, where it harms the economy, antibiotic therapy has limited value.

14.3.3 Mycobacteria

14.3.3.1 Mycobacterium marinum

Mycobacteria is a causative agent of several chronic infections in fish. Mycobacteriosis is a chronic progressive disease caused by this fast-acid bacillus, which is called nontuberculous mycobacteria (NTM).

Several NTM species, including several slow-growing and rapidly growing organisms, such as *Mycobacterium avium*, *M. chelonae*, *M. fortuitum*, and *M. marinum*, are found in association with mycobacteriosis. Among these, *M. marinum* is known to be a zoonotic fish pathogen, mostly isolated from a variety of fishes with granulomas (Gauthier & Rhodes, 2009). *Mycobacterium* spp. reside in aquaria and water supplies and cause opportunistic infections (Yanong et al., 2010). This pathogen secretes a yellow pigment only when bacteria are exposed to light and it replicates inside the host's macrophages, developing a granuloma during chronic infections (Barker et al., 1997; Tobin & Ramakrishnan, 2008). Another common disease of marine and freshwater fishes is piscine mycobacteriosis (Decostere et al., 2004). This condition also affects tropical aquarium fish and is known to be a significant cause of morbidity and mortality (Chang & Whipps, 2015).

14.3.3.1.1 Habitat

The aquatic environment acts as a reservoir of fish pathogenic mycobacteria. Several species of the group are widely disseminated around the world.

14.3.3.1.2 Morphology

The pathogenic forms of Mycobacteria are gram-positive, nonmotile, acid-fast pleomorphic rods with dimensions of about $1.5-2.0 \times 0.25-0.35\ \mu m$.

14.3.3.1.3 Culture and biochemical properties

M. marinum is a relatively slow-growing bacteria, has a generation time of 4–6 h, and requires about 2–3 weeks for the colonies to come up. It can be cultured on a special medium, namely, Brain heart infusion agar or any glycerol-based media, although for steady growth, Sauton's modified media is very appropriate (Chen et al., 1997). They secrete a yellowish pigment and form distinguishable pale-yellow colonies on solid media. The optimum temperature for growth is between 25°C and 27°C, with some exceptions as some isolates multiply well at around 37°C.

14.3.3.1.4 Epizootiology

Mycobacteriosis is a lethal disease that occurs in tropical marine and freshwater environments but is comparatively less common in fishes growing in temperate water bodies (Reichenbach-Klinke, 1972). It is reported that the transovarian passage of infection is possible specifically in viviparous species. The number of affected carrier populations has increased significantly through husbandry stress (Ramsay et al., 2009).

14.3.3.1.5 Pathogenicity and clinical signs

M. marinum induces infection in fresh and marine water fishes and, interestingly, survives in host macrophages (Arend et al., 2002; Barker et al., 1997). It causes chronic progressive disease and is considered a nontuberculous mycobacteria (Novotny et al., 2004). Non-tuberculous mycobacterial species are categorized into slow-growing and rapidly growing bacteria, whereas *M. marinum* belongs to the slow-growing category. Piscine mycobacteriosis is a widespread disease in a vast region, contaminating more than 200 species of freshwater and marine fish (Decostere et al., 2004). It is reported that *M. marinum* is the most prevalent opportunistic pathogen (Rallis & Koumantaki-Mathioudaki, 2007). Multiple symptoms associated with affected fishes include uncoordinated swimming; skin ulceration; granuloma in the different organs of fish, including liver, kidney, and spleen; and loss of weight in both fresh and marine water fish (El Amrani et al., 2010; Ferguson, 2006). The affected fishes may develop intense color and show abdominal swelling. Necroscopic studies reveal the presence of miliary tubercles in the kidney, liver, and spleen. Zebra fishes were much more susceptible to the infection.

14.3.3.1.6 Treatment and control

To control the piscine mycobacteriosis, the complete annihilation of all affected fish stock and disinfection of the holding tanks is recommended (Decostere et al., 2004). The aquaria should be kept free from *M. marinum* by using ethanol, Lysol, and sodium chlorite, as they are efficient in destroying the bacteria, while potassium peroxymonosulfate is reported to be ineffective. It is observed that cell-mediated immunity in fish play an important role in giving protection from pathogenic bacteria. Immunization with *M. salmoniphilum* mixed with Freund's adjuvant resulted in type IV delayed hypersensitivity. A DNA vaccine prepared by using Ag85A gene encodes for fibronectin-binding proteins cloned in the eukaryotic expression vector resulted in producing a humoral immune response in fish. Overall, such findings reveal that there is a possibility of eliciting protection in fish against pathogenic mycobacteria. Treatments with chloramine B or T, cycloserine, doxycycline, erythromycin, ethambutol, ethionamide, isoniazid, kanamycin, minocycline, β-lactamases, aminoglycoside antibiotics, and sulfa drugs were used in common to destroy pathogens and were found highly effective in controlling the spread of infection.

14.3.4 Renibacterium

14.3.4.1 Renibacterium salmoninarum

These are gram-positive bacilli originally assigned to Corynebacteriaceae, but factors like the molecular percentage of G + C content indicated that they should be assigned separately. *Renibacterium salmoninarum*, along with other morphologically similar opportunistic pathogens, is isolated from infected fishes and acts as a causative agent of bacterial kidney disease (BKD, also called "Dee disease") mostly reported in salmon and brown trout. The disease has been documented in different parts of the world, where it is found in rainbow trout (*Oncorhynchus mykiss*), brown trout (*Salmo trutta*), and brook trout (*Savelinus fontinalis*) (Belding & Merrill, 1935). Horizontal transmission of BKD is also reported, possibly from masu salmon. Wire-tagging seems to affect disease transmission horizontally.

14.3.4.1.1 Habitat

R. salmoninarum was first detected in Atlantic salmon in Scotland. It is possibly an obligatory Salmonidae parasite, as it was not retrieved from other fish.

14.3.4.1.2 Morphology

The coryneform bacteria is a gram-positive diplobacillus often found in pairs. It exists as small, nonmotile, asporogenous, nonacid-fast rods (0.3–1.0 × 1.0–1.5 μm); pleomorphic forms of the bacteria, metachromatic granules, and "coryneform" appearance are clearly observed in host tissues.

14.3.4.1.3 Cultural and biochemical characteristics

R. salmoninarum is a fastidious organism that requires L-cystine in the media. The KDM 2 formulation, which consists of 10% fetal calf serum and 0.1% (W/V) L-cysteine hydrochloride, is commonly used to isolate the bacteria from infected stock (Evelyn, 1997). Characteristic colonies were produced by *R. salmoninarum* on KDM2 formulated media plates after incubation at 15°C for 3 weeks. Nonpigmented, smooth, round, elevated colonies of 2-mm diameter are apparent after the incubation period. The old cultures may become extremely granular in appearance. It is found that the principal material is cysteine, which has been precipitated from the culture media. Histopathological examination of lesions revealed the presence of many aerobic gram-positive rod-shaped bacteria embedded in a crystalline solid matrix.

14.3.4.1.4 Epizootiology

BKD is widely distributed throughout the world, commonly affecting both freshwater fishes as well as salt water-cultured fishes. It is originally described from Atlantic salmon in Scotland, and has rarely been reported in wild salmonids and not at all recorded from nonsalmonid species. As bacterial kidney disease is a slowly progressing infection, fishes grow well before the disease is established in the host, unless they are subjected to other stress factors. Epizootics are also found to be seasonal as per the analysis of reported cases. The high mortality is related to either decreasing or increasing water temperature (Fryer & Sanders, 1981).

14.3.4.1.5 Pathogenicity and clinical signs

Bacterial kidney disease is a chronic progressive disease causing a low level of mortality in salmonid fishes. The pathogen can be transmitted horizontally from skin abrasions in fish or by consumption of contaminated food, or vertically from parent to progeny through eggs. It is reported clearly that intraovum infection with *R. salmoninarum* may occur, which is difficult to eliminate by disinfectant treatment of the eggs (Evelyn et al., 1984).

The affected fishes may take a few months to show visible signs of infection. BKD is one of the most difficult fish diseases to treat because of its ability to resist phagocytosis and to flourish in host cells. The infected fish are typically darker in color, with exophthalmos and hemorrhage at the base of the pectoral fins in most salmonids. Tiny distinctive raised vesicles are also found on the sides of the fish in cultured trout. The lesions are generally found in kidneys; however, they are also evident in the liver, spleen, and heart. Initially, these miliary white lesions have a red hyperemic rim but ultimately grow into large, caseous nodular granulomata. In some Pacific salmonids, large cavitations in the skeletal muscle were observed. In wild Atlantic salmon, the variation in pathology depends on environmental temperature (Smith, 1964). In farmed salmonids, the occurrence of the disease is more prevalent during the spring season, as the water temperature rises, and high mortality is usually reported in early summer.

14.3.4.1.6 Treatment and control

It is important to select a disease-free stock to control the disease, and continuous monitoring is essential to prevent disease transmission. Sulfa drugs and other antibiotic therapy have been tried, and the disease progression can be arrested by prolonged treatment. Some evidence indicates that high levels of iodine and fluorine in Atlantic salmon diets substantially decreased fatalities. *R. salmoninarum* is phylogenetically closest to *Arthrobacter* species (Wiens et al., 2008). *Arthrobacter davidanieli* is a nonpathogenic strain that is commercially used as a vaccine and can provide a significant cross-protection in Atlantic salmon, but this is not the same in the case of Pacific salmon (Salonius et al., 2005).

14.3.5 Clostridia

14.3.5.1 Clostridium botulinum

Clostridium, which develops a commensal interaction with the intestine of marine and freshwater fishes, comprises a group of strict anaerobic gram-positive bacilli, predominantly the type II strain of *C. botulinum* as a potent pathogen in fishes, usually having resistance to heat and other unfavorable conditions.

14.3.5.1.1 Habitat

The bacteria is found in diverse environments, usually isolated from decayed organic matter and in marine sediments. It is also found in association with the intestines of animals, including coho salmon and rainbow trout.

14.3.5.1.2 Morphology

The bacteria are gram-positive, motile forms in nature. They are nonacid fast, straight rods, measuring 3.4–7.5 × 0.3–0.7 μm in size. Bacterial cultures show peritrichous flagella, and form subterminal, oval-shaped endospores.

14.3.5.1.3 Cultural and biochemical characteristics

C. botulinum is a strictly anaerobic bacterium and is cultured by using infected material as inoculum that is transferred into a cooked meat medium; it needs an incubation period of about 6 days at 30°C. It forms colonies on the surface of the culture media, measuring approximately 1–3 mm in diameter. Colonies show an irregular lobate margin, raised centers, and growth appears as a semi-opaque mat on the media surface. Modest growth takes place in cooked meat broth, but to see the optimal growth of bacteria, fermentable sugars are added to the broth at an optimum temperature of 25°C–30°C. It secretes neurotoxin, hemolysin, and lecithinase enzymes but catalase/urease is not produced. It also ferments various sugars, and enzymes responsible for liquefaction are not usually produced by the type E strain of *C. botulinum*.

14.3.5.1.4 Epizootiology

A major source of *C. botulinum* is supposedly minced trash fish used as feed, mud, and bottom-living invertebrates in trout ponds (Huss & Eskilden, 1974). An outbreak of disease in a British trout farm was identified as fish botulism. The outbreak was controlled and eliminated within 1 month by conservative measures that inclue cleaning and heavy liming of contaminated ponds; intensive investigation of all ponds for immediate eradication of sick and dead fish and, banning of movement of fish outwith the farm. In Denmark, the disease was considered to be "bankruptcy disease" as it was recorded in freshwater salmonid fish in every pond, and *C. botulinum* type E was identified as the pathogen responsible for infections in all the cases.

14.3.5.1.5 Pathogenicity and clinical signs

Fish botulism has proved to be a serious threat to aquaculture. The affected fishes are generally observed to haveslightly swollen dark skin ; they swim initially on the surface and then gradually move to the bottom for some duration, and finally swim erratically to the surface again. Few pathological signs are usually found with affected fishes; other than pallor of the gills, mild abdominal edema, and an empty digestive tract, there are no obvious histopathological signs. The predominant signs of the pathology of affected fishes are basic behavioral changes and variation in skin color.

14.3.5.1.6 Treatment and control

There is no standard medication suggested for botulism in fish, but disinfection of the ponds used to culture freshwater fishes is advocated. Certain precautionary measures yield good results, including moving the uninfected stock to fresh clean areas, draining the contaminated ponds, and removing surface mud. To remove Clostridia from the gastrointestinal tract of infected trout, it is recommended that the fish should be starved for 5 days (Wenzel et al., 1971); indeed this reduces contamination with *C. botulinum* in stock.

14.4 Conclusion

There is global demand for aquaculture due to its nutritional value and ecological adaptations. Several bacterial diseases are affecting aquatic life worldwide. Many relevant pathogens can be removed by good management, which may include the use of appropriate vaccines. The novel bacterial pathogens present some major hurdles in the aquaculture industry, but several effective and economical vaccines have been developed to prevent outbreaks. Fish is a crucial source of food for humans, comprising about 60% of the world's protein intake, and approximately 60% of the underdeveloped economies derive 30% of their annual protein from fish. Still, it is clear from several studies that the ecology of fish pathogens is not yet completely understood.

The current increase in aquaculture production coincides with a growing number of infectious diseases that have a negative impact on the global aquaculture sector's growth, profitability, and sustainability. This chapter provides in-depth insight into the different bacterial pathogens associated with fish and discusses the various antibiotics used to control the pathogens, to help in effective management of diseases. For the future, a variety of modern molecular methods, including in situ and probe hybridization, microarray techniques, and restriction enzyme digestion, are entering taxonomic use and need to be employed in fish pathology. Identification of bacterial flora has pathogenic interactions with commercially significant cultured fishes and their ecological adaptations are worthy of further study.

Acknowledgement

The author MZG is thankful to Prof. Sashidhar Rao Beedu (Department of Biochemistry, UCS, Osmania University, Hyderabad) for his constant motivation, encouragement, and support.

References

Abisoye, B. F., Ojo, S. K. S., Adeyemi, R. S., & Olajuyigbe, O. O. (2011). Bacteriological assessment of some commonly sold fishes in Lagos metropolis market Nigeria. *Prime Journal of Microbiology Research*, *1*(2), 23–26.

Abowei, J., & Briyai, O. (2011). A review of some bacteria diseases in Africa culture fisheries. *Asian Journal of Medical Sciences*, *3*, 206–217.

Ahne, W., & Popp Hoffmann, R. (1982). *Pseudomonas fluorescens* as a pathogen of tench (*Tinca tinca*). *Bulletin of the European Association of Fish Pathologists*, *4*, 56–57.

Amend, D. F., & Ross, A. J. (1970). Experimental control of columnaris disease with a new nitrofuran drug P-7138. *Progressive Fish Culturist*, *32*, 19–25.

Anacker, R. L., & Ordal, E. J. (1959). Studies on the myxobacterium *Chondrococcus columnaris*. I. Serological typing. *Journal of Bacteriology*, *78*, 25–32.

Aoki, T. (1988). Drug-resistant plasmids from fish pathogens. *Microbiological Sciences*, *5*, 219–223.

Aoki, T., & Kitao, T. (1985). Detection of transferable R plasmids in strains of the fish-pathogenic bacterium *Pasteurella piscicida*. *Journal of Fish Diseases*, *8*, 345–350.

Arend, S. M., van Meijgaarden, K. E., de Boer, K., Cerda de Palou, E., van Soolingen, D., Ottenhoff, T. H. M., & van Dissel, J. T. (2002). Tuberculin skin testing and in vitro T-cell responses to ESAT-6 and culture filtrate protein 10 after infection with *Mycobacterium marinum* or *M. kansasii*. *Journal of Infectious Diseases*, *186*, 1797–1807.

Austin, B. (2011). Taxonomy of bacterial fish pathogens. *Veterinary Research*, *42*(1), 20.

Austin, B., & Austin, D. A. (1993). *Bacterial fish pathogens* (2nd ed., pp. 265–307). Chichester: Ellis Horwood Ltd.

Austin, B., & Austin, D. A. (1999). *Bacterial fish pathogens: Disease in farmed and wild fish* (3rd ed.). Switzerland: Springer International Publishing.

Austin, B., & Austin, D. A. (2007). *Bacterial fish pathogens: Disease in farmed and wild fish* (4th ed.). Goldalming: Springer-Praxis.

Austin, B., & Austin, D. A. (2016). *Bacterial fish pathogens. Disease of farmed and wild fish* (6th ed.). Switzerland: Springer International Publishing.

Barker, G. A., Smith, S. N., & Bromage, N. R. (1990). Effect of oxolinic acid on bacterial flora and hatching success rate of rainbow trout, *Oncorhynchus mykiss*, eggs. *Aquaculture (Amsterdam, Netherlands)*, *9*, 205–222.

Barker, L. P., George, K. M., Falkow, S., & Small, P. (1997). Differential trafficking of live and dead *Mycobacterium marinum* organisms in macrophages. *Infection and Immunity*, *65*(4), 1497–1504.

Barnes, A. C., Guyot, C., Hansen, B. G., Mackenzie, K., Horne, M. T., & Ellis, A. E. (2002). Resistance to serum killing may contribute to differences in the abilities of capsulate and non-capsulated isolates of *Lactococcus garvieae* to cause disease in rainbow trout (*Oncorhynchus mykiss* L.). *Fish and Shellfish Immunology*, *12*, 155–168.

Bauer, O. N., Musselius, V. A., & Strelkov Yu, A. (1973). *Diseases of pond fishes* (pp. 39–40). Jerusalem: Keter Press.

Belding, D. L., & Merrill, B. (1935). A preliminary report upon a hatchery disease of the Salmonidae. *Transactions of the American Fisheries Society*, *65*, 76–84.

Bernardet, J. F., & Bowman, J. (2006). The prokaryotes: A handbook on the biology of bacteria. Proteobacteria: Delta and Epsilon Subclasses. *Deeply Rooted Bacteria*, *7*, 481–532.

Bernardet, J. F., & Grimont, P. A. D. (1989). Deoxyribonucleic acid relatedness and phenotypic characterization of *Flexibacter columnaris* sp. nov., nom. rev., *Flexibacter psychrophilus* sp. nov., nom. rev., and *Flexibacter maritimus* Wakabayashi, Hikida and Masumura 1986. *International Journal of Systematic Bacteriology*, *39*, 346–354.

Bernardet, J. F., & Nakagawa, Y. (2006). An introduction to the family flavobacteriaceae. In M. Dworkin, S. Falkow, E. Rosenberg, K. H. Schleifer, & E. Stackebrandt (Eds.), *The prokaryotes*. New York: Springer.

Bernardet, J. F., Segers, P., Vancanneyt, M., Berthe, F., Kersters, K., & Vandamme, P. (1996). Cutting a Gordian knot: Emended classification and description of the genus *Flavobacterium*, emended description of the family *Flavobacteriaceae*, and proposal of *Flavobacterium hydatis* nom. nov. (basonym, *Cytophaga aquatilis* Strohl and Tait 1978). *International Journal of Systematic Bacteriology*, *46*(1), 128–148.

Berthe, F. C. J., Michel, C., & Bernadet, J. F. (1995). Identification of *Pseudomonas anguilliseptica* isolated from several fish species in France. *Diseases of Aquatic Organisms*, *21*, 151–155.

Blazer, V. S., Shotts, E. B., & Waltman, W. D. (1985). Pathology associated with *Edwardsiella ictaluri* in catfish Ictalurus punctatus *Rafinesque* and *Danio devario* (Hamilton-Buchanan, 1822). *Journal of Fish Biology*, *27*, 167–175.

Boomker, J., Imes, G. D., Cameron, C. M., Naudé, T. W., & Schoonbee, H. J. (1979). Trout mortalities as a result of *Streptococcus* infection. *Onderstepoort Journal of Veterinary Research*, *46*, 71–77.

Bootsma, R., & Clerx, J. P. M. (1976). Columnaris disease of cultured carp *Cyprinus carpio* L. Characteristics of the causative agent. *Aquaculture (Amsterdam, Netherlands)*, *7*, 371–384.

Bragg, R. R., & Todd, J. M. (1988). *In vitro* sensitivity to Baytril of some bacteria pathogenic to fish. *Bulletin of the European Association of Fish Pathologists*, *8*, 5.

Bullock, G. L. (1965). Characteristics and pathogenicity of a capsulated *Pseudomonas* isolated from goldfish. *Applied Microbiology*, *13*, 89–92.

Cahill, M. M. (1990). Bacterial flora of fishes: A review. *Microbial Ecology*, *19*(1), 21–41.

Chang, C. T., & Whipps, C. M. (2015). Activity of antibiotics against mycobacterium species commonly found in Laboratory Zebrafish. *Journal of Aquatic Animal Health*, *27*(2), 88–95.

Chen, S. C., Adams, A., & Richards, R. S. (1997). Extracellular products from *Mycobacterium* spp. in fish. *Journal of Fish Diseases*, *20*, 19–26.

Claucas, I. J., & Ward, A.R. (1996). *Post-harvest fisheries development: A guide to handling, preservation, processing and quality*. Charthan, Maritime, Kent ME4 4TB, UK.

Collins, M. D., Farrow, J. A. E., Phillips, B. A., & Kandler, O. (1983). *Streptococcus garvieae* sp. nov. and *Streptococcus plantarum* sp. nov. *Journal of General Microbiology*, *129*(11), 3427–3431.

Colwell, R. R., MacDonell, R. R., & De Ley, J. (1986). Proposal to recognise the family *Aeromonadaceae* fam. nov. *International Journal of Systematic Bacteriology*, *36*, 473–477.

Csaba, G., Prigli, M., Békési, L., Kovács-Gayer, E., Bajmócy, E., & Fazekas, B. (1981b). Septicaemia in silver carp (*Hypophthalmichthys molitrix*, Val.) and bighead (*Aristichthys nobilis* Rich.) caused by *Pseudomonas fluorescens*. In J. Oláh, K. Molnár, & S. Jeney (Eds.), *Fish, pathogens and environment in European polyculture* (pp. 111–123). Fisheries Research Institute, Szarvas, F. Muller.

Dalsgaard, I. (1993). Virulence mechanisms in *Cytophaga psychrophila* and other Cytophaga-like bacteria pathogenic for fish. *Annual Review of Fish Diseases*, *3*, 127–144.

Dar, G. H., Bhat, R. A., Kamili, A. N., Chishti, M. Z., Qadri, H., Dar, R., & Mehmood, M. A. (2020). *Correlation between pollution trends of freshwater bodies and bacterial disease of fish fauna. Fresh water pollution dynamics and remediation* (pp. 51–67). Singapore: Springer.

Dar, G. H., Dar, S. A., Kamili, A. N., Chishti, M. Z., & Ahmad, F. (2016). Detection and characterization of potentially pathogenic *Aeromonas sobria* isolated from fish *Hypophthalmichthys molitrix* (Cypriniformes: Cyprinidae). *Microbial Pathogenesis*, *91*, 136–140.

Dar, G. H., Kamili, A. N., Chishti, M. Z., Dar, S. A., Tantry, T. A., & Ahmad, F. (2016). Characterization of *Aeromonas sobria* isolated from fish rohu (*Labeo rohita*) collected from polluted pond. *Journal of Bacteriology & Parasitology*, *7*(3), 1–5. Available from https://doi.org/10.4172/2155-9597.1000273.

De Paola, A., Flynn, P. A., McPhearson, R. M., & Levy, S. B. (1988). Phenotypic and genotypic characterization of tetracycline- and oxytetracycline-resistant *Aeromonas hydrophila* from cultured channel catfish *(Ictalurus punctatus)* and their environment. *Applied and Environmental Microbiology*, *54*, 1861–1863.

Decostere, A., & Haesebrouck, F. (1999). Outbreak of columnaris disease in tropical aquarium fish. *The Veterinary Record*, *144*, 23–24.

Decostere, A., Hermans, K., & Haesebrouck, F. (2004). Piscine mycobacteriosis: A literature review covering the agent and the disease it causes in fish and humans. *Veterinary Microbiology*, *99*(3–4), 159–166.

Deufel, J. (1974). Prophylactic measures against bacterial diseases of salmonid fry. *Osterr Fisch*, *27*, 1–5.

Domeénech, A., Derenaáandez-Garayzábal, J. F., Pascual, C., Garcia, J. A., Cutuli, M. T., Moreno, M. A., Collins, M. D., & Dominguez, L. (1996). Streptococcosis in cultured turbot, *Scopthalmus maximus* (L.), associated with *Streptococcus parauberis*. *Journal of Fish Diseases*, *19*, 33–38.

Eissa, N., Abou, E., Elsayed., Shaheen, A., & Abbass, A. (2010). Characterization of Pseudomonas species isolated from tilapia "*Oreochromis niloticus*" in Qaroun and Wadi-El-Rayan Lakes, Egypt. Global. *Veterinaria*, *5*, 116–121.

El Amrani, M. H., Adoui, M., Patey, O., & Asselineau, A. (2010). Upper extremity *Mycobacterium marinum* infection. *Orthopaedics and Traumatology Surgery and Research*, *96*(6), 706–711.

Elliott, J. A., Collins, M. D., Pigott, N. E., & Facklam, R. R. (1991). Differentiation of *Lactococcus lactis* and *Lactococcus garvieae* from humans by comparison of whole-cell protein patterns. *Journal of Clinical Microbiology*, *29*(12), 2731–2734.

Endo, T., Ogishima, K., Hayasaka, H., Kaneko, S., & Ohshima, S. (1973). Application of oxolinic acid as a chemotherapeutic agent against infectious diseases in Fishes-I. Antibacterial activity, chemotherapeutic effects and pharmacokinetics of oxolinic acid in fishes. *Bulletin of the Japanese Society for the Science of Fish*, *39*, 165–171.

Esch, G. W., & Hazen, T. C. (1978). Thermal ecology and stress: A case history for red-sore disease in largemouth bass (*Micropterus salmoides*). In J. H. Thorpe, & J. W. Gibbons (Eds.), *Energy and environmental stress in aquatic systems*. Springfield, VA: Department of Energy symposium series no. CONF-771114. National Technical Information Service.

Esch, G. W., Hazen, T. C., Dimock, R. V., Jr., & Gibbons, J. W. (1976). Thermal effluent and the epizootiology of the ciliate Epistylis and the bacterium Aeromonas in association with centrarchid fish. *Transactions of the American Microscopical Society*, *95*(4), 687–693.

Evelyn, T. P. T. (1997). An improved growth medium for the kidney disease bacterium and some notes on using the medium. *Bulletin – Office International des Epizooties*, *87*, 511–513.

Evelyn, T. P. T., Ketcheson, J. E., & Prosperi-Porta, L. (1984). Further evidence for the presence of *Renibacterium salmoninarum* in salmonid eggs and for the failure of providone-iodine to reduce the intta-ovum infection rate in water-hardened eggs. *Journal of Fish Diseases*, *7*, 173–182.

FAO. (1997). *Review of the state of world aquaculture*. FAO Fisheries Circular No. 886, Rev. 1. Rome, Italy.

Ferguson, H. W. (2006). Spleen, thymus, Reticulo-Endotelial system, blood. In H. W. Ferguson (Ed.), *Systemic pathology of fish: A text and atlas of normal tissues in teleosts and their responses in disease* (2nd ed., pp. 199–214). London, UK: Scotian Press.

Figueiredo, H. C. P., Klesius, P. H., Arias, C. R., Evans, J., Shoemaker, C. A., Pereira, D. J., & Peixoto, M. T. D. (2005). Isolation and characterization of strains of *Flavobacterium columnare* from Brazil. *Journal of Fish Diseases*, *28*, 199–204.

Fouz, B., Biosca, E. G., & Amaro, C. (1997). High affinity iron-uptake systems in *Vibrio damsela*: Role in the acquisition of iron from transferrin. *Journal of Applied Microbiology*, *82*, 157–167.

Fouz, B., Toranzo, A. E., Biosca, E. G., Mazoy, R., & Amaro, C. (1994). Role of iron in the pathogenicity of *Vibrio damsela* for fish and mammals. *FEMS Microbiology Letters*, *121*, 181.

Fryer, J. L., & Sanders, J. E. (1981). Bacterial kidney disease of salmonid fish. *Annual Reviews in Microbiology*, *35*, 273–298.

Fujihara, M. P., & Nakatani, R. E. (1971). Antibody protection and immune responses of rainbow trout coho salmon to *Chondrococcus columnaris*. *Journal of Fisheries Research Board of Canada*, *28*, 1253–1258.

Garnjobst, L. (1945). Cytophaga columnaris in pure culture. A myxobacterium pathogenic to fish. *Journal of Bacteriology*, *49*, 113–128.

Gauthier, D. T., & Rhodes, M. W. (2009). Mycobacteriosis in fishes: A review. *The Veterinary Journal, 180* (1), 33–47.

Gosling, P. J. (1996). Pathogenic mechanisms. In B. Austin, M. Altwegg, P. J. Gosling, & P. Joseph (Eds.), *The genus Aeromonas* (pp. 245–265). Chichester: John Wiley and Sons.

Hacking, M. A., & Budd, J. (1971). Vibrio infection in tropical fish in a freshwater aquarium. *Journal of Wildlife Diseases, 7*, 273–280.

Haley, R., Davis, S. P., & Hyde, J. M. (1967). Environmental stress and Aeromonas liquefaciens in American and threadfin shad mortalities. *Progressive Fish Culturist, 29*, 193.

Hawke, J. P., & Thune, R. L. (1992). Systemic isolation and antimicrobial susceptibility of *Cytophaga columnaris* from commercially reared channel catfish. *Journal of Aquatic Animal and Health, 4*(2), 109–113.

Hettiarachchi, D. C., & Cheong, C. H. (1994). Some characteristics of *Aeromonas hydrophila* and *Vibrio* species isolated from bacterial disease outbreaks in ornamental fish culture in Sri Lanka. *Journal of the National Science Council of Sri Lanka, 22*, 261–269.

Hickey, M. E., & Lee, J. L. (2018). A comprehensive review of Vibrio (Listonella) anguillarum: Ecology, pathology and prevention. *Reviews in Aquaculture, 10*, 585–610.

Hirono, I., Masuda, T., & Aoki, T. (1996). Cloning and detection of the hemolysin gene of *Vibrio anguillarum*. *Microbial Pathogenesis, 21*, 173–182.

Holmes, P., Niccols, L. M., & Sartory, D. P. (1996). The ecology of mesophilic Aeromonas in the aquatic environment. In B. Austin, M. Altwegg, P. J. Gosling, & S. Joseph (Eds.), *The genus Aeromonas* (pp. 127–150). Chichester, UK: John Wiley and Sons, Ltd.

Hoshina, T., Sano, T., & Morimoto, Y. (1958). A Streptococcus pathogenic to fish. *Journal of the Tokyo University of Fisheries, 44*, 57–68.

Howard, S. P., Macintyre, S., & Buckley, J. T. (1996). Toxins. In B. Austin, M. Altwegg, P. J. Gosling, & P. Joseph (Eds.), *The genus Aeromonas* (pp. 267–286). Chinchester: John Wiley and Sons.

Huss, H. H., & Eskilden, U. (1974). Botulism in farmed trout caused by *Clostridium botulinum* type E. Nord. *Veterinary Medicine, 26*, 733–738.

Huys, G., Kámpfer., Albert, M. J., Kúhn, I., Denys, R., & Swings, J. (2002). *Aeromonas hydrophila* sub sp. dhakensis sub sp. nov., isolated from children with diarrhoea in Bangladesh, and extended description of *Aeromonas hydrophila* subsp. hydrophila (Chester 1901) Stanier 1943 (Approved Lists 1980). *International Journal of Systematic and Evolutionary Microbiology, 52*, 705–712.

Iranto, A., & Austin, B. (2002). Probiotics in aquaculture. *Journal of Fish Diseases, 25*, 633–642.

Johnson, P. J., & Paull, S. H. (2011). The ecology and emergence of diseases in fresh waters. *Fresh Water Biology, 56*, 638–657.

Kimura, H., & Kusuda, R. (1982). Studies on the pathogenesis of streptococcal infection in cultured yellowtails, *Seriola* spp.: Effect of crude exotoxin fractions from cell-free culture on experimental streptococcal infection. *Journal of Fish Diseases, 5*, 471–478.

Kitao, T. (1982). Erythromycin – The application of streptococcal infection in yellowtails. *Fish Pathology, 17*, 77–85.

Kitao, T., Aoki, T., Fukudome, M., Kawano, K., Wada, Y. O., & Mizuno, Y. (1983). Serotyping of *Vibrio anguillarum* isolated from diseased freshwater fish in Japan. *Journal of Fish Diseases, 6*, 175–181.

Kitao, T., Aoki, T., & Iwata, K. (1979). Epidemiological study on streptococcicosis of cultured yellowtail *(Seriola quinqueradiata)* – I. Distribution of Streptococcus sp. in sea water and muds around yellowtail farms. *Bulletin of the Japanese Society of Scientific Fisheries, 45*, 567–572.

Korun, J., Akgun-Dar, K., & Yazici, M. (2009). Isolation of *Shewanella putrefaciens* from cultured European sea bass, (*Dicentrarchus labrax*) in Turkey. *Revista de Medicina Veterinaria, 160*, 532–536.

Kusuda, R., Kawai, K., Salati, F., Banner, C. R., & Fryer, J. L. (1991). *Enterococcus seriolicida* sp. nov., a fish pathogen. *International Journal of Systematic Bacteriology, 41*(3), 406–409.

Kusuda, R., Kawai, K., Toyoshima, T., & Komatsu, I. (1976). A new pathogenic bacterium belonging to the genus streptococcus, isolated from an epizootic of cultured yellowtail. *Bulletin of the Japanese Society for the Science of Fish, 42*, 1345–1352.

Kvenberg, E. J. (1991). Non-indigenous bacterial pathogen. In R. Donn, H. Cameron, & R. Van Nostrand (Eds.), *Microbiology of marine food products* (pp. 263–291). New York: Aspen.

Larsen, J. L., & Jensen, N. J. (1977). An *Aeromonas* species implicated in ulcer-disease of the cod *(Gadus morhua). Nordisk Veterinaermedicin, 29*, 199–211.

Leadbetter, E. R. (1974). Genus II. Flexibacter Soriano 1945, 92, Lewin 1969, 192 emend. mut. char. In R. E. Buchanan, & N. E. Gibbons (Eds.), *Bergey's manual of determinative bacteriology* (8th ed., pp. 105–107). Baltimore: Williams and Wilkins.

Leung, K. Y., & Stevenson, R. M. W. (1988). Characteristics and distribution of extracellular proteases from *Aeromonas hydrophila. Journal of General Microbiology, 134*, 151–160.

Li, J., Ni, X. D., Liu, Y. J., & Lu, C. P. (2011). Detection of three virulence genes alt, ahp and aerA in *Aeromonas hydrophila* and their relationship with actual virulence to zebrafish. *Journal of Applied Microbiology, 110*, 823–830.

Li, M. F., & Flemming, C. (1967). A proteolytic pseudomonad from skin lesions of rainbow trout (*Salmo gairdnerii*). I. Characteristics of the pathogenic effects and the extracellular proteinase. *Canadian Journal of Microbiology, 13*(4), 405–416.

Li, M. F., & Traxler, G. S. (1971). A proteolytic pseudomonad from skin lesions of rainbow trout (*Salmo gairdneri*). III. Morphological studies. *Journal of the Fisheries Research Board of Canada, 28*, 104–105.

Li, Y., & Ma, Q. (2017). Iron acquisition strategies of *Vibrio anguillarum. Frontiers in Cellular and Infection Microbiology, 7*, 342.

Lipp, E. K., & Rose, J. B. (1997). The role of seafood in foodborne diseases in the United States of America. *Revue Scientifique et Technique (Paris), 16*(2), 620–640.

Locke, J. B., Aziz, R. K., Vicknair, M. R., Nizet, V., & Buchanan, J. T. (2008). Streptococcus iniae M-like protein contributes to virulence in fish and is a target for live attenuated vaccine development. *PLoS One, 3*(7), e2824.

MacInnes, J. I., Trust, T. J., & Crosa, J. H. (1979). Deoxyribonucleic acid relationships among members of the genus *Aeromonas. Canadian Journal of Microbiology, 25*(5), 579–586.

Madinabetia, I., Ohtsuka, S., Okuda, J., Iwamoto, E., Yoshida, T., Furukawa, M., Nakaoka, N., & Nakai, T. (2009). Homogeneity among *Lactococcus garvieae* isolates from striped jack, *Pseudocaranx dentex* (Bloch & Schneider), and its ectoparasites. *Journal of Fish Diseases, 32*, 901–905.

Markovic, M., Radojicic, M., Cosic, S., & Levnaic, D. (1996). Massive death of silver carp *(Hypophthalmichthys molitrix* Val.) and big head *(Aristichthys nobilis* Rich.) caused by *Pseudomonas fluorescens* bacteria. *Veterinarski. Glasnik (Srpska Akademija Nauka), 50*, 761–765.

McCarthy, D. H. (1976). Vibrio disease in eels. *Journal of Fish Biology, 8*, 317–320.

Mhango, M., Mpuchane, S. F., & Gashe, B. A. (2010). Incidence of indicator organisms, opportunistic and pathogenic bacteria in fish. *African Journal of Food, Agriculture, Nutrition and Development, 10*(10), 4202–4218.

Miller, R. M., & Chapman, W. R. (1976). Epistylis sp. and *Aeromonas hydrophila* infections in fishes from North Carolina reservoirs. *Progressive Fish Culturist, 38*, 165–168.

Mizuki, H., Whasio, S., Moita, T., Itoi, S., & Sugita, H. (2006). Distribution of the fish pathogen *Listonella anguillarum* in the Japanese flounder hatchery. *Aquaculture (Amsterdam, Netherlands), 261*, 26–32.

Morley, N. J., & Lewis, J. W. (2010). Consequences of an outbreak of columnaris disease (*Flavobacterium columnare*) to the helminth fauna of perch (*Perca fluviatilis*) in the Queen Mary reservoir, south-east England. *Journal of Helminthology, 84*, 186–192.

Nematollahi, A., Decostere, A., Pasmans, F., & Haesebrouck, F. (2003). *Flavobacterium psychrophilum* infections in salmonid fish. *Journal of Fish Diseases*, *26*(10), 563–574.

Newman, S. G. (1982). *Aeromonas hydrophila*: A review with emphasis on its role in fish diseases. In D. P. Anderson, M. Dorson, & P. H. Dubourget (Eds.), *Les antigens des Microorganismes pathogenes des poisson* (pp. 87–118). Collection Fondation Marcel Merieux.

Nieto, T. P., Santos, Y., Rodriguez, L. A., & Ellis, A. E. (1991). An extracellular acetylcholinesterase produced by *Aeromonas hydrophila* is a major lethal toxin for fish. *Microbial Pathogenesis*, *11*, 101–110.

Nieto, T. P., Toranzo, A. E., & Barja, J. L. (1984). Comparison between the bacterial flora associated with fingerling rainbow trout cultured in two different hatcheries in the north-west of Spain. *Aquaculture (Amsterdam, Netherlands)*, *42*, 193–206.

Nor-Amalina, Z., Dzarifah, M. Z., Mohd-Termizi, Y., Amal, M. N. A., Zamri-Saad, M., & Ina-Salwany, M. Y. (2017). Phenotypic and genotypic characterization of Vibrio species isolates from Epinephelus species in Selangor, Malaysia. In *Proceedings of the International Conference on Advances in Fish Health*, April 4–6. Universiti Putra Malaysia, Serdang, Selangor, Malaysia.

Novotny, L., Dvorska, L., Lorencova, A., Beran, V., & Pavlik, I. (2004). Fish: A potential source of bacterial pathogens for human beings. *Veterinary Medicine*, *49*, 343–358.

Olesen, N. J., & Vendramin, N. (2016). *20th annual workshop of the National Reference Laboratories for fish diseases*. National Veterinary Institute, Technical University of Denmark.

Ooyama, T., Hirokawa, Y., Minami, T., Yasuda, H., Nakai, T., Endo, M., Ruangpan, L., & Yoshida, T. (2002). Cell-surface properties of *Lactococcus garvieae* strains and their immunogenicity in the yellowtail *Seriola quinqueradiata*. *Diseases of Aquatic Organisms*, *51*, 169–177.

Ordal, E. J., & Rucker, R. R. (1944). Pathogenic myxobacteria. *Proceedings of the Society for Experimental Biology and Medicine*, *56*, 15–18.

Ormonde, P., Hörstedt, P., O'Toole, R., & Milton, D. L. (2000). Role of motility in adherence to and invasion of a fish cell line by *Vibrio anguillarum*. *Journal of Bacteriology*, *182*, 2326–2328.

Pacha, R. E., & Ordal, E. J. (1967). Histopathology of experimental columnaris disease in young salmon. *Journal of Comparative Pathology*, *11*, 419–423.

Park, S. B., Aoki, T., & Jung, T. S. (2012). Pathogenesis of and strategies for preventing *Edwardsiella tarda* infection in fish. *Veterinary Research*, *43*, 1–11.

Paterson, W. D. (1974). Biochemical and serological differentiation of several pigment-producing aeromonads. *Journal of the Fisheries Research Board of Canada*, *31*, 1259–1261.

Rallis, E., & Koumantaki-Mathioudaki, E. (2007). Treatment of *Mycobacterium marinum* cutaneous infections. *Expert Opinion on Pharmacotherapy*, *8*(17), 2965–2978.

Ramsay, J. M., Watral, V., Schreck, C. B., & Kent, M. L. (2009). Husbandry stress exacerbates mycobacterial infections in adult zebrafish, *Danio rerio* (Hamilton). *Journal of Fish Diseases*, *32*, 931–941.

Ransangan, J., & Mustafa, S. (2009). Identification of *Vibrio harveyi* isolated from diseased Asian Seabass Lates calcarifer by use of 16S ribosomal DNA sequencing. *Journal of Aquatic Animal Health*, *21*, 150–155.

Řehulka, J., & Minařík, B. (2007). Blood parameters in brook trout *Salvelinus fontinalis* (Mitchill, 1815), affected by columnaris disease. *Aquaculture Research*, *38*, 1182–1197.

Reichenbach-Klinke, H. H. (1972). Some aspects of mycobacterial infections in fish. *Symposia of the Zoological Society of London*, *30*, 17–24.

Roberts, R. J. (2012). *Fish pathology* (4th ed.). Hoboken, NJ: Blackwell Publishing Ltd.

Roberts, R. J., & Horne, M. T. (1978). Bacterial meningitis in farmed rainbow trout *Salmo gairdneri* Richardson affected with chronic pancreatic necrosis. *Journal of Fish Diseases*, *1*, 157–164.

Rodricks, E. G. (1991). *Indigenous pathogen: Vibrionaceae of microbiology of marine food products* (pp. 285–295). New York: Reinhold.

Rodriguez, L. A., Ellis, A. E., & Nieto, T. P. (1993). Effects of the acetylcholinesterase toxin of *Aeromonas hydrophila* on the central nervous system of fish. *Microbial Pathogenesis*, *14*, 411–415.

Rodriguez, L. A., Fernandez, A. I. G., & Nieto, T. P. (1993). Production of the lethal acetylcholinesterase toxin by different *Aeromonas hydrophila* strains. *Journal of Fish Diseases*, *16*, 73–78.

Romalde, J. L., Ravelo, C., Valdés, I., Magariños, B., Fuente, E. D. L., Martín, C. S., Avendaño-Herrera, R., & Toranzo, A. E. (2008). *Streptococcus phocae*, an emerging pathogen for salmonid culture. *Veterinary Microbiology*, *130*(1–2), 198–207.

Ross, A. J. (1972). In vitro studies with nifurpirinol (P-7138) and bacterial fish pathogens. *Progressive Fish-Culturist*, *34*, 18–20.

Ross, A. J., Martin, J. E., & Bressler, V. (1968). *Vibrio anguillarum* from an epizootic in rainbow trout (*Salmo gairdneri*) in the USA. *Bulletin Office International des Epizooties*, *69*, 1139–1148.

Rucker, R. R. (1959). Vibrio infections among marine and freshwater fishes. *Progressive Fish-Culturist*, *21*, 22–25.

Sakai, M., Atsuta, S., & Kobayashi, M. (1989a). *Pseudomonas fluorescens* isolated from the diseased rainbow trout, *Oncorhynchus mykiss*. *Kitasato Archives of Experimental Medicine*, *62*, 157–162.

Sakai, M., Atsuta, S., & Kobayashi, M. (1989b). Protective immune response in rainbow trout *Oncorhynchus mykiss*,vaccinated with ß-haemolytic streptococcal bacterin. *Fish Pathology*, *24*, 169–173.

Salonius, K., Siderakis, C., MacKinnon, A. M., & Griffiths, S. G. (2005). Use of *Arthrobacter davidanieli* as a live vaccine against *Renibacterium salmoninarum* and *Piscirickettsia salmonis* in salmonids. *Developments in Biologicals: Journal of the International Association of Biological Standardization*, *121*, 189–197.

Schaperclaus, W. (1926). Bakterium fluorescens – Infektion und Geschwulstbildungen bei Aalen mit Versuchlucktem Angelhaken. *Zufishan*, *24*, 157.

Shiomitsu, K., Kusuda, R., Osuga, H., & Munekiyo, M. (1980). Studies on chemotherapy of fish disease with erythromycin – II. Its clinical studies against streptococcal infection in cultured yellowtails. *Fish Pathology*, *15*, 17–23.

Shiraki, K., Miyamoto, F., Sato, T., Sonezaki, I., & Yano, K. (1970). Studies on a new chemotherapeutic agent nifurprazine (HB-115) against fish infectious diseases. *Fish Pathology*, *4*, 130–137.

Shoemaker, C. A., Klesius, P. H., Drennan, J. D., & Evans, J. (2011). Efficacy of a modified live *Flavobacterium columnare* vaccine in fish. *Fish and Shellfish Immunology*, *30*, 304–308.

Shotts, E., & Starliper, C. (1999). Flavobacterial diseases: Columnaris disease, cold-water disease and bacterial gill disease. In P. T. K. Woo, & D. W. Bruno (Eds.), *Fish diseases and disorders: Viral, bacterial and fungal infections* (Vol. 3, pp. 559–576). New York: CABI Publishing.

Shotts, E. B., Jr, Gaines, J. L., Jr, Martin, L., & Prestwood, A. K. (1972). Aeromonas-induced deaths among fish and reptiles in an eutrophic inland lake. *Journal of the American Veterinary Medical Association*, *161* (6), 603–607.

Smith, I. W. (1964). *The occurrence and pathology of Dee disease* (pp. 1–13). Edinburgh: Department of Agriculture and Fisheries for Scotland.

Snieszko, S. F. (1958). Columnaris disease of fishes. *US Fish and Wild-life service Fishery Leaflet*, *46*, 1–3.

Soltani, M., Shanker, S., & Munday, B. L. (1995). Chemotherapy of Cytophaga/Flexibacter-like bacteria (CFLB) infections in fish: Studies validating clinical efficacies of selected antimicrobials. *Journal of Fish Diseases*, *18*, 555–565.

Song, Y. L., Fryer, J. L., & Rohovec, J. S. (1988). Comparison of gliding bacteria isolated from fish in North America and other areas of the Pacific rim. *Fish Pathology*, *23*, 197–202.

Sonia, G. A. S., & Lipton, A. P. (2012). Pathogenicity and antibiotic susceptibility of Vibrio species isolated from the captive–reared tropical marine ornamental blue damsel fish, *Pomacentrus caeruleus* (Quoy and Gaimard, 1825). *Indian Journal of Geo-Marine Sciences*, *41*, 348–354.

Soto, E., Mauel, M. J., Karsi, A., & Lawrence, M. L. (2008). Genetic and virulence characterization of Flavobacterium columnare from channel catfish (*Ictalurus punctatus*). *Journal of Applied Microbiology*, *104*, 1302–1310.

Starliper, C. E. (2011). Bacterial cold-water disease of fishes caused by *Flavobacterium psychrophilum*. *Journal of Advanced Research*, *2*(2), 97–108.

Strohl, W. R., & Tait, L. R. (1978). *Cytophaga aquatilis* sp. nov., a facultative anaerobe isolated from the gills of freshwater fish. *International Journal of Systematic Bacteriology*, *28*(2), 293–303.

Suomalainen, L. R., Bandilla, M., & Valtonen, E. T. (2009). Immunostimulants in prevention of columnaris disease of rainbow trout, *Oncorhynchus mykiss* (Walbaum). *Journal of Fish Diseases*, *32*, 723–726.

Suomalainen, L. R., Tiirola, M., & Valtonen, E. T. (2006). Chondroitin acylase activity is related to virulence of fish pathogenic *Flavobacterium columnare*. *Journal of Fish Diseases*, *29*, 757–763.

Tobin, D. M., & Ramakrishnan, L. (2008). Comparative pathogenesis of *Mycobacterium marinum* and *Mycobacterium tuberculosis*. *Cellular Microbiology*, *10*(5), 1027–1039.

Toranzo, A. E., Barreiro, S., Casal, J. F., Figueras, A., Magarinos, B., & Barja, J. L. (1991). Pasteurellosis in cultivated gilthead bream (*Sparus auata*): First report in spain. *Aquaculture (Amsterdam, Netherlands)*, *99*, 1–15.

Toranzo, A. E., Beatriz, M., & Jesús, L. R. (2005). A review of the main bacterial fish diseases in mariculture systems. *Aquaculture (Amsterdam, Netherlands)*, *246*(1–4), 37–61.

Toranzo, A. E., Magariños, B., & Avendaño-Herrera, R. (2017). Vibriosis: *Vibrio anguillarum*, *V. ordalii* and *Aliivibrio salmonicida*. In P. T. K. Woo, & R. C. Cipriano (Eds.), *Fish viruses and bacteria: Pathobiology and protection* (pp. 314–333). Wallingford, Oxfordshire, UK: CABI Publishing.

Trust, T. J., & Barlett, K. H. (1974). Occurrence of potential pathogens in water containing ornamental fishes. *Applied Microbiology*, *28*, 35–40.

Vendrill, D., Balcazar, J. L., Zarzuela, I. R., Blas, I. D., Girones, O., & Muzquiz, J. L. (2006). *Lactococcus garvieae* in fish: A review. *Comparative Immunology, Microbiology and Infectious Diseases*, *29*(4), 177–198.

Verner-Jeffreys, D. W., Welch, T. J., Schwarz, T., Pond, M. J., & Woodward, M. J. (2009). High prevalence of multidrug-tolerant bacteria and associated antimicrobial resistance genes isolated from ornamental fish and their carriage water. *PLoS One*, *4*, 1–9.

Wagner, B. A., Wise, D. J., Khoo, L. H., & Terhune, J. S. (2002). The epidemiology of bacterial diseases in food-size channel catfish. *Journal of Aquatic Animal Health*, *14*, 263–272.

Wakabayashi, H., Huh, G., & Kimura, N. (1989). *Flavobacterium branchiophila* sp. nov., a causative agent of bacterial gill disease of freshwater fishes. *International Journal of Systematic Bacteriology*, *39*(3), 213–216.

Walczak, N., Puk, K., & Guz, L. (2017). Bacterial flora associated with diseased freshwater ornamental fish. *Journal of Veterinary Research*, *61*(4), 445–449.

Wallbanks, S., Martinez-Murcia, A. J., Fryer, J. L., Phillips, B. A., & Collins, M. D. (1990). 16s r RNA sequence determination for members of the genus Carnobacterium and related lactic acid bacteria and description of *Vagococcus salmoninarum* sp. Nov. *International Journal of Systematic Bacteriology*, *40*(3), 224–230.

Wenzel, S., Bach, R., & Müller-Prasuhn, G. (1971). Farmed trout as carriers of *Clostridium botulinum* and the cause of botulism – IV. Sources of contamination and contamination paths in fish farming and processing stations; ways to improve hygiene. *Archives für Lebensmittelhygiene*, *22*, 131.

Wiens, G. D., Rockey, D. D., Wu, Z., Chang, J., Levy, R., Crane, S., Chen, D. S., Capri, G. R., Burnett, J. R., Sudheesh, P. S., Schipma, M. J., Burd, H., Bhattacharyya, A., Rhodes, L. D., Kaul, R., & Strom, M. S. (2008). Genome sequence of the fish pathogen *Renibacterium salmoninarum* suggests reductive evolution away from an environmental *Arthrobacter ancestor*. *Journal of Bacteriology*, *190*(21), 6970–6982.

Williams, A. M., & Collins, M. D. (1990). Molecular taxonomic studies on *Streptococcus uberis* types I and II. Description of *Streptococcus parauberis* sp. nov. *Journal of Applied Bacteriology*, *68*(5), 485–490.

Xu, D. H., Zhang, D., Shoemaker, C., & Beck, B. (2019). Immune response of channel catfish (*Ictalurus punctatus*) against *Ichthyophthirius multifiliis* post vaccination using DNA vaccines encoding immobilization antigens. *Fish and Shellfish Immunology*, *94*, 308–317.

Yanong, R. P., Pouder, D. B., & Falkinham, J. O., 3rd (2010). Association of mycobacteria in recirculating aquaculture systems and mycobacterial disease in fish. *Journal of Aquatic Animal Health*, *22*(4), 219–223.

CHAPTER

Bacterial diseases in cultured fishes: an update of advances in control measures

15

Soibam Khogen Singh[1], Maibam Malemngamba Meitei[1], Tanmoy Gon Choudhary[2], Ngasotter Soibam[3], Pradyut Biswas[1] and Gusheinzed Waikhom[1]

[1]Department of Aquaculture, College of Fisheries, Central Agricultural University, Lembucherra, India [2]Department of Aquatic Animal Health, College of Fisheries, Central Agricultural University, Lembucherra, India [3]Department of Fish Processing Technology and Engineering, College of Fisheries, Central Agricultural University, Lembucherra, India

15.1 Introduction

Aquaculture intensification in the midst of higher production per unit area leads to fast emergence of diverse pathogenic bacteria in the culture environment of commercial aquaculture worldwide. It is well established that the complex interaction between the host, the pathogen, and the environment results in the occurrence of disease (Dar et al., 2016). The potential disease-causing pathogens in aquatic species comprise a wide range of both gram-positive and gram-negative bacteria and are associated with a broad range of clinical manifestations comprising swellings, ulcerations, erosions, and hemorrhagic septicemias. Many changes have been made in the sequence of disease control measures over the decade. The major issues or hindrances in developing an effective disease management perspective have mostly been due to the unusual aquatic environment, where when the host's health status is weakened, the pathogens often take advantage. In many cases, environmental deterioration, leading to stress in the cultured animals, leads to a disease outbreak (Dar et al., 2020). Some stress factors that trigger the chances of infection by opportunistic pathogens in the aquatic environment are high stocking density, poor nutritional status, nonoptimal water quality, higher microbial load, etc. Earlier, the use of antibiotics as a potential means for disease control was adapted from other sectors. However, the outcome was not sustainable and questions on antibiotic resistance and residues in the environment have long been debated. In such a situation, establishing a well-understood, sustainable approach to disease management is a prioritized research area. This chapter provides detailed updates on several control measures of bacterial diseases in aquaculture species, with suggestions on future work to be established.

15.1.1 Antibiotics residue

Antibiotics are therapeutic agents/drugs designed to hinder the growth and/or multiplication of pathogenic bacteria in an organism. The term literally means "against life." Technically, any drugs

Bacterial Fish Diseases. DOI: https://doi.org/10.1016/B978-0-323-85624-9.00019-1

having the property to terminate bacteria are antibiotics. Most antibiotics are used in aquaculture to prevent infectious diseases caused by bacteria. Antibiotic residue refers to molecules that remain in the meat of cultured animals that are treated with antibiotics. As aquaculture has become more intensified, the use of antibiotics has also risen, due to the multiplication of many bacteria (Defoirdt et al., 2011), and so antibiotic residue also increased. The emergence of antibiotic-resistant strains of bacteria is due to the indiscriminate use of antibiotics in aquaculture (Cabello et al., 2016). Also, there is an increased risk of antibiotic residue as a health hazard for consumers (Crawford, 1985), which can lead to biomagnification in higher animals. The Codex Alimentarius Commission was jointly established by the Food and Agricultural Organization (FAO) and World Health Organization (WHO) to evaluate the risk management recommendations (RMRs) and the maximum residue limits (MRLs) of antibiotic residues in foods. The European Union legislation has also established MRL values for antibiotics in fish, which is shown in Table 15.1. Different groups of antibiotics with examples of resistant pathogenic bacteria isolated from aquaculture settings are given in Table 15.2.

15.2 Preventive measures against diseases: possible outlook

An array of preventive approaches to overcome disease outbreaks in aquaculture is documented. Until a decade ago, excessive application of antibiotics and other chemicals was regular practice for preventing disease outbreaks in both cultured fish and shellfish. However, the serious threat connected with the residues and resistance developed due to antibiotics in the host as well as the environment has led to their ban in aquaculture, with the exception of selected chemicals. The quantum of chemical control strategies can thus be minimized, allowing movement towards biological strategies that are sustainable in the long run. Additionally, biosecurity measures are practical solutions in overcoming disease spread and transmission in culture periods and systems and must be seriously applied in aquaculture practices. Given the fact that the biosecurity measures are stringent measures to be adopted by the farmers, negligence and lack of governance have made this strategy ineffective, especially when the farm lacks facilities for such measures. Some important stringent quarantine measures include fish traffic control, egg disinfection, clean feed, water treatments, and disposal of carcasses. Alternative means of protection and control of bacterial pathogens are listed in the following sections.

15.2.1 Fish derived antimicrobial peptides

Recently discovered antimicrobial drugs like AMPs can now successfully address the persistent problems related to antibiotic resistance in cultured fishes. AMPs are components of eukaryotes' native immune system and are therefore not essentially susceptible to conventional mechanisms that play a role in drug resistance. The AMPs have a wide range of specificity towards pathogenic groups like bacteria, viruses, fungi, and parasites (both ecto- and endo-) (Zasloff, 2002). In general, AMPs are categorized on the basis of their biosynthesis, biological activities, or structural properties (alpha-helices, β-sheets, extended or loop structures) and have different structures, sizes, and physicochemical characteristics (Campagna et al., 2007). They exhibit amphipathic characteristics

Table 15.1 Values of MRLs (maximum residue limits) for antibiotics in fish under European Union (EU) regulations.

Pharmacologically active substance	Marker residue	MRL (μg/kg)[a]
Sulfonamides (All substances that belong to the category of sulfonamides)	Parent drug	100[b]
Diaminopyrimidine derivatives Trimethoprims	Parent drug	50
Penicillins		
Amoxicillin	Amoxicillin	50
Benzylpenicillin	Benzylpenicillin	50
Cloxacillin	Cloxacillin	300
Dicloxacillin	Dicloxacillin	300
Oxacillin	Oxacillin	300
Quinolones		
Oxolinic acid	Oxolinic acid	100
Danofloxacin	Danofloxacin	100
Difloxacin	Difloxacin	300
Enrofloxacin	Sum of enrofloxacin and ciprofloxacin	100
Flumequine	Flumequine	600 150 (salmonidae)
Sarafloxacin	Sarafloxacin	30 (salmonidae)
Macrolides		
Erythromycin	Erythromycin A	200
Tilmicosin	Tilmicosin	50
Tylosin	Tylosin A	100
Florfenicol and related compounds		
Florfenicol	Florfenicol	1000
Thiamphenicol	Thiamphenicol	50
Tetracyclines		
Chlortetracycline	Sum of parent drug and its 4-epimer	100
Oxytetracycline	Sum of parent drug and its 4-epimer	100
Tetracycline	Sum of parent drug and its 4-epimer	
Lincosamides		
Lincomycin	Lincomycin	100
Aminoglycosides		
Spectinomycin	Spectinomycin	300
Neomycin (including framycetin)	Neomycin B	500

(Continued)

Table 15.1 *(Continued)*

Pharmacologically active substance	Marker residue	MRL (μg/kg)[a]
Paramomycin	Paramomycin	500
Polymyxins		
Colistin	Colistin	150
Nitrofurans		No maximum levels can be fixed

[a]*For fin fish, these MRLs relate to "muscle and skin in natural proportions."*
[b]*Within the sulfonamide group, combined total residues for all substances should not exceed 100 μg/kg.*
Derived from Cañada-Cañada, F., Muñoz de la Peña, A., & Espinosa-Mansilla, A. (2009). Analysis of antibiotics in fish samples. Analytical and Bioanalytical Chemistry, *395, 987–1008. doi:10.1007/s00216-009-2872-z.*

with a positively charged and nonpolar face, which is the basis for their interaction with the bacterial cell membrane anionic community (Wimley & White, 1996). For clinical applications, like polymyxins in the treatment of gram-negative bacterial infections, only a few AMPs have been approved so far (Mahlapuu et al., 2016), even though new forms of antibiotics against multidrug-resistant microbes have been explored in clinical research (Hancock & Lehrer, 1998). Several peptide molecules belonging to the groups cathelicidin, defensin, and hepcidin have been isolated from teleost over the last three decades, exhibiting various biological activities. However, there exist families that are conserved explicitly to fishes, known as *piscidin* (homologous to cecropin) (Noga & Silphaduang, 2003).

Ever since the first peptide was isolated from the skin of a winter flounder (Cole et al., 1997), a good number of AMPs have been isolated from fish, mostly those living in a high microbial load aquatic environment, which have evolved as a potent innate immune system (Shabir et al., 2018). Noga and Silphaduang (2003) isolated piscidins from mast cells of hybrid striped bass (*Morone chrysops* × *M. saxatilis*) and reported three different isoforms: *piscidins 1*, *piscidins 2*, and *piscidins 3*. Later, piscidin-like peptides were reported from seabream, stored in professional phagocytic granulocyte granules (Mulero et al., 2008). *Piscidin 2* is active against fish ectoparasites (Colorni et al., 2008). Piscidins are initially synthesized as pre-pro-peptides and subsequently activated into a biologically active cationic molecule with around 22 amino acids after cleavage. However, their pore-forming behavior on the bacterial cell wall seems to be accomplished by a toroidal mechanism (Campagna et al., 2007). Subsequently, Perrin et al. (2016) suggested that transient bilayer membrane distortions are responsible for membrane disruption of very stable pores. Of recent, in the gills of hybrid striped bass, a new piscidin isoform (*piscidin 4*), having the ability to eliminate various bacterial fish pathogens, has been reported (Cañada-Cañada et al., 2009).

15.2.2 Nanotechnology-assisted delivery systems

Frequent incidences of residual remains of drug and medicinal bioactivity of traditional medications and unlicensed drugs promote environmental hazards, such as host tissue bioaccumulation that makes fish unfit for human consumption and affects nontarget species and natural microbial

Table 15.2 Different groups of antibiotics with examples of resistant pathogenic bacteria isolated from aquaculture settings used in aquaculture.

Drug class	Example	Resistance bacteria	Multiple resistance	Isolated from	Reference
Aminoglycosides	Streptomycin	*Edwardsiella ictulari*	Yes	Diseased striped catfish (*Pangasianodon hypophthalmus*), Vietnam	Dung et al. (2008)
Amphenicols	Florfenicol	*Enterobacter* spp. and *Pseudomonas* spp.	Yes	Freshwater salmon farms, Chile	Fernández Alarcón et al. (2010)
Beta-lactams	Amoxicillin	*Vibrio* spp., *Aeromonas* spp. and *Edwardsiella tarda*	Yes	Different aquaculture settings, Australia	Akinbowale et al. (2006)
Beta-lactams	Ampicillin	*Vibrio harveyi*	Yes	Shrimp farms and coastal waters, Indonesia	Teo et al. (2000)
Fluoroquinolones	Enrofloxacin	*Tenacibaculum maritimum*	Yes	Diseased turbot (*Scophthalmus maximus*) and sole (*Solea senegalensis*), Spain and Portugal	Avendaño-Herrera et al. (2008)
Macrolides	Erythromycin	*Salmonella* spp.	Yes	Marketed fish, China	Broughton and Walker (2009)
Nitrofurans	Furazolidone	*Vibrio anguillarum*	Yes	Diseased sea bass and seabream, Greece	Smith and Christofilogiannis (2007)
Nitrofurans	Nitrofurantoin	*Vibrio harveyi*	Yes	Diseased penaeid shrimp, Taiwan	Liu et al. (1997)
Quinolones	Oxolinic acid	*Aeromonas* spp., *Pseudomonas* spp. and *Vibrio* spp.	yes	Pond water, pond sediment and tiger shrimp (*Penaeus monodon*), Philippines	Tendencia and de la Peña (2001)
Sulfonamides	Sulphadiazine	*Aeromonas* spp.	yes	Diseased katla (*Catla catla*), mrigel (*Cirrhinus mrigala*), and punti (*Puntius* spp.), India	Das et al. (2009)
Tetracyclines	Tetracycline	*Aeromonas hydrophila*	yes	Water from mullet and tilapia farms, Egypt	Ishida et al. (2010)
Tetracyclines	Oxytetracycline	*Aeromonas salmonicida*	Yes	Atlantic salmon (*Salmo salar*) culture facilities, Canada	McIntosh et al. (2008)

communities. Therefore evidence indicates that alternative methods that minimize the administration of antibiotics, including natural products, are necessary. In a study on the use of nanotechnology in countering bacterial disease, nerolidol nanospheres (1.0 mL/kg diet) dietary supplementation has been reported to have potent bactericidal effects in terms of improved fish longevity and decreased brain microbial load survival. Also, they indicate an increase in the passage of nerolidol nanospheres across the blood-brain barrier, enabling the brain tissue to minimize *Streptococcus agalactiae*, a significant factor in the higher mortality of fish. Similarly, *S. agalactiae*-induced brain oxidative damage that led to the disease's pathogenesis was avoided by dietary supplementation with nerolidol nanospheres (1.0 mL/kg diet).

15.2.3 Bacterial fish vaccines

Earlier, antibiotics were used as an effective remedy for an array of bacterial fish diseases in aquaculture. Nowadays, they are little used due to the negative consequences of antibiotic resistance to different strains of bacteria. In addition, antimicrobial drugs cannot reduce bacterial infection in aquatic animals (Dar et al., 2016). With advances in research, scientists developed several fish bacterial vaccines as immune-prophylaxis. Fish vaccination is possibly the most important approach to prevent and control bacterial infection diseases, as it can immunize a large population of aquatic animals at the same time. The first reported use of a fish vaccine was Duff's killed *Aeromonas salmonicida* vaccine, which was an oral cutthroat trout vaccination (*Oncorhynchus clarkia*). A dead *Yersinia ruckeri* vaccine administered by immersion against enteric red mouth disease was the first commercially licensed vaccine for fish (Gudding & Goodrich, 2014).

15.2.3.1 Types of vaccines

15.2.3.1.1 Inactivated vaccines

Most of the licensed vaccines used in aquaculture are inactivated vaccines produced by a conventional method (Ma et al., 2019). In this type of vaccine, inactivated pathogens are used. Through multiplication and replication, the pathogens are produced in large quantities and then treated with inactivating agents like formalin, b-propiolactone (BPL), binary ethylenimine (BEI), formaldehyde, and temperature/heat that kill the microorganisms without affecting the induction of protective immunity of the vaccine candidates. Earlier, fish vaccines consisted of formalin-killed bacteria, with or without adjuvant (Taffala et al., 2013, 2014), and were delivered through immersion or injection.

The efficacy and biosafety of the vaccine depend on the conditions of cultivation, such as the media and temperature range. Most of the vaccines used in aquaculture for bacteria are cultivated in broth culture and inactivated with formalin (Toranzo et al., 2009): *Vibrio anguillarum*, *V. ordalli*, *V. salmonicida*, *Yersinia ruckeri*, and *Aeromonas salmonicida* are produced by broth fermentation followed by inactivation with formalin (Gudding et al., 1999; Shao, 2001; Toranzo et al., 2009).

15.2.3.1.2 Attenuated/live vaccines

These vaccines are chemically or genetically attenuated or the pathogen's virulence towards a particular fish species is reduced. This type of attenuated pathogen mimics a natural pathogen in the host fish and subsequently generates the immune response. They have more advantages than the killed vaccines, as they are linked with both specific and nonspecific immunity (Levine & Sztein, 2004) and also have the ability to enhance humoral and mucosal immunity (Clark & Cassidy-Hanley, 2005). At present, there are only three licensed bacterial live vaccines: arthrobactor vaccine against bacterial kidney disease (BKD), *E. ictalurii* vaccine against enteric catfish septicemia, and *Flavobacterium columnare* vaccine against columnaris disease to be directed to use in the United States. The arthrobactor has also been licensed in Canada and Chile with the brand name Renogen. The main disadvantage of this vaccine is that it can spread virulence to nontargeted species due to the residual virulence in the targeted species (Dhar & Allnutt, 2011; Salgado-Miranda et al., 2013; Shao, 2001).

15.2.3.1.3 DNA vaccine

These vaccines consist of a particular genetic material that encodes a target antigen of a specific disease. The DNA is mostly present in the plasmid and when these vaccines are injected into the host cell, they translate into immunogenic proteins (immune-stimulating antigen) and thus provide an immune response against the disease (Adams et al., 2008; Roy, 2011). But most of the DNA vaccines are used to protect against viral infection. There is a chance of reversion to virulence as the DNA sequences encode only one microbial genome, so it has to be reduced for environmental safety.

15.2.3.1.4 Recombinant vaccine

The gene sequence of the protective antigen of a pathogen is inserted into a production host with the help of recombinant technology and cultured on a large scale, from which the protective antigen is purified and used in the development of the vaccine (Adams et al., 2008). Pharos, Microtek, Bayovac 3.1 are recombinant vaccines against the disease Salmon rickettsia septicemia in salmonids. These vaccines can be delivered through intraperitoneal injection (IP), immersion, or orally through food (Table 15.3).

15.2.4 Prebiotics

An alternative method for the prevention of bacterial infection in culture fish in aquaculture is to enhance fish immunity using immune boosters or immunostimulants. Prebiotics is also one of the promising immune-stimulants in the present aquaculture industry that can be used as an alternative method of antibiotic. Prebiotics consist of a nondigestible substance that enhances the host's immunity by activating the beneficial bacteria's activity in the gut (Gibson & Roberfroid, 1995). Usually, they are referred to as the food of probiotics. There are some criteria for being a candidate for prebiotics (Gibson & Roberfroid, 1995), such as (1) it should not be hydrolyzed in the upper part of the gastrointestinal tract or absorbed, and (2) it should provide a selective growth and metabolic activation substrate for one or a small number of beneficial colon-commensal bacteria and improve the immune system. So, nondigestible carbohydrates (oligo- and polysaccharides), some lipids (both ethers and esters), and certain peptides and proteins are strong candidates as prebiotics. The use of prebiotics as an immune-stimulant in an aquaculture system will boost the immunity of the fish and thus enhance host immunity (Ringø et al., 2010; Song et al., 2014; Torrecillas et al., 2014). Fructooligosaccharides (FOSs), mannanoligosaccharides (MOSs), inulin, etc., are mostly used as fish immunostimulants.

Table 15.3 Types of bacterial vaccines applied in aquaculture.

Sl. no.	Name of vaccine	Name of disease against which vaccine is developed	Fish species
1.	*Flavobacterium columnare* vaccines	Columnaris disease	Salmonids, channel catfish, and carps
2.	Free-cell *Aeromonas hydrophila* vaccine	Dropsy	Indian major carps
3.	*Edwardsiella ictaluri* vaccine	Edwardsiellosis	Catfish
4.	*Streptococcus agalactiae* vaccine	Streptococciosis	Tilapia
5.	*Streptococcus iniae* vaccine	Streptococciosis	Tilapia
6.	Enteric red mouth disease (ERM) vaccine	Enteric red mouth disease (ERM)	Salmonids
7.	*Vibrio anguillarum-ordalii*	Vibriosis	Salmonids, rainbow trout
8.	*Aeromonas salmonicida* bacterin	Furunculosis	Salmonids
9.	*Aeromonas hydrophila* vaccine	Motile Aeromonas septicemia	Salmonids
10.	*Edwardsiella ictaluri* bacterin	Enteric septicemia	Channel catfish, Japanese flounder

15.2.4.1 Mannan oligosaccharide

MOS is a gluco-manno-protein complex extract from the *Saccharomyces cerevisiae* (yeast) cell wall (Sohn et al., 2000). Its uses are mostly documented in terrestrial animals (Benites et al., 2008; Klebaniuk et al., 2008; Yang et al., 2008). The mannose receptor (MR) is an endocytic receptor that recognizes both self-glycoprotein and microbial glycan ligands in macrophages and endothelial cells (Ringø et al., 2010). Immature cultured dendritic cells (DCs) also demonstrate MR, where it promotes the absorption of high-efficiency glycosylated antigens (Linehan et al., 2000). Mannose-containing ligands influence MR-mediated intracellular signaling cascades in DCs that can increase the production of proinflammatory cytokines. Minimal research has been done on fish using MOS as an immune stimulant (Table 15.4).

15.2.4.2 Fructooligosaccharide

Fructooligosaccharides (FOSs) are short and medium chains of β-D-fructans in which fructosyl units are bound by β-(2-1) glycosidic linkages and attached to a terminal glucose unit (Ringø et al., 2010). Naturally, they are present in various kinds of foods, such as Jerusalem artichokes, wheat, banana rye, barley, triticale, onion, garlic, and honey (Fuller & Gibson, 1998). FOSs can be used by bacteria like lactobacilli and bifidobacteria (Manning & Gibson, 2004; Sghir et al., 1998). Due to the lack of β-fructosidases in the mammalian digestive system, the β-(2-1) glycosidic bonds cannot be broken down (Teitelbaum & Walker, 2002). FOS interacts with TLR2 (Toll-like receptors), a membrane surface receptor expressed on macrophages, PMNs (polymorphonuclear leukocytes or granulocytes), and dendritic cells have resulted in immune cell activation through signal transduction pathways (Vogt et al., 2013). Some of the FOSs utilized as prebiotics are shown in Table 15.5.

Table 15.4 Mannanoligosaccharides as immune stimulant in aquaculture.

Species	Dose	Duration	Result	Reference
Atlantic salmon (*Salmo salar*)	DS 10 g/kg	4 month	11% lower routine oxygen intake, 5% lower protein and 3% higher whole-body energy concentration, and 7% higher energy retention. ↓ Neutrophil oxidative radical production and serum lysozyme activity	Grisdale-Helland et al. (2008)
Crayfish, *Cherax destructor* (yabby)	DS 0.4%	56 days	↑THC, ↑GCs, ↑ semigranular cells, ↑ Protease activity in hepatopancreas, ↑ amylase activity in the guts	Sang et al. (2011)
European sea bass (*Dicentrarchus labrax*)	DS 4 g/kg	8 weeks	↑ anterior gut mucosal and surface area folds, ↑ mucins secreting cells, ↑ ECG density, ↑ lysozyme activity	Torrecillas et al. (2011)
Japanese flounder Paralichthys olivaceus	DS 5 g/kg	56 days	↔TG, ↑ lysozyme activity, ↔ phagocytic activity	Ye et al. (2011)
Marron, *Cherax tenuimanus*	DS 0.2% and 0.4%	30 days and 112 days	↑ survivability against NH_3, *V. vimicus*, ↑THCs	Sang and Fotedar (2009)
Rainbow trout (Oncorhynchus mykiss)	DS 2000 ppm	42 days	weight gain of 9.9%, ↓ FCR, ↓Mortality, ↑ Lysozyme, ↑ APCA, ↑ CPCA ↔ serum antibody titer and bactericidal activity	Staykov et al. (2007)
Rainbow trout	DS 0.4%	12 weeks	↑ Hematocrit, ↑ macrophage activity, ↑ skin mucus volume per unit area, ↑ Survivality against *Vibrio anguillarum*	Rodriguez-Estrada et al. (2009)
Juvenile rock lobsters (*Panulirus ornatus*)	DS 0.4 %	56 days	↑ THC, ↑ GC, and ↓ bacteraemia	Sang and Fotedar (2010)

DS, *dietary supplement;* ↑, *increase/high;* ↓, *decrease/low;* ↔, *no change/not significant;* THC, *total hemocyte count;* GC, *granular cells;* ECGs, *eosinophil granulocytes,* NH_3*;* APCA, *alternate pathway of complement activation;* CPCA, *classical pathway of complete activation;* FCR, *feed conversion ratio.*

15.2.4.3 Inulin

Inulins are oligosaccharides containing β-D-fructofuranoses attached by β-2-1 linkages. They are mostly found in fruits, edible grains, and vegetables such as onions, wheat, leeks, asparagus, garlic, bananas, and artichokes (Roberfroid, 1993). Inulins bind to specific lectin-like receptors on leukocytes, inducing macrophage proliferation and thus triggering the human immune system (Causey et al., 1998; Meyer, 2008; Seifert & Watzl, 2007); also, they are beneficial in the gut microbiota (Possemiers et al., 2009) (Table 15.6).

15.2.4.4 Miscellaneous prebiotics

The other prebiotics that are used as immunostimulants are listed in Table 15.7. Some of them are GroBiotic™AE and GroBiotic®-A, which is a mixture of partially autolyzed brewers yeast, dried

Table 15.5 Fructooligosaccharides used as immunostimulant in aquaculture.

Species	Dose	Duration	Result	Reference
Atlantic salmon (*Salmo salar*)	DS 10 g/kg	4 months	Better FCR, ↔ NBT and Serum lysozyme activity, ↑ energy retained, ↓ Oxygen routine consumption	Grisdale-Helland et al. (2008)
Triangular bream (*Megalobrama terminalis*)	DS 0.3% and 0.6%	8 weeks	↑ leukocyte count, ↑ ACH50, ↑ Total serum protein and globulin, ↑ Ig M	Zhang et al. (2013)
Caspian roach (*Rutilus rutilus*) fry	DS 1%, 2%, 3%	7 weeks	At 2% and 3%, ↑ Serum Ig, ↑ Lysozyme activity, ↑ ACH50; ↑ Resistance to salinity challenge at 3%	Soleimani et al. (2012)
Japanese flounder (Paralichthys olivaceus)	DS 5.0 g/kg	56 days	↓ Lipoprotein cholesterol, ↑ Lysozyme activity, ↓ TG, ↔ Phagocytic percentage and index	Ye et al. (2011)
Red swamp crayfish (*Procambarus clarkii*)	DS 2.0, 5.0, 8.0, 10.0 g/kg	30 days	At 8.0 and 10.0 g/kg, ↑ Immune-related genes (crustin1, lysozyme, SOD, and proPO), ↑Phagocytic activity, ↑ PO, ↑ SOD, ↑ Survival against *A. hydrophila* at 5.0, 8.0, or 10.0 g/kg	Dong and Wang (2013)
Sea cucumber (*Apostichopus japonicas*)	0.25%, 0.5%	8 weeks	↑ TCC, ↑ Phagocytosis, ↑ PO, ↑ Resistance to *V. splendidus* at 0.5% FOS	Zhang et al. (2010)
Stellate sturgeon (*Acipenser stellatus*) juvenile	1%, 2%	11 weeks	At 1% FOS, ↑ total heterotrophic autochthonous bacterial and presumptive LAB levels, ↑ WBC, ↑ RBC, ↑ MCV, ↑ Hematocrit, ↑ hemoglobin, ↑ lymphocyte levels, ↑ serum lysozyme activity, ↔ respiratory burst activity	Akrami et al. (2013)

FCR, *feed conversion ratio;* NBT, *nitoblue tetrazolium;* ACH50, *alternative complement activity;* IgM, *immunoglobulin M;* TG, *triglyceride;* SOD, *superoxide dismutase;* PO, *phenol oxidase;* MCV, *mean corpuscular volume;* ↑, *increase/high;* ↓, *decrease/low;* ↔, *no change/not significant.*

fermentation products, and dairy ingredient components (Ringø et al., 2010). The main nonstarch polysaccharides found in many cereal grains are arabinoxyloligosachharides (AXOS) and are part of dietary fiber, which consists of β-(1,4)-linked D-xylopyranosyl (Ringø et al., 2010); Galactooligosaccharides (GOS) are manufactured through enzymatic treatments of lactose and consist of 2–20 molecules of galactose and glucose (Yang et al., 2008).

15.2.5 Probiotics

Probiotics are organisms and substances that contribute to intestinal microbial balance. However, the term has been redefined as a live microbial feed supplement that beneficially affects the host animal by improving its intestinal microbial balance (Fuller, 1989). First introduced probiotics. Most of the researchers used lactic acid bacteria like *Lactobacilli*, *Bifidobacteria*, Bacillus sp., and yeast. Probiotics are commonly used for growth supplements or therapeutic or prophylactic

Table 15.6 Inulin as an immunostimulant in aquaculture.

Species	Dose	Duration	Result	Reference
Juvenile great sturgeon (*Huso huso*)	Dietary supplementation of 0.0%, 1.0%, 2.0%, and 3.0%	8 weeks	↑ Inulin ↓ AP ↑ WBC; ↔ RBC, Glucose and MCH	Ahmdifar et al. (2011)
Gilthead seabream (*Sparus aurata*)	Dietary supplementation 10 g/kg	2 and 4 weeks	Serum complement activity ↑, IgM level ↑, phagocytic activity ↑, respiratory burst activity ↑	Cerezuela, Cuesta et al. (2012), Cerezuela, Guardiola et al. (2012)
Hybrid surubim (*Pseudoplatystoma* sp)	Dietary supplementation 0.5% inulin	15 days	Lactic acid bacteria ↑, *Vibrio* sp. ↓, *Pseudomonas* sp. ↓, Total Ig ↑	Mouriño et al. (2012)
Nile tilapia	Dietary supplementation of 5 g/kg insulin	1 and 2 months	Hematocrit ↑, NBT (Superoxide activity) ↑, lysozyme activity ↑, RLP↑	
Tilapia (*Tilapia aureus*) and grass carp (*Ctenopharyngodon idella*)	IP injection with 10 mg/kg	2 weeks	↑ survivability against *A. hydrophilla* and *E. tarda*	Wang and Wang (1997)
Arctic charr (*Salvelinus alpinus* L.)	DS 150 g/kg	4 weeks	↓ bacterial population and dominated by the gram-negative bacteria like *Staphylococcus*, *Streptococcus*, *Carnobacterium* and *Bacillus*	Ringø et al. (2006)
Siberian sturgeon (*Acipenser baerii*)	20 g/kg	1 month	↔ SCFA and lactate; ↑ gas production; ↓ butyrate	Mahious et al. (2006)

AP, *alkaline phosphate;* MCH, *mean corpuscular hemoglobin;* RLP, *relative level of protection;* SCFA, *short-chain fatty acid;* IgM, *immunoglobulin M;* NBT, *nitroblue tetrazolium;* ↑, *increase/high;* ↓, *decrease/low;* ↔, *no change/not significant.*

treatments against infectious diseases. Some microorganisms that can be used as probiotics in feeding as directed by Council Directive 70/524/EEC are *Bacillus cereus* var. *toyoi*, *Bacillus subtilis, Bacillus licheniformis, Enterococcus faecium, Lactobacillus casei, Lactobacillus plantarum, Lactobacillus farciminis, Lactobacillus rhamnosus, Pediococcus acidilactici, Saccharomyces cerevisiae, Streptococcus infantarius*. Probiotics enhance the immune system due to the following properties (Table 15.8):

- antagonistic action against pathogens.
- capability to produce metabolites like vitamins and enzymes.
- colonization and adhesion.

15.2.6 Synbiotic in aquaculture

The concept of synbiotic use was suggested to provide the benefits of both prebiotics and probiotics on fish immunity mainly due to their synergistic effect (Gibson & Roberfroid, 1995) and it has

Table 15.7 Other prebiotics used as immunostimulants in aquaculture.

Prebiotics	Species	Dose	Duration	Results	Reference
GroBiotic™AE	Hybrid striped bass (*Morone chrysops* × *Morone saxatilis*)	1% and 2%	7 weeks (Trial 1) and 4 weeks (Trial 2)	↑ Feed efficiency ↑ Neutrophil oxidative radical anion ↑Intracellular superoxide anion ↔ serum lysozyme ↑ Survivability against *S. iniae*	Li and Gatlin (2004)
GroBiotic®-A	Subadult hybrid striped bass (*Morone chrysops* × *M. saxatilis*)	2%	21 weeks	↑ Growth of fish ↑ Survivability against *Mycobacterium marinum*	Li and Gatlin (2004)
Dairy-yeast prebiotic	Golden shiners (Notemigonus crysoleucas)	2%	10 weeks	↓ Mortality against *F. columnare*	Sink and Lochmann (2008)
GroBiotic®-A	Red drum (Sciaenops ocellatus)	1%	3 weeks	↑ Protein, fat, and organic ADC	Burr et al. (2008)
GOS	Red drum (Sciaenops ocellatus)	1%	3 weeks	↑ Protein and organic ADC ↓ Lipid ADC	Burr et al. (2008)
AXOS	Crucian carp, Carassius auratus gibelio	50 mg/kg, 100 mg/kg and 200 mg/kg	45 days	↑ RGR and DWG ↔ Survivability At 100 ↑ mg/kg, Enzymatic activity (protease and amylase)	Xu et al. (2009)
AXOS	Juvenile Siberian sturgeon (*Acipenser baerii*)	2% in the form of AXOS-32-0.30 or AXOS-3-0.25	2 weeks	With AXOS-32-0.30, ↑ Growth performance and feed utilization ↑ Alternative hemolytic complement activity ↑ Total serum peroxidase ↑ Conc. of acetate, butyrate, and total SCFAs	Geraylou et al. (2012)
AXOS	Juvenile Siberian sturgeon (*Acipenser baerii*)	2%	4 weeks	↑ Phagocytic and respiratory burst activity of fish macrophage ↑ Colonization and/ or growth capacity of *L. lactis*	Geraylou et al. (2013)

GOS, *galactooligosaccharides;* AXOS, *arabinoxyloligosaccharides;* ↑, *increase/high;* ↓, *decrease/low;* ↔, *no change/not significant.*

Table 15.8 Probiotics application in aquatic animal nutrition.

Probiotic species	Fish species/ application mode	Beneficial effects	Reference
Carnobacterium spp.	Atlantic salmon (*Salmo salar* L.)/Feed	Inhibited *V. ordalii*, *A. salmonicida*, and *Y. ruckeri*; Reduced disease	
Lactobacillus rhamnosus	Rainbow trout (*Onchorhynchus mykiss*)/Feed	Increased resistance to *A. salmonicida*, Reduced mortality from furunculosis	
Streptococcus faecium Lactobacillus acidophilus	Nile tilapia/Feed (*Oreochromis niloticus*)	Better performance in growth and feed efficiency	
Saccharomyces cerevisae	Nile tilapia (*O. niloticus*)/Feed	Better performance in growth and feed efficiency	
Bacillus subtilis and *B. licheniformis*	Rainbow trout (*O. mykiss*)/Feed	Increased resistance to *Yersinia ruckeri*	
L. rhamnosus (JCM 1136)	Rainbow trout (*O. mykiss*)/Feed	Increased serum lysozyme and complement activities	
Live yeasts	European sea bass/Feed (*Dicentrarchus labrax*)	Better performance in growth and feed efficiency	
L. rhamnosus	Rainbow trout (*O. mykiss*)/Feed	Immune response stimulated	
B. subtilis, *L. delbriieckii*	Gilthead seabream/Feed	Stimulated cellular innate immune response	
L. lactis Sub sp. *lactis; L. sakei*; *Leuconostoc mesenteroides*	Rainbow trout (*O. mykiss*)/Feed	Stimulated phagocytosis; Enhanced the nonspecific immunity	
L. delbrueckii subsp. *delbrueckii* (AS13B)	European sea bass (*Dicentrarchus labrax* L.)/Feed	Positive effects on welfare and growth; Increased body weight	
L. rhamnosus GG	Tilapia (*O. niloticus*)/ Feed	Enhanced the growth performance and immunity	
Micrococcus luteus	Nile tilapia (*O. niloticus*)/Feed	Enhanced the nonspecific immune parameters; Improved resistance against *Edwardsiella tarda* infection	Taoka et al. (2006)
Bacillus spp. (Photosynthetic bacteria)	Common carp/Feed (*Cyprinus carpio*)	Better digestive enzyme activities; Better growth performance and feed efficiency	
Saccharomyces cerevisiae strain NCYC Sc 47 (Biosaf® Sc 47)	Rainbow trout (*O. mykiss*)/Feed	No significant effect on enzyme activities	
Saccharomyces cerevisiae var. *boulardii* CNCM I-1079 (Levucell® SB20)	Rainbow trout/Feed *O. mykiss*	Stimulated enzyme activities	
Carnobacterium divergens 6251	Atlantic salmon/Feed (*Salmo salar* L.)	*Carnobacterium divergens* is able to prevent, to some extent, pathogen-induced damage in the foregut.	
Lactobacillus delbrueckii sub sp. *bulgaricus*	Rainbow trout/Feed (*O. mykiss*)	Enhanced humoral immune response	

(Continued)

Table 15.8 ***(Continued)***

Probiotic species	Fish species/ application mode	Beneficial effects	Reference
L. rhamnosus (ATCC 53103); *B. subtilis, Enterococcus faecium*	Rainbow trout (*O. mykiss*)/Feed	Modulated cytokine production; Stimulated immune response	
Enterococcus faecium ZJ4	Nile tilapia (*O. niloticus*)/Water	Increased growth performance; Improved immune response	
Leuconostoc mesenteroides CLFP 196; *Lactobacillus plantarum* CLFP 238	Rainbow trout (*O. mykiss*)/Feed	Reduced fish mortality	
Micrococcus luteus, Pseudomonas spp.	Nile tilapia (*O. niloticus*)/Feed	Higher growth performance, survival rate and feed utilization; Enhanced fish resistance against *Aeromonas hydrophila* infection	
Saccharomyces cerevisiae (DVAQUA®)	Hybrid tilapia (*Oreochromis niloticus* ♀ × *O. aureus* ♂)/Feed	Inhibited potential harmful bacteria; Stimulated beneficial bacteria; Enhanced the nonspecific immunity; No significant effects on growth performance and feed efficiency	

become a new research area in recent times. Such products trigger the beneficial microbes' metabolic activities in the gastrointestinal (GI) tract of the host and thus improve the host immunity. There are reports that the use of synbiotics of prebiotics and probiotics is more effective than the individual use of prebiotics and probiotics. However, the attention paid towards this subject is minimal, and so the available data is still scarce. Some works are listed in Table 15.9.

15.2.7 Paraprobiotics: a new concept

Paraprobiotics is defined as "non-viable microbial cells (intact or broken) or crude cell extracts (i.e., with complex chemical composition), which, when administered (orally or topically) in adequate amounts, confer a benefit on the human or animal consumer" (Choudhury & Kamilya, 2019). It has also been named "ghost probiotics," "postbiotic," and inactivated probiotics. There have been many studies and reviews on the potential application of paraprobiotics in higher vertebrates; however, its application in aquaculture is in its beginning stage (Adams, 2010; deAlmada et al., 2016; Kataria et al., 2009). However, to address the certain negative impact of the application of live microbes as probiotics, along with assured advantages of nonviable microbes, the application of paraprobiotics in aquaculture has gained much attention recently (Choudhury & Kamilya, 2019; Singh et al., 2017; Zheng et al., 2020).

Paraprobiotics can be prepared by inactivating viable probiotic microorganisms using various inactivation methods like heat inactivation, inactivation by ultraviolet (UV) rays, ionizing radiation, high-pressure technique, and sonication (deAlmada et al., 2016). Commonly used methods of inactivation of viable probiotics for preparation of paraprobiotics for aquaculture uses are heat inactivation, formalin inactivation, UV treatment and sonication, and among these, heat inactivation is the

Table 15.9 Synbiotics application in aquaculture.

Synbiotic combination	Fish/shrimp species	Results	References
Bio-Mos® and β-1,3-D-glucan + *Pseudomonas synxantha* and *P. aeruginosa*	Western king prawns (*Penaeus latisulcatus*) juveniles	Improvement in the growth, survival, and immune response	Hai and Fotedar (2009)
Isomaltooligosaccharides + *Bacillus* OJ	*Litopenaeus vannamei* (avg. weight 1.75 g)	Synergistic effects on immune responses and disease resistance	Li et al. (2009)
Mannanoligosaccharides (MOS) + *Bacillus* spp.	European lobster (*Homarus gammarus* L.) larvae	Improved growth, survival; significant increases in microvilli length and density	Daniels et al. (2010)
Fructooligosaccharide + *Bacillus subtilis*	Sea cucumber (*Apostichopus japonicus*) (mean body weight 5.06 g)	Improved growth; synergistic effect on enhancing immunity and disease resistance	Zhang et al. (2010)
Fructooligosaccharide + *Bacillus subtilis*	Yellow croaker (*Larimichthys crocea*) juveniles	No interactions between dietary *B. subtilis* and FOS; *B. subtilis* enhanced growth, feed utilization and nonspecific immune response	Ai et al. (2011)
Biomin IMBO (Fructooligosaccharide + *Enterococcus faecium*)	Rainbow trout (*Oncorhynchus mykiss*) fingerlings	Increased growth, survival, and feeding efficiency	Mehrabi et al. (2011)
Fructooligosaccharide and mannanoligosaccharides + *Bacillus clausii*	Japanese flounder (*Paralichthys olivaceus*)	No effect on growth, reduce feed cost per unit growth, carcass composition not affected	Ye et al. (2011)
Isomaltooligosaccharide + *Bacillus licheniformis* and *B. subtilis*	*Penaeus japonicas* juveniles	Positive effect on bacterial flora and nonspecific immunity	Zhang et al. (2011)
Fructooligosaccharide + *Bacillus* TC22	Sea cucumber (*Apostichopus japonicus*) (mean body weight 4.92 g)	No effect on growth, no interaction between the two biotics	Zhao et al. (2011)
Inulin + *Weissella cibaria*	Hybrid surubim (*Pseudoplatysoma* sp.) (avg. weight 76.3 g)	Manipulation of intestinal microbiota; improved hematoimmunological parameters	Mouriño et al. (2012)
Fructooligosaccharide + *B. circulans*	*Rohu (Labeo rohita)* fingerlings	Improved growth, immunity and protection against low pH and nitrite stress	Singh, (2013); Singh et al. (2019)

most preferred method. Though the application of probiotics in aquaculture is already popularized, it is still in the early stages. Some researchers have studied the efficacy of paraprobiotics in fish and shellfish for mediating immune responses, growth, and disease resistance. There are already many research studies establishing the paraprobiotic role to stimulate the immune system in higher vertebrate models (deAlmada et al., 2016; Taverniti & Guglielmetti, 2011). Similarly, the potency of paraprobiotics that activate the immune system in in vitro conditions have also been studied in the aquaculture sector (Biswas et al., 2013; Giri et al., 2016; Kamilya et al., 2015) as well as in in vivo (Biswas et al., 2013; Dash et al., 2015; Dıaz-Rosales et al., 2006; Pan et al., 2008; Singh et al., 2017; Taoka et al., 2006) conditions. The existing literature demonstrates that paraprobiotics, either alone or in combination, can induce both cellular and molecular immunity of fish and shellfish. It has also been found that there is better performance on growth and feed utilization of fish and shellfish with the use of paraprobiotics, but there is no clear mechanism for this (Dash et al., 2015; Zheng et al., 2017).

Paraprobiotics also induce disease resistance capability against infectious diseases in fish and shellfish, though, again, the exact mechanism by which they do so is not clear. Immunostimulations may be the mechanism by which paraprobiotics increase the disease resistance of the host against pathogens (Dash et al., 2015; Irianto & Austin, 2003; Pan et al., 2008; Taoka et al., 2006). Though many works demonstrate that paraprobiotics have an effective role in immune modulation and disease resistance, similar to its viable form, that is, probiotics (Dash et al., 2015; Dawood et al., 2015; Irianto & Austin, 2003), on the other hand, studies have also shown better health benefits for probiotics compared to paraprobiotics (Munoz-Atienza et al., 2015; Taoka et al., 2006).

15.2.8 Herbal biomedicines

The application of antibiotics, vitamins, and other chemicals as a prophylactic therapy is reduced due to their residual effects. Now, researchers and farmers are showing more interest in low-cost therapeutic natural products like medicinal herbs, which have less residue. From ancient times onward, the phyto-therapy approach has been developed and its application has been well known (Citarasu, 2010). Many researchers have reported on the presence of antimicrobial properties, anti-stress, immune-stimulant, and growth promoters in herbal biomedicines.

15.2.8.1 Herbal biomedicines as growth promoters

The herbal biomedicines can be used as feed additives (Wang et al., 2015) and probiotics (Bahi et al., 2017). The herbal products stressol-I- and stressol-II-enriched *Artemia nauplii* significantly enhanced the growth of *Penaeus indicus* postlarvae (PL 10–20) and reduced the osmotic stress (Chitra, 1995). Similarly, the commercial herbal medicine Livol (IHF1000) improves digestibility in rohu (Jayaprakas & Euphrasia, 1997; Maheshappa, 1993) and other cultivable fishes, thereby promoting the growth and immune system of fishes (Shadakshari, 1993; Unnikrishnan, 1995). The papaya plant's enzyme papain improves growth, feed utilization, and protein digestibility of *P. monodon* postlarvae (Penaflorida, 1995). Francis et al. (2005) also observed the improvement in growth, feed utilization, and hematological parameters in Nile tilapia while using the commercial herbal medics from ginseng herb (Ginsana® G115). *Quillaja saponins* are also able to increase growth and decrease the metabolic rate of tilapia (Francis et al., 2005). The herbal growth promoters help enhance the production rate of proteins in the cells (Citarasu, 2010).

15.2.8.2 Herbal medicines as immune-stimulant

An immunostimulant is a chemical substance that can boost the host immune response (Anderson, 1992). For example, dietary supplementation of *Ocimum sanctum* enhances the antibody response against *A. hydrophila* (Logambal et al., 2000). Glycyrrhizin also showed increased macrophage respiratory burst activity and lymphocyte proliferative responses in rainbow trout (Jang et al., 1995; Kim et al., 1998) and the immune system of the yellowtail (Edahiro et al., 1991). This is because glycyrrhizin is a glycosylated saponin, containing one molecule of glycyrretinic acid, which has antiinflammatory and antitumor activities, mediated by its immunomodulatory activities (Wada et al., 1987; Zhang et al., 1990). Soybean protein also increases bacterial killing ability by increasing leukocyte activity when treated orally (Citarasu, 2010). Another herbal, Quil A saponin, also has the potential of in vitro bactericidal activities in rainbow trout (Grayson et al., 1987) and leukocyte migration in yellowtail (Ninomiya et al., 1995). Other herbals such as *Adathoda vasica*, *Emblica officinalis*, and *Cynodon dactylon* have immunostimulant properties that improved the immune system and reduced microbial infection in the goldfish *Carassius auratus* (Minomol, 2005). *W. somnifera*, *O. sanctum*, and *Myristica fragrans* herbs also enhanced the immunity and bactericidal activity against *Vibrio harveyi* challenge in juvenile grouper (Sivaram et al., 2004). In parallel to this, observed that extracts of five different herbal medicinal plants, *Aegle marmelos*, *C. dactylon*, *T. cordifolia*, *E. alba*, and *P. kurooa*, enhanced the biochemical, hematological, and immunological parameters in shrimp.

15.2.8.3 Herbal medicines as antibacterial agents

Antibacterial agents are groups of substances that inhibit the pathogenic activity of pathogenic bacteria by killing or reducing their metabolic activity (Kenawy, 2001). For example, methanolic extracts of herbs like *Andrographis paniculata*, *Solanum trilobatum*, and *Psoralea corylifolia* have antibacterial properties that, when enriched with *Artemia*, enhance survivability and specific growth rate (SGR) and decrease bacterial load in the *P. monodon* culture system (Citarasu et al., 2003). The butanolic extract of *W. somnifera* through *Artemia* enrichment protects shrimps from *V. parahaemolyticus* and *V. damsel* (Praseetha, 2005). Likewise, 13 bacterial pathogens have been prevented using antimicrobial Chinese herbal extracts like *Impatiens biflora*, *Stellaria aquatica*, *Artemisia vulgaris*, *Oenothera biennis*, and *Lonicera japonica* (Shangliang et al., 1990). Scientific research was started in 1992 on the antibacterial activity of guava (*Psidium guajava*) against pathogenic microbes for shrimp (Citarasu, 2010). For example, guava is reported to perform better than oxytetracycline in the prevention of luminous bacteria from black tiger shrimp (*P. monodon*) (Direkbusarakom, 2004).

15.2.8.4 Herbal biomedicines as antistress agents

Most of the herbal biomedicines contain phytochemicals, which are redox-active molecules that can inhibit oxygen anion production and scavenge free radicals (Chakraborty & Hancz, 2011; Citarasu, 2010). The antioxidant effect of herbs has been shown to be kin to that of metal ion chelators, superoxide dismutase, and xanthine oxidase inhibitors (Citarasu, 2010). Shahsavani et al. (2011) observed the reduction of lead deposition in the liver, kidney, brain, bone, and blood of common carp when fed with organosulfide allicin, a garlic clove extract. Also, a Thai herbal biomedic, *Thunbergia laurifolia* leaves, reduces the lead toxicity in *O. niloticus* (Palipoch et al., 2011). Another biomedical herb, bercoli, reduces the toxicity of carcinogenic pollutant benzo(a)-pyrene and phenol in Nile tilapia (Davila et al., 2010; Villa-Cruz et al., 2009), and its inclusion in fish feed has also contributed to the activity of tolbutamide and chlorzoxazone, which may be considered useful for detoxification (Davila et al., 2010). Wu et al. (2007) also reported the herbal extracts *Portulaca oleracea*, *A. membranaceus*, *A. paniculate*, and *Flavescent sophora* have shown antistress properties and induce immunological parameters such as SOD, serum lysozyme activity, nitric oxide synthase (NOS), and levels of total serum protein, globulin, and albumin in *C. carpio*. It is also reported that the herbal mixture is very effective in increasing stress recovery in contrast to a single herbal supplement in diets (Chakraborty & Hancz, 2011).

15.2.9 **Bacteriophage therapy**

Bacterial diseases in aquaculture are commonly treated with antibiotics, as they are easily accessible to farmers. However, indiscriminate application of antibiotics in culture systems has led to antibiotic-resistant strains of pathogens, which makes antibiotic treatment ineffective. Also, there are other negative impacts of antibiotics, which make researchers search for alternatives. As an alternative, bacteriophage therapy in aquaculture has gained much interest, as it is an eco-friendly solution to treating bacterial infections (Akmal et al., 2020; Culot, Grosset, & Gautier, 2019; Jun et al., 2016; Kim et al., 2010). The USFDA also has approved many phage-based products, like List-Shield™ from Intralytix, for treating *Listeria monocytogenes* in food/meat products and they are also recognized as safe (GRAS) and that has been another cause for the reemergence of bacteriophage research (FDA, 2013). In aquaculture, many kinds of research have been conducted for isolation and characterization of phages specific to bacterial pathogens and have tested their potential for phage therapy. Findings on phage therapy to control many bacterial diseases in fish, such as bacterial cold water disease, columnaris disease, vibriosis, edwardsiellosis, enteric septicemia of catfish, *Aeromonas hydrophila* infection, furunculosis, *Pseudomonas plecoglossicida* infection, *Pseudomonas aeruginosa* infection, streptococcosis, and lactococcosis, have been reviewed by Choudhury et al. (2017). In the case of shellfish bacterial diseases, phage therapy was extensively studied for two bacterial diseases, namely acute hepatopancreatic necrosis disease (AHPND) caused by *V. parahaemolyticus* infection and luminous vibriosis caused by *Vibrio harveyi*.

Based on this research, the strategies for the development of phage therapy in aquaculture are varied. However, the first important step in phage therapy is the determination of the disease-causing agent. Other steps for the development of phage therapy in aquaculture are isolation of potential lytic phage; large-scale culture of isolated phages; phenotypic and genotypic characterization of the isolated phages; evaluating the potency of an isolated phage for therapy; therapeutic efficacy study of an isolated phage against experimental infection in laboratory conditions followed by natural infections in field conditions; whole sequencing of a potential phage and screening for virulence genes or toxic genes presence; and finally acquiring regulatory approval and awareness for the public and scientific community for commercial uses.

Regarding the mode of application for phage therapy in aquaculture, intramuscular or intraperitoneal administration, immersion, oral administration through feed, anal intubation, and direct release of phages in culture systems have been reported (Huang & Nitin, 2019; Silva et al., 2016). Phage cocktails or different combinations with phages are a new area of interest for phage application for therapy (Choudhury et al., 2012, 2019; Duarte et al., 2018; Stalin & Srinivasan, 2017). Estimation of appropriate doses for therapy is another crucial step for developing efficient phage therapy and the existing literature describes diverse doses for phage therapy, both experimentally and in field conditions (Choudhury et al., 2017).

Bacteriophage therapy has many advantages as well as demerits. Some advantages are (1) phages are very specific to their host, meaning it kills or lyses only specific target pathogens and won't affect other bacteria (Barrow & Soothill, 1997); (2) as phages are self-replicating and self-limiting, they won't affect the environment once the purpose of phage therapy is over (Inal, 2003); (3) as phages are isolated from the environment where host/pathogenic bacteria are, the application of phages in that same environment won't harm the phages.

In addition to the advantages, some limitations of phage therapy are (1) lysogenic conversion of phage is the key limitation for phage therapy; (2) development of phage resistance by bacteria is another major drawback of bacteriophage application (Levin & Bull, 2004); (3) it is difficult to get regulatory approval for phage therapy; and (4) optimum doses and modes of application of phages are still not standardized for aquaculture.

15.3 Conclusion

Aquaculture is one of the fastest-growing food-producing industries, supplying a protein-rich diet to the ever-increasing human population. To further advance the sector's growth, disease prevention, especially bacterial diseases, is a prime concern. The use of antibiotics in fishes has received several opinions and negative concerns, for which alternative measures are being directed for control of diseases. The effectiveness and fruitfulness of novel fish vaccines have been evident. Although various vaccines have emerged to save aquatic animals from various bacterial diseases, there is still a need for vaccines against newly rising diseases in the aquaculture industry. Additionally, other biological products, such as probiotics and paraprobiotics, are effective and safe measures to combat disease. The application of prebiotics and synbiotics adapted from other animal models are quite successful in fish too. India, being a rich plant biodiversity hub, also has great potential for applying herbal biomedicines in fish disease management. Biotechnological advances in the forms of AMPs and use of novel peptides are also in line for commercial application. The future research focus should direct towards a molecular-level understanding of the disease and pathogen interaction, along with the long-term sustainability goals.

References

Adams, A., Aoki, T., Berthe, C. J., Grisez, L., & Karunasagar, I. (2008). *Recent technological advancements on aquatic animal health and their contributions toward reducing disease risks-a review. Diseases in asian aquaculture VI* (pp. 71–88). Colombo, Sri Lanka: Fish Health Section, Asian Fisheries Society.

Adams, C. A. (2010). The probiotic paradox: Live and dead cells are biological response modifiers. *Nutrition Research Reviews*, *23*, 37–46.

Ahmdifar, E., Akrami, R., Ghelichi, A., & Zarejabad, A. M. (2011). Effects of different dietary prebiotic insulin levels on blood serum enzymes, hematologic, and biochemical parameters of great sturgeon (*Huso huso*) juveniles. *Comparative Clinical Pathology*, *20*(5), 447–451.

Ai, O., Xu, H., Mai, K., Xu, W., Wang, J., & Zhang, W. (2011). Effects of dietary supplementation of *Bacillus subtilis* and fructooligosaccharide on growth performance, survival, non-specific immune response and disease resistance of juvenile large yellow croaker, *Larimichthys crocea*. *Aquaculture (Amsterdam, Netherlands)*, *317*, 155–161.

Akinbowale, O. L., Peng, H., & Barton, M. D. (2006). Antimicrobial resistance in bacteria isolated from aquaculture sources in Australia. *Journal of Applied Microbiology*, *100*(5), 1103–1113.

Akmal, M., Rahimi-Midani, A., Hafeez-ur-Rehman, M., Hussain, A., & Choi, T. J. (2020). Isolation, characterization, and application of a bacteriophage infecting the fish pathogen *Aeromonas hydrophila*. *Pathogens*, *9*(3), 215.

Akrami, R., Iri, Y., Rostami, H. K., & Mansour, M. R. (2013). Effect of dietary supplementation of fructooligosaccharide (FOS) on growth performance, survival, lactobacillus bacterial population and hemato-immunological parameters of stellate sturgeon (*Acipenser stellatus*) juvenile. *Fish & Shellfish Immunology*, *35*(4), 1235–1239.

Anderson, D. P. (1992). Immunostimulants, adjuvants, and vaccine carriers in fish: Applications to aquaculture. *Annual Review of Fish Diseases*, *2*, 281–307.

Avendaño-Herrera, R., Núñez, S., Barja, J. L., & Toranzo, A. E. (2008). Evolution of drug resistance and minimum inhibitory concentration to enrofloxacin in *Tenacibaculum maritimum* strains isolated in fish farms. *Aquaculture International*, *16*(1), 1–11.

Bahi, A., Guardiola, F. A., Messina, C. M., Mahdhi, A., Cerezuela, R., Santulli, A., & Esteban, M. A. (2017). Effects of dietary administration of fenugreek seeds, alone or in combination with probiotics, on growth performance parameters, humoral immune response and gene expression of gilthead seabream (*Sparus aurata* L.). *Fish & Shellfish Immunology*, *60*, 50–58.

Barrow, P. A., & Soothill, J. S. (1997). Bacteriophage therapy and prophylaxis: Rediscovery and renewed assessment of potential. *Trends in Microbiology*, *5*(7), 268–271.

Benites, V., Gilharry, R., Gernat, A. G., & Murillo, J. G. (2008). Effect of dietary mannan oligosaccharide from Bio-Mos or SAF-mannan on live performance of broiler chickens. *Journal of Applied Poultry Research*, *17*(4), 471–475.

Biswas, G., Korenaga, H., Nagamine, R., Takayama, H., Kawahara, S., Takeda, S., Kikuchi, Y., Dashnyam, B., Kono, T., & Sakai, M. (2013). Cytokine responses in the Japanese pufferfish (*Takifugu rubripes*) head kidney cells induced with heat–killed probiotics isolated from the Mongolian dairy products. *Fish & Shellfish Immunology*, *34*(5), 1170–1177.

Broughton, E. I., & Walker, D. G. (2009). Prevalence of antibiotic-resistant Salmonella in fish in Guangdong, China. *Foodborne Pathogens and Disease*, *6*(4), 519–521.

Burr, G., Hume, M., Neill, W. H., & Gatlin, D. M., III (2008). Effects of prebiotics on nutrient digestibility of a soybean-meal-based diet by red drum *Sciaenops ocellatus* (Linnaeus). *Aquaculture Research*, *39*(15), 1680–1686.

Cabello, F. C., Godfrey, H. P., Buschmann, A. H., & Dölz, H. J. (2016). Aquaculture as yet another environmental gateway to the development and globalisation of antimicrobial resistance. *The Lancet Infectious Diseases*, *16*(7), e127–e133.

Campagna, S., Saint, N., Molle, G., & Aumelas, A. (2007). Structure and mechanism of action of the antimicrobial peptide piscidin. *Biochemistry*, *46*, 1771–1778.

Cañada-Cañada, F., Muñoz de la Peña, A., & Espinosa-Mansilla, A. (2009). Analysis of antibiotics in fish samples. *Analytical and Bioanalytical Chemistry*, *395*, 987–1008. Available from https://doi.org/10.1007/s00216-009-2872-z.

Causey, J. L., Slain, J. L., Tangled, B. C., & Meyer, P. D. (1998). Stimulation of human immune system by inulin in vitro. In *Proceedings of Danone Conference on Probiotics and Immunity*. Germany, Bonn.

Cerezuela, R., Cuesta, A., Meseguer, J., & Esteban, M. (2012). Effects of dietary inulin and heat-inactivated *Bacillus subtilis* on gilthead seabream (*Sparus aurata* L.) innate immune parameters. *Beneficial Microbes*, *3*(1), 77–81.

Cerezuela, R., Guardiola, F. A., Meseguer, J., & Esteban, M. Á. (2012). Increases in immune parameters by inulin and *Bacillus subtilis* dietary administration to gilthead seabream (*Sparus aurata* L.) did not correlate with disease resistance to *Photobacterium damselae*. *Fish & Shellfish Immunology*, *32*(6), 1032–1040.

Chakraborty, S. B., & Hancz, C. (2011). Application of phytochemicals as immunostimulant, antipathogenic and antistress agents in finfish culture. *Reviews in Aquaculture*, *3*(3), 103–119.

Chitra, S. (1995). *Effect of feeding supplemented stresstol bioencapsulated Artemia franciscana on growth and stress tolerance-in* Penaeus indicus *postlarvae*. Doctoral dissertation, M. Phil Dissertation, MS University, Tirunelveli.

Choudhury, T. G., & Kamilya, D. (2019). Paraprobiotics: An aquaculture perspective. *Reviews in Aquaculture*, *11*(4), 1258–1270.

Choudhury, T. G., Maiti, B., Venugopal, M. N., & Karunasagar, I. (2012). Effect of total dissolved solids (TDS) and temperature on bacteriophage therapy against luminous vibriosis in shrimp. *The Israeli Journal of Aquaculture – Bamidgeh*, *64*, 761–768.

Choudhury, T. G., Maiti, B., Venugopal, M. N., & Karunasagar, I. (2019). Influence of some environmental variables and addition of r-lysozyme on efficacy of *Vibrio harveyi* phage for therapy. *Journal of Biosciences*, *44*(1), 8.

Choudhury, T. G., TharabenahalliNagaraju, V., Gita, S., Paria, A., & Parhi, J. (2017). Advances in bacteriophage research for bacterial disease control in aquaculture. *Reviews in Fisheries Science & Aquaculture (Amsterdam, Netherlands)*, *25*(2), 113–125.

Citarasu, T. (2010). Herbal biomedicines: A new opportunity for aquaculture industry. *Aquaculture International*, *18*(3), 403–414.

Citarasu, T., Venkatramalingam, K., Babu, M. M., Sekar, R. R. J., & Petermarian, M. (2003). Influence of the antibacterial herbs, *Solanum trilobatum*, *Andrographis paniculata* and *Psoralea corylifolia* on the survival, growth and bacterial load of *Penaeus monodon* post larvae. *Aquaculture International*, *11*(6), 581–595.

Clark, T., & Cassidy-Hanley, D. (2005). Recombinant subunit vaccines: Potentials and constraints. *Developmental Biology*, *121*, 153–163.

Cole, A. M., Weis, P., & Diamond, G. (1997). Isolation and characterization of pleurocidin, an antimicrobial peptide in the skin secretions of winter flounder. *The Journal of Biological Chemistry*, *272*, 12008–12013.

Colorni, A., Ullal, A., Heinisch, G., & Noga, E. J. (2008). Activity of the antimicrobial polypeptide piscidin 2 against fish ectoparasites. *Journal of Fish Diseases*, *31*, 423–432.

Crawford, L. M. (1985). The impact of residues on animal food products and human health. *Revue Scientifique et Technique de l'OIE (France)*, *4*, 669–723.

Culot, A., Grosset, N., & Gautier, M. (2019). Overcoming the challenges of phage therapy for industrial aquaculture: A review. *Aquaculture (Amsterdam, Netherlands)*, *513*, 734423.

Daniels, C. L., Merrifield, L., Boothroyd, D. P., Davies, S. J., Factor, J. R., & Arnold, K. E. (2010). Effect of dietary Bacillus spp. and mannanoligosaccharides (MOS) on European lobster (*Homarus gammarus* L.) larvae growth performance, gut morphology and gut microbiota. *Aquaculture (Amsterdam, Netherlands)*, *304*, 49–57.

Dar, G. H., Bhat, R. A., Kamili, A. N., Chishti, M. Z., Qadri, H., Dar, R., & Mehmood, M. A. (2020). *Correlation between pollution trends of freshwater bodies and bacterial disease of fish fauna. Fresh water pollution dynamics and remediation* (pp. 51–67). Singapore: Springer.

Dar, G. H., Dar, S. A., Kamili, A. N., Chishti, M. Z., & Ahmad, F. (2016). Detection and characterization of potentially pathogenic *Aeromonas sobria* isolated from fish *Hypophthalmichthys molitrix* (Cypriniformes: Cyprinidae). *Microbial Pathogenesis*, *91*, 136–140.

Das, A., Saha, D., & Pal, J. (2009). Antimicrobial resistance and in vitro gene transfer in bacteria isolated from the ulcers of EUS-affected fish in India. *Letters in Applied Microbiology*, *49*(4), 497–502.

Dash, G., Raman, R. P., PaniPrasad, K., Makesh, M., Pradeep, M. A., & Sen, S. (2015). Evaluation of paraprobiotic applicability of *Lactobacillus plantarum* in improving the immune response and disease protection in giant freshwater prawn, *Macrobrachium rosenbergii* (de Man, 1879). *Fish & Shellfish Immunology*, *43*, 167–174.

Davila, J., Marcial-Martinez, L. M., Viana, M. T., & Vazquez-Duhalt, R. (2010). The effect of broccoli in diet on the cytochrome P450 activities of tilapia fish (*Oreochromis niloticus*) during phenol exposure. *Aquaculture (Amsterdam, Netherlands)*, *304*(1–4), 58–65.

Dawood, M. A., Koshio, S., Ishikawa, M., & Yokoyama, S. (2015). Interaction effects of dietary supplementation of heat-killed *Lactobacillus plantarum* and β-glucan on growth performance, digestibility and immune response of juvenile red sea bream, *Pagrus major*. *Fish & Shellfish Immunology*, *45*(1), 33–42.

deAlmada, C. N., Almada, C. N., Martinez, R. C., & Sant'Ana, A. S. (2016). Paraprobiotics: Evidences on their ability to modify biological responses, inactivation methods and perspectives on their application in foods. *Trends in Food Science & Technology*, *58*, 96–114.

Defoirdt, T., Sorgeloos, P., & Bossier, P. (2011). Alternatives to antibiotics for the control of bacterial disease in aquaculture. *Current Opinion in Microbiology*, *14*(3), 251–258.

Dhar, A., & Allnutt, F. (2011). Challenges and opportunities in developing oral vaccines against viral diseases of fish. *Journal of Marine Science Research Development*, *1*, 2.

Dıaz-Rosales, P., Salinas, I., Rodríguez, A., Cuesta, A., Chabrillón, M., Balebona, M. C., Morinigo, M. A., Esteban, M. A., et al. (2006). Gilthead seabream (*Sparus aurata* L.) innate immune response after dietary administration of heat-inactivated potential probiotics. *Fish & Shellfish Immunology*, *20*, 482–492.

Direkbusarakom, S. (2004). Application of medicinal herbs to aquaculture in Asia. *Walailak Joural of Science and Technology*, *1*(1), 7–14.

Dong, C., & Wang, J. (2013). Immunostimulatory effects of dietary *fructooligosaccharides* on red swamp crayfish, *P. rocambarus* clarkii (Girard). *Aquaculture Research*, *44*(9), 1416–1424.

Duarte, J., Pereira, C., Moreirinha, C., Salvio, R., Lopes, A., Wang, D., & Almeida, A. (2018). New insights on phage efficacy to control *Aeromonas salmonicida* in aquaculture systems: An in vitro preliminary study. *Aquaculture (Amsterdam, Netherlands)*, *495*, 970–982.

Dung, T. T., Haesebrouck, F., Tuan, N. A., Sorgeloos, P., Baele, M., & Decostere, A. (2008). Antimicrobial susceptibility pattern of *Edwardsiella ictaluri* isolates from natural outbreaks of bacillary necrosis of *Pangasianodon hypophthalmus* in Vietnam. *Microbial Drug Resistance*, *14*(4), 311–316.

Edahiro, T., Hamaguchi, M., & Kusuda, R. (1991). Suppressive effect of glycyrrhizia against streptococcal infection promoted by feeding oxidized lipids to yellow tail *Seriola quinqueradiata*. *Suisanzoshoku*, *39*, 21–27.

FDA (Food and Drug Administration). (2013). *Center for food safety and applied nutrition GRAS notice inventory—Agency response letter*. GRAS Notice No. GRN 000435.

Fernández Alarcón, C., Miranda, C. D., Singer, R. S., López, Y., Rojas, R., Bello, H., & González Rocha, G. (2010). Detection of the floR gene in a diversity of florfenicol resistant Gram negative bacilli from freshwater salmon farms in Chile. *Zoonoses and Public Health*, *57*(3), 181–188.

Francis, G., Makkar, H. P., & Becker, K. (2005). Quillaja saponins—A natural growth promoter for fish. *Animal Feed Science and Technology*, *121*(1-2), 147–157.

Fuller, R. (1989). Probiotics in man and animals. *Journal of Applied Bacteriology*, *66*, 365–378.

Fuller, R., & Gibson, G. R. (1998). Probiotics and prebiotics: Microflora management for improved gut health. *Clinical Microbiology and Infection*, *4*(9), 477–480.

Geraylou, Z., Souffreau, C., Rurangwa, E., De Meester, L., Courtin, C. M., Delcour, J. A., & Ollevier, F. (2013). Effects of dietary arabinoxylan-oligosaccharides (AXOS) and endogenous probiotics on the growth performance, non-specific immunity and gut microbiota of juvenile Siberian sturgeon (*Acipenser baerii*). *Fish & Shellfish Immunology*, *35*(3), 766–775.

Geraylou, Z., Souffreau, C., Rurangwa, E., D'Hondt, S., Callewaert, L., Courtin, C. M., & Ollevier, F. (2012). Effects of arabinoxylan-oligosaccharides (AXOS) on juvenile *Siberian sturgeon* (*Acipenser baerii*) performance, immune responses and gastrointestinal microbial community. *Fish & Shellfish Immunology*, *33*(4), 718–724.

Gibson, G. R., & Roberfroid, M. B. (1995). Dietary modulation of the human colonic microbiota: Introducing the concept of prebiotics. *The Journal of Nutrition*, *125*(6), 1401–1412.

Giri, S. S., Sen, S. S., Jun, J. W., Park, S. C., & Sukumaran, V. (2016). Heat-killed whole-cell products of the probiotic *Pseudomonas aeruginosa* VSG2 strain affect in vitro cytokine expression in head kidney macrophages of *Labeo rohita*. *Fish & Shellfish Immunology*, *50*, 310–316.

Grayson, T. H., Williams, R. J., Wrathmell, A. B., Munn, C. B., & Harris, J. E. (1987). Effects of immunopotentiating agents on the immune response of rainbow trout, *Salmo gairdneri* Richardson, to ERM vaccine. *Journal of Fish Biology*, *31*, 195–202.

Grisdale-Helland, B., Helland, S. J., & Gatlin, D. M., III (2008). The effects of dietary supplementation with mannanoligosaccharide, fructooligosaccharide or galactooligosaccharide on the growth and feed utilization of Atlantic salmon (*Salmo salar*). *Aquaculture (Amsterdam, Netherlands)*, *283*(1-4), 163–167.

Gudding, R., & Goodrich, T. (2014). *The history of fish vaccination. Fish vaccination* (pp. 1–11). John Wiley & Sons, Ltd.

Gudding, R., Lillehaug, A., & Evensen, Ø. (1999). Recent developments in fish vaccinology. *Veterinary Immunology and Immunopathology*, *72*, 203–212.

Hai, N. V., & Fotedar, R. (2009). Comparision of the effects of the prebiotics (Bio-MOS® and β-1, 3-glucan) and the customised probiotics (*Pseudomonas synxantha* and *P. aeruginosa*) on the culture of juvenile western king prawns (*Penaeus latisulcatus* Kishinouye, 1896). *Aquaculture (Amsterdam, Netherlands)*, *289*, 310–316.

Hancock, R. E., & Lehrer, R. (1998). Cationic peptides: A new source of antibiotics. *Trends in Biotechnology*, *16*(2), 82–88.

Huang, K., & Nitin, N. (2019). Edible bacteriophage based antimicrobial coating on fish feed for enhanced treatment of bacterial infections in aquaculture industry. *Aquaculture (Amsterdam, Netherlands)*, *502*, 18–25.

Inal, J. M. (2003). Phage therapy: A reappraisal of bacteriophages as antibiotics. *Archivum Immunologiaeet Therapia Experimentalis*, *51*(4), 237–244.

Irianto, A., & Austin, B. (2003). Probiotics in aquaculture. *Journal of Fish Diseases*, *25*, 633–642.

Ishida, Y., Ahmed, A. M., Mahfouz, N. B., Kimura, T., El-Khodery, S. A., Moawad, A. A., & Shimamoto, T. (2010). Molecular analysis of antimicrobial resistance in Gram-negative bacteria isolated from fish farms in Egypt. *Journal of Veterinary Medical Science*, *72*(6), 727–734.

Jang, S. I., Marsden, M. J., Kim, Y. G., Choi, M. S., & Secombes, C. J. (1995). The effect of glycyrrhizin on rainbow trout, *Oncorhynchus mykiss* (Walbaum), leucocyte responses. *Journal of Fish Diseases*, *18*(4), 307–315.

Jayaprakas, V., & Euphrasia, C. J. (1997). Growth performance of *Labeo rohita* (Ham.) to Livol (IHF-1000), a herbal product. *Proceedings-Indian National Science Academy Part B*, *63*, 21–30.

Jun, J. W., Han, J. E., Tang, K. F., Lightner, D. V., Kim, J., Seo, S. W., & Park, S. C. (2016). Potential application of bacteriophage pVp-1: Agent combating *Vibrio parahaemolyticus* strains associated with acute hepatopancreatic necrosis disease (AHPND) in shrimp. *Aquaculture (Amsterdam, Netherlands)*, *457*, 100–103.

Kamilya, D., Baruah, A., Sangma, T., Chowdhury, S., & Pal, P. (2015). Inactivated probiotic bacteria stimulate cellular immune responses of catla, *Catla catla* (Hamilton) in vitro. *Probiotics and Antimicrobial Proteins*, *7*, 101–106.

Kataria, J., Li, N., Wynn, J. L., & Neu, J. (2009). Probiotic microbes: Do they need to be alive to be beneficial? *Nutrition Reviews*, *67*, 546–550.

Kenawy, E. R. (2001). Biologically active polymers. IV. Synthesis and antimicrobial activity of polymers containing 8-hydroxyquinoline moiety. *Journal of Applied Polymer Science*, *82*(6), 1364–1374.

Kim, J. H., Gomez, D. K., Nakai, T., & Park, S. C. (2010). Isolation and identification of bacteriophages infecting ayu *Plecoglossus altivelis* altivelis specific *Flavobacterium psychrophilum. Veterinary Microbiology*, *140*, 109–115.

Kim, K. J., Jang, S. I., Marsden, M. J., Secombes, C. J., Choi, M. S., Kim, Y. G., & Chung, H. T. (1998). Effect of glycyrrhizin on rainbow trout *Oncorhynchus mykiss* leukocyte responses. *Journal of the Korean Society for Microbiology*, *33*(3), 263–271.

Klebaniuk, R., Matras, J., Patkowski, K., & Picta, M. (2008). Effectiveness of Bio-MOS in sheep nutrition. *Annals of Animal Science*, *8*(4), 369–380.

Levin, B. R., & Bull, J. J. (2004). Population and evolutionary dynamics of phage therapy. *Nature Reviews. Microbiology*, *2*(2), 166–173.

Levine, M. M., & Sztein, M. B. (2004). Vaccine development strategies for improving immunization: The role of modern immunology. *Nature Immunology*, *5*(5), 460–464.

Li, J., Tan, B., & Mai, K. (2009). Dietary probiotic *Bacillus* OJ and isomaltooligosaccharides influence the intestine microbial populations, immune responses and resistance to white spot syndrome virus in shrimp (*Litopenaeus vannamei*). *Aquaculture (Amsterdam, Netherlands)*, *291*, 35–40.

Li, P., & Gatlin, D. M., III (2004). Dietary brewers yeast and the prebiotic Grobiotic™ AE influence growth performance, immune responses and resistance of hybrid striped bass (*Morone chrysops* × *M. saxatilis*) to *Streptococcus iniae* infection. *Aquaculture (Amsterdam, Netherlands)*, *231*(1-4), 445–456.

Linehan, S. A., Martínez-Pomares, L., & Gordon, S. (2000). Macrophage lectins in host defence. *Microbes and Infection*, *2*(3), 279–288.

Liu, P. C., Lee, K. K., & Chen, S. N. (1997). Susceptibility of different isolates of *Vibrio harveyi* to antibiotics. *Microbios*, *91*(368-369), 175.

Logambal, S. M., Venkatalakshmi, S., & Michael, R. D. (2000). Immunostimulatory effect of leaf extract of *Ocimum sanctum* Linn. in *Oreochromis mossambicus* (Peters). *Hydrobiologia*, *430*(1-3), 113–120.

Ma, J., Bruce, T. J., Jones, E. M., & Cain, K. D. (2019). A review of fish vaccine development strategies: Conventional methods and modern biotechnological approaches. *Microorganisms*, *7*(11), 569.

Maheshappa, K. (1993). *Effect of different doses of Livol on growth and body composition of rohu.* Labeo rohita *(Ham.)* (p. 59). MF Sc Thesis, University of Agricultural Science, Bangalore.

Mahious, A. S., Van Loo, J., & Ollevier, F. (2006). *Impact of the prebiotics, inulin and oligofructose on microbial fermentation in the spiral valve of Siberian sturgeon (*Acipenser baerii*).* Italy: World Aquaculture Society.

Mahlapuu, M., Håkansson, J., Ringstad, L., & Björn, C. (2016). Antimicrobial peptides: An emerging category of therapeutic agents. Frontiers in Cellular and Infection. *Microbiology (Reading, England)*, *6*, 194, –194.

Manning, T. S., & Gibson, G. R. (2004). Prebiotics. *Best Practice & Research. Clinical Gastroenterology*, *18* (2), 287–298.

McIntosh, D., Cunningham, M., Ji, B., Fekete, F. A., Parry, E. M., Clark, S. E., & Beattie, M. (2008). Transferable, multiple antibiotic and mercury resistance in Atlantic Canadian isolates of *Aeromonas salmonicida* subsp. salmonicida is associated with carriage of an IncA/C plasmid similar to the *Salmonella enterica* plasmid pSN254. *Journal of Antimicrobial Chemotherapy*, *61*(6), 1221–1228.

Mehrabi, Z., Firouzbakhsh, F., & Jafarpour, A. (2011). Effects of dietary supplementation of synbiotic on growth performance, serum biochemical parameters and carcass composition in rainbow trout (*Oncorhynchus mykiss*) fingerlings. *Journal of Animal Physioogy and Animal Nutrition*, *96*(3), 474–481. Available from https://doi.org/10.1111/j.1439-0396.2011.01167.x.

Meyer, D. (2008). Prebiotic dietary fibres and the immune system. *Agro-Food Industry Hi-Tech*, *19*, 12–15.

Minomol, M. (2005). *Culture of Gold fish Carassius auratus using medicinal plants having immunostimulant characteristics.* Doctoral dissertation, M. Phil Dissertation, MS University, India.

Mouriño, J. L. P., Do Nascimento Vieira, F., Jatobá, A. B., Da Silva, B. C., Jesus, G. F. A., Seiffert, W. Q., & Martins, M. L. (2012). Effect of dietary supplementation of inulin and W. cibaria on haemato-immunological parameters of hybrid surubim (*Pseudoplatystoma* sp). *Aquaculture Nutrition*, *18*(1), 73–80.

Mulero, I., Noga, E. J., Meseguer, J., Garcia-Ayala, A., & Mulero, V. (2008). The antimicrobial peptides piscidins are stored in the granules of professional phagocytic granulocytes of fish and are delivered to the bacteria-containing phagosome upon phagocytosis. *Developmental and Comparative Immunology*, *32*, 1531–1538.

Munoz-Atienza, E., Araújo, C., Lluch, N., Hernández, P. E., Herranz, C., Cintas, L. M., & Magadán, S. (2015). Different impact of heat-inactivated and viable lactic acid bacteria of aquatic origin on turbot (*Scophthalmus maximus* L.) head-kidney leucocytes. *Fish & Shellfish Immunology*, *44*(1), 214–223.

Ninomiya, M., Hatta, H., Fujiki, M., Kim, M., Yamamoto, T., & Kusuda, R. (1995). Enhancement of chemotactic activity of yellowtail (*Seriola quinqueradiata*) leucocytes by oral administration of *Quillaja saponin*. *Fish & Shellfish Immunology*, *5*(4), 325–328.

Noga, E. J., & Silphaduang, U. (2003). Piscidins: A novel family of peptide antibiotics from fish. *Drug News & Perspectives*, *16*, 87–92.

Palipoch, S., Jiraungkoorskul, W., Tansatit, T., Preyavichyapugdee, N., Jaikua, W., & Kosai, P. (2011). Effect of *Thunbergia laurifolia* (Linn) leaf extract dietary supplement against lead toxicity in Nile tilapia (*Oreochromis niloticus*). *World Journal of Fish and Marine Sciences*, *3*(1), 1–9.

Pan, X., Wu, T., Song, Z., Tang, H., & Zhao, Z. (2008). Immune responses and enhanced disease resistance in Chinese drum, Miichthys miiuy (Basilewsky), after oral administration of live or dead cells of Clostridium butyrium CB2. *Journal of Fish Diseases*, *31*, 679–686.

Penaflorida, V. D. (1995). Effect of papaya leaf meal on the *Penaeus monodon* post larvae. *Israeli. Journal of Aquaculture Bamidgeh*, *47*(11), 25–33.

Perrin, B. S., Jr., Fu, R., Cotten, M. L., & Pastor, R. W. (2016). Simulations of membrane-disrupting peptides II: AMP piscidin 1 favors surface defects over pores. *Biophysical Journal*, *111*, 1258–1266.

Possemiers, S., Grootaert, C., Vermeiren, J., Gross, G., Marzorati, M., Verstraete, W., & de Wiele, T. V. (2009). The intestinal environment in health and disease-recent insights on the potential of intestinal bacteria to influence human health. *Current Pharmaceutical Design*, *15*(18), 2051–2065.

Praseetha, R. (2005). *Enrichment of brine shrimp Artemia franciscana with commercial probiotics and herbal extracts and their resistance against shrimp pathogen Vibrio sp.* (Vibrio parahaemolyticus *and* Vibrio damsela). M.Phil Dissertation, Manonmaiam Sundaranar University, India.

Ringø, E., Olsen, R. E., Gifstad, T. Ø., Dalmo, R. A., Amlund, H., Hemre, G. I., & Bakke, A. M. (2010). Prebiotics in aquaculture: A review. *Aquaculture Nutrition*, *16*(2), 117–136.

Ringø, E., Sperstad, S., Myklebust, R., Mayhew, T. M., & Olsen, R. E. (2006). The effect of dietary inulin on aerobic bacteria associated with hindgut of Arctic charr (*Salvelinus alpinus* L.). *Aquaculture Research*, *37* (9), 891–897.

Roberfroid, M. (1993). Dietary fiber, inulin, and oligofructose: A review comparing their physiological effects. *Critical Reviews in Food Science & Nutrition (Burbank, Los Angeles County, Calif.)*, *33*(2), 103–148.

Rodriguez-Estrada, U., Satoh, S., Haga, Y., Fushimi, H., & Sweetman, J. (2009). Effects of single and combined supplementation of *Enterococcus faecalis*, mannan oligosaccharide and polyhydroxybutyrate acid on growth performance and immune response of rainbow trout *Oncorhynchus mykiss*. *Aquaculture Science*, *57*(4), 609–617.

Roy, P. E. (2011). *Use of vaccines in finfish aquaculture*. School of Forest Resources and Conservation, Florida Cooperative Extension Service, Institute of Food and Agricultural Sciences, University of Florida.

Salgado-Miranda, C., Loza-Rubio, E., Rojas-Anaya, E., & Garcıa-Espinosa, G. (2013). Viral vaccines for bony fish: Past, present and future. *Expert Review of Vaccines*, *12*, 567–578.

Sang, H. M., & Fotedar, R. (2009). Dietary supplementation of mannan oligosaccharide improves the immune responses and survival of marron, *Cherax tenuimanus* (Smith, 1912) when challenged with different stressors. *Fish & Shellfish Immunology*, *27*(2), 341–348.

Sang, H. M., & Fotedar, R. (2010). Effects of mannan oligosaccharide dietary supplementation on performances of the tropical spiny lobsters juvenile (*Panulirus ornatus*, Fabricius 1798). *Fish & Shellfish Immunology*, *28*(3), 483–489.

Sang, H. M., Fotedar, R., & Filer, K. (2011). Effects of dietary mannan oligosaccharide on the survival, growth, immunity and digestive enzyme activity of freshwater crayfish, *Cherax destructor* Clark (1936). *Aquaculture Nutrition*, *17*(2), e629–e635.

Seifert, S., & Watzl, B. (2007). Inulin and oligofructose: Review of experimental data on immune modulation. *The Journal of Nutrition*, *137*(11), 2563S–2567S.

Sghir, A., Chow, J. M., & Mackie, R. I. (1998). Continuous culture selection of bifidobacteria and lactobacilli from human faecal samples using fructooligosaccharide as selective substrate. *Journal of Applied Microbiology*, *85*(4), 769–777.

Shabir, U., Ali, S., Magray, A. R., Ganai, B. A., Firdous, P., Hassan, T., & Nazir, R. (2018). Fish antimicrobial peptides (AMP's) as essential and promising molecular therapeutic agents: A review. *Microbial Pathogenesis*, *114*, 50–56.

Shadakshari, G.S. (1993). *Effect of bioboost forte, Livol and Amchemin AQ on growth and body composition of common carp,* Cyprinus carpio *(Linn.)* (p. 155). MF Sc, Thesis, University of Agriculture Sciences, Bangalore.

Shahsavani, D., Baghshani, H., & Alishahi, E. (2011). Efficacy of allicin in decreasing lead (Pb) accumulation in selected tissues of lead-exposed common carp (*Cyprinus carpio*). *Biological Trace Element Research*, *142*(3), 572–580.

Shangliang, T., Hetrick, F. M., Roberson, B. S., & Baya, A. (1990). The antibacterial and antiviral activity of herbal extracts for fish pathogens. *Journal of Ocean University of Qingdao*, *2*, 53–60.

Shao, Z. J. (2001). Aquaculture pharmaceuticals and biologicals: Current perspectives and future possibilities. *Advanced Drug Delivery Reviews*, *50*, 229–243.

Silva, Y. J., Moreirinha, C., Pereira, C., Costa, L., Rocha, R. J., Cunha, Â., Gomes, N. C., Calado, R., & Almeida, A. (2016). Biological control of *Aeromonas salmonicida* infection in juvenile *Senegalese sole* (*Solea senegalensis*) with Phage AS-A. *Aquaculture (Amsterdam, Netherlands)*, *450*, 225–233.

Singh, S. K., Tiwari, V. K., Chadha, N. K., Prakash, C., Munilkumar, S., Das, P., Mandal, S. C., & Chanu, T. I. (2013). Effect of *Bacillus circulans* and fructooligosaccharide supplementationon growth and haemato-immunological function of *Labeo rohita* (Hamilton, 1822) fingerlings exposed to sub-lethal nitrite stress. *Israeli Journal of Aquaculture-Bamigedh*, *66*, 1–10.

Singh, S. K., Tiwari, V. K., Chadha, N. K., Munilkumar, S., Prakash, C., & Pawar, N. A. (2019). Effect of dietary synbiotic supplementation on growth, immune and physiological status of *Labeo rohita* juveniles exposed to low pH stress. *Fish and Shellfish Immunology*, *91*, 358–368. Available from https://doi.org/10.1016/j.fsi.2019.05.023.

Singh, S. T., Kamilya, D., Kheti, B., Bordoloi, B., & Parhi, J. (2017). Paraprobiotic preparation from *Bacillus amyloliquefaciens* FPTB16 modulates immune response and immune relevant gene expression in *Catla catla* (Hamilton, 1822). *Fish & Shellfish Immunology*, *66*, 35–42.

Sink, T. D., & Lochmann, R. T. (2008). Preliminary observations of mortality reduction in stressed, *Flavobacterium columnare*–challenged Golden Shiners after treatment with a dairy-yeast prebiotic. *North American Journal of Aquaculture*, *70*(2), 192–194.

Sivaram, V., Babu, M. M., Immanuel, G., Murugadass, S., Citarasu, T., & Marian, M. P. (2004). Growth and immune response of juvenile greasy groupers (*Epinephelus tauvina*) fed with herbal antibacterial active principle supplemented diets against *Vibrio harveyi* infections. *Aquaculture (Amsterdam, Netherlands), 237* (1-4), 9–20.

Smith, P., & Christofilogiannis, P. (2007). Application of normalised resistance interpretation to the detection of multiple low-level resistance in strains of *Vibrio anguillarum* obtained from Greek fish farms. *Aquaculture (Amsterdam, Netherlands), 272*(1-4), 223–230.

Sohn, K. S., Kim, M. K., Kim, J. D., & Han, I. K. (2000). The role of immunostimulants in monogastric animal and fish – Review. *Asian-Australasian Journal of Animal Sciences, 13*(8), 1178–1187.

Soleimani, N., Hoseinifar, S. H., Merrifield, D. L., Barati, M., & Abadi, Z. H. (2012). Dietary supplementation of fructooligosaccharide (FOS) improves the innate immune response, stress resistance, digestive enzyme activities and growth performance of Caspian roach (*Rutilus rutilus*) fry. *Fish & Shellfish Immunology, 32* (2), 316–321.

Song, S. K., Beck, B. R., Kim, D., Park, J., Kim, J., Kim, H. D., & Ringø, E. (2014). Prebiotics as immunostimulants in aquaculture: A review. *Fish & Shellfish Immunology, 40*(1), 40–48.

Stalin, N., & Srinivasan, P. (2017). Efficacy of potential phage cocktails against *Vibrio harveyi* and closely related Vibrio species isolated from shrimp aquaculture environment in the south east coast of India. *Veterinary Microbiology, 207*, 83–96.

Staykov, Y., Spring, P., Denev, S., & Sweetman, J. (2007). Effect of a mannan oligosaccharide on the growth performance and immune status of rainbow trout (*Oncorhynchus mykiss*). *Aquaculture International, 15*(2), 153–161.

Tafalla, C., Bøgwald, J., & Dalmo, R. A. (2013). Adjuvants and immunostimulants in fish vaccines: Current knowledge and future perspectives. *Fish & Shellfish Immunology, 35*(6), 1740–1750.

Tafalla, C., Bøgwald, J., Dalmo, R. A., Munang'andu, H. M., & Evensen, Ø. (2014). *Adjuvants in fish vaccines. Fish vaccination* (pp. 68–84). London: John Wiley & Sons, Ltd.

Taoka, Y., Maeda, H., Jo, J. Y., Kim, S. M., Park, S. I., Yoshikawa, T., & Sakata, T. (2006). Use of live and dead probiotic cells in tilapia *Oreochromis niloticus*. *Fisheries Science, 72*, 755–766.

Taverniti, V., & Guglielmetti, S. (2011). The immunomodulatory properties of pro- biotic microorganisms beyond their viability (ghost probiotics: Proposal of paraprobiotic concept). *Genes & Nutrition, 6*, 261–274.

Teitelbaum, J. E., & Walker, W. A. (2002). Nutritional impact of pre-and probiotics as protective gastrointestinal organisms. *Annual Review of Nutrition, 22*(1), 107–138.

Tendencia, E. A., & de la Peña, L. D. (2001). Antibiotic resistance of bacteria from shrimp ponds. *Aquaculture (Amsterdam, Netherlands), 195*(3-4), 193–204.

Teo, J. W., Suwanto, A., & Poh, C. L. (2000). Novel β-lactamase genes from two environmental isolates of *Vibrio harveyi*. *Antimicrobial Agents and Chemotherapy, 44*(5), 1309–1314.

Toranzo, A., Romalde, J., Magarinos, B., & Barja, J. (2009). Present and future of aquaculture vaccines against fish bacterial diseases. *Option Mediterraneennes, 86*, 155–176.

Torrecillas, S., Makol, A., Benítez-Santana, T., Caballero, M. J., Montero, D., Sweetman, J., & Izquierdo, M. (2011). Reduced gut bacterial translocation in European sea bass (*Dicentrarchus labrax*) fed mannan oligosaccharides (MOS). *Fish & Shellfish Immunology, 30*(2), 674–681.

Torrecillas, S., Montero, D., & Izquierdo, M. (2014). Improved health and growth of fish fed mannan oligosaccharides: Potential mode of action. *Fish & Shellfish Immunology, 36*(2), 525–544.

Unnikrishnan, G. (1995). *Effect of Livol on growth, food utilization and body composition of the Indian major carp,* Catla catla *(Ham.).* (p. 34). Doctoral dissertation, M.Sc. Dissertation, University of Kerala, India.

Villa-Cruz, V., Davila, J., Viana, M. T., & Vazquez-Duhalt, R. (2009). Effect of broccoli (*Brassica oleracea*) and its phytochemical sulforaphane in balanced diets on the detoxification enzymes levels of tilapia (*Oreochromis niloticus*) exposed to a carcinogenic and mutagenic pollutant. *Chemosphere, 74*(9), 1145–1151.

Vogt, L., Ramasamy, U., Meyer, D., Pullens, G., Venema, K., Faas, M. M., & de Vos, P. (2013). Immune modulation by different types of β2→ 1-fructans is toll-like receptor dependent. *PLoS One*, *8*(7), e68367.

Wada, T., Arima, T., & Nagashima, H. (1987). Natural killer activity in patients with chronic hepatitis treated with OK432, interferon, adenine arabinoside and glycyrrhizin. *Gastroenterologia Japonica*, *22*(3), 312–321.

Wang, J. L., Meng, X. L., Lu, R. H., Wu, C., Luo, Y. T., Yan, X., & Nie, G. X. (2015). Effects of *Rehmannia glutinosa* on growth performance, immunological parameters and disease resistance to *Aeromonas hydrophila* in common carp (*Cyprinus carpio* L.). *Aquaculture (Amsterdam, Netherlands)*, *435*, 293–300.

Wang, W. S., & Wang, D. H. (1997). Enhancement of the resistance of tilapia and grass carp to experimental *Aeromonas hydrophila* and *Edwardsiella tarda* infections by several polysaccharides. *Comparative Immunology, Microbiology and Infectious Diseases*, *20*(3), 261–270.

Wimley, W. C., & White, S. H. (1996). Experimentally determined hydrophobicity scale for proteins at membrane interfaces. *Nature Structural Biology*, *3*, 842–848.

Wu, G., Yuan, C., Shen, M., Tang, J., Gong, Y., Li, D., et al. (2007). Immunological and biochemical parameters in carp (*Cyprinus carpio*) after Qompsell feed ingredients for longterm administration. *Aquaculture Research*, *38*, 246–255.

Xu, B., Wang, Y., Li, J., & Lin, Q. (2009). Effect of prebiotic xylooligosaccharides on growth performances and digestive enzyme activities of allogynogenetic crucian carp (*Carassius auratus* gibelio). *Fish Physiology and Biochemistry*, *35*(3), 351–357.

Yang, Y., Iji, P. A., Kocher, A., Mikkelsen, L. L., & Choct, M. (2008). Effects of mannanoligosaccharide and fructooligosaccharide on the response of broilers to pathogenic *Escherichia coli* challenge. *British Poultry Science*, *49*(5), 550–559.

Ye, J. D., Wang, K., Li, F. D., & Sun, Y. Z. (2011). Single or combined effects of fructo-and mannan oligosaccharide supplements and *Bacillus clausii* on the growth, feed utilization, body composition, digestive enzyme activity, innate immune response and lipid metabolism of the Japanese flounder *Paralichthys olivaceus*. *Aquaculture Nutrition*, *17*(4), e902–e911.

Zasloff, M. (2002). Antimicrobial peptides of multi-cellular organisms. *Nature*, *415*, 389–395.

Zhang, C. N., Li, X. F., Xu, W. N., Jiang, G. Z., Lu, K. L., Wang, L. N., & Liu, W. B. (2013). Combined effects of dietary fructooligosaccharide and *Bacillus licheniformis* on innate immunity, antioxidant capability and disease resistance of triangular bream (*Megalobrama terminalis*). *Fish & Shellfish Immunology*, *35* (5), 1380–1386.

Zhang, Q., Ma, H., Mai, K., Zhang, W., Liufu, Z., & Xu, W. (2010). Interaction of dietary *Bacillus subtilis* and fructooligosaccharide on the growth performance, non-specific immunity of sea cucumber, *Apostichopus japonicus*. *Fish & Shellfish Immunology*, *29*(2), 204–211.

Zhang, Q., Tan, B., Mai, K., Zhang, W., Ma, H., Ai, Q., Wang, X., & Liufu, Z. (2011). Dietary administration of Bacillus (*B. licheniformis* and *B. subtilis*) and isomaltooligosaccharide influences the intestinal microflora, immunological parameters and resistance against *Vibrio alginolyticus* in shrimp, *Penaeus japonicus* (Decapoda: Penaeidae). *Aquaculture Research*, *42*, 943–952.

Zhang, Y. H., Yoshida, T., Isobe, K., Rahman, S. M., Nagase, F., Ding, L., & Nakashima, I. (1990). Modulation by glycyrrhizin of the cell-surface expression of H-2 class I antigens on murine tumour cell lines and normal cell populations. *Immunology*, *70*(3), 405.

Zhao, Y., Mai, K., Xu, W., Zhang, W., Ai, Q., Zhang, Y., Wang, X., & Liufu, Z. (2011). Influence of dietary probiotic *Bacillus* TC22 and Prebiotic fructooligosaccharide on growth, immune responses and disease resistance against *Vibrio splendidus* infection in sea cucumber *Apostichopus japonicas*. *Journal of Ocean University China*, *10*, 293–300.

Zheng, X., Duan, Y., Dong, H., & Zhang, J. (2017). Effects of dietary *Lactobacillus plantarum* in different treatments on growth performance and immune gene expression of white shrimp *Litopenaeus vannamei* under normal condition and stress of acute low salinity. *Fish & Shellfish Immunology*, *62*, 195–201.

Zheng, X., Duan, Y., Dong, H., & Zhang, J. (2020). The effect of *Lactobacillus plantarum* administration on the intestinal microbiota of whiteleg shrimp *Penaeus vannamei*. *Aquaculture (Amsterdam, Netherlands)*, 735331.

CHAPTER

Ulceration in fish: causes, diagnosis and prevention

16

Ishtiyaq Ahmad and Imtiaz Ahmed
Fish Nutrition Research Laboratory, Department of Zoology, University of Kashmir, Hazratbal, Srinagar, India

16.1 Introduction

The world population drastically increased during the 20th century. Although agricultural production was keeping pace with world population growth, the rate of population growth has now outstripped the agricultural production significantly. Thus there is a nutrient imbalance of essential proteins, fats, and calories in the human diet. According to the United Nation's statistics, over 900 million people in the world are malnourished. One-fourth of this number is children under the age of four or five, who suffer from acute or chronic protein energy malnutrition (PEM). Fish have become an increasingly important source of protein over the last decades in most developing nations (FAO, 2000), and the importance of fish as a source of healthy food for humans is undeniable. Fish is recognized as an extremely nutritious food due to its high protein, micronutrients, and lipids, which are rich in omega-3 unsaturated fatty acids (Ahmed & Ahmad, 2020). Compared to terrestrial sources of animal protein like beef and chicken, fish has a greater satiation effect (Uhe et al., 1992). Moreover, it serves as an affordable protein source with a wide choice of varieties and species available throughout the world (Sarma et al., 2013). Around 60% of people in developing countries mainly depend on fish for at least 30% of their animal protein supplies, while almost 80% of the population in most developed countries obtain less than 20% of their animal protein from fish. However, in some Asian countries, the intake is comparatively much higher (Delgado et al., 2003). Fish make an essential contribution to the survival and health of a significant portion of the world population, with a large share of the diet coming from fish and fishery products. However, several fish health-related issues restrain continued industry growth (Table 16.1).

Outbreaks of diseases are one of the most important concerns in the fisheries sector. It is caused by several factors related to interactions of the pathogen(s), the fish (host), and the environment (Udomkusonsri & Noga, 2005; Wedemeyer, 1996). There are numerous fish diseases in the world including bacterial (Aeromonas, Pseudomonas, Flavobacterium, Acinetobacter infections; Pękala-Safińska, 2018); fungal (Saprolegniasis, Epizootic ulcerative syndrome (EUS), Ichthyophoriasis etc. Choudhury et al., 2014); viral (Aquabirnavirus, Retrovirus, Betanodovirus, Paramyxovirus etc. Crane & Hyatt, 2011) and parasitic diseases (Whirling disease, other Myxobolus infections etc. Ahmad & Kaur, 2017, 2018; Ahmad et al., 2021; Ahmed et al., 2019; Bartholomew & Reno, 2002; Kaur & Ahmad, 2016, 2017). Out of these, EUS is considered as the most dreadful one. Fish microorganisms are generally non-hazardous as long as the fish's immune systems are not

Bacterial Fish Diseases. DOI: https://doi.org/10.1016/B978-0-323-85624-9.00018-X

Table 16.1 Ulceration in different fish species.

Disease	Locality	Species	Cause	Reference
Ulcerative disease	USA England	*Carassius auratus*	*Aeromonas salmonicida* & *A. hydrophila*	Elliot and Shotts (1980)
Ulcerative disease	Philippines	*Ophinocephalus striatus* *Clarias batrachus* etc.	*A. hydrophila*	Llobrera and Gacutan (1987)
Epizootic ulcerative syndrome	Indonesia Malaysia Singapore	*Clarias batrachus* *C. macrocephalus* *Cirrhinus mrigala*	*A. hydrophila*	Boonyaratpalin (1989)
Skin ulcer	Finland	*Platichthys flessus*	*A. salmonicida*	Wiklund and Bylund (1993)
Epizootic ulcerative syndrome	Srilanka	*Puntius sarana* *Wallago attu*	*A. hydrophila*	Pathiratne et al. (1994)
Skin ulcer	Italy	*Carcharhinus plumbeus*	*Vibrio carchariae*	Bertone et al. (1996)
Snout ulcer disease	Japan Korea	*Takifugu rubrips* *T. niphobles* etc.	Virus	Miyadai et al. (2001)
Acute ulceration	USA	*Morone saxatilis* female × *M. chrysops* male	Acute stress	Udomkusonsri et al. (2004)
Skin ulceration	USA	*M. saxatilis* male × *M. chrysops* female	*Saprolegnia*	Udomkusonsri & Noga (2005)
Skin ulceration	China	*Apostichopus japonicus*	Vibrio Photobacterium	Deng et al. (2009)
Winter ulcers	Norway	*Salmo salar*	*Moritella viscosa*	Olsen et al. (2011)
Skin ulceration	China	*Apostichopus japonicus*	Not given	Li et al. (2012)
Skin & Intestinal ulceration	Norway	*Salmo salar*	Skin- *Tenaubaculum* & *Arcobacter* Intestine- *Aliivibrio* & *Alcaligenes*	Karlsen et al. (2017)
Skin ulceration	Thailand	*Scrotum barcoo*	Ranavirus infection	Kayansamruaj et al. (2017)
Skin ulcer	Norway	*Salmo salar*	*Tenacibaculum* spp.	Olsen et al. (2017)
Skin ulceration	Belgium	*Limanda limanda*	*Vibrio tapetis* *A. salmonicida*	Vercauteren et al. (2017)
Skin ulceration	China	*Cynoglossus semilacvis*	*Photobacterium damselae* subsp. *damselae*	Shao et al. (2017)

compromised by a stressor. However, diseases persistently occur due to changes in their environment, which ultimately cause stress on fish (Pickering, 1998; Plumb, 1999; Wedemeyer, 1996). Aquaculture is continuously affected by several stressors, including environmental changes (e.g., water temperature, pH, dissolved oxygen concentration) and different management practices Dar, Dar et al., 2016; Dar, Kamili et al., 2016; Dar et al., 2020; Udomkusonsri & Noga, 2005;

Wedemeyer, 1996). All of these components can influence fish homeostasis, making them prone to a wide variety of pathogens due to which fish population is declining day by day. It has largely been accepted that environmental fluctuations as well as stressors are the main causes leading to outbreaks of infectious disease; however, the specific mechanisms responsible for these outbreaks are not fully understood. Therefore, in order to minimise the chance of diseases in freshwater as well as in cultured system, there is a dire need to have a proper knowledge about the real causes of dreadful diseases including EUS.

Among all the diseases, ulcerations are the most commonly visible abnormalities that are seen in different fish species. The precise etiologies have been determined only through continuous monitoring body injuries, bacterial and parasitic infestations and wounds, which are considered to be a main cause of the disease. During the last two decades, EUS has been reported from various parts of the world in fish affecting some part of local populations , in which EUS dominantly consist of superficial or penetrating ulcers. Among the varied disease outbreaks in fishes, ulcerative disease frequently leads to mass mortality, although some evidence of recovery in the form of scarring can also be noted in few cases. If one species is severely affected in any given locality, another species may exhibit lesser ulceration in the same environment. In this review, a general search has been made for data related to ulcerative diseases in fish fauna and its infectious agents, such as environmental stressors, viral, bacterial, or fungal, which may act as prime pathogens or secondary invaders. All these factors have been pointed out in this chapter in order to compile and authenticate data related to ulcerative syndrome, which will provide insight into the early detection of this disease, both at hatcheries and on farms.

16.2 Ulceration and its causes in different fish species

Ulceration is a common clinical sign of disease affecting different fish species, in both freshwater and marine environments. The development of ulceration has been chiefly associated with different ecological stressors, viruses, bacteria, and fungi (Sindermann & Rosenfield, 1954). A virological etiology pointed out the ulcus syndrome in cod, *Gadus morhua* (Jensen & Larsen, 1982) and ulcerations in several freshwater fish species in Southeast Asia (Dar, Dar et al., 2016; Dar, Kamili et al., 2016; Frerichs et al., 1986). Different bacterial species such as *Vibrio anguillarum*, *Aeromonas hydrophila* (Ullrich, 1992), atypical *Aeromonas salmonicida* (Wiklund & Dalsgaard, 1998), and Mycobacteria (Duijn, 1981) have been isolated from ulcerated fish. Moreover, an Aphanomyces fungus was also reported to be associated with ulcerations in different fish species (Callinan et al., 1995).

Among ulceration, skin ulcers are considered as the most significant biomarkers of contaminated or stressful environments (Bernet et al., 1999; Noga, 2000; Sindermann, 1990). In fish species, skin plays a vital role in response to ecological stressors as skin covers the whole body and remains metabolically active all the time (Iger et al., 1994; Whitear, 1986), thus epidermal damage may occur through direct contact with toxicants or indirectly due to various physiological changes (Udomkusonsri & Noga, 2005). It has been reported that the skin of hybrid striped bass and other fishes serves as a primary defense mechanism that contains two types of defense mechanisms i.e. specific and nonspecific including immunoglobulins, lectins, complement-system, lysozymes, and antimicrobial polypeptides, by which fish gets protection from the attack of such diseases (Robinette & Noga, 2001; Yano, 1996). When fishes lose their protective skin barrier, opportunistic bacteria like *Aeromonas* spp. and *Pseudomonas* spp., parasites, viruses, and water molds such as

Saprolegnia spp. can easily invade the skin and cause infection (Plumb, 1997). Infections resulting in loss of protective barrier (i.e skin) are a result of osmotic stress. Thus, acute ulceration response (AUR) might expose fishes to many opportunistic skin pathogens, which can lead to morbidity and mortality of fish facing the stress conditions.

An outbreak of the fish disease epizootic ulcerative syndrome (EUS) was first reported in 1988 and since then it has spread to most of the states in India, where around 30 genera of fishes were found to be affected in all types of water areas. It is suspected that environmental factors, including physicochemical parameters of water and anthropogenic factors such as pesticides, fertilizers, and heavy metals discharged from industries, play an important role in the outbreak of EUS. In addition, heavy metals and pesticides, other agro-chemicals, and viral, bacterial, or fungal infections, along with stress, are the other key factors that might be responsible for EUS in India.

Apart from India, several studies have been carried out by different workers in the past throughout the world, focused mainly on ulcerative diseases and their causes. Some of them are discussed in this chapter. Wiklund and Bylund (1993) reported skin ulceration in flounder, *Platichthys flesus*, from the northern Baltic Sea in Finland. Their findings showed that on the basis of correlation, males (10.7%) were found to be more affected than females (3.1%); however, larger fishes showed higher disease prevalence than smaller ones. Through bacterial examination of diseased fish, they successfully isolated a bacterium, known as cytochrome oxidase-negative, atypical *A. salmonicida* (Wiklund & Bylund, 1991). They also mentioned in their study that isolation of *A. salmonicida* was reported in the past, but it was not clear whether it came from a typical or an atypical strain or it was isolated from a diseased or a healthy specimen (Vethaak, 1992). Their results also indicated that prevalence of skin ulceration in flounder may be attributed to various factors like sex, fish size, and sampling seasons.

Lunder et al. (1995) investigated winter ulcer pathologically as well as through bacteriology in Atlantic salmon, *Salmo salar*. They also tried to establish the mechanism of transmissibility of a disease through the experimental setup. Through histological analysis, the chronic stages observed in affected fish were characterized by a severe redness of the dermis and also an involvement of intestinal muscle tissue. Their results revealed that winter ulcer is mainly caused by an injection with a *Vibrio*-like bacterium, and can be transmitted through cohabitation and injection.

Thampuran et al. (1995) conducted an exhaustive study on the EUS in different fish species such as *Channa striatus*, *Wallago attu*, and *Puntius* and reported that isolation of *Aeromonas hydrophilla* from the lesions and internal organs was dominant, which remained the main cause of this disease. Noga et al. (1998) observed that due to acute stress exposure, skin ulceration occurred in striped bass, *Morone saxatilis*, and hybrid bass, *M. saxatilis* female × *M. chrysops* male. Based on histological study, their findings revealed epithelial erosion and ulceration in affected fish species. They finally concluded that in some cases, ulceration can develop without any prior trama or before any infection.

Wiklund et al. (1999) confirmed the presence of *A. salmonicida* strains from different ulcerated fish species such as dab, *Limanda limanda*, flounder, *P. flesus*, and turbot, *Scophthalmus maximus*, from the Baltic Sea. Their findings revealed that the development of ulceration disease occurred due to the presence of *A. salmonicida*. They also reported that the preceding isolated strains were indistinguishable from those previously reported from this fish species, and presumed that ulcer disease is considered to be associated with a biotype of *A. salmonicida*. Austin et al. (2003) recovered an unusual gram-negative bacterium viz. Ultramicrobacterium ND5, *Aquaspirillum arcticum*,

Zoogloea sp., and *Janthinobacterium lividum* in ulcerated rainbow trout, *Oncorhynchus mykiss*. However, they didn't confirm whether these bacteria are responsible for ulceration syndrome or not; hence, they suggest that more studies are needed in order to determine the exact causative agent for this particular disease.

Udomkusonsri et al. (2004) made a detailed survey of the pathogenesis of AUR in hybrid striped bass. Their findings indicated that acute confinement may rapidly cause significant damage to the epidermal and ocular epithelium; however, AUR might be a primary cause of morbidity in acutely stressed fish. Altinok et al. (2006) observed the causes of ulcerative fish disease in rainbow trout. After molecular characterization based on the 16s rRNA sequence, the causative bacteria isolated from diseased fish were identified as *Pseudomonas putida*, while histological sections revealed epithelia and epithelial necrosis. They found that approximately 35% of rainbow trout died due to this disease.

Coyne et al. (2006) studied the winter ulcer disease in Atlantic salmon and observed that the root cause of this disease was the presence of *Moritella viscose*. Uniruzzaman and Chowdhury (2008) conducted a detailed survey on ulcer diseases in cultured fish of Bangladesh. They found that the majority of the fish species were visibly affected by skin ulceration disease; however, *Cirrhinus cirrhosus* and *Barbodes gonionotus* were highly affected fish species. Nsonga et al. (2010) while observing the real cause of ulceration in fish reported that this disease actually occurred during heavy rains that resulted in excess flooding, due to which pH concentration was decreased on the plains of acidic soils. They observed high mortality rates in several fish species including *Barbus* and *Clariaus* spp.

Olsen et al. (2011) studied ulcer disease in coldwater fish species and reported that winter ulcers in Atlantic salmon, *S. salar*, were mainly caused by *Moritella viscose* bacteria. Due to nonavailability and widespread use of vaccines against *M. viscosa*, winter ulcers remain a significant threat in Norway as well as in Iceland and Scotland (Benediktsdóttir et al., 1998; Bruno et al., 1998; Lunder et al., 1995). Histopathological investigation of their study revealed the presence of long, slender rods in skin ulcers. They also noticed two *Tenacibaculum* spp. in ulcerative affected fish through 16s rDNA that were also positively related to fish ulceration.

Boys et al. (2012) found an EUS in bony herring, *Nematalosa erebi*, golden perch, *Macquaria ambigua*, Murray cod, *Maccullochella peelii*, and spangled perch, *Leiopotherapon unicolor* from Murray-Darling River System, Australia. Through the polymerase chain reaction (PCR), they isolated bacteria called *Aphanomyces invadans*, which were the main cause of ulcerative syndrome, but did not found ulcer in carp, *Cyprinus carpio*. Contrary to this, Abid & Al-Hamdani, 2016 investigated ulcerated skin carp and their causative agents from Sulaimani province in Iraq. They found that the main causative agent of the skin ulceration was bacterial infection and reported nine species that were isolated from skin lesions, out of which *Pseudomonas* spp. and *Citrobacter freundii* were found dominant as compared to other bacterial species. Their results also revealed that, among infected specimens, one fish had many more complications, which may be attributed to the presence of bacterial infection along with a protozoan parasite, that is, *Chilodonella cyprinii*. An outbreak of EUS was reported in *Seranochromis robustus* from Darwendale Dam, Zimbabwe by Gomo et al. (2016). They found that fish of all ages were affected, and histopathological observations revealed that mycotic granulomas were present in all the affected fish. They finally concluded that the main cause of ulceration in *S. robustus* was due to the presence of *A. invadans* bacteria, resulting in high mortalities.

Kayansamrua et al. (2017) studied ulcerative disease in barcoo grunter, *Scrotum barcoo*, associated with ranavirus from Thailand farms. They reported for the first time largemouth bass virus (LMBV) identical to largemouth bass ulceration syndrome (LBUSV), which has been previously reported in farmed *Micropterus salmoides*, that produced ulcerative skin lesions. They found the affected fish possessed extensive hemorrhage as well as ulceration on skin and muscle. In addition to this, they observed necrosis and ulceration in other organs such as gills, spleen, and kidney through microscopic observation, which clearly indicated that LMBV was mainly responsible for the ulceration; severe mortalities occurred in farmed barcoo grunter in Thailand.

Vercauteren et al. (2017) first reported an isolation of *Vibrio tapetis* and an atypical strain of *A. salmonicida* from common dab, *Limanda limanda*, affected with skin ulcerations in the North Sea. Based on their pathological results, several ulcerative stages were observed, with epidermal and dermal tissue losses, infiltration in inflammatory tissues, and deterioration of the myofibers contiguous to the ulceration, although in varying degrees. Upon bacteriological investigation, pure cultures of *V. tapetis* were retrieved in high numbers from five fish and of *A. salmonicida* in one fish. Both agents may play a significant role in the development of the skin lesions observed in their study. Recently, Vercauteren et al. (2019) confirmed *V. tapetis* as a causative agent of ulcerative skin lesions, due to which severe mortality, with various clinical indications and sizes of the budding lesions, occurred. Zhang et al. (2019) reported severe skin ulceration in a net cage cultured around Daqing Island, China in black rockfish, *Sebastes schlegeli*. Through morphological, physiological, and biochemical observations, as well as gyrB gene sequencing, they found that the main cause of the ulceration disease was the presence of *Photobacterium damselae* subsp. *Damselae* (Fig. 16.1).

16.3 Diagnostic methods

Based on the existing literature, it is well known that ulceration in fishes causes severe mortality and is one of the major diseases responsible for fish mortality. Hence, in order to combat this dreadful disease, scientists throughout the world are trying to determine diagnostic methods for early detection. Different diagnostic methods are used in the fisheries sector, and some prominent methods are discussed in the following paragraphs.

Noga and Udomkusonsri (2002) worked on the fluorescein method, which is basically a nontoxic fluorescent dye. They observed that this method can easily and rapidly detect the presence of ulceration in fish, as they successfully tested it for diagnosis of rainbow trout, *O. mykiss*; channel catfish, *Ictalarus punctatus*; goldfish, *Carracius auratus*; and hybrid striped bass, *M. saxatalis* male × *M. chrysops* female. Ibrahem and Mesalhy (2010) also adapted the same dye for the detection of skin ulcers that are invisible in two species, that is, Nile tilapia, *Oreochromis niloticus*, and the scaleless African sharptooth catfish, *Clarias gariepinus*. Additionaly, the gross pathological and different clinical signs also serve vitally in the diagnosis of ulcerative disease in fish, as these methods mainly help in the detection of mycotic granulomas in histological sections. Different scientists have adopted this method to diagnose ulcerative disease in fish, such as goldfish ulcer disease in Atlantic salmon, *S. salar*, from *A. salmonicida* (Carson & Handlinger, 1988); and the presence of *Pseudomonas putida* infection in affected ulcerative rainbow trout (Altinok et al., 2006). Usually, an ulcerative fish shows several symptoms like loss of appetite, weight loss, and lethargy. Initially,

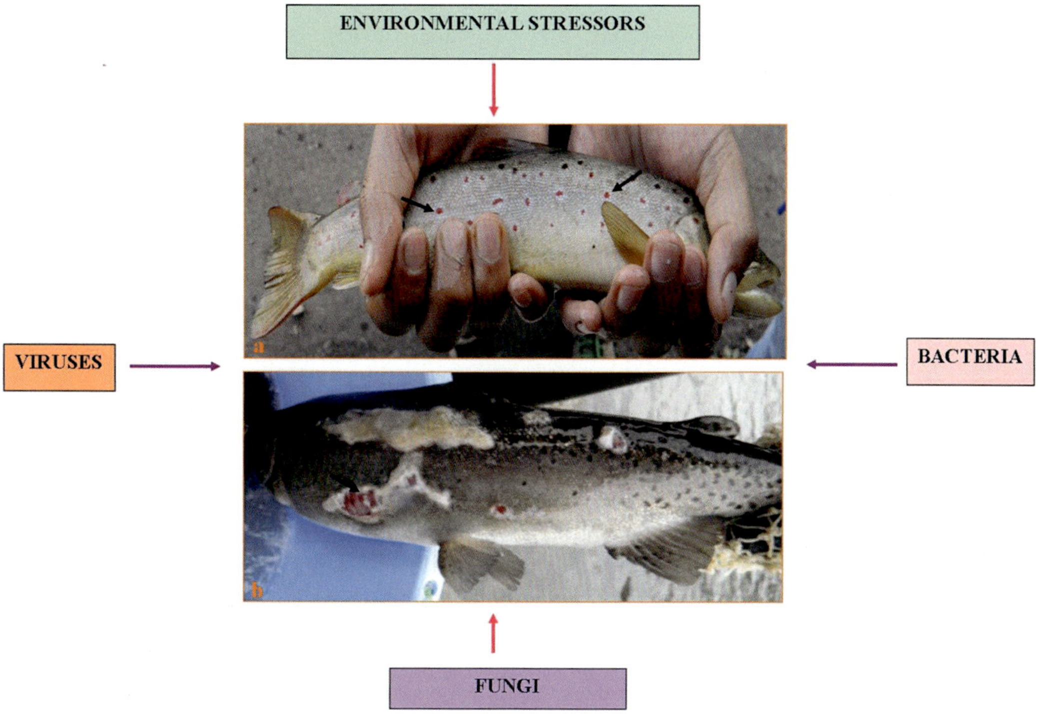

FIGURE 16.1

Cultured rainbow trout infected with ulcerative syndrome.

the affected fish shows Pin-head sized red spots appear usually on different locations of the body, which turn into small dermal ulcers during the intermediate stage. Then the skin lesions are characterized by the presence of acute hemorrhagic and necrotic ulcers on the surface of fish body. The presence of epitheloid granulomas in different organs, such as muscle, liver, kidney, etc. are the most striking diiagnostic feature of EUS (Lilley et al., 1998; Vishwanath et al., 1997). However, in some cases, infection by the presence of oomycete has been observed without the formation of granulomas in the body (Oidtmann et al., 2008; Sosa et al., 2007). Identification can also be done through a culture method, in which the oomycete is isolated and subsequently morphological and macroscopic assessment is carried out (Baruah et al., 2012; Lilley et al., 1998).

Apart from this, molecular approaches are continuously being applied for accurate diagnosis and to get valuable results. Molecular methods include electrophoretic and western blot analysis (Lilley et al., 1997), pyrolysis mass spectrometry (Lilley et al., 2001), and monoclonal antibody based detection (Miles et al., 2003; Naik et al., 2008). PCR has achieved great success in identifying fish diseases. Deng et al. (2009) isolated different bacteria from affected ulcerative sea cucumber, *Apostichopus japonicus*, by the PCR method using 16s rRNA genes. Li et al. (2012) used miRNA libraries to diagnose skin ulceration syndrome in *A. japonicas* and the sequence analysis revealed 40 conserved miRNAs found in both libraries. Another study revealed that, based on the characterization of 16s rRNA gene sequencing, six different bacteria were isolated from ulcerative *Cynoglossus semilaevis* in the Bohai Bay area of China. Finally, one more molecular method described is fluorescent peptide nucleic acid in situ hybridization (FISH). This technique was successfully tested in 50 Atlantic menhaden with typical severe ulceration disease, which provided explicit visual substantiation that hyphae in the affected ulcerative fishes were totally due to the presence of *A. invadans* (Vandersea et al., 2006).

16.4 Preventive measures

So far control of ulceration in fish has not been reported, but several measures have been established to reduce risks or minimize mortality. Liming in water is prone to have a positive effect on water parameters and is effective in reducing the infection as well as mortality (Lilley et al., 1998). It is also observed that lime applied at the rate of 100 kg/ha in an ulcerative disease prone area in the postmonsoon period just prior to the outbreak of disease has either prevented the occurrence of the disease or, if an outbreak occurred, the intensity was mild (Das, 1992). In addition to this, potassium permanganate ($KMNO_4$) dip is also recommended in order to rule out any microbial infection. Bleaching powder was also reported to be useful in healing initial lesions of ulcerative affected fishes (Sanjeevaghosh, 1992). Antibiotics, whether erythromycin, oxytetracycline, or terramycin at 60 − 100 mg/kg of feed for 7 days, cured ulcerative affected fishes (Das & Das, 1993). In addition, Robertson et al. (2005) reported that prevention of ulceration in goldfish occurred by vaccination. They concluded in their results that the use of a polyvalent vaccine by immersion and an oral booster can protect goldfish against ulcerative disease.

16.5 Conclusion

It is concluded that ulcerative disease has shown severe mortality in fish fauna and the main factors responsible for this are environmental stressors, bacteria, fungi, and viruses. Different diagnostic methods like PCR techniques, fluorescein dye method, and histopathological analysis should be taken into account to determine the actual causes of this disease. Moreover, ulceration syndrome in fish cannot be controlled fully, but some preventive measures have been reported, such as liming, bleaching powder, potassium permanganate, and some antibiotics by which reductions in ulceration were noted at initial stages. In addition, a few vaccines are also reported to reduce an ulceration effect in fish. This study will provide insight to farmers as well as government hatcheries into this disease and help them to determine early methods of detection as well.

Acknowledgements

The authors are highly grateful to the Head, Department of Zoology, University of Kashmir, Hazratbal, Srinagar, India, for providing the necessary facailites including laboratory and communications.

Funding

This study was not supported by any funding agency.

Competing interest

The authors declare that they have no competing interests.

Conflict of interest

The authors declare that there is no conflict of interest to disclose.

Data availability statement

The data used in this study is included within the manuscript.

References

Abid, O. M., & Al-Hamdani, A. (2016). Study of the causative agents of ulcerated skin lesions of carp fish ponds at sulaimani province. Basrah Journal of Veterinary Research. Iraq: College of Veterinary Medicine University of Basrah.

Ahmad, I., Ahmed, I., Reshi, Q. M., Jan, K., Gupta, A., Dar, S. A., Molnár, K., Kaur, H., Malla, B. A., & Andrabi, S. M. (2021). Morphological, histopathological and molecular characterization of *Myxobolus szekelyianus* n. sp. (Cnidaria: Myxosporea: Myxobolidae) causing acute gill disease in *Schizothorax esocinus* (Heckel, 1838) from River Jhelum of Kashmir Himalayan region, India. *Aquaculture Research, 00*, 1–13.

Ahmad, I., & Kaur, H. (2017). Redescription and histopathology of two species of myxozoans infecting gills of fingerlings of Indian major carps. *Journal of FisheriesSciences.com, 11*, 01–10.

Ahmad, I., & Kaur, H. (2018). Prevalence, site and tissue preference of myxozoan parasites infecting gills of cultured fingerlings of Indian major carps in District Fatehgarh Sahib, Punjab (India). *Journal of Parasitic Diseases, 42*, 559–569.

Ahmed, I., & Ahmad, I. (2020). Effect of dietary protein levels on growth performance, hematological profile and biochemical composition of fingerlings rainbow trout, *Oncorhynchus mykiss* reared in Indian himalayan region. *Aquaculture Reports, 16*, 100268.

Ahmed, I., Ahmad, I., Dar, S. A., Awas, M., Kaur, H., Ganie, B. A., & Shah, B. A. (2019). *Myxobolus himalayaensis* sp. nov. (Cnidaria: Myxozoa) parasiting *Schizothorax richardsonii* (Cyprinidae: Schizothoracinae) from River Poonch in North West Himalaya, India. *Aquaculture Reports, 14*, 100192.

Altinok, I., Kayis, S., & Capkin, E. (2006). *Pseudomonas putida* infection in rainbow trout. *Aquaculture (Amsterdam, Netherlands), 261*(3), 850–855. Available from https://doi.org/10.1016/j.aquaculture.2006.09.009.

Austin, D. A., Jordan, E. M., & Austin, B. (2003). Recovery of an unusual Gram-negative bacterium from ulcerated rainbow trout, *Oncorhynchus mykiss* (Walbaum), in Scotland. *Journal of Fish Diseases, 26*, 247–249.

Bartholomew, J. L., & Reno, P. W. (2002). The history and dissemination of Whirling disease. *American Fisheries Society Symposium*, *26*, 1–22.

Baruah, A., Saha, R. K., & Kamilya, D. (2012). Inter-species transmission of epizootic ulcerative syndrome (EUS) pathogen, Aphanomyces invadans and associated physiological responses. *Israeli Journal of Aquaculture – Bamidgeh*, *64*, 9.

Benediktsdóttir, E., Helgason, S., & Sigurjónsdóttir, H. (1998). Vibrio spp. isolated from salmonids with shallow skin lesions and reared at low temperature. *Journal of Fish Diseases*, *21*, 19–28.

Bernet, D., Schmidt, H., Meier, W., Burkhardt-Holm, P., & Wahli, T. (1999). Histopathology in fish: Proposal for a protocol to assess aquatic pollution. *Journal of Fish Diseases*, *22*, 25–34. Available from https://doi.org/10.1046/j.1365-2761.1999.00134.x.

Bertone, S., Gili, C., Moizo, A., & Calegari, L. (1996). *Vibrio carchariae* associated with a chronic skin ulcer on a shark, *Carcharhinus plumbeus* (Nardo). *Journal of Fish Diseases*, *19*, 429–434.

Boys, C. A., Rowland, S. J., Gabor, M., Gabor, L., Marsh, I. B., Hum, S., & Callinan, R. B. (2012). Emergence of epizootic ulcerative syndrome in native fish of the Murray–Darling river system, Australia: Hosts, distribution and possible vectors. *PLoS One*, *7*, e35568.

Bruno, D. W., Griffiths, J., Petrie, J., & Hastings, T. S. (1998). *Vibrio viscosus* in farmed Atlantic salmon *Salmo salar* in Scotland: Field and experimental observations. *Diseases of Aquatic Organisms*, *34*, 161–166.

Callinan, R. B., Paclibare, J. O., Bondad-Reantaso, M. G., Chin, J. C., & Gogolewski, R. P. (1995). Aphanomyces species associated with epizootic ulcerative syndrome (EUS) in the Philippines and red spot disease (RSD) in Australia: Preliminary comparative studies. *Diseases of Aquatic Organisms*, *21*, 233–238.

Carson, J., & Handlinger, J. (1988). Virulence of the aetiological agent of goldfish ulcer disease in Atlantic salmon, *Salmo salar* L. *Journal of Fish Diseases*, *11*, 471–479.

Choudhury, T. G., Singh, S. K., Parhi, J., Barman, D., & Das, B. S. (2014). Common fungal diseases of fish: A Review. *Environment & Ecology*, *32*, 450–456.

Coyne, R., Smith, P., Dalsgaard, I., Nilsen, H., Kongshaug, H., Bergh, O., & Samuelsen, O. (2006). Winter ulcer disease of post-smolt Atlantic salmon: An unsuitable case for treatment? *Aquaculture (Amsterdam, Netherlands)*, *253*, 171–178.

Dar, G. H., Bhat, R. A., Kamili, A. N., Chishti, M. Z., Qadri, H., Dar, R., & Mehmood, M. A. (2020). *Correlation between pollution trends of freshwater bodies and bacterial disease of fish fauna. Fresh water pollution dynamics and remediation* (pp. 51–67). Singapore: Springer.

Dar, G. H., Dar, S. A., Kamili, A. N., Chishti, M. Z., & Ahmad, F. (2016). Detection and characterization of potentially pathogenic *Aeromonas sobria* isolated from fish *Hypophthalmichthys molitrix* (Cypriniformes: Cyprinidae). *Microbial Pathogenesis*, *91*, 136–140.

Dar, G. H., Kamili, A. N., Chishti, M. Z., Dar, S. A., Tantry, T. A., & Ahmad, F. (2016). Characterization of *Aeromonas sobria* isolated from fish rohu (*Labeo rohita*) collected from polluted pond. *Journal of Bacteriology & Parasitology*, *7*(3), 1–5. Available from https://doi.org/10.4172/2155-9597.1000273.

Das, M. K. (1992). *Status of research on the fish disease epizootic ulcerative syndrome in India*. Consultation on Epizootic Ulcerative Syndrome (EUS) vis a vis the Environment and People. International Collective in Support of Fish Workers. 25–26 May, 1992, Trivandrum, India.

Das, M. K., & Das, R. K. (1993). A review of the fish disease epizootic ulcerative syndrome in India. *Environment & Ecology.*, *11*, 134–145.

Delgado, C. L., Wada, N., Rosegrant, M. W., Meijer, S., & Ahmed, M. (2003). *Fish to 2020: Supply and demand in changing global markets* (p. 226) Washington, DC: International Food Policy Research Institute, WorldFish Center.

Deng, H., He, C. B., Zhou, Z. C., Liu, C., Tan, K. F., Wang, N. B., Jiang, B., Gao, X. G., & Liu, W. D. (2009). Isolation and pathogenicity of pathogens from skin ulceration disease and viscera ejection syndrome of the sea cucumber *Apostichopus japonicus*. *Aquaculture (Amsterdam, Netherlands)*, *287*(1), 18–27.

van Duijn, C. (1981). Tuberculosis in fishes. *Journal of Small Animal Practice*, *22*, 391–411.

Elliot, D. G., & Shotts, E. B. (1980). Aetiology of an ulcerative disease in goldfish *Carassius auratus* (L): microbial examination of diseased fish from seven location. *Journal of Fish Diseases*, *3*, 133.

FAO. (2000). *The State of Food Insecurity in the World 2000*. Rome.

Frerichs, G. N., Millar, S. D., & Roberts, R. J. (1986). Ulcerative rhabdovirus in fish in South-East Asia. *Nature*, *322*, 216.

Gomo, C., Hanyire, T., Makaya, P. V., & Sibanda, S. (2016). Outbreak of epizootic ulcerative Syndrome (EUS) in *Seranochromis robustus* fish species in Darwendale dam, Zimbabwe. *African Journal of Fisheries Science*, *4*, 204–205.

Ibrahem, M. D., & Mesalhy, S. (2010). Determining the safety and suitability of fluorescein dye for characterization of skin ulcerations in cultured Nile tilapia (*Oreochromis niloticus*) and African sharptooth catfish (*Clarias gariepinus*). *Journal of Advanced Research*, *1*, 361–366.

Iger, Y., Lock, R. A. C., Jenner, H. A., & Wendelaar Bonga, S. E. (1994). Cellular responses in the skin of carp (*Cyprinus carpio*) exposed to copper. *Aquatic Toxicology*, *29*(1-2), 49–64. Available from https://doi.org/10.1016/0166-445x(94)90047-7.

Jensen, N. J., & Larsen, J. L. (1982). The ulcus-syndrome in cod (*Gadus morhua*) IV. Transmission experiments with two viruses isolated from cod and *Vibrio anguillarum*. *Nordisk Veterinærmedicin*, *34*, 136–142.

Karlsen, C., Ottem, K. F., Brevik, Ø. J, Davey, M., Sørum, H., & Winther-Larsen, H. C. (2017). The environmental and host-associated bacterial microbiota of Arctic seawater-farmed Atlantic salmon with ulcerative disorders. *Journal of Fish Diseases*. Available from https://doi.org/10.1111/jfd.12632.

Kaur, H., & Ahmad, I. (2016). Morphological description of *Myxobolus markiwi* n. sp. (Cnidaria: Myxosporea: Myxozoa) infecting gills of fingerlings of aquaculture ponds from Punjab. *Species*, *17*, 141–149.

Kaur, H., & Ahmad, I. (2017). A report on two new myxozoan parasites infecting gills of fingerlings of Indian major carps cultured in nursery ponds in Punjab (India). *Journal of Parasitic Diseases*, *41*, 987–996.

Kayansamruaj, P., RangsicholL, A., DonG, H. T., Rodkhum, C., Maita, M., Katagiri, T., & Pirarat, N. (2017). Outbreaks of ulcerative disease associated with ranavirus infection in barcoo grunter, *Scortum barcoo* (McCulloch & Waite). *Journal of Fish Diseases*, *40*(10), 1341–1350. Available from https://doi.org/10.1111/jfd.12606.

Li, C., Feng, W., Qiu, L., Xia, C., Su, X., Jin, C., Zhou, T., Zeng, Y., & Li, T. (2012). Characterization of skin ulceration syndrome associated microRNAs in sea cucumber *Apostichopus japonicus* by deep sequencing. *Fish & Shellfish Immunology*, *33*, 436–441.

Lilley, J. H., Beakes, G. W., & Hetherington, C. S. (2001). Characterization of *Aphanomyces invadans* isolates using pyrolysis mass spectrometry (PyMS). *Mycoses*, *44*, 383–389.

Lilley, J. H., Callinan, R. B., Chinabut, S., Kanchanakhan, S., MacRae, I. H., & Phillips, M. J. (1998). *Epizootic ulcerative syndrome (EUS) technical handbook*. Bangkok: Aquatic Animal Health Research Institute.

Lilley, J. H., Thompson, K. D., & Adams, A. (1997). Characterization of *Aphanomyces invadans* by electrophoretic and Western blot analysis. *Diseases of Aquatic Organisms*, *30*, 187–197.

Llobrera, A. T., & Gacutan, R. Q. (1987). *Aeromonas hydrophila* associated with ulcerative disease epizootic in Laguna de Bay, Philippines. *Aquaculture*, *67*, 273–278.

Lunder, T., Evensen, O., Holstad, G., & Hastein, T. (1995). Winter ulcer in the Atlantic salmon *Salmo salar*. Pathological and bacteriological investigations and transmission experiments. *Diseases of Aquatic Organisms*, *23*, 39–49.

Miles, D. J. C., Thompson, K. D., Lilley, J. H., & Adams, A. (2003). Immunofluorescence of the epizootic ulcerative syndrome pathogen, *Aphanomyces invadans*, using a monoclonal antibody. *Diseases of Aquatic Organisms*, *55*, 77–84.

Miyadai, T., Kitamura, S. I., Uwaoku, H., & Tahara, D. (2001). Experimental infection of several fish species with the causative agent of Kuchijirosho (snout ulcer disease) derived from the tiger puffer *Takifugu rubripes*. *Diseases of Aquatic Organisms*, *47*, 193–199.

Naik, M. G., Rajesh, K. M., Sahoo, A. K., & Shankar, K. M. (2008). Monoclonal antibody-based detection of *Aphanomyces invadans* for surveillance and prediction of epizootic ulcerative syndrome (EUS) outbreak in fish. In M. G. Bondad-Reantaso, C. V. Mohan, M. Crumlish, & R. P. Subasinghe (Eds.), *Diseases in Asian Aquaculture VI, Fish Health Section* (pp. 157–168). Manila, Philippines: Asian Fisheries Society.

Noga, E. J. (2000). Skin ulcers in fish: *Pfiesteria* and other etiologies. *Toxicologic Pathology*, *28*, 807–823.

Noga, E. J., Botts, S., Yang, M. S., & Avtalian, R. (1998). Acute stress causes skin ulceration in Stripped bass and hybrid bass (*Morone*). *Vetrinary Pathology*, *35*, 102–107.

Noga, E. J., & Udomkusonsri, P. (2002). Fluorescein: A rapid, sensitive, nonlethal method for detecting skin ulceration in fish. *Veterinary Pathology*, *39*, 726–731.

Nsonga, A., Mfitilodze, W., & Samui, K. (2010). Epidemiology of Epizootic Ulceration Syndrome on fish of the Zambezi river basin: A case study for Zambia. *Second RUFORUM Biennial Meeting Entebbe, Uganda*.

Oidtmann, B., Steinbauer, P., Geiger, S., & Hoffmann, R. W. (2008). Experimental infection and detection of *Aphanomyces invadans* in European catfish, rainbow trout and European eel. *Disease of Aquatic Organisms*, *82*, 195–207.

Olsen, A., Nilsen, H., Sandlund, N., Mikkelsen, H., Sørum, H., & Colquhoun, D. (2011). *Tenacibaculum* sp. associated with winter ulcers in sea-reared Atlantic salmon *Salmo salar*. *Diseases of Aquatic Organisms*, *94*(3), 189–199. Available from https://doi.org/10.3354/dao02324.

Olsen, A. B., Gulla, S., Steinum, T., Colquhoun, D. J., Nilsen, H. K., & Duchaud, E. (2017). Multilocus sequence analysis reveals extensive genetic variety within *tenacibaculum* spp. associated with ulcers in sea-farmed fish in Norway. *Veterinary Microbiology*, *205*, 39–45. Available from https://doi.org/10.1016/j.vetmic.2017.04.028.

Pathiratne, A., Widanapathirana, G. S., & Chandrakanti, W. H. S. (1994). Association of *Aeromonas hydrophila* with epizootic ulcerative syndrome (EUS) of freshwater fish in Sri Lanka. *Journal of Applied Ichthyology*, *10*, 204–208.

Pękala-Safińska, A. (2018). Contemporary threats of bacterial infections in freshwater fish. *Journal of Veterinary Research*, *62*, 261–267.

Pickering, A. D. (1998). Stress responses of farmed fish. In K. D. Black, & A. D. Pickering (Eds.), *Biology of farmed fish* (pp. 222–255). Boca Raton, FL: CRC Press LLC.

Plumb, J. A. (1997). Infectious diseases of striped bass. In R. M. Harrell (Ed.), *Striped bass and other Morone culture* (pp. 271–313). Amsterdam: Elsevier Science BV.

Plumb, J. A. (1999). *Health maintenance and principal microbial diseases of cultured fished* (p. 328) Ames, Iowa: Iowa State University Press.

Robertson, P. A. W., Austin, D. A., & Austin, B. (2005). Prevention of ulcer disease in goldfish by means of vaccination. *Journal of Aquatic Animal Health*, *17*, 203–209.

Robinette, D. W., & Noga, E. J. (2001). Histone-like protein: A novel method for measuring stress in fish. *Diseases of Aquatic Organisms*, *44*, 97–107.

Sanjeevaghosh, D. (1992). *Socio-economic impact of epizootic ulcerative syndrome on the inland fisherfolk of Kerela*. Consultation on Epizootic ulcerative Syndrome (EUS) vis-a-vis the Environment and People. International Collective in Support of Fish Workers. 25–26 May 1992. Trivandrum, India.

Sarma, D., Akhtar, M. S., Das, P., Das, P., Shahi, N., Yengkokpam, S., Debnath, D., Alexander, C., & Mahanta, P. C. (2013). Nutritional quality in terms of amino acid and fatty acid of five coldwater fish species: Implications to human health. *National Academy Science Letter*, *36*(4), 385–391.

Sindermann, C., & Rosenfield, A. (1954). Diseases of fishes of the western North Atlantic. I. Diseases of the sea herring (Clupea harengus). *Research Bulletin of the Department of Sea and Shore Fisheries*, *18*, 1–23.

Shao, P., Yong, P., Zhou, W., Sun, J., Wang, Y., Tang, Q., Ren, S., Zun, W., Zhao, C., Xu, Y., & Wang, X. (2017). First isolation of *Photobacterium damselae* subsp. damselae from half-smooth tongue sole suffering from skin-ulceration disease. *Aquaculture*, *511*, 734208.

Sindermann, C. J. (1990). *Principal diseases of marine fish and shellfish, . Diseases of marine fish* (Vol. 1). New York: Academic Press.

Sosa, E. R., Landsberg, J. H., Stephenson, C. M., Forstchen, A. B., Vandersea, M. W., & Litaker, R. W. (2007). *Aphanomyces invadans* and ulcerative mycosis in estuarine and freshwater fish in Florida. *Journal of Aquatic Animal Health*, *19*, 14–26. Available from https://doi.org/10.1577/H06-012.1.

Thampuran, N., Surendran, P. K., Mukundan, M. K., & Gopakumar, K. (1995). Bacteriological studies on fish affected by epizootic ulcerative syndrome (EUS) in Kerala India. *Asian Fisheries Science*, *8*, 103–111.

Udomkusonsri, P., & Noga, E. J. (2005). The acute ulceration reponse (AUR): A potentially widespread and serious cause of skin infection in fish. *Aquaculture*, *246*, 63–67.

Udomkusonsri, P., Noga, E. J., Nancy, A., & Monteiro-Riviere, N. A. (2004). Pathogenesis of acute ulceration response (AUR) in hybrid striped bass. *Diseases of Aquatic Organisms*, *61*, 199–213.

Uhe, A. M., Collier, G. R., & O'Dea, K. (1992). A comparison of the effects of beef, chicken and fish protein on satiety and amino acid profiles in lean male subjects. *Journal of Nutrition*, *122*, 467–472.

Ullrich, S. (1992). *Bakterielle Fischkrankheiten in Untereider und Unterelbe und ihre Beeinflussung durch Umweltfaktoren* (115 pp). PhD/Doctoral thesis, Berichte aus dem Institut fur Meereskunde an der Christian-Albrechts-Universitat, Kiel Nr. 223, Germany.

Uniruzzaman, M., & Chowdhury, M. B. R. (2008). Ulcer diseases in cultured fish in Mymensingh and surrounding districts. *The Bangladesh Veterinarian.*, *25*, 40–49.

Vandersea, M. W., Litaker, R., Yonnish, B., Sosa, E., Landsberg, J. H., Pullinger, C., Moon-Butzin, P., Green, J., Morris, J. A., Kator, H., Noga, E. J., & Tester, P. A. (2006). Molecular assays for detecting *Aphanomyces invadans* in ulcerative mycotic fish lesions. *Applied and Environmental Microbiology*, *72*, 1551–1557.

Vercauteren, M., De Swaef, E., Declercq, A., Bosseler, L., Gulla, S., Balboa, S., Romalde, J. L., Devriese, L., Polet, H., Boyen, F., Chiers, K., & Decostere, A. (2017). First isolation of *Vibrio tapetis* and an atypical strain of *Aeromonas salmonicida* from skin ulcerations in common dab (*Limanda limanda*) in the North Sea. *Journal of Fish Diseases*, *41*, 1–7. Available from https://doi.org/10.1111/jfd.12729.

Vercauteren, M., De Swaef, E., Declercq, A. M., Polet, H., Aerts, J., Ampe, A., Romalde, J. L., Haesebrouck, F., Devriese, L., Decostere, A., & Chiers, K. (2019). Scrutinizing the triad of *Vibrio tapetis*, the skin barrier and pigmentation as determining factors in the development of skin ulcerations in wild common dab (*Limanda limanda*). *Veterinary Research*, *50*, 41–53.

Vethaak, A. D. (1992). Diseases of flounder (*Platichthys flesus* L.) in the Dutch Wadden Sea, and their relation to stress factors. *Netherland Journal of Sea Research*, *29*, 257–272.

Vishwanath, T. S., Mohan, C. V., & Shankar, K. M. (1997). Mycotic granulomatosis and seasonality are the consistent features of epizootic ulcerative syndrome (EUS) of fresh and brackish water fishes of Karnataka, India. *Asian Fisheries Science*, *10*, 155–160.

Wedemeyer, G. A. (1996). *Physiology of fish in intensive culture systems* (p. 232) New York: Chapman & Hall.

Whitear, M. (1986). The skin of fishes including cyclostomes – Epidermis. In J. Bereiter-Hahn, A. G. Matoltsy, & K. S. Richards (Eds.), *Biology of the integument* (Vol. 2, pp. 8–38)). New York, NY: Springer-Verlag.

Wiklund, T., & Bylund, G. (1991). A cytochrome oxidase negative bacterium (presumptively an atypical *Aeromonas salmonicida*) isolated from ulcerated flounders (*Platichthys flesus* (L.)) in the northern Baltic Sea. *Bulletin of the European Association of Fish Pathologist*, *11*(2), 74.

Wiklund, T., & Bylund, G. (1993). Skin ulcer disease of flounder *Platichthys flesus* in the northern Baltic Sea. *Diseases of Aquatic Organisms*, *17*, 165–174.

Wiklund, T., & Dalsgaard, I. (1998). Occurrence and significance of atypical *Aeromonas salmonicida* in non-salmonid and salmonid fish species: A review. *Diseases of Aquatic Organisms*, *32*, 49–69.

Wiklund, T., Tabolina, I., & Bezgachina, T. V. (1999). Recovery of atypical *Aeromonas salmonicida* from ulcerated fish from the Baltic Sea. *ICES Journal of Marine Science*, *56*, 175–179.

Yano, T. (1996). The nonspecific immune system: Humoral defense. In W. S. Hoar, D. J. Randall, & A. P. Farrell (Eds.), *The fish immune system: Organism, pathogen, and environment* (pp. 150–157). San Diego, CA: Academic Press.

Zhang, Z., Yu, Y. X., Wang, K., Wang, Y. G., Jiang, Y., Liao, M. J., & Rong, X. J. (2019). First report of skin ulceration caused by *Photobacterium damselae* subsp. *damselae* in net-cage cultured black rockfish (*Sebastes schlegeli*). *Aquaculture (Amsterdam, Netherlands)*, *503*, 1–7.

CHAPTER

Application of probiotic bacteria for the management of fish health in aquaculture

17

Sandip Mondal[1], Debashri Mondal[2], Tamal Mondal[3] and Junaid Malik[4]

[1]School of Pharmaceutical Technology, School of Medical Sciences, ADAMAS University Kolkata, India [2]Department of Zoology, Raiganj University, University Road, College Para, Raiganj, India [3]Department of Botany, Hiralal Mazumdar Memorial College for Women, College Para, Dakshineswar, Kolkata, India [4]Department of Zoology, Government Degree College, Bijbehara, India.

17.1 Introduction

Aquaculture is the cultivation of various aquatic organisms, both marine and freshwater, using different management procedures to enhance production of the cultivated stock. Aquaculture has become a significant economic activity worldwide. Currently, the contributions of aquaculture to food production, industrial raw materials, medical applications, and and using marine species for the ornamental trade have risen exponentially.

With the growth and intensification of aquaculture production, aquatic animals are subject to the frequent occurrence of adverse environments and disease-related complications, as well as environmental deterioration, resulting in major financial casualties. In recent years, antibiotic use has decreased due to the complex environmental problems it causes in the ecosystem, such as the proliferation of antibiotic-resistant bacterial strains. Incorporating antibiotics into the cultured organisms, in addition to removing pathogenic microbes, often destroys bacteria that are advantageous to the same individuals. Also, the accumulation of certain compounds is not healthy for humans, who are the ultimate consumers, as health risks are found to be linked with antibiotics use. Hence the production of nonantibiotic means of treatment is a critical aspect of aquaculture health management, with a view toward the current dietary trends of consuming whole natural foods, in support of a healthier and longer life. Similarly, protection of the environment has changed over time, with consumers giving preference to commercial entities with an economic mission of environmental sustainability, also important in the aquaculture industry.

In this context, for several decades the implementation of probiotics has been explored in fish cultivation primarily for commercial purposes. Browdy (1998) said that probiotics are among the most remarkable innovations that have emerged as biofriendly agents of disease prevention. Probiotics can be applied in the aquatic ecosystem to enhance microbial resistance and improve feeding quality, growth rate, and survival of the cultured species (Browdy, 1998). In addition, they have no unwanted side effects on treated organisms (Dar, Dar et al., 2016; Dar, Kamili et al., 2016; Dar et al., 2020; Huynh et al., 2017; Mohapatra et al., 2013). This review highlights the present

Bacterial Fish Diseases. DOI: https://doi.org/10.1016/B978-0-323-85624-9.00024-5

knowledge concerning the application of probiotics for growth enhancement, fish disease management, and water quality management in aquaculture.

17.2 Probiotics definition

Probiotics are beneficial living bacteria found in some foods or supplements. They can provide numerous health benefits. The term *probiotic* was first used to define compounds synthesized by a microorganism that promotes the development of others (Lilly & Stillwell, 1965). The word probiotic derives from the Greek words *pro bios*, which means "for life" (Chalas et al., 2016). Dr. Elie Metchnikoff (1908), in the year 1905, first explained the effects of certain bacteria among farmers who consumed pathogens in milk and showed that health could be improved by enhancing our intestinal microflora and replacing pathogenic microorganisms with beneficial ones. Up to 1965, Lilly and Stillwell (1965) modified the original word *probiotika* and introduced the term *probiotic*. Afterward, Sperti changed the idea of cell extracts that promote bacteria load. Presently, the term probiotic represents those substances having a different function than that of antibiotics, generated from a microorganism and prolonging the logarithmic development process in other organisms (Sperti, 1971).

According to the World Health Organization (WHO) and the Food and Agriculture Organization (FAO), probiotics are living microorganisms that are used orally with certain significant health benefits for the host (Akhter et al., 2015). Seeing the difference between the aquatic ecosystem and that of terrestrial species, an updated concept was proposed by Merrifield et al. (2010) for probiotics in aquaculture, as a probiotic organism may be a living, dead, or part of a microbial cell delivered through food or the rearing water, supporting the host through enhancing disease tolerance and overall health (Hai, 2015). The probiotics include various forms of bacteria, bacteriophages, microalgae, and yeast, commonly using a water routine or feed substitute in aquaculture. Several commercially available probiotics are currently available with mono- or multistrains.

17.3 Routes of administration

Probiotics may be administered with feed, according to Abidi (Rehna, 2003), and cod liver oil or egg as a binder; the most common commercial preparations comprise either *Saccharomyces cerevisiae* or *Lactobacillus* sp. FAO and WHO mentioned in their strategies that probiotic organisms essentially resist the gastric juices and bile, as they must have the ability to multiply and inhabit the gastrointestinal area. They need to be active, secure, and retain their efficacy for the product's shelf life (Senok et al., 2005). Probiotics may be applied through injection also. Experimentally, the probiotic delivery of *Micrococcus luteus* by intraperitoneal injection to *Oreochromis niloticus* had a fatality rate of just 25%, compared to 90% for *Pseudomonas* using the same path (Khattab et al., 2005). Of all the probiotic delivery routes in aquaculture, applying it via the farming water is the only approach accessible to all kinds of fish. Through the water, probiotic delivery may also affect the fish's safety by enhancing the water quality, as the probiotics alter the bacterial composition of water and sediments (Ashraf, 2000; Venkateswara, 2007).

17.4 Significant factors governing the advantages of probiotic form of administration

The length of therapy (contact time), the dosage, and the probiotics source are significant factors that may influence efficacy (Cabello, 2006; Van Hai, 2015). However, the key factors that can influence the effects of water-based probiotics are described here.

1. *Temperature*: Temperature tends to be an essential aspect in the medicinal operation of probiotics in aquaculture systems when applied by immersion. The probiotic strain *Aeromonas media* A199's maximal antagonistic activity toward saprolegniosis infection was accomplished by a small increase in water temperature (Lategan et al., 2004).
2. *Dose*: The effective dose of the probiotics employed through specific application to the water medium shall be determined based on the weight of the fish handled and the amount of water. The effects on developmental efficiency and immune response in *Perca fluviatilis* L. of commercial probiotics composed of a combination of *B. licheniformis* and *Bacillus pumilus* were tested on larvae. The use of the bacteria probiotics via water did not result in substantial growth, survival, or immunity. Applying probiotics with this operational method may not have been adequate to stimulate their digestive processes (Suzer et al., 2008). Similarly, *Lactobacillus* spp. was immediately applied. Like a probiotic for the gilthead bream tank water (*Sparus aurata* L.), the larvae, development, and gastrointestinal enzyme production did not improve dramatically relative to probiotic administration in live fish (Jahangiri & Esteban, 2018).
3. *Inoculation times*: Inoculation must be repeated during the treatment cycle to ensure a healthy probiotic population at a final measured concentration. In cases when the probiotics are injected into the water, the time will depend on how long it takes for a given probiotic to vanish from the column of water. Because of this, this period should be measured before treatment begins.
4. *Age of treated fishes*: In initial larval phases, probiotic administration can more efficiently control the gut microbiota, since the digestive tract isn't fully formed and the established microbiota represents that of the feed and rearing water. Therefore initial probiotic therapies are strongly recommended for full benefit via rearing water. In the initial postfertilization of cod, administration by water column (105−107 CFU/mL) of two probiotic strains (*Arthrobacter* sp. and *Enterococcus* sp.) may regulate the endogenous microbiota and support the growth and survival of larvae (Lauzon et al., 2010).
5. *Salinity*: Salinity is believed to reduce probiotic bacteria survival and restrict their seawater use. Lactic acid bacteria (LAB) half-lives in seawater (35 gL/b) were stated to be between 3 and 21 h (20−23°C) (Gatesoupe, 2008). This small range of practicality could end in lower probiotic colonization in fish species that live in higher salinity environments. Yet, eight lactic acid bacteria extracted from fish and seafood for turbot probiotics in vitro, and in vivo analysis showed that all isolates could live in seawater for 7 days (at 18°C) (Muñoz-Atienza et al., 2014). It seems as though salinity and temperature have a concomitant effect on the advantageous effects of probiotics. On the other hand, it has to be noted that in marine aquaculture, the direct addition of probiotics to the water column has generally been considered more successful than other methods (Villamil et al., 2010). Apart from the factors mentioned previously, it seems as though the effects of probiotics are widely diverse among different

species of fish (Jahangiri & Esteban, 2018), although there is not enough evidence to substantiate this.

17.5 Rationale for use of probiotics in aquaculture

Researchers require a detailed understanding of the digestive tract as an ecosystem to identify the function and future importance of probiotics to hydrobiont safety and well-being. The recent years of increased scientific activities in this area have verified the significance of microbials in the digestive system. The hydrobiont digestive tract is an open structure that continuously comes into contact with the natural atmosphere—water. Compared with water, the digestive tract is a much richer nutrient environment and more suitable for most bacterial growth. Gastrointestinal bacteria (GIT) are involved in nutritional decomposition, supplying pharmacologically important products like sugars, vitamins, and amino acids to the macroorganisms (Sugita et al., 1997; Thompson et al., 1999; Verschuere et al., 2000).

Bairagi et al. (2002) evaluated GIT-associated aerobic bacteria in nine species of freshwater fish. They concluded that digestive enzymes were produced by selected bacterial strains, thereby promoting food consumption and digestion. Probiotics applied to the diet also have the same beneficial characteristics (Wang, 2007; Yanbo & Zirong, 2006). The digestive tract microflora have been established; limited hydrobionts play a significant part in the confrontation against contagious diseases, since they produce antibacterial resources that prevent pathogenic bacteria from penetrating the organism. Probiotics were originally meant to preserve or to restore an effective interaction between healthy and pathogenic microbes that form the microbiota of the gastrointestinal tract or the surface mucus of hydrobionts. A few bacteria also have antiviral properties and are employed as candidate probiotics. While the precise process by which bacteria achieve this is not understood, experimental studies suggest that chemical and biological suppression of viruses, such as samples of aquatic algae and extracellular bacteria, may occur. Mucosal surfaces attachment and colonization are potential defense mechanisms against pathogenic agents for binding sites and nutrients (Westerdahl et al., 1991) or immune regulation.

17.6 Selection criteria for probiotics

Probiotics' primary function is to create or sustain an equilibrium between favorable and detrimental bacteria naturally found in the intestine or gut of fish (Chauhan & Singh, 2019; Thirumurugan & Vignesh, 2015). Effective probiotics should have the following qualities (AFRC, 1989; Merrifield et al., 2010; Ouwehand et al., 1999):

1. Probiotics should have beneficial effects, against various pathogenic bacteria, on the fish's growth, development, and defense.
2. The probiotic bacteria should not negatively affect the host.
3. The genetically stable probiotics do not have the potential for drug resistance.
4. They may exhibit the following properties as a useful feed:

a. Tolerance to bile and acid.

b. Resistance to digestive fluids.
c. Attach to the surfaces of digestive system.
d. Aversion of pathogens.
e. Immunity enhancement.
f. Greater motility in the stomach.
g. Endurance in mucus.
h. Synthesis of vitamins and enzymes.
5. They should have good sensory properties, fermentation operation, cryo-drying resistance, and feed survival through packing and storage processes.

To determine their potential as ideal probiotics, bacteria extracted from various sites are subjected to multiple steps of sampling. The screening method includes gram staining, indexing, in-vitro assessment of hostile characteristics, acid resistance, bile resistance, drug sensitivity, and biofilms formation. Effective compliance with all requirements qualifies them as a possible probiotic suitable for aquaculture use.

17.7 Probiotics formulation and commercialization

Although the definition does not specify exact quantities, elevated amounts of active microorganisms for probiotics are suggested (Gatesoupe, 1999). Therefore high viability maintenance during planning and storing poses specific difficulties and may be considered a significant roadblock in industrial probiotics development. This phenomenon is chiefly true for "tech adaptive" strains (such as particular LABs) with the impact that the most widely marketed probiotics are typically only resilient.

Most liquid/frozen probiotic products need storage and delivery refrigeration, thus adding cost and frequent aquaculture applications. Probiotic bacteria resilience and healthy cell counts differ according to strain and producer (Schillinger, 1999). It is necessary to display the effective sustainability of the bacteria in products during their shelf life to retain trust in the probiotic materials used for aquaculture. Over the last 20 years, the understanding of how probiotic products are processed and formulated has grown. Most probiotic experiments are performed in vitro, and even without the implementation of uniformity criteria.

The genus *Bacillus* is a gram-positive rod that produces a single endospore (spore), which is a rare case within probiotic bacteria. *Bacillus* spp. *B. subtleis*, *B. cereuses*, *B. coagulans*, *B. clouseii*, *B. megaterium*, and *B. licheniformis* are used as probiotics (Oggioni et al., 2003). These powders or solutions of distilled water formulations are incredibly tolerated by spores because of the spore climate's physical and biological features and long lifespan. The price of growing spores for aquaculture in comparison to the production of distilled components is also low.

The prospect of using these species as probiotics is made much more possible by distinct structures, comprising the unique plasmids obtainable for genetic engineering of *Bacillus subtilis* (Driks, 1999). Nevertheless, as a basic guideline, the plasmids' existence is not a justification for disregarding the strain as a possible probiotic. However, this additional chromosomal DNA's role in producing phenotypes important to technical and probiotic elements should be evaluated.

The consistency is also important to ensure probiotic potency and their potential to deliver benefits in the host's final product preparation. Probiotic colonies must also preserve their probiotic properties after manufacturing and survive in adequate amounts throughout their shelf life to be viable and provide their beneficial effects. It is well documented that probiotic viability is determined by different variables, including species, strain biotype, water storage, hydrogen ion concentrations (pH), mechanical friction, osmotic strength, and oxygen. Therefore particular care and methods are required throughout the probiotic development procedure. Diverse methods have been suggested, including oxygen-impermeable tanks, two-step fermentation, pressure tolerance, the introduction of micronutrients like amino acids and peptides, and microencapsulation to improve the resilience of these probiotics to adverse environments.

There is growing attention on probiotics as an ecologically sustainable substitute used in both methodological and theoretical situations (Soccol et al., 2010). As per a recent study by Grand View Research, Inc., the international probiotics demand is expected to hit USD 77.09 billion by 2025 (Narayana et al., 2020), observing a compound annual growth rate of 6.9%. There are currently numerous marketable probiotic formulations containing one or more live microorganisms developed to enhance aquatic organism production. Probiotics should be used as a food supplement applied either to or combined with food in the culture tank.

In addition to the laboratory treatment of bacteria, some items are now available for commercial use. One of the first consumer products to be focused on *Bacillus* isolated extracted commercial formulations such as Biostart to research the influence of inoculum concentration during the processing of cultured catfish. The usage of marketable *Bacillus* spp. probiotic strains improved the consistency and viability of pond-grown shrimp (Moriarty, 1998).

Marketed products are usually sold in powder or liquid formats, and new technologies for development have been created. In the fermentation procedures, the emphasis was on optimization conditions for fermentation to improve probiotic viability and versatility, enhancing performance (Lacroix & Yildirim, 2007). The manufacture is usually done in batch cultures because of the complexity of continuous processing in large-scale systems (Soccol et al., 2010).

Recently, systems for the immobilization of probiotics have been developed, in particular using microencapsulation. Microbial cells of large density are encapsulated in a colloidal matrix using alginate, chitosan, carboxymethylcellulose, or pectin to secure the microorganisms biologically and chemically. Approaches widely intended for probiotic microencapsulation are extrusion, emulsion, starch adhesion, and spray drying (Rokka & Rantamäki, 2010). Aimed at extension to aquaculture, Rosas-Ledesma et al. (2012) have efficiently encapsulated *Shewanella putrefaciens* cells into calcium alginate, showing the persistence of compressed probiotic cells via the sole gastrointestinal tract.

Alginate matrix encapsulation defends bacteria from low pH and gastrointestinal enzymes; this defense allows the discharge of the probiotic without harm to the intestine (Morinigo et al., 2008). The benefits of freeze-dried industrial preparations lie in preservation and shipping. Conditions for the execution of reconstituting these formulations, such as degree of hydration, temperature, and solution osmolarity, are nevertheless critical for ensuring the bacteria's viability.

It is crucial to note that these items must offer health value to the host; to this point, confined microorganisms can withstand storage conditions and stay viable and stable in the digestive tract of the aquatic species to increase production (Dar, Dar et al., 2016; Dar, Kamili et al., 2016; Martínez Cruz et al., 2012). In the opinions of the producers, these preparations are safe to use and useful for the maintenance of the well-being of aquatic animals.

17.8 Classification of probiotics in aquaculture

Probiotics are classified according to their use, administration, function, etc.

17.8.1 Commercial form

The aquatic probiotics are commonly sold in two primary forms, dry and wet. Dry forms are more stable and are paired with host water or feed. Liquid probiotic materials are usually specifically combined with food or placed in the tank, generally used in hatcheries for chickens. The liquid forms of probiotics show stronger and more beneficial outcomes because of their reduced concentration as compared to dried probiotics (Nageswara & Babu, 2006).

17.8.2 Mode of administration

Depending on the way they are treated, the aquatic probiotics can be further classified into two groups. The first one involves combining probiotic bacteria with feed supplements to improve beneficial intestinal bacteria. The second class includes applying probiotics in water directly; therefore fish can absorb the available nutrients in the aqueous system and prevent pathogen propagation. These two types of probiotics have been used in aquaculture with finfish and shrimp (Sahu et al., 2008).

17.8.3 Based on derivation

Putative probiotics refers to probiotics derived from various organic sources such as gastrointestinal tract, liver, gill, gonads, intestine, and certain inner organs. In comparison, the (nonputative) commercial sources include those previously manufactured and commercially accessible in the marketplace. Bacillus, lactobacillus, and the genus *Bifidobacterium* are the most commonly used probiotic microorganisms (Hai, 2015).

17.8.4 Depending upon the function

Probiotics are classified into three categories according to their actions against pathogens:

1. *Antibacterial*: Several probiotics employed in aquaculture are known for their antibacterial properties against recognized pathogenic agents. The probiotic *Lactococcus lactis* RQ516 shows an inhibitory effect when given to tilapia against *Aeromonas hydrophila. Leuconostoc mesenteroides* can hinder the pathogens of fish present in Nile tilapia (Zhou et al., 2010). *Bacillus subtilis* significantly decreases the motile total of Coliforms, Aeromonads, and Pseudomonads present in aquarium fish, according to studies. Many species of Lactobacilli extracted from inside the intestine of *Anguilla*, *Labeo rohita*, *Clarias orientalis*, *Oreochromis*, and *Puntius carnaticus* have shown potent antimicrobials combating *Aeromonas* and *Vibrio* species.
2. *Antiviral*: Probiotics' antiviral function has developed a reputation in recent years. However, the precise mechanism by which probiotic bacteria exhibit antiviral activity is still unclear.

The in-vitro studies show that virus inhibition can happen via the bacteria's extracellular enzyme secretion. Usage of *Lactobacillus* as a probiotic, whether as a single strain or in combination with Sporolac, has led in the case of *Paralichthys olivaceus* (oil flounders) to less susceptibility to viral lymphocytes disease. The antiviral activity against IHNV was shown by the species *Aeromonas*, *Corynebacterium*, *Pseudomonas*, and *Vibrio*.

3. *Antifungal*: Very few studies on the probiotic antifungal function have been published. The cultivated water of the *Anguilla australis Aeromonas* strain A199 (eel) possessed a strong inhibition property against *Saprolegnia* species (Lategan & Gibson, 2003). In another study, in *Oncorhynchus mykiss* (rainbow trout), increased tolerance for saprolegniasis has been shown by the species *Pseudomonas* M162, *Pseudomonas* M174, and *Janthinobacterium* M169. Nurhajati et al. (Nurhajati et al., 2012) showed the inhibitory potential of *Lactobacillus Plantarum* FNCC 226 against *Saprolegnia parasitica* in catfish (*Pangasius hypothalamus*).

17.9 Use of probiotics

17.9.1 Probiotics as a growth enhancer

Probiotics have been shown experimentally to increase fish growth. Probiotic treatments have increased marine organisms' growth rates, their effectiveness in feeding by affecting digestive enzyme cycles, and their longevity. In reality, probiotics enhanced feed digestibility by enhancing digestive enzymes, for example, lyase alginate, protease, and amylase. Probiotics work effectively through extracellular enzymes, such as proteases, carbohydrolases, and lipases, and the production of growth factors. Vibrio midae SY9 improved digestive protease function, absorption level protein digestion, and *Haliotis midae* growth rate. The highest growth rate and best feed utilization ratio for tilapia species such as *Oreochromis niloticus* were observed using probiotic *Micrococcus luteus* (Yassir et al., 2002). Thus in fish aquaculture, *M. luteus* is known as a growth promoter. A probiotic formulation culminated in producing vital nutrients, such as biotin, fatty acids, and vitamin B12. Probiotics may serve as a supplemental food supply or be added to food digestion, as bacteria are essential foods in typical environments of deposit-feeding holothurians. Worldwide, different research studies have been ongoing using probiotics as a growth enhancer. Table 17.1 provides information about research on the usage of probiotics as a growth enhancer for fish.

17.9.2 Probiotics for disease management

In aquaculture, probiotics or their derivatives for health purposes have been described as beneficial. This involves microbial nutrients that keep bacteria from spreading throughout the gastrointestinal tract, external layers, and cultured species (Verschuere et al., 2000). The effectiveness of these advantageous organisms is accomplished by improving the immune system of cultivated species, reducing their susceptibility to disease, or developing inhibitors that deter pathogenic organisms from manifesting infection in the host. It was noticed that streptococcus, which is one of the biggest problems in tilapia communities, has been controlled after the administration of a probiotic with artificial feed (Utami & Suprayudi, 2015).

Table 17.1 Use of probiotics as a growth enhancer of fish.

Probiotic	Test fish	Activity	Reference
Shewanella putrefaciens Pdp11	*Solea senegalensis*	Regulating the microbiota of digestive tract, increment in growth performance	Ashraf (2000)
Lac. pentosus BD6, *Lac. fermentum* LW2, *Bacillus subtilis* E20, *Saccharomyces cerevisiae* P13	Asian seabass	Augmenting growth performance	Lin et al. (2017)
Lactobacillus plantarum	*Oreochromis niloticus*	Improving growth performance and survival rate	Meidong et al. (2017)
Lactobacillus casei	*Keureling* (*Tor tambra*) fish	Increase the rate of growth and feed production	Muchlisin et al. (2017)
Lactobacillus plantarum	Tilapia	Regulating some hematological parameters and enhancing the growth performance	Yamashita et al. (2017)
Lactobacillus rhamnosus	*Pagrus major*	Promoting growth	Dawood et al. (2015)
Enterococcus casseliflavus	*Oncorhynchus mykiss*	Favoring growth performance	Safari et al. (2016)
Enterococcus faecalis	*Oncorhynchus mykiss*	Augmenting growth	Rodriguez-Estrada et al. (2013)
"Ecotec" (*Lactobacillus acidophilus, Streptococcus thermophiles, STY-31, LA-5, Bifidobacterium, BB-12, and Lactobacillus delbrueckii* ssp. *Bulgaricus, LBY-27*)	*Labeo rohita* juveniles	Increase in growth and survival	Kanwal and Tayyeb (2019)
Baker's yeast, *Saccharomyces cerevisiae*	*Mystus cavasius*	Promoting growth and survival	Banu et al. (2020)

Probiotics not only help in disease management but also augment innate immune responses. The ability of probiotics to increase the number of lymphocytes, macrophages, granulocytes, and erythrocytes of various fish is well known (Anand et al., 2015; Irianto & Austin, 2002; Kumar et al., 2008). The immune systems of fish larvae, shrimp, and other invertebrates are less evolved than those of the adult stage, and for their resistance to infection they are dependent mainly on non-specific immune responses (Verschuere et al., 2000). The application of probiotics promotes rainbow trout's immunity by inducing phagocyte growth, enhanced bacterial killing, and immunoglobulin development (Nikoskelainen et al., 2003). The probiotic *Lactobacillus fermentum* LbFF4 extracted from fermented Nigerian food ("fufu") and *L. plantarum* LbOGI from the "Ogi" beverage may induce immunity in *Clarias gariepinus* (Burchell) to certain selected bacterial pathogens (Ogunshe & Olabode, 2009). It is known that both live probiotics and dead probiotics have health benefits and can be used to change biological reactions (Adams, 2010). The probiotics will thus shield the host from numerous diseases caused by pathogenic microorganisms (Akhter et al., 2015).

17.9.2.1 Modes of action

The valuable actions of probiotics on the defensive system of the host's gastrointestinal tract have a vital role in the treatment of inflammation and prevention of different diseases (Azimirad et al., 2016; Modanloo et al., 2017). Considering the potential probiotic effect in vivo, a distinction must be made between the strain's inherent ability to positively affect the host and its success in achieving and sustaining itself at the position where the action is to be carried out. For example, if the host does not ingest the strain, it is meaningless if siderophore development or inhibiting molecules are of adequate quantity and under GI conditions. This is significant, because Prieur (1981) showed that the bivalve *Mytilus edulis* selectively ingested and digested microbes. Likewise, if probiotics cannot successfully multiply in the intestine after being consumed, significant results cannot appear unless they are routinely added into the diet. Therefore potential means of action essentially call for the selected probiotics to get to where probiotic outcomes are expected. The modes are described in the following paragraphs.

17.9.2.1.1 Production of bacteriocidal substances

Microbial communities can discharge biochemical materials with a bactericidal or bacteriostatic impact on other microbial species, which may change interpopulation associations by affecting chemical or accessible energy rivalry (Fredrickson & Stephanopoulos, 1981; Pybus et al., 1994). In the host surface or culture media, bacteria produce inhibitory substances, thus serving as an obstacle to propagating (optimistic) pathogenic substances. Bacteriologic antibacterial activity, in particular, is related to the following: antibiotic development (145), lysozymes, siderophores, hydrogen peroxide, proteases, and alteration of its pH properties through organic acid processing (126), both individually and in combination.

Lactic acid bacteria develop compounds that inhibit the development of other microorganisms such as bacteriocins (136). Because lactic acid bacteria naturally constitute only a small portion of the intestinal microbiota and, although they may be commonly regarded as nonpathogenic, this can be resolved by suppressing closely associated organisms from lactic acid bacteria by producing bacteriocin, which effectively contributes to the higher health status of their organisms.

In amensalism, which can happen between bacterial species, compounds apart from bacteriocins and antibiotics have been proposed to play a role. Strand B-10–31, isolated from the Japanese nearshore seawater, formed a monastatin alkaline protease inhibitor. The purified and diluted monastatin in vitro test showed inhibitory activity to suppress the fish pathogen *A. hydrophila* protease and *V. anguillarum* thiol protease.

Probiotics are bacterial alternatives to contaminants in aquaculture (Decamp et al., 2008; Heo et al., 2013). While the mechanism of action through which probiotics cause antibacterial effects still remains to be determined, several studies suggested that probiotics develop antibiotic compounds (Moriarty, 1998); pH decrease after the production of organic acid would also inhibit the growth of pathogenic bacteria (Ma et al., 2009). *Bacillus licheniformis* and *B. pumilus* were reported to have antibacterial activity by Ramesh et al. (2015). Pumilus is immune to low pH and elevated bile levels. An inhibitor of *Vibrio alginolyticus* in whiteleg prawns was also shown in a further study of *Bacillus licheniformis* (Ferreira et al., 2015). The *Lactobacillus* spp. is confirmed. Common probiotics synthesize diacetyl, hydroperoxide, short-chain fatty acids, and bactericidal proteins (Faramarzi et al., 2011), reinforcing both the immune response and disease tolerance

(Gram et al., 1999). Thus, through developing antibiotic compounds, probiotics may shield marine animals from pathogen threats.

17.9.2.1.2 Competition for nutrients

Rivalries for nutrients, or usable energy, can determine how various communities of microbials coexist within a similar environment (Fredrickson & Stephanopoulos, 1981). Nutrient rivalry is regarded as one of the pathways by which probiotics prevent pathogenic agents (Ringø et al., 2016). In aquaculture ecosystems, the microbial ecology is typically dominated by heterotrophs acting as carbon and energy sources for organic substrates. Basic knowledge of the factors regulating microbiota composition in aquaculture systems is needed for manipulation. However, this information is not widely accessible, and so one needs to focus on an observational approach. Previous research has shown that iron rivalry is a vital component in marine bacteria. For their growth, most bacteria require iron. However, iron is scarce in animal tissues and body fluids (Verschuere et al., 2000). The bacteria get the quantity of iron required for their development from the iron-binding siderophores. There is a clear link between the development of siderophores and specific diseases (Gram et al., 1999).

17.9.2.1.3 Competition for binding sites

Competition over gastrointestinal sites and other tissue interfaces is one potential method to avoid invasion by pathogens. Capability to bind to the surface wall and enteric mucus is considered to be essential for bacteria survival in fish intestines (Westerdahl et al., 1991). Because bacterial adhesion to the tissue surface is necessary throughout the initial stages of pathogenic infection (Krovacek et al., 1987), the first probiotic impact could be the contest for binding receptors with pathogens (Montes, 1993). Adhesion may be unspecific, depending on physicochemical or particular factors, like adhesive molecules on the epithelial outer membrane of adherent bacteria and receptor molecules (Salminen et al., 1996). Mutual exclusion is suggested as an operation mode of probiotics in the deterrence of pathogens (Sorroza et al., 2012), accomplished by probiotic settlement in the GI mucosal epithelium (Korkea-Aho et al., 2012). Various forms of surface factors have indicated participation in a probiotic association with intestinal epithelial cells and mucus that inhibits pathogen colonization (competitive exemption) per se. This can contribute specifically to associations in adhesion receptor systems (Montes, 1993) that could antagonize (Luis-Villaseñor et al., 2011) and limit the invasion of pathogens (Chabrillón et al., 2005). This phenomenon explicitly illustrates the ability to prescribe probiotics as a supplement for antibiotics and other elements (Cheng et al., 2014). Electrostatic interactions, passive forces, steric forces, hydrophobic and lipoteichoic acids have been reported to be among the factors that cause probiotic adhesion to the binding position (Wilson et al., 2011). Westerdahl et al. (1991) reported that rivalry following mucosal surfaces' occupation for attachment sites and nutrients might be a potential functional mode for the defensive implications of probiotics combating pathogens.

17.9.2.1.4 Immunomodulation

The innate immune responses are the nonspecific immune response of the first line to defend host species from infection, involving various cells and pathways. It has been documented that probiotics can affect nonspecific elements of the immune system, such as mononuclear phagocytes (monocytes, macrophages), natural killer cells (NKs), and polymorphonuclear leukocytes (neutrophils). Immune stimulants are chemicals that stimulate animals' immune systems and increase their resistance to viral, bacterial, fungal, and parasite infections (Raa, 1996). However, currently, it is not evident whether probiotic bacteria can positively affect the immune response of cultured aquatic species, but this cannot be excluded a priori from this system. Previous studies have indicated improvements in monocytes (Aly et al., 2008), leukocytes (Korkea-Aho et al., 2012), granulocytes, macrophages, erythrocytes, and lymphocytes in numerous fish after probiotic therapy (Nayak et al., 2007). For example, rainbow trout that consumed *Clostridium butyricum* demonstrated improved tolerance to vibriosis by impacting phagocytic action in leukocytes (Sakai et al., 1995). Often, *Vibrio* species and *Bacillus* have joint administration. In young white shrimps, the growth efficiency, survival, and resistance to white spot syndrome virus and *V. harveyi* were beneficial (Hoseinifar et al., 2018). The authors contributed to the advancement of phagocytosis and antibacterial action in this defense, and thus immunomodulation. Probiotics are also useful bacteria capable of inhibiting infections and host immune system control. The main microbe-associated molecular patterns (MAMPs) of probiotics include peptidoglycanS-layer protein A (Slpa), lipoteichoic acid (LTA), exopolysaccharide (EPS) flagellin, and microbial nucleic acids. Some pattern recognition receptors or PRRs can detect the MAMPs and cause signaling cascades that can lead to cytokines, chemokines, and other results. These receptors are used for identifying patterns. In recent years, interest has been growing in determining the mechanism of action of probiotics on the immune system of the intestine. The investigators examined the potential association between TLR-mediated signaling probiotic identification and bowel immunity stimulation in this context. For example, the TLR2 signaling pathway was stated to have been involved in identifying the *Psychrobacter* sp. SE6 probiotic and subsequent activation of the immune reaction in *Epinephelus coioides*, the grouper (Sun et al., 2014) (Table 17.2).

17.9.2.1.5 Quorum sensing disruption

Quorum sensing (QS) is a bacterial regulating system responsible for monitoring the production of different biological macromolecules such as cell density-dependent virulence factors. Bacteria control the gene that expresses this mechanism by generating, releasing, and recognizing small signal molecules (Chu et al., 2014) named auto-inducers. This mechanism is used by many bacteria to interact and control a range of physiological processes (Miller & Bassler, 2001). Disruption of the pathogen QS mechanism, a new antiinfective approach, has been proposed for aquaculture.

Since N-Acyl homoserine lactones (AHLs) are the critical family of auto-inducers used in gram-negative bacteria, their biodegradation proves to be an important way to disrupt QS. Sp. *Bacillus* QSI-1 is an effective quorum quencher of the fish pathogen *A. hydrophila* on virulence factor development and biofilm formation (Chu et al., 2014). Sp. *Bacillus* QSI-1 reduced AHL accumulation but did not impact the growth of *A. hydrophila* strain YJ-1 when co-cultured. The QSI-1 supernatant was found to exhibit substantial inhibition of protease development (83.9%), hemolytic function (77.6%), and biofilm-forming (77.3%) in the YJ-1 strain.

Bacillus species QSI-1 significantly reduced the pathogenicity of the *A. hydrophila* YJ-1 strain in zebrafish (*Danio rerio*) in biocontrol studies. Fish fed with QSI-1 were found to have 80.8% relative survival. The findings showed that AHLs destroying bacteria should be seen as an alternative to antibiotics for bacterial fish diseases in aquaculture. Probiotic bacteria such as the strains *Lactobacillus*, *Bifidobacterium*, and *Bacillus cereus* destroy the signal molecules of pathogenic

Table 17.2 Use of probiotics for disease management and augmentation of the immune system of fish.

Probiotic	Test fish	Activity	Reference
Bacillus licheniformis (TSB27) *Lactobacillus thuringiensis* *Bacillus plantarum* *Bacillus subtilis* (B46).	*Sparus aurata* L.	Boosting the immune system	Bahi et al. (2017)
Carnobacteria inhibens	Atlantic salmon, Rainbow trout	Reduced mortalities by defending with *A. salmonicida*, *Vibio ordalii*, *Yersinia ruckeri*	Robertson et al. (2000)
Bacillus subtilis and *Bacillus licheniformis* (BioPlus2B)	Trout	Increased survival rate against *Y. ruckeri*	Raida et al. (2003)
Bacillus subtilis	Indian major carp	Control of infection, against *A. hydrophila*	Kumar et al. (2006)
Bacillus subtilis	Rainbow trout	Increased survival rate against *Aeromonas*	Newaj-Fyzul et al. (2007)
Bacillus subtilis	Grouper	Enhance the relative survival percentages against *Streptococcus* sp.	Liu et al. (2012)
Bacillus circulans	Catla catla	Enhanced the immune response and therefore survival against *A. hydrophila*	Bandyopadhyay and Mohapatra (2009)
Lactobacillus acidophilus	African catfish	Reduced mortalities against *Staphylococcus xylosus*, *A. hydrophila*, and *Streptococcus agalactiae*	Al-Dohail et al. (2011)
Lactobacillus sakei	Rock bream	A nonsignificant decrease in the cumulative mortality against *Edwardsiella tarda*	Harikrishnan et al. (2011)
Lactococcus lactis	Olive flounder	Activated the innate immune system and protection against pathogen infection against *Streptococcus iniae*	Kim et al. (2013)
Flavobacterium sasangense	Common carp	Enhance immune response and disease resistance against *A. hydrophila*	Chi et al. (2014)
Pseudomonas aeruginosa	Zebrafish	Protect fish by inhibiting biofilm formation and enhancing defense mechanisms against *Vibrio parahaemolyticus*	Vinoj et al. (2015)
Shewanella xiamenensis	Grass carp	Improved disease resistance against *A. hydrophila*	Wu et al. (2015)
Enterococcus casseliflavus	Rainbow trout	Improve growth performance and enhance disease resistance against *Streptococcus iniae*	Safari et al. (2016)
Lac. pentosus BD6, *Lac. fermentum* LW2, *Bacillus subtilis* E20, *Saccharomyces cerevisiae* P13	Asian seabass	Improving disease resistance of Asian seabass against *A. hydrophila*	Lin et al. (2017)

(*Continued*)

Table 17.2 Use of probiotics for disease management and augmentation of the immune system of fish. *Continued*

Probiotic	Test fish	Activity	Reference
Bacillus sp. MVF1	*Labeo rohita*	Decreasing susceptibility to disease	Park et al. (2017)
Bacillus subtilis, *Bacillus licheniformis*	Juvenile rainbow trout	Increasing resistance against *A. salmonicida*	Park et al. (2017)
Vibrio lentus	*Dicentrarchus labrax*	Protecting against vibriosis caused by *V. harveyi* in seabass larvae	Schaeck et al. (2017)
Enterococcus casseliflavus	*Oncorhynchus mykiss*	Favoring disease resistance by inmunomodulation	Safari et al. (2016)
Bacillus sp. *Pediococcus* sp.	*Solea senegalensis*	Improving protection against pathogen outbreaks	Batista et al. (2015)
Lactobacillus plantarum (LP20)	*Seriola dumerili*	Improving immune response and stress	Dawood et al. (2015)
Lactobacillus mesenteroides SMM69 *Weissella cibaria* P71	*Scophthalmus maximus* L.	Acting as antimicrobial agent against the turbot pathogens *T. maritimum* and *V. splendidus*	Muñoz-Atienza et al. (2014)
Bacillus subtilis Bacillus licheniformis Bacillus sp. *Pediococcus* sp.	*Oreochromis* sp.	Increasing resistance to *S. agalactiae*	Ng et al. (2014)
Enterococcus faecalis	*Oncorhynchus mykiss*	Stimulating immune system and protecting against diseases	Rodriguez-Estrada et al. (2013)
Saccharomyces cerevisiae	Nile tilapia (*Oreochromis niloticus*)	Activating the immune responses (lysozyme activity and phagocytosis) that help in improving resistance against pathogens	Dawood et al. (2020)

bacteria by enzymatic secretion or auto-inducer antagonist development (Brown, 2011). Medellin-Peña et al. (2007) demonstrated that *Lactobacillus acidophilus* secretes a molecule inhibiting the QS or interfering with *Escherichia coli* O157 gene bacterial transcription (Medellin-Peña et al., 2007).

17.9.3 Probiotics for water quality management in aquaculture

Water plays a crucial role in the lives of all organisms, but especially in all aquatic species. Characteristics of water affect the survival, reproduction, and growth of fish and other aquatic organisms. The hydrobiological parameters and distribution of nutrients have a significant impact on aquatic organisms. Water quality analysis helps to detect whether the water is suitable for human consumption, fish health, etc., which is one of the criteria associated with outbreaks of fish diseases in aquatic organisms. The main concern in aquaculture is water quality improvement, avoiding organic, nitrogen, ammonia, and nitrite wastes. High levels of these compounds can cause severe harm and significant mortality (Das et al., 2017). The oxidizing bacteria of ammonia

(ammonia to nitrite) and oxidizing nitrate bacteria (nitrite to nitrate) will turn these toxic substances into safer forms (Qi et al., 2009). Probiotic bacteria have been argued to be an ecological biocontrol or bioremediation agent for sustainable aquaculture production (Ibrahem, 2015). The decreased growth of algae, decreased organic load, increased concentration of nutrients, increased population of beneficial bacteria, inhibition of potential pathogens, and increased concentration of dissolved oxygen are probiotic effects (Ibrahem, 2015).

Among the different routes of probiotic application in aquaculture, supplementation of rearing water is the only method applicable for all ages of fish. Administration of probiotics through feeding (dry feed) or injection has some limitations during early larval stages, due to the immature digestive tracts in that stage of fish development and the high level of stress.

On the other hand, from the beginning day of hatching in incubators, probiotics can be directly administered to the rearing water. Commercialized probiotics (Remus, Avecom, Ghent, Belgium) can be directly applied to the water containing the larvae of cod (*Gadus morhua* L.), enriching rotifer unregulated growth-related proteins and downregulated proteins related to stress (Sveinsdóttir et al., 2009). According to the findings, *L. plantarum* consisted of 70% of the microbial community in the cod larvae intestine. The probiotic *L. plantarum* was inoculated at the rearing water stage. was inoculated at the rearing water stage. (Strøm & Ringø, 1993). Overall, the most beneficial mode of administration of probiotics in cod larviculture was administration through the rearing water (Lauzon et al., 2014).

Studies have documented gram-positive bacteria (*Bacillus* species) as probiotics to boost the water quality. The bacterial community of *Bacillus* is more effective in converting organic matter into CO_2 than gram-negative bacteria. It is proposed that fish farmers should reduce the accumulation of dissolved and particulate organic material during the growing season, while maintaining high rates of probiotics in production ponds. What is more, this can support phytoplankton production. Nonetheless, this theory could not be tested using one or more *Bacillus* species, *Nitrobacter*, *Pseudomonas*, *Enterobacter*, *Cellulomonas*, and *Rhodopseudomonas*, during the cultivation of shrimp or channel catfish. Except for the nitrates, published evidence for improving the water quality is therefore limited. Use of the *Bacillus* sp. has been documented to result in improved water safety, longevity, and development rates, as well as improved juvenile *Penaeus monodon* health status and decreased pathogenic Vibrio species (Dalmin et al., 2001).

Boyd acknowledged the positive impact of probiotics on organic matter decomposition and decreased phosphate and nitrogen compound amounts (Boyd & Massaut, 1999). Aerobic denitrifying bacteria are deemed ideal candidates in aquaculture waters to reduce nitrate or nitrite to N_2. Many bacteria have been extracted in tanks for shrimp farming to that purpose. Some of the already known denitrifying bacteria are *Acinetobacter*, *Arthrobacter*, *Bacillus*, *Cellulosimicrobium*, *Halomonas*, *Microbacterium*, *Paracoccus*, *Pseudomonas*, *Sphingobacterium*, and *Stenotrophomas*.

17.10 Safety and evaluation of probiotics

The basic principle that probiotics can have multiple beneficial effects in aquaculture has been proven beyond question. This principle's application to health and disease in hydrobionts has already shown some promising results (Olafsen, 2001), even though still in its infancy. However, previously ignored safety factors for the production and promotion of probiotics are now being considered (Courvalin, 2006). Safety is the state of certainty that an entity will not inflict harmful consequences under specified circumstances.

New species and more complex strains of probiotic bacteria are continually being discovered. These new probiotic colonies cannot be expected to share the historical stability of established or typical strains. New strains should be thoroughly tested and examined for both safety and effectiveness before integrating them into products. Furthermore, current producers of probiotics should submit molecular methods for ensuring the accurate detection of the types of bacteria present in their products, for quality assurance and safety.

Different bacteria belonging to the genera *Streptococcus*, *Lactococcus*, *Vagococcus*, and *Carnobacterium* can be linked to an increasing number of diseases that have emerged with the worldwide growth of aquaculture (Ringø & Gatesoupe, 1998). Therefore the safety profile of a possible probiotic strain in the selection process is of vital importance. This research may include assessing strain tolerance to a broad range of different antibiotic groups such as tetracyclines, quinolones, and macrolides, and subsequent evidence of nontransmission of drug-resistance gene virulence plasmids (Moubareck et al., 2005). The end-product composition should also be considered, as to whether it causes adverse effects in specific subjects or eliminates the positive effects. A greater understanding of the possible pathways by which probiotic species may induce adverse effects will help create successful assays that predict which strains may not be safe for probiotic products. Enhanced awareness will also strengthen recommendations for using particular drugs and help define conditions under which probiotics can be supervised closely. However, current molecular methods should also be applied to ensure that the probiotic organisms used in aquaculture are adequately established for consistency and safety compliance.

Probiotics used in the food industry have historically been considered safe, and in reality no human hazards have been reported, remaining the best evidence of their safety (Saxelin et al., 1996). In principle, probiotics could be blamed for four kinds of side effects in susceptible people: bacterial infections, harmful metabolic activity, unnecessary immune suppression, and gene spread. However, no hard evidence has been found regarding the safety of aquaculture products; *Penaeus monodon* cultures have been documented in Asia and, more recently, in Latin America with bacterial white spot syndrome (BWSS), in farms with repeated use of *Bacillus subtilis* related probiotics. The spots are similar to those produced in white spot viral syndrome (WSS), a deadly disease that spreads rapidly and causes mass mortality in shrimp cultures (Wang et al., 1999), as proven by Wang et al. (2000).

In 2000, BWSS was a nonsystemic infection of *P. monodon*, and lesions usually vanished after molting. Cultures having this disease are still stable and typically develop without substantial mortality. However, it is of considerable concern that most farmers are unable to differentiate between BWSS and WSS. In the event of doubt, farmers are urged to send samples to a confirmatory testing laboratory.

Also, because certain aquaculture products are eaten raw or half-cooked, the question has been asked whether residual probiotics in the end user could induce some infection. Shakibazadeh et al. (2011) evaluated the capacity of danger for humans through *Shewanella* probiotic algae in shrimp farms. Studies in mice given up to 1036 CFU were conducted to achieve the LD50 value for *Shewanella* algae, demonstrating the probiotics' efficacy in animals. Based on their observations, these authors noted that using *Shewanella* algae is healthy for shrimp users, such as those working in manufacturing plants and farms.

Since there has been no scientific consensus to ensure probiotics quality and protection, the FAO and WHO acknowledged the need to establish recommendations for a comprehensive approach to testing probiotics in food to substantiate their health claims. Based on scientific evidence (Pineiro & Stanton, 2007), field experts developed a working group to suggest standards and probiotics assessment methods, and *Guidelines for the Evaluation of Probiotics in Food* was published, providing guidance on assessing health and nutrient properties of probiotics in foods.

The working group reported no pathogenic or virulent properties present in lactobacilli, bifidobacteria, or lactococci, although they noted that certain lactobacilli strains were identified with unusual cases of bacteremia under certain circumstances. However, with increasing the use of lactobacillus in probiotics, the prevalence of the number of unusual cases of bacteremia does not increase. An enterococcus was also identified that can exhibit virulence characteristics; thus it is not intended for human consumption as a probiotic (FAO, 2006). While the *Guidelines* are not based on aquaculture products, they provide a basis for conducting studies to determine the safety of probiotics in this field.

To date, it has not been possible to identify particular virulence determinants or pathogenicity of the probiotic microorganisms tested using animal models, including mice, rats, and fish (Lahtinen et al., 2009), to show their overall safety. It is necessary, however, to continue research using three approaches: (1) analyzing the inherent properties of probiotic strains, (2) researching their pharmacokinetics (survival, intestinal function, dosage reaction, and mucosal recovery), and (3) recognizing the relationships between the microorganism and the host.

17.11 Research gaps and future research plans

Usage of probiotics as substitutes for antibiotics and chemicals has led to enormous benefits in aquaculture, so this achievement deserves further research to examine relevant but less studied problems. One such problem is the efficacy of various methods of administration. This problem can be seen as a critical factor in the new probiotics generation in the aquaculture industry. Using probiotics straight into the water column in aquaculture systems as a possible method has been less investigated than the other implementation methods, though some studies point to the high efficacy of probiotic administration in this manner.

In addition, because this approach is more successful in marine environments due to higher probiotic absorption by treated fish (owing to intense drinking activity in these habitats), further research should be carried out on probiotics through the water in different marine fish species. It appears that further studies on the environmental safety of these additives are needed (Chinabut & Puttinaowarat, 2005). The least-studied problems regarding the administration of probiotics as a water additive involve the relationship between the quality and quantity of probiotics required to regulate ammonia nitrogen in aquatic environments (Gross et al., 2004).

Similarly, in finfish aquaculture, considering the beneficial effects of yeasts as a probiotic, no information is available on their rearing-water administration. The application of various probiotics in biofloc systems has proven successful in improving shrimp quality (Yuniasari & Ekasari, 2010). No studies to examine the effects of biofloc technology combined with probiotics in finfish aquaculture have been performed, to our knowledge. A study on this route of administration may include numerous instruments for molecular biotechnology.

DNA microarrays are commonly used for the assessment of immune responses as well as dietary effects in fish (Murray et al., 2010; Yuniasari & Ekasari, 2010), but knowledge of transcriptomic effects induced by the administration of probiotics via water is minimal; further research should therefore concentrate on this new method to gain a better understanding of the effectiveness of probiotic application as a water additive.

Some experiments have shown that the use of chosen probiotics may be an effective way to protect marine animals from diseases. Farmers, though, cannot anticipate when the initiation of disease will occur or schedule probiotic feeding in the weeks before infection. If the initiation of disease has already occurred, additional analysis of the treatment results is then needed. It is noted that a successful probiotics screen plays a vital role in the collection of suitable probiotics in aquaculture as positive findings in vitro.

There are times where the in vivo results are not known (Kesarcodi-Watson et al., 2008). Additionally, the safety of probiotic health benefits is also unclear. In marine conditions, the existence of live probiotics is unclear (Newaj-Fyzul et al., 2014). While there is no evidence supporting short-term cyclic probiotic feeding techniques, this strategy is believed to prevent overstimulating of an immune response, thus retaining a degree of immunostimulation or defense.

Studies related to dosage are minimal and somewhat inconsistent. We will require further inquiries before presenting guidance with some degree of trust. Moreover, overdose or excessive probiotic administration causes immunosuppression of the hosts' continuous responses. While there is not much evidence of sustained probiotic administration in aquaculture, Sakai (Sakai, 1999) disproves the theory that long-term exposure to immune stimulants causes immunosuppression in aquatic organisms.

More studies are required on the effects, or even mortality, of probiotics, when they are administered over a long period and with an indiscriminate dosage.

17.12 Conclusion

This chapter shows that the application of probiotics in aquaculture as an alternative to chemicals and antibiotics has proven useful in promoting profitable aquaculture, because probiotics have the potential to boost the quality of the water, increase stress tolerance, produce high-quality livestock, and provide varied potential benefits to fish health, among other advantages. Further research is required into the most effective routes of probiotic administration. In addition, new research must be carried out on new biotechnological processes that contribute to mass production and application of cost-effective probiotics on an industrial scale. To transfer the laboratory results to industry, we must overcome some non-minor gaps, such as the legal permission involved when working with living organisms for human consumption. Probiotics are a welcome addition to the mix of disease prophylaxis solutions in aquaculture, but the research and science behind this field are still very much in the developmental stages. It seems likely that the use of probiotics will gradually increase and, if validated through rigorous scientific investigation and wide application, it may prove to be a boon for the aquaculture industry.

References

Adams, C. A. (2010). The probiotic paradox: Live and dead cells are biological response modifiers. *Nutrition Research Reviews*, *23*, 37–46. Available from https://doi.org/10.1017/S0954422410000090.

Afrc, R. F. (1989). Probiotics in man and animals. *Journal of Applied Bacteriology, 66*, 365–378.

Akhter, N., Wu, B., Memon, A. M., & Mohsin, M. (2015). Probiotics and prebiotics associated with aquaculture: A review. *Fish & Shellfish Immunology, 45*, 733–741. Available from https://doi.org/10.1016/j.fsi.2015.05.038.

Al-Dohail, M. A., Hashim, R., & Aliyu-Paiko, M. (2011). Evaluating the use of *Lactobacillus acidophilus* as a biocontrol agent against common pathogenic bacteria and the effects on the haematology parameters and histopathology in African catfish *Clarias gariepinus* juveniles. *Aquaculture Research, 42*, 196–209. Available from https://doi.org/10.1111/j.1365-2109.2010.02606.x.

Aly, S. M., Mohamed, M. F., & John, G. (2008). Effect of probiotics on the survival, growth and challenge infection in Tilapia nilotica (*Oreochromis niloticus*). *Aquaculture Research., 39*, 647–656. Available from https://doi.org/10.1111/j.1365-2109.2008.01932.x.

Anand, P. S. S., Kohli, M. P., Roy, S. D., Sundaray, J. K., Kumar, S., Sinha, A., Pailan, G. H., & Sukham, Mk (2015). Effect of dietary supplementation of periphyton on growth, immune response and metabolic enzyme activities in *P. enaeus* monodon. *Aquaculture Research, 46*, 2277–2288. Available from https://doi.org/10.1111/are.12385.

Ashraf, A. (2000). *Probiotics in fish farming. Evaluation of a candidate bacterial mixture. Vattenbruksinstitutionen* (pp. 1–18). Report 19, Umea'2000, Ph. Licentiate Thesis.

Azimirad, M., Meshkini, S., Ahmadifard, N., & Hoseinifar, S. H. (2016). The effects of feeding with synbiotic (*Pediococcus acidilactici* and fructooligosaccharide) enriched adult Artemia on skin mucus immune responses, stress resistance, intestinal microbiota and performance of angelfish (*Pterophyllum scalare*). *Fish & Shellfish Immunology, 54*, 516–522. Available from https://doi.org/10.1016/j.fsi.2016.05.001.

Bahi, A., Guardiola, F., Messina, C., Mahdhi, A., Cerezuela, R., Santulli, A., Bakhrouf, A., & Esteban, M. (2017). Effects of dietary administration of fenugreek seeds, alone or in combination with probiotics, on growth performance parameters, humoral immune response and gene expression of gilthead seabream (*Sparus aurata* L.). *Fish & Shellfish Immunology, 60*, 50–58. Available from https://doi.org/10.1016/j.fsi.2016.11.039.

Bairagi, A., Ghosh, K. S., Sen, S. K., & Ray, A. K. (2002). Enzyme producing bacterial flora isolated from fish digestive tracts. *Aquaculture International, 10*, 109–121. Available from https://doi.org/10.1023/A:1021355406412.

Bandyopadhyay, P., & Mohapatra, P. K. D. (2009). Effect of a probiotic bacterium *Bacillus circulans* PB7 in the formulated diets: On growth, nutritional quality and immunity of *Catla catla* (Ham.). *Fish Physiology and Biochemistry, 35*, 467–478. Available from https://doi.org/10.1007/s10695-008-9272-8.

Banu, M. R., Akter, S., Islam, M. R., Mondol, M. N., & Hossain, M. A. (2020). Probiotic yeast enhanced growth performance and disease resistance in freshwater catfish gulsa tengra, *Mystus cavasius*. *Aquaculture Reports, 16*, 100237. Available from https://doi.org/10.1016/j.aqrep.2019.100237.

Batista, S., Ramos, M., Cunha, S., Barros, R., Cristóvão, B., Rema, P., Pires, M., Valente, L., & Ozório, R. (2015). Immune responses and gut morphology of Senegalese sole (*Solea senegalensis*, Kaup 1858) fed monospecies and multispecies probiotics. *Aquaculture Nutrition, 21*, 625–634. Available from https://doi.org/10.1111/anu.12191.

Boyd, C. E., & Massaut, L. (1999). Risks associated with the use of chemicals in pond aquaculture. *Aquacultural Engineering, 20*, 113–132. Available from https://doi.org/10.1016/S0144-8609(99)00010-2.

Browdy, C. L. (1998). Recent developments in penaeid broodstock and seed production technologies: Improving the outlook for superior captive stocks. *Aquaculture (Amsterdam, Netherlands), 164*, 3–21.

Brown, M. (2011). Modes of action of probiotics: Recent developments. *Journal of Animal and Veterinary Advances, 10*, 1895–1900. Available from https://doi.org/10.3923/javaa.2011.1895.1900.

Cabello, F. C. (2006). Heavy use of prophylactic antibiotics in aquaculture: A growing problem for human and animal health and for the environment. *Environmental Microbiology, 8*, 1137–1144. Available from https://doi.org/10.1111/j.1462-2920.2006.01054.x.

Chabrillón, M., Rico, R., Arijo, S., Diaz-Rosales, P., Balebona, M., & Moriñigo, M. (2005). Interactions of microorganisms isolated from gilthead sea bream, *Sparus aurata* L., on *Vibrio harveyi*, a pathogen of farmed Senegalese sole, *Solea senegalensis* (Kaup). *Journal of Fish Diseases*, *28*, 531–537. Available from https://doi.org/10.1111/j.1365-2761.2005.00657.x.

Chalas, R., Janczarek, M., Bachanek, T., Mazur, E., Cieszko-Buk, M., & Szymanska, J. (2016). Characteristics of oral probiotics – A review. *Current Issues in Pharmacy and Medical Sciences*, *29*, 8–10. Available from https://doi.org/10.1515/cipms-2016-0002.

Chauhan, A., & Singh, R. (2019). Probiotics in aquaculture: A promising emerging alternative approach. *Symbiosis*, *77*, 99–113. Available from https://doi.org/10.1007/s13199-018-0580-1.

Cheng, G., Hao, H., Xie, S., Wang, X., Dai, M., Huang, L., & Yuan, Z. (2014). Antibiotic alternatives: The substitution of antibiotics in animal husbandry? *Frontiers in Microbiology*, *5*, 217. Available from https://doi.org/10.3389/fmicb.2014.00217.

Chi, C., Jiang, B., Yu, X.-B., Liu, T.-Q., Xia, L., & Wang, G.-X. (2014). Effects of three strains of intestinal autochthonous bacteria and their extracellular products on the immune response and disease resistance of common carp, *Cyprinus carpio*. *Fish & Shellfish Immunology*, *36*, 9–18. Available from https://doi.org/10.1016/j.fsi.2013.10.003.

Chinabut, S., & Puttinaowarat, S. (2005). The choice of disease control strategies to secure international market access for aquaculture products. *Developments in Biologicals*, *121*, 255.

Chu, W., Zhou, S., Zhu, W., & Zhuang, X. (2014). Quorum quenching bacteria *Bacillus* sp. QSI-1 protect zebrafish (*Danio rerio*) from *Aeromonas hydrophila* infection. *Scientific Reports*, *4*, 5446. Available from https://doi.org/10.1038/srep05446.

Courvalin, P. (2006). Antibiotic resistance: The pros and cons of probiotics. *Digestive and Liver Disease*, *38*, S261–S265. Available from https://doi.org/10.1016/s1590-8658(07)60006-1.

Dalmin, G., Kathiresan, K., & Purushothaman, A. (2001). Effect of probiotics on bacterial population and health status of shrimp in culture pond ecosystem. *Indian Journal of Experimental Biology*, *39*(9), 939–942.

Dar, G. H., Bhat, R. A., Kamili, A. N., Chishti, M. Z., Qadri, H., Dar, R., & Mehmood, M. A. (2020). *Correlation between pollution trends of freshwater bodies and bacterial disease of fish fauna. Fresh water pollution dynamics and remediation* (pp. 51–67). Singapore: Springer.

Dar, G. H., Dar, S. A., Kamili, A. N., Chishti, M. Z., & Ahmad, F. (2016). Detection and characterization of potentially pathogenic *Aeromonas sobria* isolated from fish *Hypophthalmichthys molitrix* (Cypriniformes: Cyprinidae). *Microbial Pathogenesis*, *91*, 136–140.

Dar, G. H., Kamili, A. N., Chishti, M. Z., Dar, S. A., Tantry, T. A., & Ahmad, F. (2016). Characterization of *Aeromonas sobria* isolated from fish rohu (*Labeo rohita*) collected from polluted pond. *Journal of Bacteriology & Parasitology*, *7*(3), 1–5. Available from https://doi.org/10.4172/2155-9597.1000273.

Das, S., Mondal, K., & Haque, S. (2017). A review on application of probiotic, prebiotic and synbiotic for sustainable development of aquaculture. *Growth*, *14*, 15.

Dawood, M. A., Eweedah, N. M., Khalafalla, M. M., Khalid, A., El Asely, A., Fadl, S. E., Amin, A. A., Paray, B. A., & Ahmed, H. A. (2020). Saccharomyces cerevisiae increases the acceptability of Nile tilapia (*Oreochromis niloticus*) to date palm seed meal. *Aquaculture Reports*, *17*, 100314. Available from https://doi.org/10.1016/j.aqrep.2020.100314.

Dawood, M. A., Koshio, S., Ishikawa, M., & Yokoyama, S. (2015). Effects of partial substitution of fish meal by soybean meal with or without heat-killed *Lactobacillus plantarum* (LP20) on growth performance, digestibility, and immune response of amberjack, *Seriola dumerili* juveniles. *BioMed Research International*, *2015*, 514196. Available from https://doi.org/10.1155/2015/514196.

Decamp, O., Moriarty, D. J., & Lavens, P. (2008). Probiotics for shrimp larviculture: Review of field data from Asia and Latin America. *Aquaculture Research*, *39*, 334–338. Available from https://doi.org/10.1111/j.1365-2109.2007.01664.x.

Driks, A. (1999). Bacillus subtilis spore coat. *Microbiology and Molecular Biology Reviews*, *63*, 1–20.

FAO. (2006). *Guidelines for the evaluation of probiotics in food.* Report of a Joint FAO/WHO Working Group, London, Ontario, Canada, 30 April to 1 May 2002. Probiotics in Food: Health and Nutritional Properties and Guidelines For Evaluation. (pp. 1-56). Available at: https://www.who.int/foodsafety/fs_management/en/probiotic_guidelines.pdf

Faramarzi, M., Kiaalvandi, S., & Iranshahi, F. (2011). The effect of probiotics on growth performance and body composition of common carp (*Cyprinus carpio*). *Journal of Animal and Veterinary Advances*, *10*, 2408–2413. Available from https://doi.org/10.3923/javaa.2011.2408.2413.

Ferreira, G. S., Bolívar, N. C., Pereira, S. A., Guertler, C., do Nascimento Vieira, F., Mouriño, J. L. P., & Seiffert, W. Q. (2015). Microbial biofloc as source of probiotic bacteria for the culture of *Litopenaeus vannamei*. *Aquaculture (Amsterdam, Netherlands)*, *448*, 273–279. Available from https://doi.org/10.1016/j.aquaculture.2015.06.006.

Fredrickson, A., & Stephanopoulos, G. (1981). Microbial competition. *Science (New York, N.Y.)*, *213*, 972–979. Available from https://doi.org/10.1126/science.7268409.

Gatesoupe, F. J. (1999). The use of probiotics in aquaculture. *Aquaculture (Amsterdam, Netherlands)*, *180*, 147–165. Available from https://doi.org/10.1016/S0044-8486(99)00187-8.

Gatesoupe, F.-J. (2008). Updating the importance of lactic acid bacteria in fish farming: Natural occurrence and probiotic treatments. *Journal of Molecular Microbiology and Biotechnology*, *14*, 107–114. Available from https://doi.org/10.1159/000106089.

Gram, L., Melchiorsen, J., Spanggaard, B., Huber, I., & Nielsen, T. F. (1999). Inhibition of *Vibrio anguillarum* by *Pseudomonas fluorescens* AH2, a possible probiotic treatment of fish. *Applied and Environmental Microbiology*, *65*, 969–973. Available from https://doi.org/10.1128/aem.65.3.969-973.1999.

Gross, A., Abutbul, S., & Zilberg, D. (2004). Acute and chronic effects of nitrite on white shrimp, *Litopenaeus vannamei*, cultured in low-salinity brackish water. *Journal of the World Aquaculture Society*, *35*, 315–321. Available from https://doi.org/10.1111/j.1749-7345.2004.tb00095.x.

Hai, N. V. (2015). The use of probiotics in aquaculture. *Journal of Applied Microbiology*, *119*, 917–935. Available from https://doi.org/10.1111/jam.12886.

Harikrishnan, R., Kim, M.-C., Kim, J.-S., Balasundaram, C., & Heo, M.-S. (2011). Protective effect of herbal and probiotics enriched diet on haematological and immunity status of *Oplegnathus fasciatus* (Temminck & Schlegel) against *Edwardsiella tarda*. *Fish & Shellfish Immunology*, *30*, 886–893. Available from https://doi.org/10.1016/j.fsi.2011.01.013.

Heo, W.-S., Kim, Y.-R., Kim, E.-Y., Bai, S. C., & Kong, I.-S. (2013). Effects of dietary probiotic, *Lactococcus lactis* subsp. lactis I2, supplementation on the growth and immune response of olive flounder (*Paralichthys olivaceus*). *Aquaculture (Amsterdam, Netherlands)*, *376*, 20–24. Available from https://doi.org/10.1016/j.aquaculture.2012.11.009.

Hoseinifar, S. H., Sun, Y.-Z., Wang, A., & Zhou, Z. (2018). Probiotics as means of diseases control in aquaculture, a review of current knowledge and future perspectives. *Frontiers in Microbiology*, *9*, 2429. Available from https://doi.org/10.3389/fmicb.2018.02429.

Huynh, T.-G., Shiu, Y.-L., Nguyen, T.-P., Truong, Q.-P., Chen, J.-C., & Liu, C.-H. (2017). Current applications, selection, and possible mechanisms of actions of synbiotics in improving the growth and health status in aquaculture: A review. *Fish & Shellfish Immunology*, *64*, 367–382. Available from https://doi.org/10.1016/j.fsi.2017.03.035.

Ibrahem, M. D. (2015). Evolution of probiotics in aquatic world: Potential effects, the current status in Egypt and recent prospectives. *Journal of Advanced Research*, *6*, 765–791. Available from https://doi.org/10.1016/j.jare.2013.12.004.

Irianto, A., & Austin, B. (2002). Probiotics in aquaculture. *Journal of Fish Diseases*, *25*, 633–642. Available from https://doi.org/10.1046/j.1365-2761.2002.00422.x.

Jahangiri, L., & Esteban, M. Á. (2018). Administration of probiotics in the water in finfish aquaculture systems: A review. *Fishes*, *3*, 33. Available from https://doi.org/10.3390/fishes3030033.

Kanwal, Z., & Tayyeb, A. (2019). Role of dietary probiotic Ecotec in growth enhancement, thyroid tuning, hematomorphology and resistance to pathogenic challenge in *Labeo rohita* juveniles. *Journal of Applied Animal Research*, *47*, 394–402. Available from https://doi.org/10.1080/09712119.2019.1650050.

Kesarcodi-Watson, A., Kaspar, H., Lategan, M. J., & Gibson, L. (2008). Probiotics in aquaculture: The need, principles and mechanisms of action and screening processes. *Aquaculture (Amsterdam, Netherlands)*, *274*, 1–14. Available from https://doi.org/10.1016/j.aquaculture.2007.11.019.

Khattab, Y., Shalaby, A., & Abdel-Rhman, A. (2005). Use of probiotic bacteria as growth promoters, antibacterial and their effects on physiological parameters of *Oreochromis niloticus*. In: *Proceedings of International Symposium on Nile Tilapia in Aquaculture*.

Kim, D., Beck, B. R., Heo, S.-B., Kim, J., Kim, H. D., Lee, S.-M., Kim, Y., Oh, S. Y., Lee, K., & Do, H. (2013). *Lactococcus lactis* BFE920 activates the innate immune system of olive flounder (*Paralichthys olivaceus*), resulting in protection against *Streptococcus iniae* infection and enhancing feed efficiency and weight gain in large-scale field studies. *Fish & Shellfish Immunology*, *35*, 1585–1590. Available from https://doi.org/10.1016/j.fsi.2013.09.008.

Korkea-Aho, T., Papadopoulou, A., Heikkinen, J., Von Wright, A., Adams, A., Austin, B., & Thompson, K. (2012). Pseudomonas M162 confers protection against rainbow trout fry syndrome by stimulating immunity. *Journal of Applied Microbiology*, *113*, 24–35. Available from https://doi.org/10.1111/j.1365-2672.2012.05325.x.

Krovacek, K., Faris, A., Ahne, W., & Månsson, I. (1987). Adhesion of *Aeromonas hydrophila* and *Vibrio anguillarum* to fish cells and to mucus-coated glass slides. *FEMS Microbiology Letters*, *42*, 85–89. Available from https://doi.org/10.1111/j.1574-6968.1987.tb02304.x.

Kumar, R., Mukherjee, S., Ranjan, R., & Nayak, S. (2008). Enhanced innate immune parameters in *Labeo rohita* (Ham.) following oral administration of *Bacillus subtilis*. *Fish & Shellfish Immunology*, *24*, 168–172. Available from https://doi.org/10.1016/j.fsi.2007.10.008.

Kumar, R., Mukherjee, S. C., Prasad, K. P., & Pal, A. K. (2006). Evaluation of *Bacillus subtilis* as a probiotic to Indian major carp *Labeo rohita* (Ham.). *Aquaculture Research*, *37*, 1215–1221. Available from https://doi.org/10.1111/j.1365-2109.2006.01551.x.

Lacroix, C., & Yildirim, S. (2007). Fermentation technologies for the production of probiotics with high viability and functionality. *Current Opinion in Biotechnology*, *18*, 176–183. Available from https://doi.org/10.1016/j.copbio.2007.02.002.

Lahtinen, S., Boyle, R., Margolles, A., & Gueimonde, M. (2009). *Safety assessment of probiotics. Prebiotics and probiotics science and technology*. New York: Springer Science & Business Media, LLC. Available from https://doi.org/10.1007/978-0-387-79058-9_31.

Lategan, M., & Gibson, L. (2003). Antagonistic activity of Aeromonas media strain A199 against *Saprolegnia* sp., an opportunistic pathogen of the eel, *Anguilla australis* Richardson. *Journal of Fish Diseases*, *26*, 147–153. Available from https://doi.org/10.1046/j.1365-2761.2003.00443.x.

Lategan, M., Torpy, F., & Gibson, L. (2004). Control of saprolegniosis in the eel *Anguilla australis* Richardson, by Aeromonas media strain A199. *Aquaculture (Amsterdam, Netherlands)*, *240*, 19–27. Available from https://doi.org/10.1016/j.aquaculture.2004.04.009.

Lauzon, H., Gudmundsdottir, S., Steinarsson, A., Oddgeirsson, M., Petursdottir, S., Reynisson, E., Bjornsdottir, R., & Gudmundsdottir, B. (2010). Effects of bacterial treatment at early stages of Atlantic cod (*Gadus morhua* L.) on larval survival and development. *Journal of Applied Microbiology*, *108*, 624–632. Available from https://doi.org/10.1111/j.1365-2672.2009.04454.x.

Lauzon, H. L., Pérez-Sánchez, T., Merrifield, D. L., Ringø, E., & Balcázar, J. L. (2014). *Probiotic applications in cold water fish species. Aquaculture nutrition: Gut health, probiotics and prebiotics* (pp. 223–252). John Wiley & Sons, Ltd. Available from http://doi.org/10.1002/9781118897263.ch9.

Lilly, D. M., & Stillwell, R. H. (1965). Probiotics: Growth-promoting factors produced by microorganisms. *Science (New York, N.Y.)*, *147*, 747–748. Available from https://doi.org/10.1126/science.147.3659.747.

Lin, H.-L., Shiu, Y.-L., Chiu, C.-S., Huang, S.-L., & Liu, C.-H. (2017). Screening probiotic candidates for a mixture of probiotics to enhance the growth performance, immunity, and disease resistance of Asian seabass, *Lates calcarifer* (Bloch), against *Aeromonas hydrophila*. *Fish & Shellfish Immunology*, *60*, 474–482. Available from https://doi.org/10.1016/j.fsi.2016.11.026.

Liu, C.-H., Chiu, C.-H., Wang, S.-W., & Cheng, W. (2012). Dietary administration of the probiotic, Bacillus subtilis E20, enhances the growth, innate immune responses, and disease resistance of the grouper, *Epinephelus coioides*. *Fish & Shellfish Immunology*, *33*, 699–706. Available from https://doi.org/10.1016/j.fsi.2012.06.012.

Luis-Villaseñor, I. E., Macías-Rodríguez, M. E., Gómez-Gil, B., Ascencio-Valle, F., & Campa-Córdova, Á. I. (2011). Beneficial effects of four Bacillus strains on the larval cultivation of *Litopenaeus vannamei*. *Aquaculture (Amsterdam, Netherlands)*, *321*, 136–144. Available from https://doi.org/10.1016/j.aquaculture.2011.08.036.

Ma, C.-W., Cho, Y.-S., & Oh, K.-H. (2009). Removal of pathogenic bacteria and nitrogens by *Lactobacillus* spp. JK-8 and JK-11. *Aquaculture (Amsterdam, Netherlands)*, *287*, 266–270. Available from https://doi.org/10.1016/j.aquaculture.2008.10.061.

Martínez Cruz, P., Ibáñez, A. L., Monroy Hermosillo, O. A., & Ramírez Saad, H. C. (2012). Use of probiotics in aquaculture. *ISRN Microbiology*, *2012*. Available from https://doi.org/10.5772/50056.

Medellin-Peña, M. J., Wang, H., Johnson, R., Anand, S., & Griffiths, M. W. (2007). Probiotics affect virulence-related gene expression in *Escherichia coli* O157: H7. *Applied and Environmental Microbiology*, *73*, 4259–4267. Available from https://doi.org/10.1128/AEM.00159-07.

Meidong, R., Doolgindachbaporn, S., Sakai, K., & Tongpim, S. (2017). Isolation and selection of lactic acid bacteria from Thai indigenous fermented foods for use as probiotics in tilapia fish *Oreochromis niloticus*. *Aquaculture, Aquarium, Conservation & Legislation*, *10*, 455–463.

Merrifield, D. L., Dimitroglou, A., Foey, A., Davies, S. J., Baker, R. T., Bøgwald, J., Castex, M., & Ringø, E. (2010). The current status and future focus of probiotic and prebiotic applications for salmonids. *Aquaculture (Amsterdam, Netherlands)*, *302*, 1–18. Available from https://doi.org/10.1016/j.aquaculture.2010.02.007.

Metchnikoff, E. (1908). *The prolongation of life*. GP Putnam's Sons. Available from http://doi.org/10.1038/077289b0.

Miller, M. B., & Bassler, B. L. (2001). Quorum sensing in bacteria. *Annual Reviews in Microbiology*, *55*, 165–199. Available from https://doi.org/10.1146/annurev.micro.55.1.165.

Modanloo, M., Soltanian, S., Akhlaghi, M., & Hoseinifar, S. H. (2017). The effects of single or combined administration of galactooligosaccharide and *Pediococcus acidilactici* on cutaneous mucus immune parameters, humoral immune responses and immune related genes expression in common carp (*Cyprinus carpio*) fingerlings. *Fish & Shellfish Immunology*, *70*, 391–397. Available from https://doi.org/10.1016/j.fsi.2017.09.032.

Mohapatra, S., Chakraborty, T., Kumar, V., DeBoeck, G., & Mohanta, K. (2013). Aquaculture and stress management: A review of probiotic intervention. *Journal of Animal Physiology and Animal Nutrition*, *97*, 405–430. Available from https://doi.org/10.1111/j.1439-0396.2012.01301.x.

Montes, A. (1993). The use of probiotics in food-animal practice. *Veterinary Medicine*, *88*, 282–289.

Moriarty, D. (1998). Control of luminous Vibrio species in penaeid aquaculture ponds. *Aquaculture (Amsterdam, Netherlands)*, *164*, 351–358. Available from https://doi.org/10.1016/S0044-8486(98)00199-9.

Morinigo, M., Sánchez, V., & Martınez, T. (2008). *Encapsulation of a bacterial fish probiiotic in alginate beads: Protective effect under in vitro simulations of fish gastric conditions*. In: International Conference on Fish Diseases and Fish Immunology, pp. 6–9. Reykjavik, Iceland: Iceland University.

Moubareck, C., Gavini, F., Vaugien, L., Butel, M., & Doucet-Populaire, F. (2005). Antimicrobial susceptibility of bifidobacteria. *Journal of Antimicrobial Chemotherapy*, *55*, 38–44. Available from https://doi.org/10.1093/jac/dkh495.

Muchlisin, Z., Nazir, M., Fadli, N., Adlim, M., Hendri, A., Khalil, M., & Siti-Azizah, M. (2017). Efficacy of commercial diets with varying levels of protein on growth performance, protein and lipid contents in carcass of *Acehnese mahseer*, Tor tambra. *Iranian Journal of Fisheries Sciences*, *16*, 557–566.

Muñoz-Atienza, E., Araújo, C., Magadán, S., Hernández, P. E., Herranz, C., Santos, Y., & Cintas, L. M. (2014). In vitro and in vivo evaluation of lactic acid bacteria of aquatic origin as probiotics for turbot (*Scophthalmus maximus* L.) farming. *Fish & Shellfish Immunology*, *41*, 570–580. Available from https://doi.org/10.1016/j.fsi.2014.10.007.

Murray, H. M., Lall, S. P., Rajaselvam, R., Boutilier, L. A., Blanchard, B., Flight, R. M., Colombo, S., Mohindra, V., & Douglas, S. E. (2010). A nutrigenomic analysis of intestinal response to partial soybean meal replacement in diets for juvenile Atlantic halibut, *Hippoglossus hippoglossus*, L. *Aquaculture (Amsterdam, Netherlands)*, *298*, 282–293. Available from https://doi.org/10.1016/j.aquaculture.2009.11.001.

Nageswara, P., & Babu, D. (2006). Probiotics as an alternative therapy to minimize or avoid antibiotics use in aquaculture. *Fishing Chimes*, *26*, 112–114.

Narayana, S. K., Mallick, S., Siegumfeldt, H., & van den Berg, F. (2020). Bacterial flow cytometry and imaging as potential process monitoring tools for industrial biotechnology. *Fermentation*, *6*, 10. Available from https://doi.org/10.3390/fermentation6010010.

Nayak, S., Swain, P., & Mukherjee, S. (2007). Effect of dietary supplementation of probiotic and vitamin C on the immune response of Indian major carp, *Labeo rohita* (Ham.). *Fish & Shellfish Immunology*, *23*, 892–896. Available from https://doi.org/10.1016/j.fsi.2007.02.008.

Newaj-Fyzul, A., Adesiyun, A. A., Mutani, A., Ramsubhag, A., Brunt, J., & Austin, B. (2007). Bacillus subtilis AB1 controls Aeromonas infection in rainbow trout (*Oncorhynchus mykiss*, Walbaum). *Journal of Applied Microbiology*, *103*, 1699–1706. Available from https://doi.org/10.1111/j.1365-2672.2007.03402.x.

Newaj-Fyzul, A., Al-Harbi, A., & Austin, B. (2014). Developments in the use of probiotics for disease control in aquaculture. *Aquaculture*, *431*, 1–11. Available from https://doi.org/10.1016/j.aquaculture.2013.08.026.

Ng, W.-K., Kim, Y.-C., Romano, N., Koh, C.-B., & Yang, S.-Y. (2014). Effects of dietary probiotics on the growth and feeding efficiency of red hybrid tilapia, *Oreochromis* sp., and subsequent resistance to *Streptococcus agalactiae*. *Journal of Applied Aquaculture*, *26*, 22–31. Available from https://doi.org/10.1080/10454438.2013.874961.

Nikoskelainen, S., Ouwehand, A. C., Bylund, G., Salminen, S., & Lilius, E.-M. (2003). Immune enhancement in rainbow trout (*Oncorhynchus mykiss*) by potential probiotic bacteria (*Lactobacillus rhamnosus*). *Fish & Shellfish Immunology*, *15*, 443–452. Available from https://doi.org/10.1016/S1050-4648(03)00023-8.

Nurhajati, J., Aryantha, I., & Indah, D. (2012). The curative action of *Lactobacillus plantarum* FNCC 226 to *Saprolegnia parasitica* A3 on catfish (*Pangasius hypophthalamus* Sauvage). *International Food Research Journal*, *19*.

Oggioni, M. R., Ciabattini, A., Cuppone, A. M., & Pozzi, G. (2003). Bacillus spores for vaccine delivery. *Vaccine*, *21*, S96–S101. Available from https://doi.org/10.1016/s0264-410x(03)00207-x.

Ogunshe, A. A., & Olabode, O. P. (2009). Antimicrobial potentials of indigenous *Lactobacillus strains* on gram-negative indicator bacterial species from *Clarias gariepinus* (Burchell.) microbial inhibition of fish-borne pathogens. *African Journal of Microbiology Research*, *3*, 870–876. Available from https://doi.org/10.5897/AJMR.9000082.

Olafsen, J. A. (2001). Interactions between fish larvae and bacteria in marine aquaculture. *Aquaculture (Amsterdam, Netherlands)*, *200*, 223–247. Available from https://doi.org/10.1016/S0044-8486(01)00702-5.

Ouwehand, A., Kirjavainen, P., Grönlund, M.-M., Isolauri, E., & Salminen, S. (1999). Adhesion of probiotic micro-organisms to intestinal mucus. *International Dairy Journal*, *9*, 623–630. Available from https://doi.org/10.1016/S0958-6946(99)00132-6.

Park, Y., Lee, S., Hong, J., Kim, D., Moniruzzaman, M., & Bai, S. C. (2017). Use of probiotics to enhance growth, stimulate immunity and confer disease resistance to Aeromonas salmonicida in rainbow trout (*Oncorhynchus mykiss*). *Aquaculture Research.*, *48*, 2672–2682. Available from https://doi.org/10.1111/are.13099.

Pineiro, M., & Stanton, C. (2007). Probiotic bacteria: Legislative framework—Requirements to evidence basis. *The Journal of Nutrition*, *137*, 850S–853S. Available from https://doi.org/10.1093/jn/137.3.850S.

Prieur, D. (1981). Experimental studies of trophic relationships between marine bacteria and bivalve molluscs. *Kieler Meeresforsch(Sonderh)*, *5*, 376–383.

Pybus, V., Loutit, M., Lamont, I., & Tagg, J. (1994). Growth inhibition of the salmon pathogen *Vibrio ordalii* by a siderophore produced by *Vibrio anguillarum* strain VL4355. *Journal of Fish Diseases*, *17*, 311–324. Available from https://doi.org/10.1111/j.1365-2761.1994.tb00227.x.

Qi, Z., Zhang, X.-H., Boon, N., & Bossier, P. (2009). Probiotics in aquaculture of China—Current state, problems and prospect. *Aquaculture (Amsterdam, Netherlands)*, *290*, 15–21. Available from https://doi.org/10.1016/j.aquaculture.2009.02.012.

Raa, J. (1996). The use of immunostimulatory substances in fish and shellfish farming. *Reviews in Fisheries Science*, *4*, 229–288. Available from https://doi.org/10.1080/10641269609388587.

Raida, M., Larsen, J., Nielsen, M., & Buchmann, K. (2003). Enhanced resistance of rainbow trout, *Oncorhynchus mykiss* (Walbaum), against *Yersinia ruckeri* challenge following oral administration of *Bacillus subtilis* and *B. licheniformis* (BioPlus2B). *Journal of Fish Diseases*, *26*, 495–498. Available from https://doi.org/10.1046/j.1365-2761.2003.00480.x.

Ramesh, D., Vinothkanna, A., Rai, A. K., & Vignesh, V. S. (2015). Isolation of potential probiotic *Bacillus* spp. and assessment of their subcellular components to induce immune responses in Labeo rohita against *Aeromonas hydrophila*. *Fish & Shellfish Immunology*, *45*, 268–276. Available from https://doi.org/10.1016/j.fsi.2015.04.018.

Rehna, A. (2003). *Use of probiotics in larval rearing of mew candidate species*. Aquaculture Asia April 8 (Vol. VIII No. 2). Available at: https://www.researchgate.net/publication/253199241_Use_of_Probiotics_in_larval_rearing_of_new_candidate_species.

Ringø, E., & Gatesoupe, F.-J. (1998). Lactic acid bacteria in fish: A review. *Aquaculture (Amsterdam, Netherlands)*, *160*, 177–203. Available from https://doi.org/10.1016/S0044-8486(97)00299-8.

Ringø, E., Zhou, Z., Vecino, J. G., Wadsworth, S., Romero, J., Krogdahl, Å., Olsen, R., Dimitroglou, A., Foey, A., & Davies, S. (2016). Effect of dietary components on the gut microbiota of aquatic animals. A never-ending story? *Aquaculture Nutrition*, *22*, 219–282. Available from https://doi.org/10.1111/anu.12346.

Robertson, P., O'Dowd, C., Burrells, C., Williams, P., & Austin, B. (2000). Use of *Carnobacterium* sp. as a probiotic for Atlantic salmon (*Salmo salar* L.) and rainbow trout (*Oncorhynchus mykiss*, Walbaum). *Aquaculture (Amsterdam, Netherlands)*, *185*, 235–243. Available from https://doi.org/10.1016/S0044-8486(99)00349-X.

Rodriguez-Estrada, U., Satoh, S., Haga, Y., Fushimi, H., & Sweetman, J. (2013). Effects of inactivated *Enterococcus faecalis* and *Mannan oligosaccharide* and their combination on growth, immunity, and disease protection in rainbow trout. *North American Journal of Aquaculture*, *75*, 416–428. Available from https://doi.org/10.1080/15222055.2013.799620.

Rokka, S., & Rantamäki, P. (2010). Protecting probiotic bacteria by microencapsulation: Challenges for industrial applications. *European Food Research and Technology*, *231*, 1–12. Available from https://doi.org/10.1007/s00217-010-1246-2.

Rosas-Ledesma, P., León-Rubio, J. M., Alarcón, F. J., Moriñigo, M. A., & Balebona, M. C. (2012). Calcium alginate capsules for oral administration of fish probiotic bacteria: Assessment of optimal conditions for encapsulation. *Aquaculture Research*, *43*, 106–116. Available from https://doi.org/10.1111/j.1365-2109.2011.02809.x.

Safari, R., Adel, M., Lazado, C. C., Caipang, C. M. A., & Dadar, M. (2016). Host-derived probiotics *Enterococcus casseliflavus* improves resistance against *Streptococcus iniae* infection in rainbow trout (*Oncorhynchus mykiss*) via immunomodulation. *Fish & Shellfish Immunology*, *52*, 198–205. Available from https://doi.org/10.1016/j.fsi.2016.03.020.

Sahu, M. K., Swarnakumar, N., Sivakumar, K., Thangaradjou, T., & Kannan, L. (2008). Probiotics in aquaculture: Importance and future perspectives. *Indian Journal of Microbiology*, *48*, 299–308. Available from http://doi.org/10.1007/s12088-008-0024-3.

Sakai, M. (1999). Current research status of fish immunostimulants. *Aquaculture (Amsterdam, Netherlands)*, *172*, 63–92. Available from https://doi.org/10.1016/S0044-8486(98)00436-0.

Sakai, M., Yoshida, T., Atsuta, S., & Kobayashi, M. (1995). Enhancement of resistance to vibriosis in rainbow trout, *Oncorhynchus mykiss* (Walbaum), by oral administration of *Clostridium butyricum* bacterin. *Journal of Fish Diseases*, *18*, 187–190. Available from https://doi.org/10.1111/j.1365-2761.1995.tb00276.x.

Salminen, S., Isolauri, E., & Salminen, E. (1996). Clinical uses of probiotics for stabilizing the gut mucosal barrier: Successful strains and future challenges. *Antonie Van Leeuwenhoek*, *70*, 347–358. Available from https://doi.org/10.1007/bf00395941.

Saxelin, M., Rautelin, H., Salminen, S., & Mäkelä, P. (1996). Safety of commercial products with viable Lactobacillus strains. *Infectious Diseases in Clinical Practice*, *5*, 331–335.

Schaeck, M., Reyes-López, F. E., Vallejos-Vidal, E., Van Cleemput, J., Duchateau, L., Van den Broeck, W., Tort, L., & Decostere, A. (2017). Cellular and transcriptomic response to treatment with the probiotic candidate *Vibrio lentus* in gnotobiotic sea bass (*Dicentrarchus labrax*) larvae. *Fish & Shellfish Immunology*, *63*, 147–156. Available from https://doi.org/10.1016/j.fsi.2017.01.028.

Schillinger, U. (1999). Isolation and identification of lactobacilli from novel-type probiotic and mild yoghurts and their stability during refrigerated storage. *International Journal of Food Microbiology*, *47*, 79–87. Available from https://doi.org/10.1016/s0168-1605(99)00014-8.

Senok, A., Ismaeel, A., & Botta, G. (2005). Probiotics: Facts and myths. *Clinical Microbiology and Infection*, *11*, 958–966. Available from https://doi.org/10.1111/j.1469-0691.2005.01228.x.

Shakibazadeh, S., Saad, C., Christianus, A., Kamarudin, M., Sijam, K., & Sinaian, P. (2011). Assessment of possible human risk of probiotic application in shrimp farming. *International Food Research Journal*, *18*, 433–437.

Soccol, C. R., Vandenberghe, L. Pd. S., Spier, M. R., Medeiros, A. B. P., Yamaguishi, C. T., Lindner, J. D. D., Pandey, A., & Thomaz-Soccol, V. (2010). The potential of probiotics: A review. *Food Technology and Biotechnology*, *48*, 413–434.

Sorroza, L., Padilla, D., Acosta, F., Román, L., Grasso, V., Vega, J., & Real, F. (2012). Characterization of the probiotic strain *Vagococcus fluvialis* in the protection of European sea bass (*Dicentrarchus labrax*) against vibriosis by *Vibrio anguillarum*. *Veterinary Microbiology*, *155*, 369–373. Available from https://doi.org/10.1016/j.vetmic.2011.09.013.

Sperti, G. S. (1971). *Probiotics*. AVI Publishing Company.

Strøm, E., & Ringø, E. (1993). *Changes in the bacterial composition of early developing cod,* Gadus morhua *(L.) larvae following inoculation of* Lactobacillus plantarum *into the water. Physiological and biochemical aspects of fish development* (pp. 226–228). Bergen, Norway: University of Bergen, https://doi.org/10.1128%2Fmmbr.64.4.655-671.2000.

Sugita, H., Kawasaki, J., & Deguchi, Y. (1997). Production of amylase by the intestinal microflora in cultured freshwater fish. *Letters in Applied Microbiology*, *24*, 105–108. Available from https://doi.org/10.1046/j.1472-765X.1997.00360.x.

Sun, Y.-Z., Xia, H.-Q., Yang, H.-L., Wang, Y.-L., & Zou, W.-C. (2014). TLR2 signaling may play a key role in the probiotic modulation of intestinal microbiota in grouper Epinephelus coioides. *Aquaculture (Amsterdam, Netherlands)*, *430*, 50–56. Available from https://doi.org/10.1016/j.aquaculture.2014.03.042.

Suzer, C., Çoban, D., Kamaci, H. O., Saka, Ş., Firat, K., Otgucuoğlu, Ö., & Küçüksari, H. (2008). *Lactobacillus* spp. bacteria as probiotics in gilthead sea bream (*Sparus aurata*, L.) larvae: Effects on growth performance and digestive enzyme activities. *Aquaculture (Amsterdam, Netherlands)*, *280*, 140–145. Available from https://doi.org/10.1016/j.aquaculture.2008.04.020.

Sveinsdóttir, H., Steinarsson, A., & Gudmundsdóttir, Á. (2009). Differential protein expression in early Atlantic cod larvae (*Gadus morhua*) in response to treatment with probiotic bacteria. *Comparative Biochemistry and Physiology Part D: Genomics and Proteomics*, *4*, 249–254. Available from https://doi.org/10.1016/j.cbd.2009.06.001.

Thirumurugan, R., & Vignesh, V. (2015). *Probiotics: Live boon to aquaculture. Advances in marine and brackishwater aquaculture* (pp. 51–61). Springer.

Thompson, F. L., Abreu, P. C., & Cavalli, R. (1999). The use of microorganisms as food source for *Penaeus paulensis* larvae. *Aquaculture (Amsterdam, Netherlands)*, *174*, 139–153. Available from https://doi.org/10.1016/S0044-8486(98)00511-0.

Utami, D. A. S., & Suprayudi, M. A. (2015). Administration of microencapsulated probiotic at different doses to control streptococcosis in tilapia (*Oreochromis niloticus*). *Microbiology Indonesia*, *9*, 17–24. Available from https://doi.org/10.5454/mi.9.1.3.

Van Hai, N. (2015). Research findings from the use of probiotics in tilapia aquaculture: A review. *Fish & Shellfish Immunology*, *45*, 592–597. Available from https://doi.org/10.1016/j.fsi.2015.05.026.

Venkateswara, A. (2007). Bioremediation to restore the health of aquaculture. *Pond Ecosystem Hyderabad*, *500*, 1–12.

Verschuere, L., Rombaut, G., Sorgeloos, P., & Verstraete, W. (2000). Probiotic bacteria as biological control agents in aquaculture. *Microbiology and Molecular Biology Reviews*, *64*, 655–671, https://doi.org/10.1128%2Fmmbr.64.4.655-671.2000.

Villamil, L., Figueras, A., Planas, M., & Novoa, B. (2010). Pediococcus acidilactici in the culture of turbot (*Psetta maxima*) larvae: Administration pathways. *Aquaculture (Amsterdam, Netherlands)*, *307*, 83–88. Available from https://doi.org/10.1016/j.aquaculture.2010.07.004.

Vinoj, G., Jayakumar, R., Chen, J.-C., Withyachumnarnkul, B., Shanthi, S., & Vaseeharan, B. (2015). N-hexanoyl-L-homoserine lactone-degrading *Pseudomonas aeruginosa* PsDAHP1 protects zebrafish against *Vibrio parahaemolyticus* infection. *Fish & Shellfish Immunology*, *42*, 204–212. Available from https://doi.org/10.1016/j.fsi.2014.10.033.

Wang, Y., Hassan, M., Shariff, M., Zamri, S., & Chen, X. (1999). Histopathology and cytopathology of white spot syndrome virus (WSSV) in cultured *Penaeus monodon* from peninsular Malaysia with emphasis on pathogenesis and the mechanism of white spot formation. *Diseases of Aquatic Organisms*, *39*, 1–11. Available from https://doi.org/10.3354/dao039001.

Wang, Y., Lee, K., Najiah, M., Shariff, M., & Hassan, M. (2000). A new bacterial white spot syndrome (BWSS) in cultured tiger shrimp *Penaeus monodon* and its comparison with white spot syndrome (WSS) caused by virus. *Diseases of Aquatic Organisms*, *41*, 9–18. Available from https://doi.org/10.3354/dao041009.

Wang, Y.-B. (2007). Effect of probiotics on growth performance and digestive enzyme activity of the shrimp *Penaeus vannamei*. *Aquaculture (Amsterdam, Netherlands)*, *269*, 259–264. Available from https://doi.org/10.1016/j.aquaculture.2007.05.035.

Westerdahl, A., Olsson, J. C., Kjelleberg, S., & Conway, P. L. (1991). Isolation and characterization of turbot (*Scophtalmus maximus*)-associated bacteria with inhibitory effects against *Vibrio anguillarum*. *Applied and Environmental Microbiology*, *57*, 2223–2228. Available from https://doi.org/10.1128/aem.57.8.2223-2228.1991.

Wilson, B., Salyers, A., Whitt, D., & Winkler, M. (2011). *Bacterial pathogenesis: A molecular approach*. Washington, DC: American Society for Microbiology. Available from http://doi.org/10.1128/9781555816162.

Wu, Z.-Q., Jiang, C., Ling, F., & Wang, G.-X. (2015). Effects of dietary supplementation of intestinal autochthonous bacteria on the innate immunity and disease resistance of grass carp (*Ctenopharyngodon idellus*). *Aquaculture (Amsterdam, Netherlands)*, *438*, 105–114. Available from https://doi.org/10.1016/j.aquaculture.2014.12.041.

Yamashita, M., Pereira, S., Cardoso, L., de Araujo, A., Oda, C., Schmidt, É., Bouzon, Z., Martins, M., & Mouriño, J. (2017). Probiotic dietary supplementation in Nile tilapia as prophylaxis against streptococcosis. *Aquaculture Nutrition*, *23*, 1235–1243. Available from https://doi.org/10.1111/anu.12498.

Yanbo, W., & Zirong, X. (2006). Effect of probiotics for common carp (*Cyprinus carpio*) based on growth performance and digestive enzyme activities. *Animal Feed Science and Technology*, *127*, 283–292. Available from https://doi.org/10.1016/j.anifeedsci.2005.09.003.

Yassir, R., Adel, M., & Azze, A. (2002). Use of probiotic bacteria as growth promoters, antibacterial and the effect on physiological parameters of *Orechromis niloticus*. *Journal of Fish Diseases*, *22*, 633–642.

Yuniasari, D., & Ekasari, J. (2010). Nursery culture performance of *Litopenaeus vannamei* with probiotics addition and different C/N Ratio under laboratory condition. *Hayati Journal of Biosciences*, *17*, 115–119. Available from https://doi.org/10.4308/hjb.17.3.115.

Zhou, X., Wang, Y., Yao, J., & Li, W. (2010). Inhibition ability of probiotic, *Lactococcus lactis*, against A. hydrophila and study of its immunostimulatory effect in tilapia (*Oreochromis niloticus*). *International Journal of Engineering, Science and Technology*, *2*, 73–80. Available from https://doi.org/10.4314/ijest.v2i7.63743.

CHAPTER

Efficacy of different treatments available against bacterial pathogens in fish

18

Younis Ahmad Hajam[1], Rajesh Kumar[2], Raksha Rani[1], Preeti Sharma[1] and Diksha[1]

[1]*Division Zoology, Department of Biosciences, Career Point University, Hamirpur, India* [2]*Department of Biosciences, Himachal Pradesh University, Shimla, India*

18.1 Introduction

Fish represent a significant food for humans, and are also used for environmental research purposes. Every year, more environmental pollutants are found, which is worrisome due to the relationship between toxic sources of pollution and disease. Aquatic toxins can be of special concern, as water bodies may act as a means of accumulation or spread of various toxins. Additional methods are required for better detection and risk evaluation of health-related toxins in water ecosystems (Dar et al., 2016b, 2016a, 2020; Ravi et al., 2007). At the immature stage of the fish life cycle, diseases cause fish mortality.

Most of the disease outbreaks in aquaculture occur due to stress, such as pollution, predators, etc. When a pathogen is able to establish an infectious disease, the impact on fish species can be problematic. Disease from pathogens can be a viral infection, bacterial infection, or fungal infection. Fishes are more susceptible to disease when environmental pollution, such as chemicals or drugs, is present, and can also be adversely affected or even killed by other microorganisms and some parasites.

Many diseases and conditions, mainly in freshwater, affect fish: columnaris, also referred to as cottonmouth, gill infections, ich, swelling, tail and fin rot, fungal diseases, pop-eye and cloudy eye, swim bladder disease, lice and nematode worm infestations, water quality-induced diseases, alimentary blockages, anorexia, chilodonella, ergasilus, TB, glugea, henneguya, hexamita, head and lateral line erosion disease, injuries, leeches in aquariums, lymphocystis, marine velvet, neon tetra disease, and many others. The most significant causes of death are due to bacterial diseases in aquaculture, such as *Streptococcus agalactiae*, *Lactococcus garvieae*, *Enterococcus faecalis*, *Aeromonas hydrophila*, and *Yersinia ruckeri*.

In the whole world, about 1/3 food source is agriculture management or manufacturing, which might play a major role in facing the subsequent need (Ravi et al., 2007). Pisciculture may lead to as independent or absolute farming due to the enlargement of manufacturing in farms. The recent intensive growth of fish-farming production has led to lowered immunity, causing mass fatalities or also reducing the number of fishes, leading to major financial losses (Harper & Wolf, 2009). Fish infections are cured by different treatments including drugs and chemicals. Chemicals have led to greater production in fish farming (Table 18.1). Sanitation management of bacterial infection in aquafarms is used in parallel with other measures, through specific pathogen-free brood stocks,

Bacterial Fish Diseases. DOI: https://doi.org/10.1016/B978-0-323-85624-9.00013-0

Table 18.1 Different diseases in fishes.

Sr. no.	Disease condition	Pathogens involved	Symptoms
1.	Columnaris disease	*Flavobacterium columnare*	Hemorrhagic and ulcerative lesions on fins, head, and back, which may look yellow to orange due to bacterial growth and pigmentation
2.	Tail rot and fin rot	*Pseudomonas* spp., *Cytophaga* spp.	Erosions, discoloration, and disintegration of fins and tails
3.	Bacterial gill disease or gill rot or environmental disease	*Flavobacterium branchiophilum, Cytophaga* spp., *Flexibacter* spp.	Gasping, lethargic, gills look discolored with trapped materials, secondary fungal infection
4.	Aeromoniasis or motile Aeromonas septicemia	*Aeromonas hydrophila, A. veronii bv. Sobria, A. sobria*	Hemorrhagic and ulcerative lesions on skin fins, head, exophthalmia
5.	Edwardsiellosis or Edwardsiella septicemia	*Edwardsiella tarda*	Ulcerative abscesses in internal organs, hemorrhagic ulcers on skin, fins and body, rectal protrusion
6.	Vibriosis	*Vibrio anguillarum, V. parahaemolyticus, V. alginolyticus*	Ulcerative abscesses in internal organs, hemorrhagic ulcers on skin, fins, and body
7.	Eye disease	*Aeromonas liquefaciens, Staphylococcus aureus,* various other bacteria	Cataract of eyes, affect cornea, eyeball becomes putrefied
8.	Pseudomoniasis/ Pseudomonas septicemia	*Pseudomonas* sp. *Pseudomonas fluorescens*	Hemorrhagic lesions on skin, fins, tail

with good sanitization, feed optimization, good water quality, and good health control. Pathogen prevention measures (used on vehicles, staff, equipment, and visitors) are employed to prevent pathogen entry into the farms. Also, the water can be treated with UV radiation to avoid water contamination. This is mainly used in hatcheries and land-based recycling units in which a lower water level can be processed (Almeida et al., 2009) (Fig. 18.1).

18.2 Bacterial infections occurring in freshwater fish

18.2.1 *Aeromonas* infections

Analysis shows that a number of diseases are recognized year after year in certain fish farms in Poland. There are many reasons behind this; most of the diseases are due to motile *Aeromonas*, such as *A. hydrophila*, *A. sobria*, or *A. caviae*. The symptoms depend upon the disease type. First, skin ulcers and gill or fin injuries are observed in MAI (motile Aeromonas infection). This can change into a systematic infectious disease, MAS (motile Aeromonas septicemia). In salmonids, psychrophilic *Aeromonas*, *Aeromonas salmonicida*, may cause furunculosis, leading to skin ulcers. Due to these, death rates have increased (up to 80%) (Austin & Austin, 2016). A species that

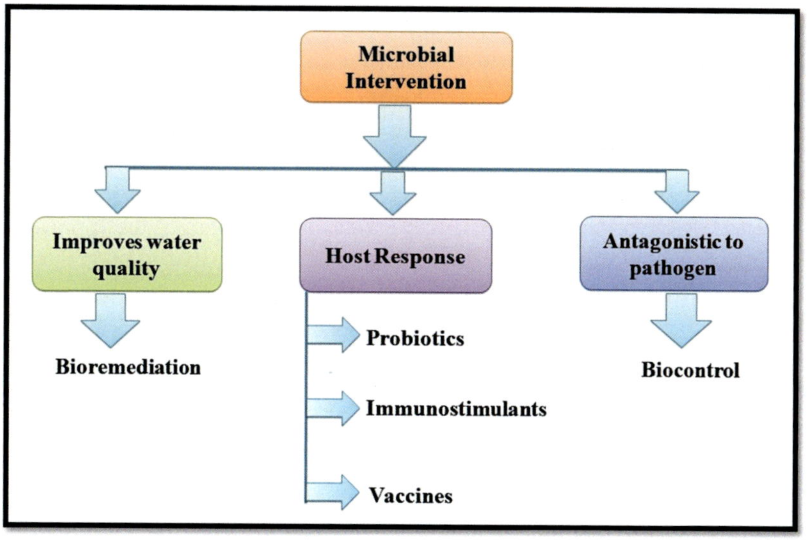

FIGURE 18.1

Different treatments used for fish diseases.

commonly found in any condition may display to bacteria, that is, *Aeromonas* species. Some conditions are risky, such as bacterial interchange occurring during fish transport. *Aeromonas* spp. can play a role in the physical flora of fish intestines (Austin & Austin, 2016), and they can cause disorders in *Cyprinus carpio* L. as seen in data from over 5 years ago in Poland, in which deaths of fish were observed (Kozińska et al., 2014). Also, infections can occur from multiple pathogen species, such as *Pseudomonas* species, *S. putrefaciens*, *Acinetobacter* species, and *S. maltophilia*.

18.2.2 *Pseudomonas* infections

The species *Pseudomonas* is dispersed worldwide, representing a great number of microorganisms. All psychrophilic bacteria grow in cold conditions, and they are the dominant microflora. The bacteria *Aeromonas* and the *P. fluorescens* group are significant, and can be identified in infected skin or fins at above 10°C. The increase in infections may cause mutations to form new genera, allowing the bacteria to change, based on the situation. *Pseudomonas* species can cause cold water strawberry disease in rainbow trout (*Oncorhynchus mykiss*) and *Tinca tinca*. Whole body disease along with the usual condition of septic infection has been detected in *Carassius carassius* and *Carassius gibelio* (Ahne et al., 1982; Csaba et al., 1984).

18.2.3 Flavobacterium infections

In aquatic habitats, the *Flavobacterium* species is found as a native species and can be a part of gill microbiota in fish. Three species of *Flavobacterium* can cause flavobacteriosis: *F. columnare* is an etiologic pathogen that causes columnaris disease; *F. branchiophilum* spp. cause bacterial gill disease; *F. psychrophilum* spp. are related to cold water disease and rainbow trout fry syndrome. Cyprinidae are a family of freshwater fish also referred to as cryprinids. The species *F. columnare* and *F. branchiophilum* have given rise to some health issues. However, *F. psychrophilum* spp. also cause deadly infections but usually are isolated from salmonids (Lehmann et al., 1991). These are acute diseases in which the death rate can reach 50%, but in critical situations they can cause about 80% death rate in fish communities. In the previous 10 years, these bacteria species have been isolated from salmonids or cyprinids in Polish fish farms. In these farms the diseases are detected through clinical identification of disease signs in particular species (Kozińska & Pekala, 2007), primarily outbreaks of cold water disease in Poland in rainbow trout.

18.2.4 Acinetobacter infections

Normally, infectious diseases originate because of particular bacterial microorganisms. The *Acinetobacter* species is distributed throughout the environment, which also includes water bodies. Recently in Poland, these microorganisms were isolated from trout and carp. Generally, the infection is detected in a distinct period, usually in the month of May or September (Kozinska & Pekala, 2004). The laboratory work detects scale loss, nonpigmented body, injury in gills, and eye obstruction in afflicted trout. Also, loss of blood and blockage in the gills are observed in afflicted trout or in autopsies of these fish species. The disease syndrome has a death rate of 5%–20% (Kozinska & Pekala, 2004). *Acinetobacter* species cause diseases that are commonly varied from supplementary bacterial diseases, that is, *Aeromonas* and *Chryseobacterium* species, but the inactive pathogen study included the *Acinetobacter* group. *Acinetobacter* species usually are considered a transporter of disinfectant genes. They play a significant role in increasing antibiotic resistance in the environment (Manchanda et al., 2010).

18.2.5 Shewanella putrefaciens infections

For more than 10 years, *Shewanelloses* have been known as serious infectious diseases in freshwater aquacultured fish. *Shewanella putrefaciens* is a halobacterium that is a significant microorganism in the putrefaction, or decay, process. It mainly spoils fish stored in cold temperatures, but also affects chicken and beef as well (Borch et al., 1996). *Shewanella putrefaciens* has also been isolated from marine and brackish water, as well as marine fish (Al-Harbi & Uddin, 2005). The first bacteria of this type to be isolated from freshwater infected fish were described by Kozinska and Pekala (2004). Disease developed very fast in different species, both ornamental and cultured fish, including *Cyprinus carpio* L., *Oncorhynchus mykiss*, *Anguilla anguilla*, *Salmo trutta* m. *trutta*, *Hypophthalmichthys molitrix*, *Coregonus lavaretus*, *Sander lucioperca*, *Leuciscus idus*, *Rutilus rutilus*, *Brachydanio rerio*, *Pelvicachromis taeniatus*, *Heterandria formosa*, and *Cyprinus carpio* L. (Qin et al., 2012; Rusev et al., 2016). The disorders in cultured fish were noticed in the spring season when the water temperature went up to 7°C–10°C (Dar et al., 2016b, 2016a; Pękala et al., 2015). Common clinical signs in afflicted fish are dark skin, lethargy, and skin ulcers. In postmortem exams, kidney and spleen hemorrhage was observed. In different species the death rate ranged from 40% to 50% (Peralta et al., 2012; Qin et al., 2012).

18.2.6 Fish infection with gram-positive bacteria

Recently, various diseases caused by gram-positive bacteria in *Oncorhynchus mykiss* suddenly increased in Poland (Kozińska et al., 2014). Two species of gram-positive bacteria, *Lactococcus garviae* and *Streptococcus iniae*, are of particular importance. Both of these species cause serious illness in both freshwater and marine fish, especially *Oncorhynchus mykiss* or the oncorhynchus group, eels, and fish of the catfish and tilapia family (Austin & Austin, 2016). The source of bacterial infections can be from water or sediments (Park et al., 1998). Though bacteria are present in the environment and in the fish all year, the diseases mainly occur in the summer season when temperatures rise to 18°C–25°C. The conditions of the aquatic environment and temperature are considered to be the most important factors affecting the appearance of disease (Austin & Austin, 2016). The clinical signs of infection of *L. garviae* and *Streptococcus iniae* are similar, depending upon the fish species. They both cause exophthalmos. In addition, darkened skin, accumulation of fluid in body tissue, and hemorrhage are seen. Dissection (necropsy) shows hemorrhage in the swim bladder, liver, spleen, or kidney, and also inflammation in the stomach. The disease characteristics are similar and are detected easily at the initial stage. Then the fish show nervous whirling movements as the disease develops, from encephalitis or meningitis (Eldar & Ghittino, 1999; Eldar et al., 1994). In Poland *streptococcosis* was first recognized in salmonids in 2010 (Grawiński, 2010). During summer each year since then, *streptococcosis* has appeared in salmonid farms.

18.3 Emerging potential pathogens of freshwater fish

For the past 10 years in Poland, new bacterial infectious diseases have been detected. All these diseases represent emerging infections. These include new, previously undefined diseases as well as old diseases with new features. These new features may include the introduction of a disease to a new location or a new population, or new clinical features (Okamura & Feist, 2011). For instance, in 2004, shewanellosis infection first appeared as a critical health issue in carp and rainbow trout (Kozinska & Pekala, 2004), and then spread to other fish species in other countries (Pękala et al., 2015; Qin et al., 2012; Rusev et al., 2016). These emerging infectious diseases may lead to critical health issues in some fish species. Recently, *Plesiomonas shigelloides* and *Stenotrophomonas maltophilia* are gram-negative bacteria that have been frequently isolated from both diseased and healthy fish.

18.3.1 Infections due to Plesiomonas shigelloides

Plesiomonas shigelloides is a bacterium that belongs to the Enterobactiaceae family and acts as a pathogen for both animals as well as humans. Until relatively recently, disease issues observed in fishes related to *P. shigelloides* have been few. However, high mortality in salmonids was caused by some of them, affecting up to 40% of the stock (Cruz et al., 1986; Vladik & Vitovec, 1974). Cachexia of fish and redness of the anus were clinical symptoms observed, and the presence of "punctate petechiae" was reported based on post-mortem examination of the fish peritoneum and exudative fluid was also present in the body cavity. Growth of *P. shigelloides* in monoculture and infection with both *Flavobacterium* sp. and *Aeromonas hydrophila* were detected in the bacterial examination of samples collected from diseased fish (Cruz et al., 1986; Vladik & Vitovec, 1974). In African catfish (*Heterobranchus bidorsalis*), eels (*Anguilla anguilla*), and sturgeon (*Acipenser sturio*) the presence of *P. shigelloides* was also detected (Klein et al., 1993). Again, clinical signs of infection included redness of anus, petechial hemorrhage in the peritoneum, and ascites, similar to those seen in rainbow trout. However, it should be noted that *P. shigelloides* is considered to be part of the microbiota of fish intestines (Vandepitte et al., 1980).

18.3.2 Infections due to Stenotrophomonas maltophilia

Stenotrophomonas maltophilia is a bacterium that can cause infection in freshwater fishes. This microorganism is present everywhere in the sediments of the bottom surface, in both freshwater and salt water (Dungan et al., 2003; Juhnke & des Jardin, 1989). *S. maltophilia* was detected in industrial and agricultural soils in terrestrial environments (Sturz et al., 2001) and also isolated from plant tissues (Taghavi et al., 2009). *S. maltophilia* is associated with the degradation of xenobiotics compounds and thus it plays a very important role in the processes of biological purification (Dubey & Fulekar, 2012). This bacterium has the ability to produce phytohormones, and due to this ability it is used to promote plant growth and as a biological agent for fighting various plant pathogens (Peralta et al., 2012). *S. maltophilia* bacteria are resistant to a number of antibacterial agents and thus this species is termed a multidrug-resistant bacterium. It is commonly related to respiratory diseases in humans (Brooke, 2012).

The most important fish diseases reported in the literature for *S. maltophilia* are those in African catfish (*Heterobranchus bidorsalis*) and channel catfish (*Ictalurus punctatus*) (Abraham et al., 2016; Geng et al., 2010). Various symptoms such as lethargy, depigmentation of the skin, focal hemorrhages and petechiae, as well as edema in the body cavity, were observed during clinical examination of the fish. Congestion of internal organs, petechiae on their surface, and intestines filled with gases were also observed in post-mortem studies, with 20% mortality rate of the fish stock. Geng et al. (2010) defined these conditions as "infectious intussusception syndrome." In Poland, *S. maltophilia* has been frequently isolated from internal organs and skin of fish showing nonspecific symptoms of disease, often along with many and varied microflora.

18.3.3 Infections due to *Kocuria rhizophila*

Kocuria rhizophila is another species of bacteria that causes infections in fish. This is a new emerging bacterium that acts as a pathogen for fish such as salmonids (Pękala et al., 2018). *Kocuria* spp. are recognized as a parasite in mammals and have been isolated from "marine sediments, chicken meat, freshwater, or food" (Becker et al., 2008). Kim et al. (2007) isolated *Kocuria rhizophila* from the gut of trout, where they formed the microfloral physiology. Fishes infected with *Kocuria rhizophila* showed 50% mortality of the stock and pathological changes in both outer and inner organs were also observed. Clinical symptoms like "exophthalmia, swollen abdomen, increased skin melanisation as well as skin petechiae and focal lesions" were observed in "moribund fish" (or those fish with severe sublethal symptoms). Intestinal inflammation, congestion of liver, and hemorrhages in the tail muscles (mainly in the caudal part) were observed in post-mortem examinations (Pękala et al., 2018). By viewing the available literature and information collected with regard to the impacts on health of fish from *Kocuria* spp., it should be assumed that this bacterium can act as a pathogen to fish, depending on conditions.

18.3.4 Infections caused by myxobacteria

Myxobacteria are pathogens of fish and various diseases are caused by agents of this group, in soil as well as water. There are many species of myxobacteria that can serve as fish pathogens. If the water temperature increases to 20°C, then fish can die due to columnaris disease (Dubos & Davis, 1946). Also, at lower temperatures, various psychrophilic forms have been reported. It is suggested that treatment with 1:2000 copper sulfate can be used only in initial phases of the diseases, while the infection is still minor.

Many surface disinfectants have been tested to control these bacteria, but none of these were able to control myxobacterioses (Rucker et al., 1954). However, the most promising surface disinfectant is pyridylmercuric acetate. The chemotherapeutic effects of many sulfonamides for the control of myxobacterioses disease in fingerlings of *Oncorhynchus tschawytscha* (Chinook salmon) were first explored by Snieszko (1954). From the sulfonamides tested, sulfamethazine gave the results but these results were not as effective. Sulfathiazole gave no effect and sulfanilamide in food was refused by the fishes after a few days of treatment. In the first week of treatment, all sulfonamides experimented on were processed by mouth in the proportion of 12 g/100 lbs of fish per day for a week, and during the next week, the dosage was reduced by half. The treatment with sulfonamides had merely a provisional effect and infection due to myxobacterioses increased just after the therapy was stopped. It was reported that the sulfamethazone was absorbed at a more gradual rate than that of sulfamerazine in fingerlings of *Salvelinus fontinalis* (brook trout) retained at temperatures of 12°C–13°C and the sulfathiazole could barely be spotted in the tissue of trout. Hence, these consequences are in agreement with the observations (Snieszko & Friddle, 1951). Sulfanilamide was poisonous in nature due to its prompt absorption and subsequent high tissue levels. The sulfadiazine acts on this disease significantly as in *Salvelinus fontinalis*. A treatment of sulfamerazine and sulfadiazine suggested by Slater was given to the fingerlings of salmon of various species and trout in which myxobacterioses disease developed at temperature ranges from 16°C–21°C. This treatment helped to reduce the mortality rate among *Oncorhynchus kisutch* (coho salmon) and *Oncorhynchus mykiss* (rainbow trout). On the other hand, it had no effect on that of *O. tshawytscha* (Chinook salmon) and results with cutthroat trout were also unsure. It is possible that species of fish that showed no response to therapy had not absorbed sufficient sulfamerazine to accumulate a protective concentration in the tissues. To confirm the tissue levels of sulfonamides in these two species, it is hard to believe that the drug resistance developed by pathogens could change with the host. A study revealed that sulfamerazine is simply absorbed by rainbow trout (Snieszko & Friddle, 1951). Rainbow trout also respond to sulfonamide treatment of myxobacterioses. Hence, it was observed that fishes reared at lower temperature developed no myxobacterioses; the temperature was less than 15°C and it appeared meaningful, so that the most appropriate method for prevention of this disease could be to nurture the fishes at lower temperature. If fish have to be reared at higher temperature, then the carriers of infections should be removed from the water supply or the water should be free from any infectious agents. Routine chlorination along with dichlorination of the water source may be applied when waterborne fish diseases are creating heavy losses. Hence, it may be supposed that the effectual treatment with sulfonamides promisingly provide strong consequences, although the presence of pathogens and high temperature of water may allow permanent infection. Surprisingly, a federal trout hatchery at LaCrosse, Wisconsin that obtains chlorinated city water, which is dechlorinated before it reach the fish, is basically free from communicable trout diseases.

18.4 Treatment of bacterial pathogens in fish

Disease prevention is more important than treatment to stop and overturn the disease processes once they have started. Fish immunization to protect fish from bacterial diseases has been carried out for years, with mixed results. Some alternatives to antibiotics have been proposed by researchers such as vaccines (Kurath, 2008), antibiotic substitutes (Dorrington & Gomez-Chiarri, 2008), and probiotics use (Kesarcodi-Watson et al., 2008). Bacteriocinogenic bacterial strains are emerging as an important applicant for an alternative, because bacteriocin has been used as a substitute for antibiotics (Joerger and Zhu 2003), while bacteria are used as potential probiotics (Gillor et al., 2008).

18.4.1 Bacteriocins

Bacteriocins are peptides, proteinaceous compounds, synthesized by ribosomes in some bacteria. These inhibits the growth of other similar or closely related bacterial strains (Gillor et al., 2005; Joerger and Zhu 2003). In the communities of microbes, the function of bacteriocins has not been fully determined. Bacteriocins may act as anticompetitor compounds, facilitating an invasion of a strain into an established community of microbes (Lenski & Riley, 2002; Riley & Gordon, 1999; Riley & Wertz, 2002), or act as molecules that play an important role in communication between bacterial groups (Gillor et al., 2008).

Bacteriocins are basically toxins produced by bacteria to stop the growth of their related bacterial strains. Further, most of the classified bacteriocins are detected in coastal aquacultures. However, it is also expected that the bacteriocin-like inhibitory substance, that is, BLIS-400, isolated from *Vibrio mediterranei* can inhibit hemorrhagic septicemia, vibrio septicemia, and ulcer disease due to *A. hydrophila* and *Vibrio* sp. (Carraturo et al., 2014). Similarly, a novel bacteriocin-like substance, also called BLIS, from *Vibrio* sp. NM10, BLIS AP8 from *L. casei* AP8, and bacteriocins such as repressive H5 from *Lactobacillus plantarum* H5 (Ghanbari et al., 2013) could help to inhibit hemorrhagic septicemia and vibrio septicemia bacterial diseases, although the mechanism of action is not yet clear. Pscicocin V1a and Pscicocin V1b detected from *Carnobacterium piscicola* or CS526 from *Carnobacterium piscicola* V1 (Bhugaloo-Vial et al., 1996) correspondingly can inhibit hemorrhagic septicemia disease transmitted by *Pseudomonas* sp. It is possible that the Phocaecin PI80 bacteriocin detected from the gut microbiota of *Fenneropenaeus indicus*, which is also known as Indian white shrimp (Kumar & Arul, 2009), could obstruct Vibrio septicemia caused by *Vibrio* sp. *Equulites elongatus* or spot nape pony fish and Pacific sturgeon or sturgeon fish (*Acipenseridae*) are efficacious fish species having microbiota capable of producing BLIS and could obstruct various bacterial diseases such as *Vibrio* septicemia, hemorrhagic septicemia, and bacterial gill diseases in aquaculture. It was observed that bacteriocin-releasing bacteria isolated from freshwater fish show action against the most common fish bacteria *A. hydrophila* (Banerjee et al., 2013; Giri et al., 2011; Vijayabaskar & Somasundaram, 2008), but the mechanism of the same is still unidentified. Additionally, wide-ranging study is needed to find the scientific evidence and knowledge to prevent diseases in aquaculture and to combat the losses and depletions in aquaculture production.

18.4.2 Fish gut microbiota

A native microbiota may have some advantages within the intestine of a host, such as nutrient metabolism, colonizing resistance, protective action in the case of a pathogen, etc. (Denev et al., 2000; Guarner & Malagelada, 2003). It affects physiological activities, anatomical activity, and/or immunological growth of the host (Rawls et al., 2004). It involves *A. hydrophila*, *Acinetobacter*, *Bacillus*, *Flavobacterium*, *Pseudomonas* characterized as Enterobacteriaceae, as well as anaerobic bacteria from *Bacteroides*, *Clostridium*, or *Fusobacterium* groups (Huber et al., 2004; Kim et al., 2007).

Different lactic acid bacteria, that is, *Lactobacillus*, *Lactococcus*, *Streptococcus*, *Leuconostoc*, and *Carnobacterium* spp., reside in the gastrointestinal tract (Nikoskelainen et al., 2001). Thus, well-developed bacterial flora have a crucial role to play in the generation of pathogen immunological processes. (Salminen et al., 2005). It helps to control infections of microorganisms in fish, that is, *furunculosis*, *columnaris*, *streptococcosis*, or can be associated with antiinfectious effects (Gutowska et al., 2004; Saha et al., 2006; Skrodenyte-Arbaciauskiene et al., 2006; Sugita & Ito, 2006). Recently, the gastrointestinal tract has emerged as an antimicrobial promoter. Compounds such as Thuricin CD, abp118, and microcin C7 from bacteria present in the intestine, that is, *B. thuringiensis*, *L. salivarius*, *E. coli*, having a narrow range of microorganism targets, generally act as beneficial biomedicine.

18.5 Treatment with beneficial gram-negative and gram-positive bacteria

Gram-negative bacteria commonly known to be pathogenic to fish, such as *Aeromonas* species, *Flavobacterium* species, *Pseudomonas* species, and *Shewanella putrefaciens* are replaced by other species, which until now have not been known to be virulent or even conditionally pathogenic to fish. (Jiang et al., 2012). Enterobacteriaceas, the largest family of gram-negative bacteria, gave rise to the synthesized ribosomal peptides called microcins (Zschüttig et al., 2012). EcN (*E. coli* Nissle) 1917 is a probiotic strain in which microcins M and H47 have been observed (Patzer et al., 2003).

In gram-positive bacteria, *Bacillus* spp. (S11) can protect against various infections (Rengpipat et al., 2000). *B. subtilis* was isolated from *catla* (major carp) introducing conflict in *P. fluorescens* (a pathogen) (Ghosh et al., 2007). Different species, such as *L. plantarum*, *L. fermentum*, and *Lactococcus lactis* are probiotics that inhibit some pathogens, such as *A. hydrophila*, *A. salmonicida*, *V. anguillarum*, or *Y. ruckeri* through attachment to intestine secretions (Balcázar et al., 2008; Denev et al., 2009). Different strains such as D14 (*Paenibacillus* sp.), B11 (*Staphylocooccus cohnii*), D12 (*Paenibacillus barcinonensis*), E28 (*B. megaterium*) formed in red tilapia have resistance potential against bacterial pathogens *A. hydrophila* ATCC 35654, *Vibrio alginolyticus* ATCC

33839, and *A. salmonicida. Paenibacillus barcinonensis* strain D12 and *Paenibacillus* sp. strains D14 showed strong antagonistic capability.

18.6 Bioremediation (improving water quality)

Gram-positive bacteria (*Bacillus* spp.) are more useful than gram-negative bacteria to improve water quality, as they convert organic matter into CO_2 or microorganism biofuel (Balcázar et al., 2006). In addition, NH_3 and NO_2 toxicity can be removed by introducing bacterial culture into the aquatic environment. Also, pH, oxygen, ammonia, hydrogen sulfide, and temperature were found to be present in breeding water at acceptable levels when probiotics had been added. The improvement of water quality and the fish environment is termed "bioremediation" (Mohapatra et al., 2013).

18.6.1 Disinfectants

In fish farms, chemicals are often employed, depending upon the organism types, different stages of the life cycle, method of culturing, and the people who are using the chemicals (Gomez-Gil et al., 2000). Some of the chemicals or drugs used to prevent bacterial infection are $KMnO_4$ (5 mg/L), C_6H_6O (5%), NaClO (1%), iodine solution, $C_5H_8O_2$ and CH_2O; $C_{23}H_{25}N_2$ and $CuSO_4$ are also used, but overdosing can lead to toxicity (Bornø & Colquhoun, 2009). Antibiotics are used on a large scale in aquaculture (Bruun et al., 2000), because many contagious or infectious diseases can be prevented or treated in fish farms by the use of antibiotics; however, they have certain drawbacks like the expense, short time of protection and need for multiple doses, antibiotic-resistant strains of bacteria, and the problem of large amounts of toxins remaining in seafood and being consumed (Miranda & Zemelman, 2002). Also, fewer types of antibiotics are available for the treatment of fishes (van der Waaij & Nord, 2000).

18.6.2 Prebiotics

Prebiotics are nondigestible food components consisting of MOS, that is, mannan oligosaccharides, derived from the yeast cell wall. MOS excite the intestinal health-promoting bacteria and due to this the bacterial pathogen rate is less in aquaculture treated with prebiotics (Sohn et al., 2000). According to Rodrigues-Estrada et al. (2008), health supplements with MOS led to better growth, hemolytic activity, and phagocytic activity, increasing fish survival in a challenge with *V. anguillarum*. In the case of *Oncorhynchus mykiss*, when the MOS diet was fed, then growth was significantly increased, as well as antibody titer and lysozyme activities (Staykov et al., 2007). Prebiotics and probiotics together form synbiotics, which enhance development and reproduction of the microbial flora in the digestive tract. *E. faecalis* and MOS are regulated in goldfish to increase the beneficial immune responses and it is live, as compared to *V. anguillarum*. Synbiotic feeding has better results than prebiotic or probiotic approaches (Gatlin & Peredo, 2012). Fig. 18.2 shows the action of prebiotic bacteria.

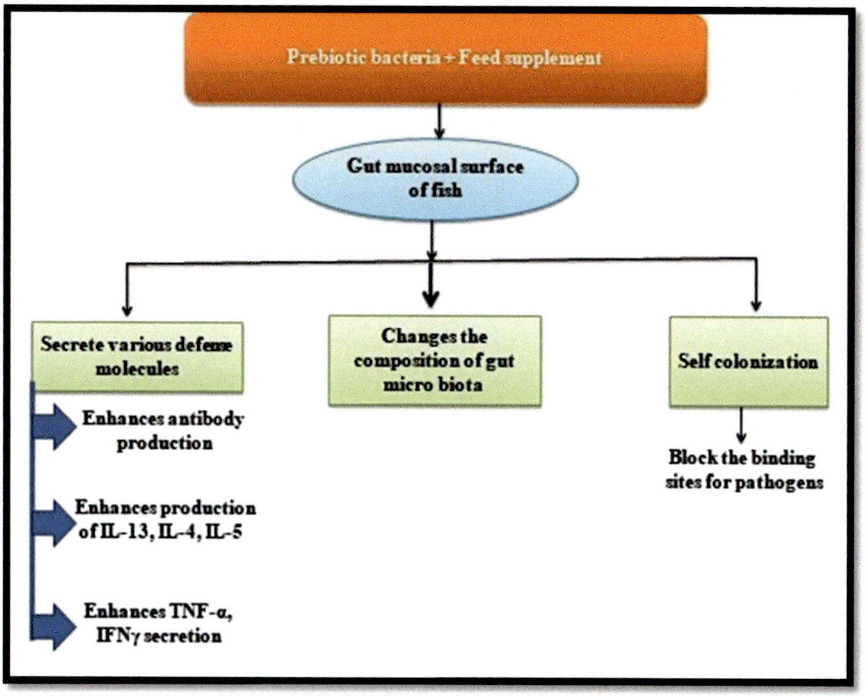

FIGURE 18.2

The prebiotic bacteria.

18.7 Vaccination

Treatment using conventional vaccines uses killed bacteria (bacterins) from broth culture of a particular strain undergoing inactivation of formalin, with bacterins including both bacterial cells and extracellular products (Soliman et al., 2019). There are increasing problems with vaccines, for example, the need for booster doses that stimulate just the humoral immune response, and the short length of protection (Dahiya et al., 2010). Hence, many researchers are searching for better and more ecofriendly treatment approaches (Dahiya et al., 2010).

18.7.1 Biovaccines (living attenuated vaccines)

There are several types of biovaccines. Live attenuated vaccines are not inactivated, but their virulence is lowered genetically (Adams et al., 2008). Giving a live vaccine can initiate an immune response in the host for a short period. Living attenuated vaccines play a significant role in fish aquaculture. The live bacterial vaccine was first used in aquaculture in 1990 (Sun et al., 2010). This type of weakened vaccine is a prototype of infection. In particular, it induces cellular immune response. In addition, it can stimulate humoral and mucosal immunity (Clark & Cassidy-Hanley, 2005).

18.7.2 Encapsulated oral vaccine

18.7.2.1 Live feeds bioencapsulated vaccine

In this type of vaccine, encapsulation is applied to stop antigens from escaping from the pellets or to protect them from the acidic environment in the stomach of fish. Bioencapsulated feed is mainly for fish fry. Live feeds such as Artemia, copepods, and rotifers are included with the vaccine suspension and fed to fish (Lin et al., 2005). The vaccines are released by the bioencapsulated feed into the fish alimentary canal, which is a superior method for vaccine releasing. It lessens fish handling and as a result the fish stress is reduced. It also appears to be appropriate for immunization of groups of fishes. "For effective oral vaccine delivery, the antigen should not be subject to digestive hydrolysis and should be absorbed well for inducing a protective immune response" (Vandenberg, 2004).

18.7.2.2 Nanobioencapsulated vaccine

Recently researchers have been giving more attention to the use of nanoparticles (NPs). "Nanoparticles (NPs) are adjuvant and efficient delivery systems in fish vaccine development due to their nano size." A cellular endocytosis mechanism grasps the nanoparticles, which allow easy uptake of cell antigens and increase their ability to be presented (Vinay et al., 2018).

18.8 Immunomodulation

Probiotics, a combination of live beneficial yeast and bacteria, can accelerate the immune system (fight against infection), regulating the action of lymphocytes, which play an important role in the immune system. They also increase the lysozymes, complement, and antimicrobial peptides (Mohapatra et al., 2013). In humans and animals, probiotics plays an important role in modulating the immune system (Galdeano & Perdigon, 2006). Recent studies showed that probiotics regulate growth and protect against disease. Immunological studies show that different probiotics are effective and can regulate teleost and fish immunity in both in vitro and in vivo conditions (Aly, 2008). Monocytes and macrophages interact with probiotics, stimulating NK cells to enhance immune responses. In in vivo and in vitro conditions, proliferation of B-lymphocytes is stimulated in fish. Some probiotics can enhance the number of RBCs and WBCs in fishes and other higher vertebrates. Probiotics interact with "phagocytic cells" and "polymorph nuclear leukocytes" to enhance the level of immunoglobulin, as immunoglobulin level elevated by probiotics supplementation is reported in fishes and other higher vertebrates (Nayak, 2010). Song et al. (2006) reported high globulin in the mucosa layer of the skin of Miichthysmiiuy due to *Clostridium butyricum*. This group of probiotics, commonly known as lactic acid bacteria, decreases the level of immunoglobulin in fish. Due to supplementation with probiotics, *Lactobacillus rhamnosus* increased the level of immunoglobulin in some fishes, such as *Oncorhynchus mykiss* (Nikoskelainen et al., 2003).

18.9 Bacteriophage therapy

Bacteriophage therapy is another very important biological control method for bacterial pathogens in aquaculture. This therapy is risk-free for humans as well as animals. In the field of biotechnology and medical sciences dealing with prevention of bacterial disease, treatment, rapid detection of disease, and biological control, these phages are widely used (Haq et al., 2012). Furthermore, phages are very specific and can only infect bacterial cells with receptors on the surface of the cells that match the phages (Kutter & Sulakvelidze, 2004). Currently, these phages have gained interest due to their use in biological control of fish pathogens, as no drug residues are linked with such treatment (Jun et al., 2016; Silva et al., 2016). Depending on the isolation and identification of the bacteriophages that will specifically kill the desired pathogens, their use for therapeutic purposes in aquaculture has great potential to control diseases (Higuera et al., 2013).

18.10 Phage therapy dosage

Accurate phage therapy dosage is necessary for effective therapy. Different dosages have been reported in in vitro and in vivo conditions, and the treatment may not be cost effective if a very high concentration of phage is required. Research should be focused on finding those phages with a high infection rate at lower dosages as well as a high replication rate (Rong et al., 2014). Currently, the ability of phage therapy to treat bacterial infections in aquaculture has made great progress (Higuera et al., 2013; Karunasagar et al., 2007; Nakai & Park, 2002; Nakai et al., 1999; Park et al., 2000; Shivu et al., 2007). To authorize it as a marketable treatment, however, a precise assessment of this approach is required. Though studies have tackled the use of phages to manage against vibriosis in aquaculture (Vinod et al., 2006), current research is needed to assess the capacity to limit vibriosis in fish larvae production. Assessment of the suitable bacteriophages, phage handover method, and life cycle stage, such as eggs, larvae, juveniles, or adult fish throughout which the phage therapy is applied, become the main factors in the accomplishment of phage-mediated control of *Vibrio* sp. in aquaculture.

There are various factors and conditions needed to determine the steps for phage therapy in aquaculture, such as host range, latent period, burst size, survival in the environment, and efficiency of bacterial inactivation. Two double standard DNA phages were assessed in a study that detected *V. parahaemolyticus* and *V. anguillarum* transmission of disease to the discussed three hosts, *V. parahaemolyticus*, *V. anguillarum*, and *A. salmonicida*, showing strong ability to deactivate the pathogenic *Vibrio* species assessed. Even so, phage VP-2 indicated greater ability to incapacitate *V. anguillarum* than phage VA-1 and serve as superior as acquired whenever this phage was consumed to disable possess host. Phage VP-2 effectively treated larval fish diseased with *V. anguillarum*, but defeating arcs analyses determined that, 8 h after phage add-on, certain bacteria continued feasible and could regrow, as seen in preceding studies (Barrow et al., 1998). The majority of the resistant bacteria were phage-resistant mutants; however, it has been shown that the virulent bacteria which flattered repellent to phage infectivity are not as much of fit or evade their morbific assets (Capparelli et al., 2010; Filippov et al., 2011). This arises mainly because the cell exterior constituents, such as LPS and proteins, imitate receptors for phage adsorption and could also act as virulence factors. When comparable conditions arise in vivo, then its significance revealing that the bacterial inhabitants would affect a nonlethal state or nonvirulent to fish larvae. Alterations in such receptors to acquire endurance to the phage could diminish pathogenicity (Capparelli et al., 2010; Filippov et al., 2011; Wagner & Waldor, 2002) and, accordingly, bacteria regrowth after phage treatment could be affected by limited or no outcomes for fish larvae. More analyses are necessary to perceive mutations in surface bacterial molecules of unaffected bacteria after phage therapy, as they can behave as phage receptors and, perhaps at the same time, as virulence factors.

It was also suggested that phages reduce the bacterial level sufficiently to be eradicated by the fish immune system by attained reaction (Levin & Bull, 2004), but this does not happen in the case of fish larvae since they do not have the ability to develop the particular acquired immunity (Vadstein, 1997).

References

Abraham, T. J., Paul, P., Adikesavalu, H., Patra, A., & Banerjee, S. (2016). *Stenotrophomonas maltophilia* as an opportunistic pathogen in cultured African catfish *Clarias gariepinus* (Burchell, 1822). *Aquaculture (Amsterdam, Netherlands)*, *450*, 168–172.

Adams, A., Aoki, I., Berthe, C. J., Grisez, L., & Karunasagar, I. N. D. R. A. N. I. (2008). *Recent technological advancements on aquatic animal health and their contributions toward reducing disease risks – A review. Diseases in Asian Aquaculture VI* (pp. 71–88). Colombo, Sri Lanka: Fish Health Section, Asian Fisheries Society.

Ahne, W., Popp, W., & Hoffmann, R. (1982). *Pseudomonas fluorescens* as a pathogen for tench. *Bulletin of the European Association of Fish Pathologists*, *2*, 56–57.

Al-Harbi, A. H., & Uddin, N. (2005). Bacterial diversity of tilapia (*Oreochromis niloticus*) cultured in brackish water in Saudi Arabia. *Aquaculture (Amsterdam, Netherlands)*, *250*, 566–572.

Almeida, A., Cunha, Â., Gomes, N., Alves, E., Costa, L., & Faustino, M. A. (2009). Phage therapy and photodynamic therapy: Low environmental impact approaches to inactivate microorganisms in fish farming plants. *Marine Drugs*, *7*, 268–313.

Aly, M. (2008). *Real time detection of lane markers in urban streets. 2008 IEEE Intelligent Vehicles Symposium* (pp. 7–12). The Netherlands: IEEE, Eindhoven.

Austin, B., & Austin, D. A. (2016). *Aeromonadaceae representative (*Aeromonas salmonicida*). In* Bacterial fish pathogens (pp. 215–321). Cham: Springer.

Balcázar, J. L., De Blas, I., Ruiz-Zarzuela, I., Cunningham, D., Vendrell, D., & Múzquiz, J. L. (2006). The role of probiotics in aquaculture. *Veterinary Microbiology*, *114*, 173–186.

Balcázar, J. L., Vendrell, D., de Blas, I., Ruiz-Zarzuela, I., Muzquiz, J. L., & Girones, O. (2008). Characterization of probiotic properties of lactic acid bacteria isolated from intestinal microbiota of fish. *Aquaculture (Amsterdam, Netherlands)*, *278*, 188–191.

Banerjee, S. P., Dora, K. C., & Chowdhury, S. (2013). Detection, partial purification and characterization of bacteriocin produced by *Lactobacillus brevis* FPTLB3 isolated from freshwater fish. *Journal of Food Science and Technology*, *50*, 17–25.

Barrow, P., Lovell, M., & Berchieri, A. (1998). Use of lytic bacteriophage for control of experimental *Escherichia coli* septicemia and meningitis in chickens and calves. *Clinical and Diagnostic Laboratory Immunology*, *5*, 294–298.

Becker, K., Rutsch, F., Uekötter, A., Kipp, F., König, J., Marquardt, T., & von Eiff, C. (2008). *Kocuria rhizophila* adds to the emerging spectrum of micrococcal species involved in human infections. *Journal of Clinical Microbiology*, *46*, 3537–3539.

Bhugaloo-Vial, P., Dousset, X., Metivier, A., Sorokine, O., Anglade, P., Boyaval, P., & Marion, D. (1996). Purification and amino acid sequences of piscicocins V1a and V1b, two class IIa bacteriocins secreted by *Carnobacterium piscicola* V1 that display significantly different levels of specific inhibitory activity. *Applied and Environmental Microbiology*, *62*, 4410–4416.

Borch, E., Kant-Muemansb, M. L., & Blixt, Y. (1996). Bacterial spoilage of meat products and cured meat. *International Journal of Food Microbiology*, *33*, 103–120.

Bornø, G., & Colquhoun, D. (2009). Classical furunculosis (in Norwegian). Fact Sheet. Norwegian Veterinary Institute. Available at: https://stim.no/tjenester/fiskehelsetjenester/diagnostikk/.

Brooke, J. S. (2012). Stenotrophomonas maltophilia: An emerging global opportunistic pathogen. *Clinical Microbiology Reviews*, *25*, 2–41.

Bruun, M. S., Schmidt, A. S., Madsen, L., & Dalsgaard, I. (2000). Antimicrobial resistance patterns in Danish isolates of *Flavobacterium psychrophilum*. *Aquaculture (Amsterdam, Netherlands)*, *187*, 201–212.

Capparelli, R., Nocerino, N., Lanzetta, R., Silipo, A., Amoresano, A., Giangrande, C., & Iannaccone, M. (2010). Bacteriophage-resistant *Staphylococcus aureus* mutant confers broad immunity against staphylococcal infection in mice. *PLoS One*, *5*, e11720.

Carraturo, A., Raieta, K., Kim, J., & Russo, G. L. (2014). Antibacterial activity of phenolic compounds derived from *Ginkgo biloba* sarcotestas against food-borne pathogens. *Microbiology Research Journal International*, *4*, 18–27.

Clark, T. G., & Cassidy-Hanley, D. (2005). Recombinant subunit vaccines: Potentials and constraints. *Developments in Biologicals*, *121*, 153–163.

Cruz, J. M., Saraiva, A., Eiras, J. C., Branco, R., & Sousa, J. C. (1986). An outbreak of *Plesiomonas shigelloides* in farmed rainbow trout, Salmo gairdneri Richardson, in Portugal. *Bulletin of the European Association of Fish Pathologists (Denmark)*, *6*, 20–22.

Csaba, G., Prigli, M., Kovacs-Gayer, E., Bekesi, L., Bajmocy, E., & Fazekas, B. (1984). Septicaemia in silver carp (*Hypophthalmichthys molitrix* Val.) and bighead carp (*Aristichthys nobilis* Rich.) caused by *Pseudomonas fluorescens*. *Symposia Biologica Hungarica*, *23*, 75–84.

Dahiya, T. P., Kant, R., & Sihag, R. C. (2010). Use of probiotics as an alternative method of disease control in aquaculture. *Biosphere*, *2*, 52–57.

Dar, G. H., Bhat, R. A., Kamili, A. N., Chishti, M. Z., Qadri, H., Dar, R., & Mehmood, M. A. (2020). *Correlation between pollution trends of freshwater bodies and bacterial disease of fish fauna. Fresh water pollution dynamics and remediation* (pp. 51–67). Singapore: Springer.

Dar, G. H., Dar, S. A., Kamili, A. N., Chishti, M. Z., & Ahmad, F. (2016a). Detection and characterization of potentially pathogenic *Aeromonas sobria* isolated from fish *Hypophthalmichthys molitrix* (Cypriniformes: Cyprinidae). *Microbial Pathogenesis*, *91*, 136–140.

Dar, G. H., Kamili, A. N., Chishti, M. Z., Dar, S. A., Tantry, T. A., & Ahmad, F. (2016b). Characterization of *Aeromonas sobria* isolated from fish rohu (*Labeo rohita*) collected from polluted pond. *Journal of Bacteriology & Parasitology*, *7*(3), 1–5. Available from https://doi.org/10.4172/2155-9597.1000273.

Denev, S., Beev, G., Staykov, Y., & Moutafchieva, R. (2009). Microbial ecology of the gastrointestinal tract of fish and the potential application of probiotics and prebiotics in finfish aquaculture. *International Aquatic Research*, *1*, 1–29.

Denev, S. A., Suzuki, I., & Kimoto, H. (2000). Role of Lactobacilli in human and animal health. *Nihon Chikusan Gakkaiho*, *71*, 549–562.

Dorrington, T., & Gomez-Chiarri, M. (2008). Antimicrobial peptides for use in oyster aquaculture: effect on pathogens, commensals, and eukaryotic expression systems. *Journal of Shellfish Research*, *27*(2), 365–373.

Dubey, K. K., & Fulekar, M. H. (2012). Chlorpyrifos bioremediation in *Pennisetum rhizosphere* by a novel potential degrader *Stenotrophomonas maltophilia* MHF ENV20. *World Journal of Microbiology and Biotechnology*, *28*, 1715–1725.

Dubos, R. J., & Davis, B. D. (1946). Factors affecting the growth of tubercle bacilli in liquid media. *The Journal of Experimental Medicine*, *83*, 409.

Dungan, R. S., Yates, S. R., & Frankenberger, W. T., Jr (2003). Transformations of selenate and selenite by *Stenotrophomonas maltophilia* isolated from a seleniferous agricultural drainage pond sediment. *Environmental Microbiology*, *5*, 287–295.

Eldar, A., Bejerano, Y., & Bercovier, H. (1994). *Streptococcus shiloiand* and *Streptococcus difficile*: Two new streptococcal species causing a meningoencephalitis in fish. *Current Microbiology*, *28*, 139–143.

Eldar, A. A., & Ghittino, C. (1999). *Lactococcus garvieae* and *Streptococcus iniae* infections in rainbow trout *Oncorhynchus mykiss*: Similar, but different diseases. *Diseases of Aquatic Organisms*, *36*, 227–231.

Filippov, A. A., Sergueev, K. V., He, Y., Huang, X. Z., Gnade, B. T., Mueller, A. J., & Nikolich, M. P. (2011). Bacteriophage-resistant mutants in *Yersinia pestis*: Identification of phage receptors and attenuation for mice. *PLoS One*, *6*, e25486.

Galdeano, C. M., & Perdigon, G. (2006). The probiotic bacterium *Lactobacillus casei* induces activation of the gut mucosal immune system through innate immunity. *Clinical and Vaccine Immunology*, *13*, 219–226.

Gatlin III, D. M., & Peredo, A. M. (2012). *Prebiotics and probiotics: Definitions and applications*. SRAC Publication No. 4711.

Geng, Y., Wang, K., Chen, D., Huang, X., He, M., & Yin, Z. (2010). *Stenotrophomonas maltophilia*, an emerging opportunist pathogen for cultured channel catfish, *Ictalurus punctatus*, in China. *Aquaculture (Amsterdam, Netherlands)*, *308*, 132–135.

Ghanbari, M., Jami, M., Kneifel, W., & Domig, K. J. (2013). Antimicrobial activity and partial characterization of bacteriocins produced by lactobacilli isolated from Sturgeon fish. *Food Control*, *32*, 379–385.

Ghosh, S., Sinha, A., & Sahu, C. (2007). *Isolation of putative probionts from the intestines of Indian major carps. Israeli Journal of Aquaculture*.

Gillor, O., Etzion, A., & Riley, M. A. (2008). The dual role of bacteriocins as anti-and probiotics. *Applied microbiology and biotechnology*, *81*(4), 591–606.

Gillor, O., Nigro, L. M., & Riley, M. A. (2005). Genetically engineered bacteriocins and their potential as the next generation of antimicrobials. *Current pharmaceutical design*, *11*(8), 1067–1075.

Giri, S. S., Sukumaran, V., Sen, S. S., Vinumonia, J., Banu, B. N., & Jena, P. K. (2011). Antagonistic activity of cellular components of potential probiotic bacteria, isolated from the gut of Labeorohita, against *Aeromonas hydrophila*. *Probiotics and Antimicrobial Proteins*, *3*, 214–222.

Gomez-Gil, B., Roque, A., & Turnbull, J. F. (2000). The use and selection of probiotic bacteria for use in the culture of larval aquatic organisms. *Aquaculture (Amsterdam, Netherlands)*, *191*, 259–270.

Grawiński, E. (2010). Less notorious diseases of salmonids in Northern Poland territory. *ŻycieWeterynaryjne*, *85*, 522–528.

Guarner, F., & Malagelada, J. R. (2003). Bacterial flora of the digestive tract. *Gastroenterologiá y Hepatologiá*, *26*, 1.

Gutowska, M. A., Drazen, J. C., & Robison, B. H. (2004). Digestive chitinolytic activity in marine fishes of Monterey Bay, California. *Comparative Biochemistry and Physiology Part A: Molecular & Integrative Physiology*, *139*, 351–358.

Haq, I. U., Chaudhry, W. N., Akhtar, M. N., Andleeb, S., & Qadri, I. (2012). Bacteriophages and their implications on future biotechnology: A review. *Virology Journal*, *9*, 1–8.

Harper, C., & Wolf, J. C. (2009). Morphologic effects of the stress response in fish. *Ilar Journal*, *50*, 387–396.

Higuera, G., Bastías, R., Tsertsvadze, G., Romero, J., & Espejo, R. T. (2013). Recently discovered *Vibrio anguillarumphages* can protect against experimentally induced vibriosis in Atlantic salmon, Salmo salar. *Aquaculture (Amsterdam, Netherlands)*, *392*, 128–133.

Huber, I., Spanggaard, B., Appel, K. F., Rossen, L., Nielsen, T., & Gram, L. (2004). Phylogenetic analysis and in situ identification of the intestinal microbial community of rainbow trout (Oncorhynchus mykiss, Walbaum). *Journal of applied microbiology*, *96*(1), 117–132.

Jiang, H. X., Tang, D., Liu, Y. H., Zhang, X. H., Zeng, Z. L., Xu, L., & Hawkey, P. M. (2012). Prevalence and characteristics of β-lactamase and plasmid-mediated quinolone resistance genes in *Escherichia coli* isolated from farmed fish in China. *Journal of Antimicrobial Chemotherapy*, *67*, 2350–2353.

Joerger, R. D., & Zhu, X. Y. (2003). Composition of microbiota in content and mucus from cecae of broiler chickens as measured by fluorescent in situ hybridization with group-specific, 16S rRNA-targeted oligonucleotide probes. *Poultry science*, *82*(8), 1242–1249.

Juhnke, M. E., & des Jardin, E. (1989). Selective medium for isolation of *Xanthomonas maltophilia* from soil and rhizosphere environments. *Applied and Environmental Microbiology*, *55*, 747–750.

Jun, J. W., Han, J. E., Tang, K. F., Lightner, D. V., Kim, J., Seo, S. W., & Park, S. C. (2016). Potential application of bacteriophage pVp-1: Agent combating *Vibrio parahaemolyticus* strains associated with acute hepatopancreatic necrosis disease (AHPND) in shrimp. *Aquaculture (Amsterdam, Netherlands)*, *457*, 100–103.

Karunasagar, I., Shivu, M. M., Girisha, S. K., Krohne, G., & Karunasagar, I. (2007). Biocontrol of pathogens in shrimp hatcheries using bacteriophages. *Aquaculture (Amsterdam, Netherlands)*, *268*, 288–292.

Kesarcodi-Watson, A., Kaspar, H., Lategan, M. J., & Gibson, L. (2008). Probiotics in aquaculture: the need, principles and mechanisms of action and screening processes. *Aquaculture*, *274*(1), 1–14.

Kim, D. H., Brunt, J., & Austin, B. (2007). Microbial diversity of intestinal contents and mucus in rainbow trout (*Oncorhynchus mykiss*). *Journal of Applied Microbiology*, *102*(6), 1654–1664.

Klein, B., Kleingeld, D., & Bohm, K. (1993). From samples of cultured fish in Germany. *Bulletin of the European Association of Fish Pathologists*, *13*, 70.

Kozinska, A., & Pekala, A. (2004). First isolation of *Shewanella putrefaciens* from freshwater fish – A potential new pathogen of fish. *Bulletin European Association of Fish Pathologists*, *24*, 189–193.

Kozińska, A., & Pękala, A. (2007). Various cases of flavobacteriosis in trout and carp cultured in Poland. *Medycyna Weterynaryjna*, *63*, 858–863.

Kozińska, A., Paździor, E., Pękala, A., & Niemczuk, W. (2014). *Acinetobacter johnsonii* and *Acinetobacter lwoffii* – The emerging fish pathogens. *Bulletin of the Veterinary Institute in Pulawy*, *58*, 193–199.

Kumar, R., & Arul, V. (2009). Purification and characterization of phocaecin PI80: an anti-listerial bacteriocin produced by Streptococcus phocae PI80 Isolated from the gut of Peneaus indicus (Indian white shrimp). *Journal of microbiology and biotechnology*, *19*(11), 1393–1400.

Kurath, G. (2008). Biotechnology and DNA vaccines for aquatic animals. *Revue scientifique et technique-Office international des épizooties*, *27*(1), 175.

Kutter, E., & Sulakvelidze, A. (2004). *Bacteriophages: Biology and applications* (p. 528) Boca Raton, FL: CRC Press.

Lehmann, J. D. F. J., Mock, D., Stürenberg, F. J., & Bernardet, J. F. (1991). First isolation of *Cytophaga psychrophila* from a systemic disease in eel and cyprinids. *Diseases of Aquatic Organisms*, *10*, 217–220.

Lenski, R. E., & Riley, M. A. (2002). Chemical warfare from an ecological perspective. *Proceedings of the National Academy of Sciences*, *99*(2), 556–558.

Levin, B. R., & Bull, J. J. (2004). Population and evolutionary dynamics of phage therapy. *Nature Reviews. Microbiology*, *2*, 166–173.

Lin, J. H., Yu, C. C., Lin, C. C., & Yang, H. L. (2005). An oral delivery system for recombinant subunit vaccine to fish. *Developments in Biologicals*, *121*, 175–180.

Manchanda, V., Sanchaita, S., & Singh, N. P. (2010). Multidrug resistant acinetobacter. *Journal of Global Infectious Diseases*, *2*, 291.

Miranda, C. D., & Zemelman, R. (2002). Bacterial resistance to oxytetracycline in Chilean salmon farming. *Aquaculture (Amsterdam, Netherlands)*, *212*, 31–47.

Mohapatra, S., Chakraborty, T., Kumar, V., DeBoeck, G., & Mohanta, K. N. (2013). Aquaculture and stress management: A review of probiotic intervention. *Journal of Animal Physiology and Animal Nutrition*, *97*, 405–430.

Nakai, T., & Park, S. C. (2002). Bacteriophage therapy of infectious diseases in aquaculture. *Research in Microbiology*, *153*, 13–18.

Nakai, T., Sugimoto, R., Park, K. H., Matsuoka, S., Mori, K. I., Nishioka, T., & Maruyama, K. (1999). Protective effects of bacteriophage on experimental *Lactococcus garvieae* infection in yellowtail. *Diseases of Aquatic Organisms*, *37*, 33–41.

Nayak, S. K. (2010). Probiotics and immunity: A fish perspective. *Fish & Shellfish Immunology*, *29*, 2–14.

Nikoskelainen, S., Ouwehand, A., Salminen, S., & Bylund, G. (2001). Protection of rainbow trout (*Oncorhynchus mykiss*) from furunculosis by *Lactobacillus rhamnosus*. *Aquaculture (Amsterdam, Netherlands)*, *198*, 229–236.

Nikoskelainen, S., Ouwehand, A. C., Bylund, G., Salminen, S., & Lilius, E. M. (2003). Immune enhancement in rainbow trout (*Oncorhynchus mykiss*) by potential probiotic bacteria (*Lactobacillus rhamnosus*). *Fish & Shellfish Immunology*, *15*, 443–452.

Okamura, B., & Feist, S. W. (2011). Emerging diseases in freshwater systems. *Freshwater Biology*, *5*, 627–637.

Park, K. H., Kato, H., Nakai, T., & Muroga, K. (1998). Phage typing of *Lactococcus garvieae* (formerly *Enterococcus seriolicida*) a pathogen of cultured yellowtail. *Fisheries Science*, *64*, 62–64.

Park, S. C., Shimamura, I., Fukunaga, M., Mori, K. I., & Nakai, T. (2000). Isolation of bacteriophages specific to a fish pathogen, *Pseudomonas plecoglossicida*, as a candidate for disease control. *Applied and Environmental Microbiology*, *66*, 1416–1422.

Patzer, S., Baquero, M. R., Bravo, D., Moreno, F., & Hantke, K. (2003). The colicin G, H and X determinants encode microcins M and H47, which might utilize the catecholatesiderophore receptors FepA, Cir, Fiu and IroN. *Microbiology (Reading, England)*, *149*, 2557–2570.

Pękala, A., Kozińska, A., Paździor, E., & Głowacka, H. (2015). Phenotypical and genotypical characterization of *S. hewanellaputrefaciens* strains isolated from diseased freshwater fish. *Journal of Fish Diseases*, *38*, 283–293.

Pękala, A., Paździor, E., Antychowicz, J., Bernad, A., Głowacka, H., Więcek, B., & Niemczuk, W. (2018). *Kocuria rhizophila* and *Micrococcus luteus* as emerging opportunist pathogens in brown trout (*Salmo trutta* Linnaeus, 1758) and rainbow trout (*Oncorhynchus mykiss* Walbaum, 1792). *Aquaculture (Amsterdam, Netherlands)*, *486*, 285–289.

Peralta, K. D., Araya, T., Valenzuela, S., Sossa, K., Martínez, M., Peña-Cortés, H., & Sanfuentes, E. (2012). Production of phytohormones, siderophores and population fluctuation of two root-promoting rhizobacteria in *Eucalyptus globulus* cuttings. *World Journal of Microbiology and Biotechnology*, *28* (5), 2003–2014.

Qin, L., Zhang, X., & Bi, K. (2012). A new pathogen of gibel carp *Carassius auratus* gibelio – *Shewanella putrefaciens*. *Wei sheng wu xue bao Acta Microbiologica Sinica*, *52*, 558–565.

Ravi, A. V., Musthafa, K. S., Jegathammbal, G., Kathiresan, K., & Pandian, S. K. (2007). Screening and evaluation of probiotics as a biocontrol agent against pathogenic Vibrios in marine aquaculture. *Letters in Applied Microbiology*, *45*, 219–223.

Rawls, S. M., Baron, D. A., Gomez, T., Jacobs, K., & Tallarida, R. J. (2004). Pronounced hypothermic synergy between systemic baclofen and NOS inhibitor. *European Journal of Pharmacology*, *50*, 271–272.

Rengpipat, S., Rukpratanporn, S., Piyatiratitivorakul, S., & Menasaveta, P. (2000). Immunity enhancement in black tiger shrimp (*Penaeus monodon*) by a probiont bacterium (Bacillus S11). *Aquaculture (Amsterdam, Netherlands)*, *191*, 271–288.

Rodrigues-Estrada, U., Satoh, S., Haga, Y., Fushimi, H., & Sweetman, J. (2008). Studies of the effects of mannan-oligosaccharides, Enterococcus faecalis, and poly hydrobutyric acid as immune stimulant and growth promoting ingredients in rainbow trout diets. In *5th World Fisheries Congress, Yokohama, Japan, October* (pp. 20–25).

Riley, M. A., & Gordon, D. M. (1999). The ecological role of bacteriocins in bacterial competition. *Trends in microbiology*, *7*(3), 129–133.

Riley, M. A., & Wertz, J. E. (2002). Bacteriocin diversity: ecological and evolutionary perspectives. *Biochimie*, *84*(5-6), 357–364.

Rong, R., Lin, H., Wang, J., Khan, M. N., & Li, M. (2014). Reductions of *Vibrio parahaemolyticus* in oysters after bacteriophage application during depuration. *Aquaculture (Amsterdam, Netherlands)*, *418*, 171–176.

Rucker, R. R., Earp, B. J., & Ordal, E. J. (1954). Infectious diseases of Pacific salmon. *Transactions of the American Fisheries Society*, *83*, 297–312.

Rusev, V., Rusenova, N., Simeonov, R., & Stratev, D. (2016). *Staphylococcus warneri* and *Shewanella putrefaciens* coinfection in Siberian sturgeon (*Acipenser baerii*) and Hybrid sturgeon (*Huso huso* x *Acipenser baerii*). *Journal of Microbiology & Experimentation*, *3*, 00078.

Saha, S., Roy, R. N., Sen, S. K., & Ray, A. K. (2006). Characterization of cellulase-producing bacteria from the digestive tract of tilapia, *Oreochromis mossambica* (Peters) and grass carp, *Ctenopharyngodon idella* (Valenciennes). *Aquaculture Research*, *37*, 380–388.

Sakai, D., Mochida, J., Iwashina, T., Hiyama, A., Omi, H., Imai, M., Nakai, T., Ando, K., & Hotta, T. (2006). Regenerative effects of transplanting mesenchymal stem cells embedded in atelocollagen to the degenerated intervertebral disc. *Biomaterials*, *27*(3), 335–345.

Salminen, S. J., Gueimonde, M., & Isolauri, E. (2005). Probiotics that modify disease risk. *The Journal of Nutrition*, *135*, 1294–1298.

Shivu, M. M., Rajeeva, B. C., Girisha, S. K., Karunasagar, I., Krohne, G., & Karunasagar, I. (2007). Molecular characterization of *Vibrio harveyi* bacteriophages isolated from aquaculture environments along the coast of India. *Environmental Microbiology*, *9*, 322–331.

Silva, Y. J., Moreirinha, C., Pereira, C., Costa, L., Rocha, R. J., Cunha, Â., & Almeida, A. (2016). Biological control of *Aeromonas salmonicida* infection in juvenile Senegalese sole (*Solea senegalensis*) with Phage AS-A. *Aquaculture (Amsterdam, Netherlands)*, *450*, 225–233.

Skrodenyte-Arbaciauskiene, V., Sruoga, A., & Butkauskas, D. (2006). Assessment of microbial diversity in the river trout *Salmo trutta* fario L. intestinal tract identified by partial 16S rRNA gene sequence analysis. *Fisheries Science*, *72*, 597–602.

Snieszko, S. F. (1954). Therapy of bacterial fish diseases. *Transactions of the American Fisheries Society*, *83*, 313–330.

Snieszko, S. F., & Friddle, S. B. (1951). Tissue levels of various sulfonamides in trout. *Transactions of the American Fisheries Society*, *80*, 240–250.

Sohn, K. S., Kim, M. K., Kim, J. D., & Han, I. K. (2000). The role of immunostimulants in monogastric animal and fish – Review. *Asian-Australasian Journal of Animal Sciences*, *13*, 1178–1187.

Soliman, W. S., Shaapan, R. M., Mohamed, L. A., & Gayed, S. S. (2019). Recent biocontrol measures for fish bacterial diseases, in particular to probiotics, bio-encapsulated vaccines, and phage therapy. *Open Veterinary Journal*, *9*, 190–195.

Song, Z. F., Wu, T. X., Cai, L. S., Zhang, L. J., & Zheng, X. D. (2006). Effects of dietary supplementation with *Clostridium butyricum* on the growth performance and humoral immune response in *Miichthys miiuy*. *Journal of Zhejiang University. Science. B*, *7*, 596–602.

Staykov, Y., Spring, P., Denev, S., & Sweetman, J. (2007). Effect of a mannan oligosaccharide on the growth performance and immune status of rainbow trout (*Oncorhynchus mykiss*). *Aquaculture International*, *15*, 153–161.

Sturz, A. V., Matheson, B. G., Arsenault, W., Kimpinski, J., & Christie, B. R. (2001). Weeds as a source of plant growth promoting rhizobacteria in agricultural soils. *Canadian Journal of Microbiology*, *47*(11), 1013–1024.

Sugita, H., & Ito, Y. (2006). Identification of intestinal bacteria from Japanese flounder (*Paralichthys olivaceus*) and their ability to digest chitin. *Letters in Applied Microbiology*, *43*, 336–342.

Sun, Y., Liu, C. S., & Sun, L. (2010). Isolation and analysis of the vaccine potential of an attenuated *Edwardsiella tarda* strain. *Vaccine*, *28*, 6344–6350.

Taghavi, S., Garafola, C., Monchy, S., Newman, L., Hoffman, A., Weyens, N., & van der Lelie, D. (2009). Genome survey and characterization of endophytic bacteria exhibiting a beneficial effect on growth and development of poplar trees. *Applied and Environmental Microbiology*, *75*, 748–757.

Vadstein, O. (1997). The use of immunostimulation in marine larviculture: Possibilities and challenges. *Aquaculture (Amsterdam, Netherlands)*, *155*, 401–417.

van der Waaij, D., & Nord, C. E. (2000). Development and persistence of multi-resistance to antibiotics in bacteria; an analysis and a new approach to this urgent problem. *International Journal of Antimicrobial Agents*, *16*, 191–197.

Vandenberg, G. W. (2004). Oral vaccines for finfish: Academic theory or commercial reality? *Animal Health Research Reviews*, *5*, 301.

Vandepitte, J., Van Damme, L., Fofana, Y., & Desmyter, J. (1980). *Edwardsiella tarda* and *Plesiomonas shigelloides*. Their role as diarrhea agents and their epidemiology. *Bulletin de la Societe de Pathologie Exotique et de ses Filiales*, *73*, 139–149.

Vijayabaskar, P., & Somasundaram, S. T. (2008). Isolation of bacteriocin producing lactic acid bacteria from fish gut and probiotic activity against common fresh water fish pathogen *Aeromonas hydrophila*. *Biotechnology (Reading, Mass.)*, *7*, 124–128.

Vinay, T. N., Bhat, S., Gon Choudhury, T., Paria, A., Jung, M. H., Shivani Kallappa, G., & Jung, S. J. (2018). Recent advances in application of nanoparticles in fish vaccine delivery. *Reviews in Fisheries Science & Aquaculture*, *26*, 29–41.

Vinod, M. G., Shivu, M. M., Umesha, K. R., Rajeeva, B. C., Krohne, G., Karunasagar, I., & Karunasagar, I. (2006). Isolation of *Vibrio harveyi* bacteriophage with a potential for biocontrol of luminous vibriosis in hatchery environments. *Aquaculture (Amsterdam, Netherlands)*, *255*, 117–124.

Vladik, P., & Vitovec, J. (1974). Plesiomonas shigelloides in rainbow troup septicemia. *Veterinarni Medicina*, *19*, 297–301.

Wagner, P. L., & Waldor, M. K. (2002). Bacteriophage control of bacterial virulence. *Infection and Immunity*, *70*(8), 3985–3993.

Zschüttig, A., Zimmermann, K., Blom, J., Goesmann, A., Pöhlmann, C., & Gunzer, F. (2012). Identification and characterization of microcin S, a new antibacterial peptide produced by probiotic *Escherichia coli* G3/10. *PLoS One*, *7*, e33351.

CHAPTER 19

Summary of economic losses due to bacterial pathogens in aquaculture industry

Juan José Maldonado-Miranda[1], Luis Jesús Castillo-Pérez[2], Amauri Ponce-Hernández[3] and Candy Carranza-Álvarez[1]

[1]*Professor of Faculty of Professional Studies Huasteca Zone, Autonomous University of San Luis Potosi, San Luis Potosi, Mexico* [2]*Multidisciplinary Graduate Program in Environmental Sciences, Autonomous University of San Luis Potosi, San Luis Potosi, Mexico* [3]*Student of Graduate Studies and Research Center, Faculty of Chemistry, Autonomous University of San Luis Potosi, San Luis Potosi, Mexico*

19.1 Introduction

The global aquaculture industry has had rapid growth, as in recent decades (1980–2020) it has contributed significantly to food production. According to the Food and Agriculture Organization (FAO) of the USDA (United States Department of Agriculture, 2018), fish production globally reached approximately 171 million tons in 2016, 47% of which came from aquaculture production. In the same year, the first sale of fishery production reached a total value of $362 billion USD, out of which $232 billion belonged to aquaculture production. Nowadays, aquaculture is a very dynamic food sector, showing an annual growth rate of 6%. Some reports indicate that by 2025 aquaculture will generate profits exceeding $208.9 billion USD. Organisation for Economic Co-operation and Development (OECD) and FAO (2019) project that aquaculture will have a higher fish production than the wild catch sector by 2028. However, despite the high impact this industry has had on food production, there have been serious threats to its economic growth due to the various diseases that attack farmed fish.

Diseases caused mainly by the action of bacteria, parasites, viruses, and fungi have a significant economic impact on fish production (Rodger, 2016). According to Mishra et al. (2017) the growing variety of diseases has resulted in the total loss in the aquaculture sector worldwide of more than $6 billion USD per year. Bacteria-caused diseases are responsible for serious losses in the production of hatcheries and aquaculture farms, mainly because bacteria have the ability to survive in the aquatic environment regardless of their hosts (Dar, Dar, et al., 2016; Dar, Kamili, et al., 2016; Dar et al., 2020; Jayaprakashvel & Subramani, 2019). Bacterial diseases in fish are enhanced when the temperature of the aquatic environment is warm, physicochemical conditions and microbial quality are inadequate, the nutritional status of the medium is poor, and the storage density is high (Pridgeon & Klesius, 2012).

Bacterial Fish Diseases. DOI: https://doi.org/10.1016/B978-0-323-85624-9.00023-3

19.2 Principal fish species produced in the aquaculture industry worldwide

Aquaculture production is defined as cultivated fish and crustaceans extracted from marine and inland waters and marine tanks. Aquaculture plays a key role in many emerging economies worldwide, due to its potential to contribute to increasing food production, as it helps to reduce the pressure on fishery resources. This indicator is measured in tons and USD (OECD, 2020).

Global fish (fish, crustaceans, mollusks, and other aquatic animals, excluding aquatic mammals, reptiles, seaweeds, and other aquatic plant production) is estimated to have reached about 179 million tons in 2018 with a total first sale value estimated at $401 billion USD, out of which 82 million tons, valued at $250 billion USD, came from aquaculture production. Of the overall total, 156 million tons were used for human consumption, equivalent to an estimated annual supply of 20.5 kg/capita (FAO, 2020). Aquaculture accounted for 46% of the total production and 52% of fish for human consumption.

Fig. 19.1 shows the main "food fish" (fish destined for human consumption) worldwide in 2016, highlighting the production of *Ctenopharyngodon idellus* (Grass carp) with a total of 6068 and 5704 thousand tons in 2016 and 2018, respectively.

Fig. 19.2 shows aquaculture production by continent in thousands of tons, among which the production of fin fish stands out over crustaceans, mollusks, and other aquatic animals. Asia had the highest production in 2018, followed by Africa, the Americas, and Europe (Fig. 19.2). Within Asian countries, China is the largest producer, with 35% of the global fish production in 2018 (FAO, 2018).

19.3 Principal causes of economic loss in the aquaculture industry

There are a number of environmental and health factors in the aquaculture industry that can affect and trigger diseases in fish, causing considerable economic losses to producers. According to Thrusfield (1995), diseases in aquaculture can cause two types of economic losses: (1) disease control costs and (2) reduced production, resulting in an economic loss for the producer and consumer. Income economic loss can be calculated as the difference in income that producers earn after experiencing an outbreak of disease and the income they would have earned if the fish had grown without disease. The profit is affected by the expenses of disease control, which include the costs to prevent disease and the costs to prepare farming tanks (Israngkura and Sae-Hae, 2002). These economic losses are estimated between 10% and 15% of total production. Economic losses increase in the world's highest production species, that is, the higher the production volumes, the greater the handling and incidence of diseases in farming tanks or ponds.

Fig. 19.3 shows economic losses in aquaculture production during the period from 2010 to 2018 for the main food-fish species, including *Ctenopharyngodon idellus*, *Hypophthalmichthys molitrix*, and *Cyprinus carpio*.

Due to the economic impacts that occur globally, it is very important to carry out socioeconomic impact assessments of bacterial diseases on aquaculture, with the intention of implementing appropriate strategies according to the type and characteristic of each pathogen. Such assessments can be used as a guide to determine the proper investment on food-fish disease control (Israngkura &

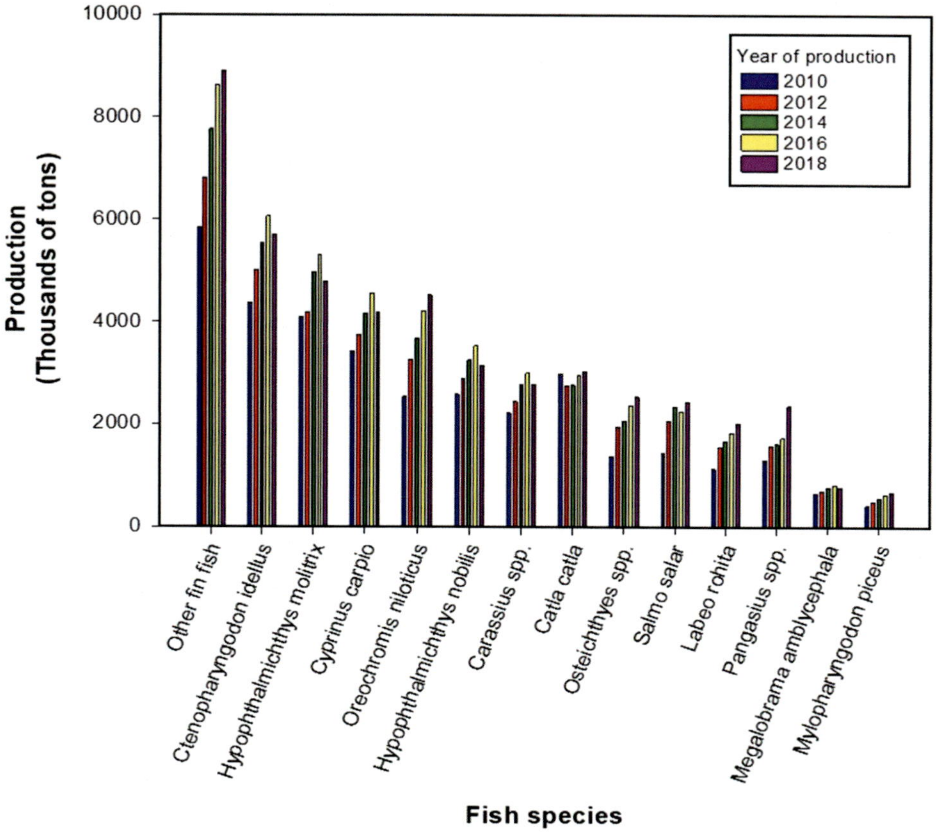

FIGURE 19.1

Principal food fish worldwide from 2010 to 2018.

Sae-Hae, 2002). Unfortunately, carrying out such assessments is an arduous task, due to the complexities around them (Mohan & Bhatta, 2002). According to Murray et al. (2016), economic estimates of disease impacts have been supported by expert opinions and restricted calculations; very few have employed systematic and clear methods. In many cases, even if systematic approaches are available, the lack of data restricts the realization of detailed economic assessments. It is important to mention that opinions about economic impacts will be variable in each country, and even among aquaculture companies within the same region, because this will depend on the needs of the different stakeholders (Brooks et al., 2014). However, it is necessary to consider that the occurrence of diseases in fish is the result of the incidence of one or more biological, physicochemical, nutritional, and density-dependent factors. Fig. 19.4 details said factors.

Table 19.1 presents global economic losses for the aquaculture industry caused by different types of pollutants: organic, inorganic, and biological agents. Economic losses depend on the type of pollutant, which is likely associated with the environmental conditions and economic activities of each region; even within the same country, pollutants vary from region to region (Table 19.1).

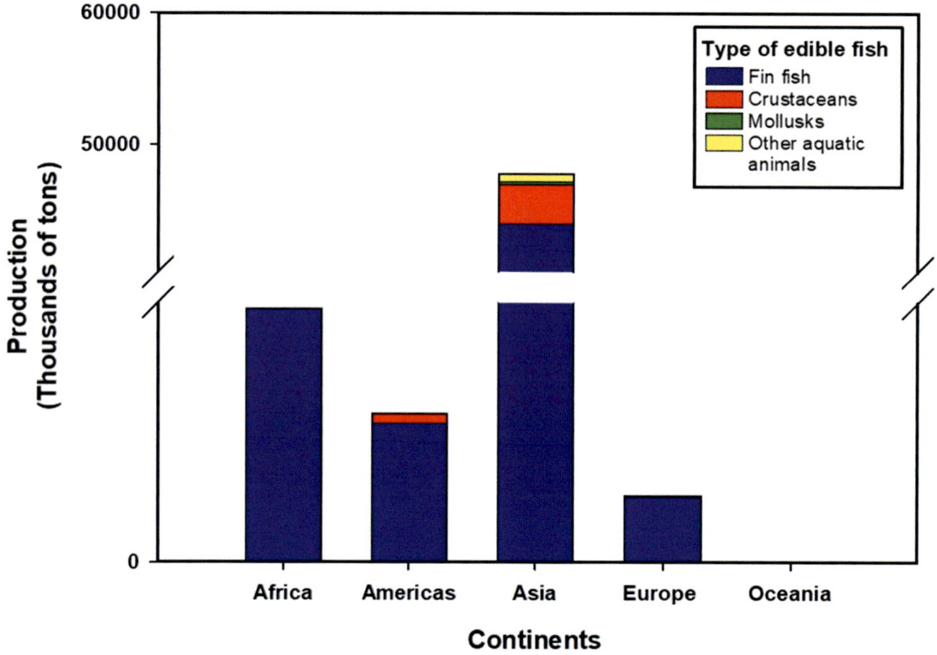

FIGURE 19.2

Production of edible fish per continent.

19.4 Pathogens that causes economic loss in the aquaculture industry

Although the aquaculture industry generates significant gains and is a profitable activity, there are considerable economic losses due to attacks of pathogenic bacteria. According to Kuebutornye et al. (2020), most bacterial diseases in aquaculture are caused by the gram-negative genera *Aeromonas*, *Vibrio*, *Edwardsiella*, *Flavobacterium*, *Franciella*, *Photobacterium*, *Piscirickettsia*, *Tenacibaculum*, *Yersinia*, *Acinetobacter*, *Pseudomonas*, and *Mycobacterium*; and gram-positive genera such as *Streptococcus*, *Lactococcus*, and *Clostridium* (Pridgeon & Klesius, 2012). These bacteria are the cause of the main diseases presented in fish: (1) pseudomoniasis (*Pseudomonas* sp.); (2) aeromoniasis or ascitis (*Aeromonas hydrophila* and *Aeromonas salmonic*); (3) botulism (*Clostridium botulinum*); (4) streptococcosis (*Streptococcus* sp.); and (5) *Flavobacterium columnare* (Abowei & Briyai, 2011). Fig. 19.5 shows the principal pathogenic bacteria and the global economic losses they cause.

Moreover, a study published by Wei (2002) indicates that bacterial sepsis caused by pathogens such as *Aeromonas hydrophila*, *Yersinia ruckeri*, and *Vibrio fluvialis* had a major impact on Chinese aquaculture in the years 1990–92; these bacteria hit almost all regions and provinces of China and had a considerable effect on fish of the species *Hypophthalmichthys* and *Carassius*. According to Wei (2002), the annual economic loss was greater than $120 million USD. For this reason, scientific research methods were applied to reduce the economic loss to about $72.4 million USD.

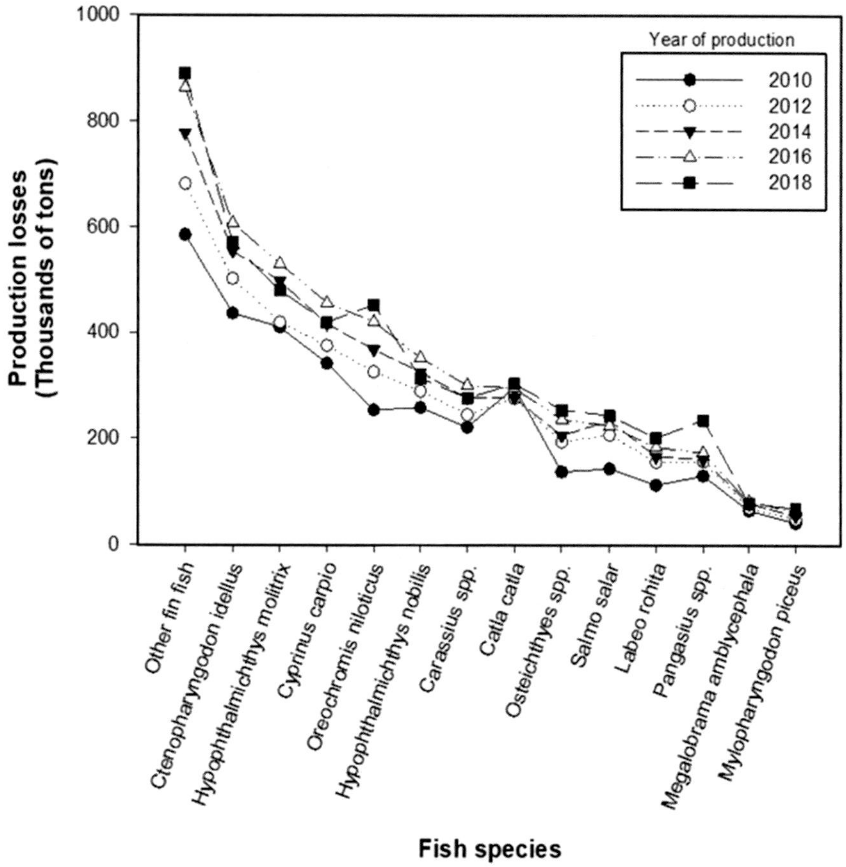

FIGURE 19.3

Economic loss for the principal species of food fish worldwide.

That same year Thi-Van et al. (2002) detailed the socioeconomic impact of red spot disease (RSD) on aquaculture in different regions of North Vietnam over several years (1983–98). This report did not identify the causal agent of the disease; however, by linking the symptoms mentioned by the farmers with the information shown in the FAO letter (2009), it can be concluded that the RSD was caused by the bacterium *Vibrio anguillarum*. Thi-Van et al. (2002) reported that in 1988 the Da River experienced severe losses of 80% in the production of herbivorous carp due to RSD. By 1992, in the Hoa Binh province, the carp production loss was 100%, and in 1994, RSD caused a mortality rate of 70%–80% in the Tuyen Quang province. Overall, in North Vietnam, during the period 1995–97, 80%–90% of the ponds were affected by RSD, and the estimated economic losses for that period were greater than $500,000 USD.

According to Gracía-Mendoza (2017), this same bacterium (*V. anguillarum*) is an important causal agent of diseases with an economic impact in the fish culture of the *Seriola* genus. Some

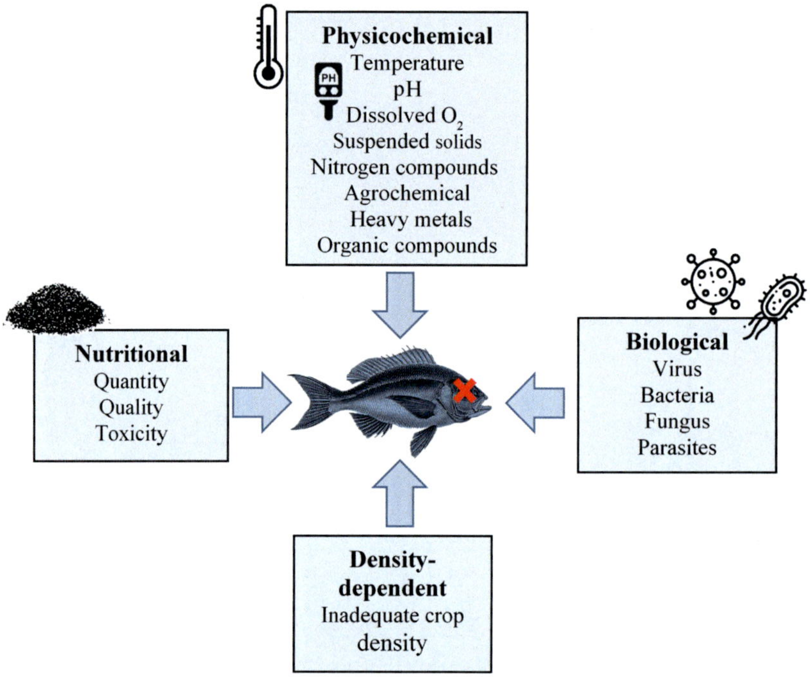

FIGURE 19.4

Principal factors that lead to disease in the aquaculture industry.

reports indicate that in the Japanese aquaculture industry alone, losses per year have reached $200 million USD, including expenses for disease treatments (Nelson, 1999).

On the other hand, a study by Peterman (2019) indicated that commercial catfish production was affected by diseases caused by the action of virulent *Hydrophila*, *Aeromonas*, *Edwardsiella ictaluri*, and *Flavobacterium columnare*. This resulted in a significant economic loss in the United States, since catfish production is the leading aquaculture industry in that country. In total, $16.9 million USD were lost in the production season in 2016 in Eastern Mississippi, out of which $10 million was attributed to catfish disease, $4.3 million to production loss, $1.9 million invested in medicated food, $0.6 million to the use of chemicals, and $0.1 to other various expenses. In addition, the USDA reported that columnaris disease, caused by *Flavobacterium columnare* bacteria, affects fish species such as catfish, rainbow trout, tilapia, sport fish, bait fish, and ornamental fish, costing the aquaculture industry an economic loss of between $40 million and $50 million USD annually. In addition, the USDA indicates that the bacterium *Aeromonas hydrophila* has significantly affected the catfish industry in sums up to $10 million USD annually in fish production losses and treatment costs.

On the other hand, the Chilean aquaculture industry stands out for its high production of salmonid species, such as Atlantic salmon (*Salmo salar*), coho salmon (*Oncorhynchus kisutch*), and rainbow trout (*Oncorhynchus mykiss*). However, salmonid rickettsial septicemia (SRS) has been the main problem to be treated in salmonid aquaculture farms, where production losses from infectious

Table 19.1 Summary of economic losses due to different types of contamination in aquaculture industry.

Region	Country	Continent	Type of contamination	Relevant notes	References
Lakes of Pátzcuaro Michoacán	Mexico	Americas	Agrochemical dumping	Pollution leads to low fish catch	Gaspar-Dillanes et al. (2002)
Coastal areas of Veracruz state	Mexico	Americas	Heavy metal contamination	Pollution affects commercial species such as *Centropomus undecimalis*, *Gerres* sp., *Carcharhinus limbatus* and *Rhizoprionodon terranovae*	Vázquez-Botello et al. (2002)
Magdalena and Meta River basins	Colombia	Americas	Hydrocarbons and persistent organochlorine compounds	These pollutants are the result of chemicals that are banned around the world but are still used in Colombia	Mancera-Rodríguez and Álvarez-León (2005)
Mojana region and wetlands in the south of the Department of Bolivar	Colombia	Americas	Chemical contamination, especially heavy metals	Chemical pollution is the most dangerous in the aquatic ecosystems of Colombia	Mancera-Rodríguez and Álvarez-León (2006)
Fish farms in different regions of Brazil: Dourados, Rio de Janeiro, Itambaracá, Itaju, Arealva, Porto Ferreira, Guaíra, Santa Fé do Sul, Palmital and Jaboticabal	Brazil	Americas	Bacterial contamination	Twelve species of occurrence in Brazilian fish farms were detected and a diagnostic method for bacterial pathogens in fish farms was proposed	Sebastião et al. (2015)
Central and Southwestern Uganda	Uganda	Africa	Bacterial and parasitic contamination	The outbreaks of diseases caused by bacterial pathogens such as *Flavobacterium columnare*, *Aeromonas* sp., *Edwardsiella* sp., *Pseudomonus* sp., *Streptococcus* sp., *Staphylococcus* sp.,	Walakira et al. (2014)

(Continued)

Table 19.1 Summary of economic losses due to different types of contamination in aquaculture industry. *Continued*

Region	Country	Continent	Type of contamination	Relevant notes	References
				production of *euryhaline* fish between Mediterranean countries and bacterial contamination origin has become one of the major agents of economic losses	Yiagnisis and Athanassopoulou (2011)
Three fish farms in north-eastern Poland	Poland	Europe	Zearalenone mycotoxin	Accumulation of this mycotoxin in the ovaries may be a concern for the aquaculture industry	Woźny et al. (2013)

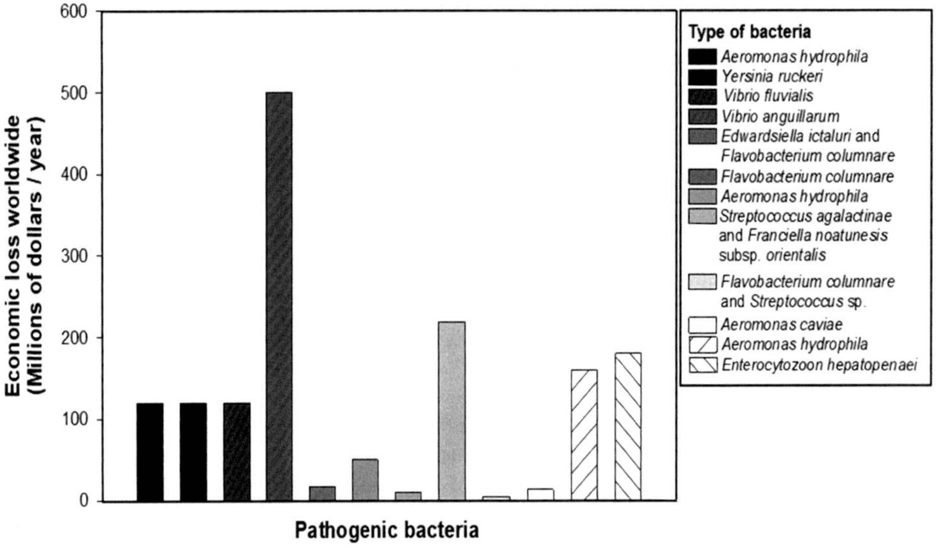

FIGURE 19.5

Economic losses due to bacterial pathogens in aquaculture industry.

diseases reach $700 million USD per year and remain uncontrolled, even though 30 vaccines and antimicrobials are available. In order to reduce the ravages of this problem, the Chilean authorities have been motivated to employ various strategies such as diversifying aquaculture, and helping and strengthening small- to medium-scale cultivation of native species with high export potential (Flores-Kossack et al., 2020).

In a study conducted by Junior et al. (2020), etiological agents causing injuries in tilapia fillets of the Nile River were identified, resulting in economic losses to fish farmers and farms in Brazil. The results of several tests indicated that the agents causing these lesions in the tilapia fillets were the bacteria *Streptococcus agalactinae* and *Franciella noatunesis* sub. *orientalis*. The losses caused by these bacteria were estimated to be numerically significant. To explain this, the authors considered that 1 kg of tilapia in Brazil in 2017 had an average value of $7.30 USD. Therefore, an average loss of 30 tons of steak per month can be translated into an economic loss of $219,000 USD. However, according to Junior et al. (2020), further study is needed to determine a more accurate estimate of economic losses.

In addition, a study carried out by Tavares-Dias and Martins (2017) in the same country (Brazil) estimated the costs of economic losses due to fish mortality caused by various diseases. The study indicated that the direct losses are mainly caused by parasites and bacteria. However, they do not consider secondary losses caused by other factors (e.g., medicines, use of chemicals, food, among other miscellaneous costs); productivity, loss of growth, and food conversion rates were also not considered. Among the significant economic losses caused by bacterial diseases in fish, there was a loss of $4.0 million USD in the production of Nile tilapia, *Oreochromis niloticus*, due to the attack of various bacteria such as *Aeromonas* spp., *Pseudomonas fluorescens*, *Flavobacterium columnare*, and *Streptococcus* sp. (Kubitza et al., 2013). In addition, there are reports on this same species of fish being affected by *Aeromonas caviae*, causing an economic loss of $13,000 USD (Dar et al., 2016; Martins et al., 2008).

Furthermore, the bacterium *Aeromonas hydrophila* caused a loss in the production of surubim hybrid fish, reaching an amount of $160,000 USD (Silva et al., 2012). Finally, there are records of economic loss caused by the bacterium *Citrobacter freundii* in the production of *Pseudoplatystoma reticulatum* fish; however, the cost of production loss is not shown (Pádua et al., 2014). In all cases analyzed, the authors indicate that more reliable data on the type of pathogen and causative disease is needed in order to perform a reliable economic analysis. Table 19.2 highlights some cases where economic losses are reported in the aquaculture industry due to the attack of pathogenic bacteria in different host fish species (Table 19.2).

19.5 Identification of bacterial diseases in fish farms

In fish farms, diseases can occur due to the interaction of various environmental or manipulation variables, as well as the presence of pathogens and suboptimal conditions both nutritional and immunological of the organisms in cultivation. In this environment, fish cohabit or are infected with numerous pathogens without contracting the disease; this situation is established by a balance between the resistance of the host (fish) and the virulence of the pathogen (harmful). This condition is broken when there are important stress factors enough for the fish to develop the disease.

The disease may be acute or severe, and without apparent physical effects or noticeable physical alterations. In some cases, knowledge of the normal behavior and external anatomy of the fish can help identify the presence of diseases on a fish farm. In order to identify the presence

Table 19.2 Economic losses due to bacterial pathogens in aquaculture industry.

Host species	Bacterium	Estimated loss (in million USD)	References
Hypophthalmichthys molitrix, *Hypophthalmichthys nobilis* and *Carassius carassius*	*Aeromonas hydrophila*, *Yersinia ruckeri*, and *Vibrio fluvialis*	120	Wei (2002)
Ctenopharyngodon idella	*Vibrio anguillarum*	0.5	Thi-Van et al. (2002)
Oreochromis niloticus	*Aeromonas* spp., *Pseudomonas fluorescens*, *Flavobacterium columnare*, and *Streptococcus* sp.	4	Kubitza et al. (2013)
Oreochromis niloticus	*Aeromonas caviae*	0.013	Martins et al. (2008)
Hybrid surubim	*Aeromonas hydrophila*	0.16	Silva et al. (2012)
Pseudoplatystoma reticulatum	*Citrobacter freundii*	Not reported	Pádua et al. (2014)
Seriola spp.	*V. anguillarum*	200	Gracía-Mendoza (2017)
Ictalurus punctatus	*A. hydrophila*	10	USDA (2018)
I. punctatus, *Oncorhynchus mykiss*, sport fish, bait fish, and ornamental fish	*Flavobacterium columnare*	40–50	USDA (2018)
Catfish *I. punctatus* and Hybrid catfish	*Aeromonas hydrophila* virulentas, *Edwardsiella ictaluri*, and *F. columnare*	16.9	Peterman (2019)
Oncorhynchus kisutch, *Oncorhynchus mykiss*, and *Salmo salar*	*Piscirickettsia salmonis*	700	Flores-Kossack et al. (2020)
Oreochromis niloticus	*Streptococcus agalactinae* and *Franciella noatunesis* sub. *Orientalis*	0.219	Junior et al. (2020)

of diseases, it is necessary to observe motor behavior and performance, or the physical appearance of the fish; but it is also important to rely on different scientific and traditional methods, which allow analysis of both the origin and effects of the biological or chemical agents causing the disease.

An example within physical characteristics is the pigmentation as a factor that is defined in some species; in some fish of the genus catfish (Fig. 19.6A), the coloration is darker than the color presented in tilapia (Fig. 19.6B), whose skin has a lighter coloration and a greater presence of scales. Traditionally, some farmers consider a healthy fish to be one with soft skin with no flaking and no bruising with discharge, besides having bright eyes and a transparent cornea (Fig. 19.6C).

Fig. 19.7 presents the steps to follow for the identification of diseases on aquatic farms. Constant monitoring is essential in order to prevent a new disease in a group of fish from multiplying and affecting all other fish in the pond. Economic losses will depend on the type and level of effects presented.

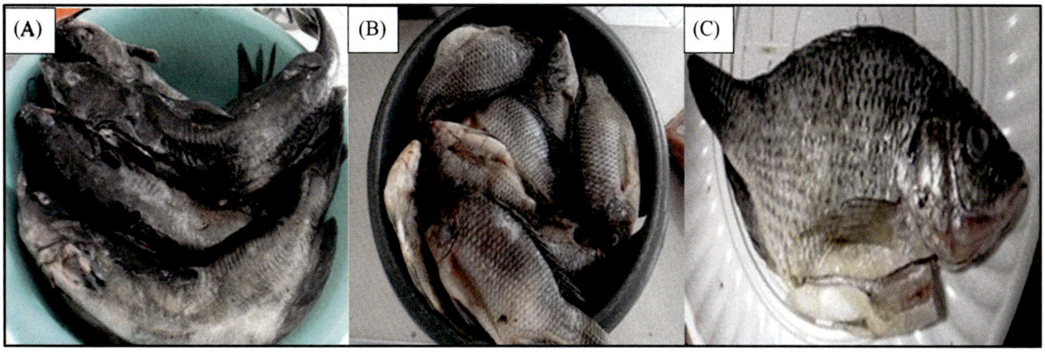

FIGURE 19.6

Physical appearance of different food fish. (A) Catfish genus, (B) Several tilapia fish, and (C) Appearance of a healthy tilapia.

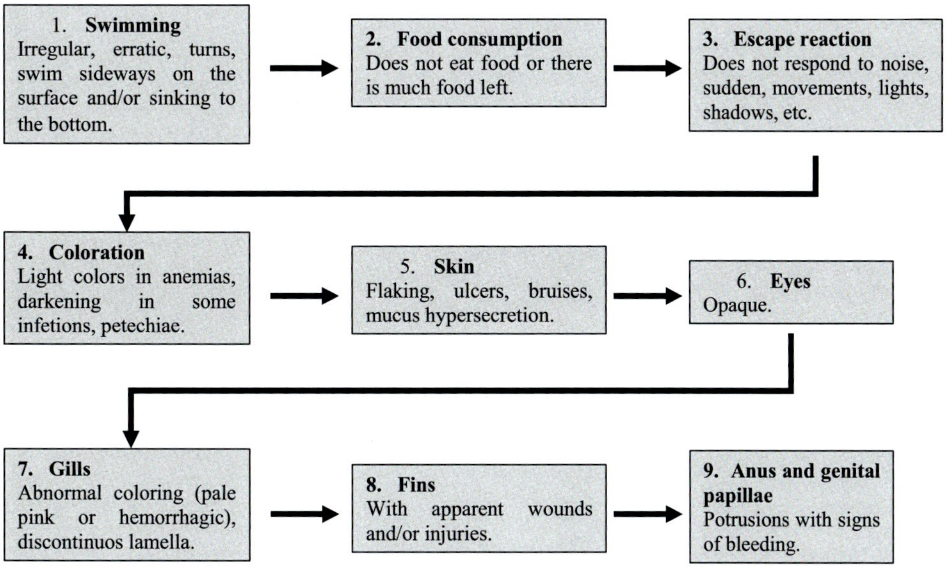

FIGURE 19.7

Steps to follow in order to identify disease in fish tanks or ponds (Balbuena-Rivarola, 2011).

19.6 Analysis of a fish farm system in Huasteca Potosina, Mexico

In aquaculture farms, fish are subject to a density (number of organisms per unit area or volume) much higher than what they would find in their natural environment, which causes greater stress because of the competition for space, food, oxygen, etc. The more intensive (the greater the number of fish per unit area) the fish farm, the greater the stress caused by said competition and the greater the likelihood of disease being generated, if adequate management is not available. Therefore, the fish farmer should seek to maintain the appropriate conditions of aquaculture management, as well as having an emergency plan to carry out corrective measures in a timely manner, and thus reduce the occurrence of diseases in fish. In aquaculture farms, it is essential to take into account the following effects: (1) water quantity and quality, (2) cleaning of tanks or ponds, (3) density of fish stock, (4) water treatment, (5) fish feeding, (6) environmental conditions, (7) type of culture, and (8) size of tanks or ponds.

A field trip was carried out in order to analyze the management practices on an aquaculture farm in a region of Huasteca Potosina, Mexico, where fish are farmed in both tanks, and in a pond and river. Fish farmers have an adequate disease management plan and identification system, which

FIGURE 19.8

System of aquaculture in Huasteca Potosina, Mexico. (A) Water pumping system, (B) Sanitization of the tank (C) water distribution, and (D) fingerling stage (3–5 cm fish).

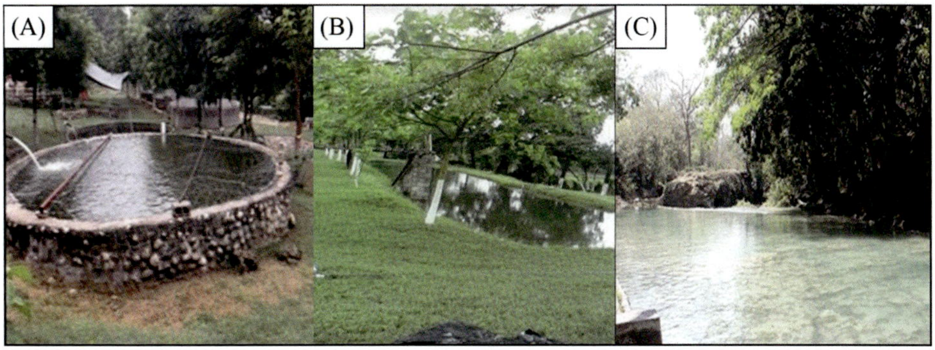

FIGURE 19.9

Types of fish farms in Huasteca Potosina, Mexico. (A) Stock in a traditional tank, (B) Free fish in a pond, and (C) Aquaculture in cage systems in a river.

has allowed them to reduce losses to only 5% of the total production per year. The practices they carry out include constant monitoring and cleaning of the tanks and pond, monthly evaluation of the water quality including physicochemical analysis of water in specialized laboratories, the pumping and oxygenation of the ponds, proper management while stocking, and the maintenance of an adequate density of fish. Fig. 19.8 shows the system of fish farming in a tank.

In traditional fish farming systems, sufficient water must be available in the tanks and ponds, in terms of quantity and quality. The required volumes should be appropriate, and the tanks and ponds refilled when there is water evaporation, especially when external temperatures are high. In addition, it is necessary to empty and clean the tanks and ponds when the water quality is low. In fish farming, water pumping is usually carried out to maintain adequate oxygen levels, as well as being a means for removing the waste generated by the fish themselves, through circulation of water from the bottom. In addition, this activity is carried out when environmental temperatures exceed optimal levels. These activities help the fish inside the tank or pond to remain healthy and maintain an adequate defense system against pathogens that may enter the aquatic environment. In addition, the water quality analyses performed consist of physicochemical (pH, conductivity, hardness, whole solids, total alkalinity) and microbiological analyses, which allow for decisions to be made before any problems affecting production arise.

In this traditional fish farming system, alternate fish cultures start in tanks, and then are moved into a pond as the fish grow larger, and the last stage is carried out in cage systems in rivers (Fig. 19.9), which reduces production costs, as pumping and water change is reduced. Sustainable fish production systems are implemented in which fish waste is used as food for chicken farms. This process is a productive cycle that favors the economy of producers in rural communities.

19.7 Conclusions

In food fish production in the aquaculture industry, bacteria are the most important groups of pathogenic organisms and the cause of the largest economic losses worldwide. The most common bacteria and most likely to be identified by the symptoms they generate correspond to the gram-negative genera *Flexibacter columnaris*, *Aeromonas*, and *Pseudomonas*, and the gram-positive genera *Streptococcus* and *Clostridium*, which are the causes of the main diseases presented in fish. Timely identification of biological agents and environmental factors that can promote the spread of disease in fish can significantly reduce economic losses in the aquaculture industry. For this reason, it is important to have basic health practices in fish management, as well as sustainable systems in both rural and industrial production systems.

References

Abowei, J. F. N., & Briyai, O. F. (2011). A review of some bacteria diseases in Africa culture fisheries.

Balbuena-Rivarola, E. D. (2011). *Manual básico de sanidad piscícola. Ministerio de Agricultura y Ganadería-Viceministerio de Ganadería*. Paraguay: Food and Agriculture Organization of the United Nations.

Balebona, M. C., Zorrilla, I., Moriñigo, M. A., & Borrego, J. J. (1998). Survey of bacterial pathologies affecting farmed gilt-head sea bream (*Sparus aurata* L.) in southwestern Spain from 1990 to 1996. *Aquaculture (Amsterdam, Netherlands)*, *166*(1-2), 19–35. Available from https://doi.org/10.1016/S0044-8486(98)00282-8.

Brooks, V. J., Hernández-Jover, M., Cowled, B., Holyoake, P. K., & Ward, M. P. (2014). Building a picture prioritisation of exotic diseases for the pig industry in Australia using multi-criteria decision analysis. *Preventive Veterinary Medicine*, *113*, 103–117. Available from https://doi.org/10.1016/j.prevetmed.2013.10.014.

Chowdhury, M. B. R. (1998). Involvement of aeromonads and pseudomonads in diseases of farmed fish in Bangladesh. *Fish Pathology*, *33*, 247–254. Available from https://doi.org/10.3147/jsfp.33.247.

Dar, G. H., Bhat, R. A., Kamili, A. N., Chishti, M. Z., Qadri, H., Dar, R., & Mehmood, M. A. (2020). *Correlation between pollution trends of freshwater bodies and bacterial disease of fish fauna. Fresh water pollution dynamics and remediation* (pp. 51–67). Singapore: Springer.

Dar, G. H., Dar, S. A., Kamili, A. N., Chishti, M. Z., & Ahmad, F. (2016). Detection and characterization of potentially pathogenic *Aeromonas sobria* isolated from fish *Hypophthalmichthys molitrix* (Cypriniformes: Cyprinidae). *Microbial Pathogenesis*, *91*, 136–140.

Dar, G. H., Kamili, A. N., Chishti, M. Z., Dar, S. A., Tantry, T. A., & Ahmad, F. (2016). Characterization of *Aeromonas sobria* isolated from fish rohu (*Labeo rohita*) collected from polluted pond. *Journal of Bacteriology & Parasitology*, *7*(3), 1–5. Available from https://doi.org/10.4172/2155-9597.1000273.

El Morabit, A., García-Márquez, S., & Santos, Y. (2004). Is sea lamprey a potential source of infection with *Aeromonas salmonicida* for wild and farmed fish? *Bulletin-European Association of Fish Pathologists*, *24*, 100–103.

FAO (Food and Agriculture Organizatión of the United Nations). (2018). *Part 1: World review in The State of World Fisheries and Aquaculture 2018 - Meeting the sustainable development goals* (pp. 2–75). Rome.

FAO (Food and Agriculture Organizatión of the United Nations). (2009). *Cultured Aquatic Species Information Programme* Mugil cephalus. Cultured Aquatic Species Fact Sheets. In: Saleh, M.A. (Eds.) Departamento de Pesca y Acuicultura de la FAO.

FAO (Food and Agriculture Organizatión of the United Nations). (2020). *The State of World Fisheries and Aquaculture 2020. Sustainability in action*. Rome: Food and Agriculture Organizatión of the United Nations. Available from https://doi.org/10.4060/ca9229en.

Flores-Kossack, C., Montero, R., Köllnerb, B., & Maisey, K. (2020). Chilean aquaculture and the new challenges: Pathogens, immune response, vaccination and fish diversification. *Fish & Shellfish Immunology*, *98*, 52–67. Available from https://doi.org/10.1016/j.fsi.2019.12.093.

Gaspar-Dillanes, M. T., Rojas-Carrillo, P., Fernández-Méndez, J. I., Toledo, D. R., & M.P. (2002). Lago de Pátzcuaro. In M. P. Cisneros-Mata, L. Beléndez-Moreno, E. Zárate-Becerra, M. T. Gaspar-Dillanes, L. C. López-González, C. Saucedo-Ruiz, & J. Tovar-Ávila (Eds.), *Sustentabilidad y pesca responsable en México* (pp. 796–820). Sagarpa, México: Instituto Nacional de la Pesca.

Gracía-Mendoza, M. E. (2017). *Calidad bacteriológica del agua en sistemas de mantenimiento de reproductores de* Seriola lalandi. Baja California. México: CICESE, Centro de Investigación Científica y de Educación Superior de Ensenada.

Israngkura, A., & Sae-Hae, S. (2002). A review of the economic impacts of aquatic animal disease. *FAO fisheries technical*, 253–286.

Jayaprakashvel, M., & Subramani, R. (2019). Implications of quorum sensing and quorum quenching in aquaculture health management. In P. Bramhachari (Ed.), *Implication of quorum sensing and biofilm formation in medicine, agriculture and food industry*. Singapore: Springer. Available from https://doi.org/10.1007/978-981-32-9409-7_18.

Johansen, L. H., Jensen, I., Mikkelsen, H., Bjørn, P. A., Jansen, P. A., & Bergh, Ø. (2011). Disease interaction and pathogens exchange between wild and farmed fish populations with special reference to Norway. *Aquaculture (Amsterdam, Netherlands)*, *315*, 167–186.

Junior, J. A. F., Gomes Leal, C. A., Ferreira de Oliveira, T., Alvarenga-Nascimento, K., S. A. de Macêdo, J. T., & Ocampos Pedroso, P. M. (2020). Anatomopathological characterization and etiology of lesions on Nile tilapia fillets (*Oreochromis niloticus*) caused by bacterial pathogens. *Aquaculture (Amsterdam, Netherlands)*, 526. Available from https://doi.org/10.1016/j.aquaculture.2020.735387.

Kubitza, F., Campos, J. L., Ono, E. A., & Istchuk, P. L. (2013). Panorama da piscicultura no Brasil: A sanidade na piscicultura, do ponto de vista dos produtores e técnicos. *Panorama da Aqüicultura*, *23*, 16–25.

Kuebutornye, F. K. A., Abarike, E. D., Lu, Y., Hlordzi, V., Sakyi, M. E., Afriyie, G., Wang, Z., Li, Y., & Xie, C. X. (2020). Mechanisms and the role of probiotic Bacillus in mitigating fish pathogens in aquaculture. *Fish Physiology and Biochemistry*, *46*, 819–841. Available from https://doi.org/10.1007/s10695-019-00754-y.

Mancera-Rodríguez, N. J., & Álvarez-León, R. (2005). Estado del conocimiento de las concentraciones de hidrocarburos y residuos organoclorados en peces dulceacuícolas de Colombia. *Dahlia - Rev Aso Col Ictiólogos*, *8*, 89–103.

Mancera-Rodríguez, N. J., & Álvarez-León, R. (2006). Estado del conocimiento de las concentraciones de mercurio y otros metales pesados en peces dulceacuícolas de Colombia. *Acta Biológica Colombiana*, *11*, 3–23.

Martins, M. L., Miyazaki, D. M. Y., & Mouriño, J. L. P. (2008). *Aeromonas caviae* durante surto de mortalidade em tilápia do Nilo e suplementação com vitamina C na dieta. *Boletim do Instituto de Pesca, Sao Paulo*, *34*, 585–590.

Maurya, P. K., Malik, D. S., Yadav, K. K., Kumar, A., Kumar, S., & Kamyab, H. (2019). Bioaccumulation and potential sources of heavy metal contamination in fish species in River Ganga basin: Possible human health risks evaluation. *Toxicology Reports*, *6*, 472–481. Available from https://doi.org/10.1016/j.toxrep.2019.05.012.

Mishra, S. S., Das, R., Choudhary, P., Debbarma, J., Sahoo, S. N., Giri, B. S., Rathore, R., Kumar, A., Mishra, C. K., & Swain, P. (2017). Present status of fisheries and impact of emerging diseases of fish and shellfish in Indian aquaculture. *Journal of Aquatic Research and Marine Science*, *1*, 5–26.

Mohan, C. V., & Bhatta, R. (2002). Social and economic impacts of aquatic animal health problems on aquaculture in India. In J. R. Arthur, M. J. Phillips, R. P. Subasinghe, M. B. Reantaso, & I. H. MacRae (Eds.), *Primary Aquatic Animal Health Care in Rural, Small-scale, Aquaculture Development* (pp. 63–75). FAO, Fish. Tech. Pap. No. 406.

Murray, S., Kohli, G.S., Harwood, D.T., Boulter, M., (2016). Safeguarding seafood consumers in new South wales from Ciguatera fish poisoning. Fisheries Research and Development Corporation.

Murray, A. G., Wardeh, M., & McIntyre, K. M. (2016). Using the Hindex to assess disease priorities for salmon aquaculture. *Preventive Veterinary Medicine*, *126*, 199–207. Available from https://doi.org/10.1016/j.prevetmed.2016.02.007.

Nelson, D. R. (1999). *Maritimes protecting the health of farmed fish. Maritimes* (p. 41)) University of Rhode Island.

OECD. (2020). *Aquaculture production (indicator)*. Organisation for Economic Co-operation and Development. Available from http://doi.org/10.1787/d00923d8-en.

OECD & FAO. (2019). *OECD-FAO agricultural outlook 2019-2028*. Rome: OECD Publishing, Paris/Food and Agriculture Organization of the United Nations. Available from https://doi.org/10.1787/agr_outlook-2019-en.

Pádua, S. B., Marques, D. P., Sebastião, F. A., Pilarski, F., Martins, M. L., & Ishikawa, M. M. (2014). Isolation, characterization and pathology of Citrobacter freundii infection in native Brazilian catfish Pseudoplatystoma. *Brazilian Journal of Veterinary Pathology*, *7*, 151–157.

Park, S. I. (2009). Disease control in Korean aquaculture. *Fish Pathology*, *44*, 19–23. Available from https://doi.org/10.3147/jsfp.44.19.

Peterman, M. A. (2019). Direct economic impact of fish diseases on the East Mississippi catfish industry. *North American Journal of Aquaculture*, *81*, 222–229. Available from https://doi.org/10.1002/naaq.10090.

Pietsch, C. (2020). Risk assessment for mycotoxin contamination in fish feeds in Europe. *Mycotoxin Research*, *36*, 41–62. Available from https://doi.org/10.1007/s12550-019-00368-6.

Pridgeon, J. W., & Klesius, P. H. (2012). Major bacterial diseases in aquaculture and their vaccine development. *CAB Reviews*, *7*, 1–6. Available from https://doi.org/10.1079/PAVSNNR20127048.

Rodger, H. D. (2016). Fish disease causing economic impact in global aquaculture. In A. Adams (Ed.), *Fish vaccines. Birkhäuser advances in infectious diseases*. Basel: Springer. Available from https://doi.org/10.1007/978-3-0348-0980-1_1.

Sebastião, F. A., Furlan, L. R., Hashimoto, D. T., & Pilarski, F. (2015). Identification of bacterial fish pathogens in Brazil by direct colony PCR and 16S rRNA gene sequencing. *Advances in Microbiology*, *5*, 409. Available from https://doi.org/10.4236/aim.2015.56042.

Silva, B. C., Mouriño, J. L. P., Vieira, F. N., Jatobá, A., Seiffert, W. A., & Martins, M. L. (2012). Haemorrhagic septicaemia in the hybrid surubim (*Pseudoplatystoma corruscans x Pseudoplatystomac fasciatum*) caused by *Aeromonas hydrophila*. *Aquaculture Research*, *43*, 908–916. Available from https://doi.org/10.1111/j.1365-2109.2011.02905.x.

Tavares-Dias, M., & Martins, M. L. (2017). An overall estimation of losses caused by diseases in the Brazilian fish farms. *Journal of Parasitic Diseases*, *41*, 913–918. Available from https://doi.org/10.1007/s12639-017-0938-y.

Thi-Van, P., Van-Khoa, L., Thi-Lua, D., Van-Van, K., & Thi-Ha, N. (2002). The impacts of red spot disease on small-scale aquaculture in northern Vietnam. In J. R. Arthur, M. J. Phillips, R. P. Subasinghe, M. B. Reantaso, & I. H. MacRae (Eds.), *Primary aquatic animal health care in rural, small-scale, aquaculture development* (pp. 165–176). FAO, Fish. Tech. Pap. No. 406.

Thrusfield, M. (1995). Veterinary epidemiology, 2nd Edn. 479 Blackwell Science.

USDA (United States Department of Agriculture). (2018). *Guarding against threats to fish health*. AgResearch Magazine (vol. 66).

Vázquez-Botello, A., Barrera, G., G. Díaz, G., Ponce, V., G. Villanueva-Fragoso, S., & Wong Ch, I. (2002). Contaminación marina y costera. In A. Guzmán, C. P. Quiroga B., L. Díaz, C. C. Fuentes, C. M. D. Contreras, L. Silva, & G. (Eds.), *La pesca en Veracruz y sus perspectivas de desarrollo* (pp. 97–111). México: Instituto Nacional de la Pesca, Sagarpa y Universidad Veracruzana.

Walakira, J., Akoll, P., Engole, M., Sserwadda, M., Nkambo, M., Namulawa, V., Kityo, G., Musimbi, S., Abaho, I., Kasigwa, H., Mbabazi, D., Kahwa, D., Naigaga, I., Birungi, D., Rutaisire, J., & Majalija, S. (2014).

Common fish diseases and parasites affecting wild and farmed Tilapia and catfish in Central and Western Uganda. *Uganda Journal of Agricultural Sciences*, *15*, 113–125.

Wanja, D. W., Mbuthia, P. G., Waruiru, R. M., Mwadime, J. M., Bebora, L. C., Nyaga, P. N., & Ngowi, H. A. (2019). Bacterial pathogens isolated from farmed fish and source pond water in Kirinyaga County, Kenya. *International Journal of Fisheries and Aquatic Studies*, *7*, 34–39.

Wei, Q. (2002). Social and economic impacts of aquatic animal health problems in aquaculture in China. In J. R. Arthur, M. J. Phillips, R. P. Subasinghe, M. B. Reantaso, & I. H. MacRae (Eds.), *Primary aquatic animal health care in rural, small-scale, aquaculture development* (pp. 55–61). FAO, Fish. Tech. Pap. No. 406.

Woźny, M., Obremski, K., Jakimiuk, E., Gusiatin, M., & Brzuzan, P. (2013). Zearalenone contamination in rainbow trout farms in north-eastern Poland. *Aquaculture (Amsterdam, Netherlands)*, *416–417*, 209–211. Available from https://doi.org/10.1016/j.aquaculture.2013.09.030.

Yiagnisis, M., & Athanassopoulou, F. (2011). Bacteria isolated from diseased wild and farmed marine fish in Greece. In F. Aral (Ed.), *Recent advances in fish farms*. IntechOpen. Available from https://doi.org/10.5772/1122.

Zhang, L., & Wong, M. H. (2007). Environmental mercury contamination in China: sources and impacts. *Environment International*, *33*, 108–121. Available from https://doi.org/10.1016/j.envint.2006.06.022.

Index

Note: Page numbers followed by "*f*" and "*t*" refer to figures and tables, respectively.

Printed in the United States
by Baker & Taylor Publisher Services